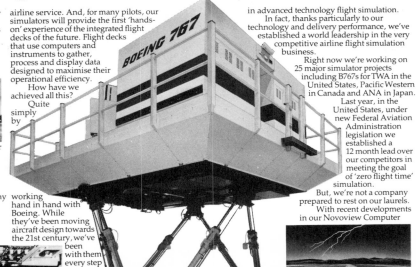
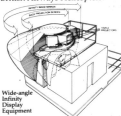

JANE'S AVIONICS
1982-83

Alphabetical list of advertisers

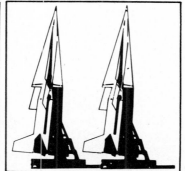

Classified list of advertisers

The companies advertising in this year's publication have informed us that they are involved in the fields of manufacture indicated below:-

Accelerometers
Rockwell International

Accelerometers for Inertial Navigation
Ferranti

Accident Recorders
SFIM

Actuators
Rockwell International

Advanced Airborne HF Communications
Rockwell International

AFCS
Ferranti
SFIM

Airborne Battery Chargers
Ferranti

Airborne Communication Systems
Rockwell International

Airborne Data Handling Systems
Ferranti

Airborne Digital Computer Systems
Ferranti

Airborne Systems Calibration Equipment
Ferranti

Airborne Weapon Systems Simulation
Ferranti

Aircraft Instruments
Ferranti

Air Data Computers
Rockwell International

Air Data Flight Instruments
Ferranti
Thomson-CSF

Amplifiers
Rockwell International

Analysers
Rockwell International

Antennas
Racal Avionics
Rockwell International
SMA
Thomson-CSF

Altimeters
Rockwell International

Artificial Horizons
Ferranti

ASW Helicopter Trainers
Ferranti

ASW
CAE Electronics
Ferranti
SFIM
Thomson-CSF

ATC Airborne Equipment
Rockwell International

Attack Radar
Ferranti
SMA
Thomson-CSF

Attenuators, Avionic
Rockwell International

Attitude/Heading Flight Instruments
Ferranti
SFIM
Thomson-CSF

Autopilot Approach Monitors
Ferranti

Autopilots
Ferranti
Rockwell International
SFIM

Autostability
Ferranti

Avionics
Rockwell International

Cad/Cam Systems
Rockwell International

Calibration Equipment & Services
Rockwell International

Cockpit/Flight Deck Displays
Ferranti
Racal Avionics
Thomson-CSF

Cockpit Management Systems
Rockwell International

Combined Map & Electronic Displays
Ferranti

Computers
Ferranti
Racal Avionics
Thomson-CSF

Current and Voltage Sensors
Ferranti

Designator Lasers
Ferranti
Thomson-CSF

Direction Indicators
Ferranti

Electronic Displays
Rockwell International

Electronic Flight Displays
Rockwell International

Electronic Warfare (ECM)
Ferranti
Thomson-CSF

Electronic Warfare (COMINT)
Thomson-CSF

Electronic Warfare (ELINT)
Ferranti
Thomson-CSF

Electronic Warfare (ESM)
Thomson-CSF

Electronic Warfare (SIGINT)
Thomson-CSF

Encryption & Scrambling Equipment
Rockwell International

Engine/Thrust Management—Health Monitoring
Racal Avionics

Fatigue Recorders
SFIM

Flight Data Recorders
SFIM

Flight Directors
Rockwell International

Flight Following
Offshore Navigation

Flight Performance Management
Racal Avionics
Rockwell International
SFIM

Flight Simulation
CAE Electronics
Rediffusion Simulation
Thomson-CSF

Flight Simulation for Helicopters & Fixed-Wing Aircraft
CAE Electronics

Flight Simulation Components eg Motion System/Control Force
CAE Electronics

Flight Testing Services
Rockwell International

FLIR
Ferranti

Gyroscopes
Ferranti

Height Radar
Ferranti

Helicopter Radars
Ferranti

Helicopter Stabilisation Systems
Ferranti

Helicopter Weapon Sights
Ferranti

Helmet Displays
Thomson-CSF

High Performance HF Radio
Rockwell International

Horizon Gyro Units
Ferranti

HUD
Ferranti
Thomson-CSF

Hybrid Microcircuits
Ferranti

Inertial Navigation Digital
Ferranti

Lightweight Airborne HF Transceivers
Rockwell International

Linescan Radar
Thomson-CSF

Low-Light TV
Ferranti
Thomson-CSF

Marked-Target Seeker Lasers
Ferranti
Thomson-CSF

Marker Beacons
Rockwell International

Military Standard 1553 Data-Bus Equipment
Ferranti

Missile-Guidance TV
Ferranti

Mission Planning Equipment
Ferranti

Moving Map Displays
Ferranti
Racal Avionics
Thomson-CSF

Nav/Attack Systems
Ferranti

Navigation/Nav-Com (ADF)
Racal Avionics

Navigation/Nav-Com (Decca/Dectra)
Racal Avionics

Navigation/Nav-Com (DME)
Racal Avionics
Thomson-CSF

Navigation/Nav-Com (Doppler)
Racal Avionics

Navigation/Nav-Com (GNAV)
Racal Avionics

Navigation/Nav-Com (INS)
SFIM

Navigation/Nav-Com (LORAN)
Offshore Navigation
Racal Avionics

Navigation/Nav-Com (Magnetic Compass)
SFIM

Navigation/Nav-Com (Moving Map)
Ferranti
Racal Avionics
Thomson-CSF

Navigation/Nav-Com (Omega/VLF)
Racal Avionics

Navigation/Nav-Com (RNAV)
Racal Avionics

Navigation/Nav-Com (Tacan)
Racal Avionics
Rockwell International
Thomson-CSF

Navigation/Nav-Com (VOR/ILS)
Racal Avionics
Thomson-CSF

Navigation/Nav-Com (Vortac)
Thomson-CSF

Portable Data Storage & Transfer
Ferranti

Radar Systems
SMA

Radio Communications (Command/Data Links)
Offshore Navigation
Thomson-CSF

Radio Communication (HF)
Rockwell International

Radio Communications (Integrated Management)
Thomson-CSF

Radio Communications (Marker-Beacon Receivers)
Racal Avionics

Radio Communications (Ranger Lasers)
Ferranti

Radio Communication (UHF)
Italtel

Radio Communication (VHF)
Italtel

Satellite Navigation
Rockwell International

Selcal
Rockwell International

Simulation Systems
Rockwell International

Stall Warning Systems
Ferranti

Static Invertors
Ferranti

Stores Management
Rockwell International

Surveillance Radar
Ferrant
Racal Avionics
SMA
Thomson-CSF

Tactical Command Systems
Ferranti

Tactical Control Systems
Ferranti

Telemetry
Offshore Navigation

Transformer Rectifier Units
Ferranti

Turn & Slip Indicators
Ferranti

UHF Airborne Transceivers
Rockwell International

UHF Line-of-Sight/Satellite Transceivers
Rockwell International

Ultra Compact VHF-UHF Multimode Systems
Rockwell International

Vertical & Azimuth Gyros
Ferranti

VHF
Rockwell International

VLF
Rockwell International

What does nearly every military aircraft in the free world have in common?

AN/ARC-171(V)
UHF Line-of-sight/satellite transceiver

AN/ARC-190(V)
Advanced Airborne HF
Communications, 400 Watts

Brussels (2) 242-4048 • Cairo 989358 • Frankfurt (Rodgau 6) (6106) 671 • Hong Kong (5) 274-321 • Kuala Lumpur (3) 482-251 • London (1) 759-9911
Manila (2) 8186689 • Mexico City (905) 533-1846 • Lilydale, Victoria (3) 726-0766 • Toulouse (61) 711-141 • Rio de Janeiro (21) 286-8296
Riyadh (1) 476-9060 • Rome (6) 919-5271 • Seoul (2) 74-8470 • Tokyo (3) 265-8804 • Toronto (416) 757-1101

Collins Telecommunications Products.

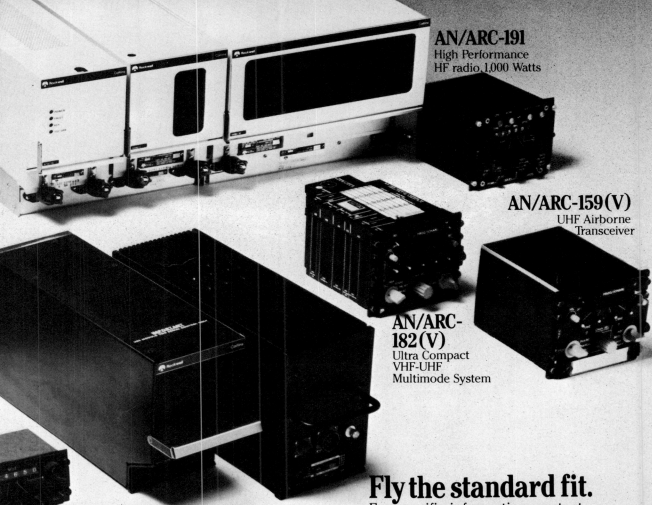

AN/ARC-191
High Performance
HF radio,1,000 Watts

AN/ARC-159(V)
UHF Airborne
Transceiver

AN/ARC-182(V)
Ultra Compact
VHF-UHF
Multimode System

AN/ARC-174(V)
Lightweight Airborne HF
Transceiver, 100 Watts

Fly the standard fit.

For specific information, contact
your nearest Collins representative
(see international sales listing, below).
Or write Business Development,
Collins Telecommunications Products
Division, Defense Electronics Operations,
Rockwell International, 855 35th St., N.E.,
Cedar Rapids, Iowa 52498, U.S.A.

Rockwell International

...where science gets down to business

Some idea of the complexity of airborne electronic equipment is given by this study of a combined laser and infra-red target detection and ranging set built by Hughes Aircraft Corporation for the US Navy's Grumman A-6 Intruder. The system 'sees' through darkness, smoke and haze to pinpoint surface targets and provide the guidance information that will accurately deliver laser-guided or conventional weapons launched by the carrier-borne aircraft

JANE'S AVIONICS

FIRST EDITION

EDITED BY
MICHAEL WILSON BSc CEng FBIS MRAeS

1982-83

JANE'S YEARBOOKS

"Jane's" is a registered trade mark

Copyright © 1982 by Jane's Publishing Company Limited

Published by Jane's Publishing Company Limited, 238 City Road, London EC1V 2PU, England

ISBN 0 86720-611-X

Distributed in Canada, the Philippines and the USA and its dependencies by
Science Books International Inc, 51 Sleeper Street, Boston, Massachusetts 02210, USA

Ferranti Avionics.
A vital part of airborne defence.

Today, airborne defence demands the most advanced technology of many kinds. It demands the wide ranging avionics technology Ferranti has, together with the unique Ferranti experience in applying it.

Ferranti is contributing strongly to worldwide airborne defence with airborne radar and laser equipment, Head Up Displays, navigation and weapon delivery systems and cockpit displays.

Ferranti is supplying artificial horizons and other instruments, aircraft electrical equipment and helicopter autostabilisers, and is in the lead with airborne computers and microprocessors, and air defence systems.

Ferranti can supply a total avionics suite to meet all airborne defence needs.

Ferranti systems and equipment are used in Tornado, Harrier, Jaguar, Mitsubishi F-1, F/A18 Hornet and other leading military aircraft around the world. It's the surest proof of their supremacy.

Ferranti plc.
Bridge House, Gatley, Cheadle, Cheshire SK8 4HZ

FERRANTI
Selling technology

Contents

ONI-7000 LORAN-C NAVIGATION SYSTEM

The **ONI-7000** is the only Loran-C system to provide coast-to-coast U.S. coverage as well as Transatlantic and Transpacific capability using existing stations. In addition operation throughout Alaska, the Mediterranean, and much of Canada, Mexico, Europe and the Caribbean is possible using this advanced design.

The **ONI-7000** incorporates features to reduce workload including flight planning mode, automatic waypoint sequencing, prestored routes, and automatic magnetic variation correction.

The **ONI-7000** provides all-altitude/all-weather operation with non-precision approach accuracy at a fraction of the price of other long range navigation systems.

FLITE-TRAK

FLITE-TRAK airborne equipment provides the telemetry of aircraft position and altitude automatically at selected intervals on integral or existing aircraft radios.

FLITE-TRAK dispatcher terminals incorporate a CRT display of aircraft positions with computer generated map/route overlays and permanent storage of flight plans and tracks.

FLITE-TRAK equipment adds a new dimension of safety and efficiency to fleet aircraft operations.

Write or phone today for details.

OFFSHORE NAVIGATION, INC.

5728 Jefferson Highway, P.O. Box 23504, New Orleans, Louisiana 70183, U.S.A.
Phone (504) 733-6790, Cable "OFFNAV", Telex 058-381

Foreword

There could hardly be a more propitious time to launch a book on aviation electronics. Technology just budding in the early 1970s is coming into full flower, with developments in some areas progressing so fast that the challenge is how to introduce them in the most beneficial way. With such advances in micro-electronics, information processing and displays, the stage has now for the first time been reached that virtually any demand can be met technically. Whether the technology on offer is bought depends on other and more subjective things such as funds available, economic trade-off, industrial agreements, and perhaps nowadays even trades union pressure.

In the military field the technology revolution has matured in such advanced aircraft as the McDonnell Douglas F/A-18 Hornet multi-purpose fighter and Europe's tri-national Tornado bomber, both just now going into service; Dassault's Mirage 2000 fighter, in which the delta-wing formula long since dismissed by other manufacturers is being given a new lease of life by the marriage of electronics and aerodynamics; and the General Dynamics F-16 fighter, blooded by the Israeli Air Force in its June 1981 attack on the Iraqi nuclear reactor at Osirak. Electronics also played a part in the UK's successful attempt to regain the Falkland Islands from Argentine possession in 1982, and supremely so in the highly orchestrated Israeli land/sea/air invasion of the Lebanon to annihilate the PLO in the same year.

Current military aircraft have a voracious appetite for electronics equipment: the content of the 3600-plus F-16s and F/A-18s ordered by the world's air forces since June 1975 alone amounts, in round figures, to $US15 000 million.

Perhaps even more topical are events in the international air transport field. As this book closes for press the first of the so-called new generation transports was being handed over to its initial customer. The medium-range Boeing 767, intended replacement for a host of obsolescent Boeing 707s and Douglas DC-8s around the world, was launched by a United Airlines order in July 1978, first flew in September 1981, was certificated in August 1982, and was scheduled for its inaugural service the following month. Its formidable European rival, the Airbus Industrie A.310, begins commercial service with Air France and Lufthansa in early 1983. Pacing the A.310 is another new-generation Boeing transport, the short-haul 757, that will be going to its first and so far principal customers, Eastern Airlines and British Airways, in January 1983.

All three aircraft carry completely new digital processing and display equipment that, together with quiet, low-consumption engines, new low-drag aerodynamics and computer-designed long-life structures, will cut the cost of ownership down to maybe half that of the aircraft they will replace. They bring with them as standard items many of the systems that in previous generations of aircraft have been regarded as optional extras: Category III autopilot and flight instruments, and performance management, for example. Healthy order books — at least for this stage in the career of any new aeroplane — reflect customers' confidence in the new designs and their ability to make money, despite the appalling financial situations of most airlines in the wake of rising fuel costs and by low passenger demand.

Not everybody is pleased, however. The ubiquitous chip, having deprived many people of their jobs on the ground, is now doing the same thing in the sky and one of the remaining flight-crew members, the flight engineer, is inexorably set for extinction, doomed in the United States by order of the president himself.

Again topical, as we close this first edition, is the news that the world's first 'zero flight time' commercial aircraft flight simulator has been approved by the Federal Aviation Authority (FAA), America's certification authority, as a means of converting pilots on to new types and enabling them to become proficient on them, all without leaving the ground.

The term 'avionics' unfortunately has no precise definition, and is sometimes used to describe equipment far removed from what is generally understood to be covered by the word. For the purpose of this book the term is defined as operational systems or equipment (but not components) designed specially for piloted fixed-wing aircraft and helicopters, airships and balloons, and drones and remote-piloted vehicles (RPVs). Flight simulators qualify for inclusion. They are really pseudo aircraft, with the same equipment and behaviour, certificated in the same manner, and capable of providing the same workload and — at times — cause

for adrenalin flow as the real thing. Systems for missiles and weapons are excluded, except for one or two special cases, as is flight-test instrumentation, which is non-operational and does not generally conform to aeronautical operational standards.

The industry is so large and varied that a selective approach to this directory has been necessary, the aim being to cover the most significant, recent, or widely used equipment. Categorisation sometimes has not been easy, but in general conforms with commonly accepted usage.

With specialisation increasing all the time, it is clear that a large proportion of the professional electronic engineers who form the backbone of the industry have little more than a modest appreciation of the 'real-world' significance of their work. Accordingly a particular effort has been made to give some prominence to the applications and aircraft themselves. Likewise, they and engineers coming into the industry from college or university or other, non-aviation fields, are often unaware of the background to their field, and so it was felt worthwhile to outline in an introductory article the development of specialised electronics for aviation use.

Current developments in navigation, for example, stem from the 1930s, when great skill and not a little 'lore' was required of pilots to find their way in bad weather or at night with the safety and reliability needed to set the infant air-transport industry on a credible footing. The size of the navigation section in this book gives some idea of the effort still being channelled into aerial sure-footedness.

The world avionics market has grown steadily since the early post-Second World War years, with a notable depression between 1969 and 1973. Growth accelerated in 1977 as the new micro-processor-based technology emerged from the laboratory and became frozen into brochures and engineering specifications. Very roughly, world sales for the year 1981/82 will have amounted to about $US20 000 million. Overall leader by far is the USA, though it is impossible to quantify relative sales because of the widely varying methods of classifying equipment.

Projections for the rest of the decade are optimistic, too, as military and commercial users increasingly adopt the capabilities of the new technologies now emerging. Advanced, digital electronic equipment can so augment the effectiveness of combat aircraft that smaller fleets are now needed to accomplish a given task, or alternatively the same budget can buy a greater potency. Furthermore, the new generation of fighters and bombers can assimilate evolutionary improvements or innovations without extensive rework in a way that their non-digital predecessors never could. Thus it is becoming customary to have one or two 'mid-term refits' during the life of an aircraft in order to upgrade its effectiveness. On the commercial side, airlines that have bought the McDonnell Douglas DC-9 Super 80 and Boeing 737 Advanced 200 are already having a foretaste of what benefits the new transports will be bringing in their train; much greater reliability, lower maintenance costs, and smaller and lighter black boxes operating on less electrical power.

Equipment on the new civil and military aircraft is becoming increasingly integrated, so that individual systems are tending to lose their distinctive identities as they 'talk to', or provide information for, other perhaps totally different equipment. Thus the weather radar displays, once dedicated to painting precipitation maps, are now often used to show flight-manual or navigation information, so that their traditional job has become just one of several modes.

While technically and economically beneficial, the integration philosophy can make life difficult for airline customers who may want to make their own choice of equipment, for example to maintain commonality with systems already in use or to protect national industry. National and international technical standards such as those of ARINC define the operational, dimensional and signal interface characteristics so as to ensure interchangeability between equipment of different manufacturers and compatibility with other equipment. However, the airframe builder is responsible for the choice of equipment, and re-certification to suit individual customers' needs may be difficult or impossible and certainly expensive. This has been very much the case with the Boeing 757 where the manufacturer has adopted Henry Ford's dictum, 'You can have any colour you like so long as it's black'.

Free enterprise and a naturally competitive environment,

together with a huge domestic market and the responsibility of providing most of the equipment to defend the Western world, have ensured that the US industry remains technically dominant overall. Names such as Collins and Bendix evoke respect in every part of the world, not least in the Soviet bloc countries. However the European star is rising, and has begun to pose a threat to the traditional US suppliers. Competition is principally from the UK, in which a number of progressive companies such as Marconi Avionics, Smiths Industries, and Rediffusion have learned their way through complicated US civil and military procurement policies, to the benefit of export order books. This is fortunate, since UK airlines and defence forces are using fewer and fewer 'national' aeroplanes, so that the home market has decreased. France, too, has emerged as a supplier of world stature, its partisan and technically literate administrative and procurement machine having made great and successful efforts to establish the electronics industry as a national asset. West Germany's electronics industry, gaining momentum in the 1960s and 1970s, has been slowed by lack of money and so is unable to make the advances that it would like. However, research and technology demonstrator programmes such as the VAK 191B v/STOL fighter of the mid-1960s and the concurrent Dornier Do31 v/STOL transport gave development impetus to a number of companies, notably in the area of flight controls. Tornado and the Alpha Jet strike fighter have been the principal recent markets. Italy is further behind, again through lack of money for research and development, but has considerable expertise in a number of areas. Japan has a formidable reputation in consumer electronics, but has not really broken through into avionics. The Soviet Union is more of a nucleus in relation to its allies than is the USA to Europe, and so provides virtually the entire design and manufacturing capacity. The USSR is reckoned generally to be some five years behind equivalent US technology in the avionics field, with perhaps smaller gaps in particular military areas.

The air transport system is one of the world's major industries, and one of the most technologically based. The introduction of the four Western wide-body airliners (the term is used to denote aircraft substantially larger and heavier than Boeing 707s and Douglas DC-8s, and with two aisles in the passenger cabin) in the 1970s brought a host of refinements and, in particular, a newcomer: inertial navigation, or IN. (Strictly speaking, IN made its debut in 1968 aboard 707s of American Airlines, which had been contracted to ferry supplies over the long-haul US West Coast to South-East Asia routes, entirely out of reach of ground-based navigation aids.)

The reliability of these first-generation commercial IN systems has been outstanding. Captain Laurie Taylor, a former senior pilot with British Airways, notes in an operational appraisal of the Boeing 747 and its equipment: "In my experience it is the only aeronautical innovation within the past 30 years that has lived up to the claims made for it in terms of accuracy, usefulness, reliability, and reduced crew workload". Only old-timers who have grown up in the flying game to share a widespread mistrust of 'electrics' can really grasp the magnitude of this compliment. For a number of airlines it sounded the death-knell of the navigator.

However, IN equipment is expensive to buy and maintain. A single set may cost $US100 000 to 120 000 and certification authorities require a minimum of two sets for over-water operation away from the coverage of ground-based navaids. Sometimes circumstances may demand three, for example operators with long, polar routes where compass readings may be unreliable for hours at a time. Maintenance costs at perhaps $2 an hour are not particularly low, high-grade test equipment is needed, and also skilled labour. Operators therefore casting around for other solutions to the navigation problem have come up with one that may catch on: a mixture of IN and omega or vlf being developed jointly by Sperry and Canadian Marconi.

A number of IN manufacturers include omega and/or vlf (very low frequency) equipment in their product range, and systems have been certificated for over-water use as the sole navigational aid. While mechanically simpler than IN, these radio-based aids do not provide attitude or heading, and some operators are concerned that coverage of the earth's surface is not 100 per cent yet and that propagation anomalies still exist. Nevertheless they have caught on in a big way.

Since the wide-bodies came into service, another type of equipment has been introduced: the IRS, or inertial reference system, and it has become standard on the Boeing 767 and 757. While not providing the comprehensive information and commands of an IN, the IRS is much more than a half-way house, deriving basic position, acceleration and speed. One of the reasons for the cost of IN is its mechanical complexity. In an effort to eliminate in IRS equipment the large number of precision electro-mechanical components — gimbals and bearings, gyros and synchros, torque motors, slip-rings and resolvers — industry has introduced a new method of sensing linear and angular accelerations known as strapdown technology, using so-called laser gyros. It makes use of phase differences between two beams of laser light travelling round a closed circuit in a solid-state optical system. The devices (one for each axis) are solidly mounted in standard boxes, along with associated electronics that provide the computing facilities for transforming attitude and acceleration information from 'box' into aircraft axes.

The prestige of being chosen to provide the world's first production laser-gyro IRS equipment for commercial aircraft goes to Honeywell, which has tooled up to build 400 sets for the 767 and 757. There should be little or no technical risk involved by virtue of the company's very long research into these devices culminating with flight trials beginning in 1974. Honeywell claims that the laser-gyro IRS will cost significantly less than any other combination of equipment giving the same performance and capability.

The flight-decks of the three 'new-generation' aeroplanes: 767, 757 and A.310, are instantly distinguished from those of their forebears by the presence of television-type instrument displays. Each aircraft has six such displays: two for each pilot, giving flight guidance, one showing engine performance and health, and one a general-purpose display that can show 'pages' from the flight manual, or other information that may be needed. Front-runners among suppliers at the present time are Collins, chosen by Boeing for its two new transports, and France's Thomson-CSF, which is aboard the A.310.

These are indeed impressive devices, and are now being proposed for virtually all new commercial aircraft or projects, including helicopters. The advantages are the elimination of the intricate mechanical elements associated with the two principal flight instruments, and the facility of being able to change the format of one of them — the horizontal situation indicator — to expand the display or include other information such as weather maps.

Head-up displays (hud) are the recognised centre of information in military aircraft, and in June Marconi Avionics handed over the 1000th hud to be built for the multi-national F-16 programme. The event was a gratifying sign that top European companies with heavy investments in technology and high-grade staff can still fight the US giants and win on their home ground. The UK company has sold overseas no less than £500 million of aviation electronics in the past decade. In May the same firm delivered to the US Air Force the first of a new type of hud, using holographic techniques to provide the much wider field of view needed by the new generation of highly manoeuvrable fighters and missiles.

One of the most noteworthy and sustained issues in recent years has been the question of crew numbers. With the navigator gone, ousted by the introduction of automated systems, particularly inertial navigation, the standard complement on most aircraft has been two pilots and a flight engineer. For many years airlines have been trying to dispense with the engineer, and indeed the Boeing 737 twin-jetliner of the late 1960s turned out to be a battleground over this issue. But now declining revenues as a result of world-wide recession, deregulation, and disproportionate rises in fuel costs on the one hand and the need to justify the adoption of automated flight and navigation equipment on the other has forced a showdown, and the United States president himself set up a task force to enquire into the matter. In June 1981 this task force reported its conclusion: aircraft designed for operation by crews of two were safe to fly that way.

Pressure from trades unions had compelled Boeing in the early stage of 767 design to abandon the two-crew configuration for the traditional layout, and consequently there was less reason to adopt the new electronics. The company was obliged by airline insistence to reverse its conservative approach, however, and shielded by the presidential task force recommendations, decided in December 1981 that it would certificate both two- and three-man crew layouts, the new configuration having priority. Despite a cost increase, at the time of writing virtually all 767 customers have elected to have the new flight deck. In a supreme management and engineering effort, the very difficult mid-stream changeover was accomplished and the aeroplane certificated in the time originally scheduled for the three-man layout.

The 1973 oil crisis that obliged airlines to seek more profitable ways of operating their fleets, and the realisation that cheap fuel had vanished for ever, stimulated industry to see how burgeoning electronic capabilities could be harnessed specifically to save fuel. In hard figures, the annual fuel bill for older aeroplanes such as the 707 and DC-8 has risen from 21 per cent of the direct operating cost in 1967 to 55 per cent by 1981, and projections show that it could rise to 60 per cent. This savage penalty has caused many of these early-1960s transports to be retired, and given rise to a new family of black boxes for the wide-bodies that provides the flight-crew with information on the best speeds, heights, and routes, or commands the aircraft through the autopilot to fly in accordance with them.

The technology has grown up so rapidly — the first system was introduced about 1977 — that terminology has not kept pace and each manufacturer, it seems, has a different name for the same thing. There are however two basic varieties of equipment: the performance management system, computing best speeds and heights, and the flight management system (FMS) in which the capability is expanded to provide lateral (ie navigation) information or commands. In British terminology flight management is regarded as taking place within one or two dedicated units, while in American thinking it embraces a host of aircraft operation and navigation systems.

In the short time it has been flying, the new equipment has built up a good reputation for itself. Systems on the 727, 737 and 747 transports with far less capability than those being built for the new aircraft have been turning in fuel savings of 1.5 to 4.5 per cent on routes from 2000 down to 400 miles. Again, the equipment is not cheap: a dual FMS for the 747 can cost $US400 000. But figures are relative; British Airways says that a 1 per cent fuel economy on the jumbo can save $250 000 a year. While FMS is not essential to safety, dual installations are obligatory if the calculated fuel savings are to be translated into permissible smaller fuel uplift at the beginning of a flight, resulting in even further savings.

The increasing capability of the new aircraft and their systems calls for more extensive training, and the flight simulation industry is also benefiting as more and more of the proficiency-maintenance load is put on to these devices. The same computing technology that is helping to justify the huge investment in the wide-bodies and new-generation aircraft is being used to improve the performance and realism of flight simulators. Pressure from the airlines and some weakening of resolve on the part of national regulatory bodies have led to proportionately increasing amounts of training being carried out on these devices, and the big news of the year was the FAA's certification of a Boeing 727 simulator by Canadian company CAE Industries for Phase III operation. This means that it can be used not only for standard proficiency training, but also to convert pilots to that aircraft from other types without actually having to fly a 727.

The rationale for the increasing use of simulators may be summarised as improved safety and crew efficiency stemming from a more comprehensive training schedule, substantial fuel savings, reduced aircraft maintenance, the release of aircraft for line operations instead of non-revenue-earning training, and better community relations owing to the lower total noise. In the past the real-life simulation of emergency procedures, such as handling on the approach to land with asymmetric engine power, has not infrequently led to the very accidents students were being taught to avoid. In simulators which accurately duplicate aircraft characteristics emergency and recovery procedures can be practised and re-run, stopped at critical junctures for discussion, and if need be deliberately allowed to degrade to the point that in real life would have resulted in an accident.

The factor that has permitted these devices to be certificated with progressively greater responsibility is the increasing degree of realism being attained. This is partly due to the use of more computing power, better mathematical models, more realistic displays, and not least to more accurate measurements of actual aircraft behaviour, particularly in emergencies. With all these advantages, it is unsurprising to note that one market survey puts the value of the simulator industry between 1981 and 1985 at no less than $US7000 million.

The digital revolution introduced by the US military in the early 1960s and now being adopted by the new transports and some refurbishment of older ones was a huge step, and the arrival of single-chip micro-processors finally made it possible to do 'on the spot' computing, thus settling an old argument: central or distributed processing? Nevertheless it did not do much to ease the mechanical problems of handling terminations on equipment (plugs and sockets) and the growing size and weight of cable looms as more and more data was being passed between increasingly exotic equipment. The weight of cable on Britain's first-generation digital aeroplane, TSR.2, was around five tons, or some 20 per cent of the bomber's empty weight. Not only weight, but finding room for bulky cable looms in increasingly small airframes has become a major headache for designers and production workers.

A major advance in digital technology therefore was the advent of the data-bus, in which the digital information from a piece of equipment is run along a single pair of wires, being multiplexed so that each signal is sampled many times a second. At a stroke this technique cuts down on size and weight of cable looms, and simplifies the problem of introducing new equipment. US industry in conjunction with a government agency (the US Air Force Avionics Laboratory at Wright-Patterson AFB, Ohio) began work on data-buses in 1969, and the first aircraft to benefit was the General Dynamics YF-16 LWF (Light Weight Fighter) technology demonstrator. This aircraft — significantly, it had a small airframe — went on to become the top-selling F-16 fighter. Data-bus technology has been applied to all major civil and military aircraft since that time, and is being continuously improved. Other applications are being sought or introduced, eg in highly automated vehicles such as tanks and ships, and in early warning or defensive ground-based installations.

Digital technology brings with it the associated needs for adequate software. The provision of software is now a major industry and an expensive item since it is so labour-intensive. It is said that development of the software for the Sperry flight-management system chosen for the Boeing 767 and 757 took no fewer than 250 man-years. Just as with equipment, software has to be certificated by national regulatory bodies such as the US Federal Aviation Authority or Britain's Civil Aviation Authority. Experience so far indicates that there is still some distance to go before changes or modifications to software implemented by aircraft manufacturers can be accepted with confidence by users. One airline complains that virtually every change has a fault somewhere that has gone undetected through perhaps months of work at the manufacturer's, but has shown up within days of introduction to service. It is not unusual to spent $US100 000 on a program chip, only to find that it does not work, so that the airline ends up having to do the debugging, a job which is not only time-consuming but which the user is legally not permitted to do.

The problem is largely one of common-mode failures — the same fault affecting all three supposedly independent channels in an automatic landing system, for example. And the cause? Largely the software engineer, whose obscure fault in a program infects all equipment or channels employing that program. French policy is to employ dissimilar redundancy, though this is naturally more expensive because the software, or perhaps at least part of it, has to be designed twice.

The old computer engineer's adage 'Garbage in — garbage out' loses none of its force merely because airborne equipment is so much more expensive than its ground-based counterpart, and two examples suffice to show the fallibility of the finest computer of all — the human brain. First, a civil incident. The Royal Commission Report on the 1979 Air New Zealand DC-10 crash in the Antarctic puts the cause of the accident, in which all 257 people aboard died, as the use of incorrect navigation co-ordinates to program the wide-body's IN system. Secondly, a military incident. A McDonnell Douglas F/A-18 Hornet tasked for a test flight during 1981 failed to recover from a spin and crashed into the water, the pilot ejecting after methodically going through the approved recovery sequence. The subsequent inquiry found that the flight-control system had not been programmed to respond to the very high rates of rotation set up under these conditions. For the new 'digital' aircraft, adequate software management is going to be essential, and is one of the greatest challenges to be faced.

From an avionics point of view, the fight between British and Argentine forces for possession of the Falkland Islands appears to have been a relatively simple affair, a ding-dong match between Royal Air Force and Royal Navy Harriers and Sea Harriers, and Argentine Air Force A-4 Skyhawks and Mirage IIIs, without any great technical sophistication.

Not so in the Israeli efforts to dislodge the Palestinian fighters from their Beirut stronghold. Here the Israeli Air Force displayed its supreme skill in modern air warfare, integrating RPV (remote

FOREWORD

piloted vehicle) surveillance and targeting, stand-off jamming by means of dedicated aircraft, and precision bombing accuracy.

The little Scout RPV, an indigenous design by Israel Aircraft Industries, was able, through an electro-optical sensor, to monitor the preparation and despatch of Syrian fighters on their airfields, and to transmit pictures back to Israeli positions. Meanwhile electronic warfare Boeing 707s equipped with comprehensive ecm equipment jammed communication links between Syrian ground controllers and their fighters. At the same time Grumman E-2C Hawkeye command and control aircraft provided the early-warning of approaching fighters. Thanks to the efforts of intrepid television camera crews on the ground, the viewing world was able for the first time actually to see electronic warfare technology in action, in the form of flares ejected by Israeli fighter-bombers (principally Phantoms) to deceive PLO and Syrian anti-aircraft missiles.

What viewers saw was just the tip of the electronic-warfare iceberg. For security reasons, EW is a very secretive industry, but clearly a very profitable one for the firms concerned. Loral, one of the foremost names in the business, for example, predicts worldwide demand of $US5300 million in 1983, a 39 per cent increase over the 1980 figure of $3800 million.

In terms of quantity production, the ASPJ (advanced self-protective jammer) programme by ITT and Westinghouse is expected to be the most significant such Western effort of the decade, worth in total around $US1500 million at 1978 prices. The level of technology and effort needed to develop this system was early on recognised to be so high that even top US companies, acting alone, were unlikely to be able to take it on. Accordingly the joint US Air Force and Navy steering organisation encouraged multi-company consortia to bid for the contract.

Production began in June 1982, the target cost of $350 000 per set, even with a potential 4000-aircraft market — giving some idea of the complexity involved.

Acknowledgements
My grateful thanks to the many people who have helped with material, advice, and enthusiasm in the production of this first issue of *Jane's Avionics* to get the new brainchild airborne, so to speak. The Jane's people, especially Sidney Jackson, Valerie Passmore, Ken Harris and Ania Warne, have been outstandingly helpful and patient. At a critical juncture Mike Hirst, Mike Witt, Richard Whitaker, and Don Parry pitched in to improve the thrust/weight ratio, giving unstintingly of their time and skills.

Electronics calls the tune

Civilisation has always been equated with communication. The ability to communicate instantly by voice or picture, and personally in only a little longer time, permits mankind to exchange ideas, decide actions, and form bonds with those who in earlier times would have been ignored or looked on as potential enemies. In well under 50 years this capability has obliged nations previously concerned only with their own parochial affairs to recognise that they are responsible members of a global village. The twin keys that more than any others have opened the doors to this change are electronics and aviation.

The idea of a village connotes a common language or method of communication, a continuing and easy dialogue among members of the community, a unity of purpose, and a fellowship that underpins a caring and compassionate society. Electronics and aviation have smoothed a path along which humanity may travel as far as it wishes toward these desirable goals.

This path-smoothing began in earnest after the Second World War, when air transport got into its stride. It was greatly extended in the early 1960s, when the first jet airliners began to shrink the globe, and again a decade later when the wide-body transports broadened the way with low-cost travel available almost universally. In the wake of the politicians who by their personal statesmanship create frameworks of international agreement came business executives and then tourists to 'flesh out' the skeletal structures of mutual declarations with commerce and human relationships.

On the obverse side of the coin is the darker reality that technology — an ethically neutral activity — has conferred on humanity an awesome power to destroy, a capability greater than anything that could have been imagined half a century ago. Sadly, the technology that now gives people an easier and more enjoyable life, culminating for many with package holidays in benign climates once as remote as the moon, has largely been developed under pressure of the old maxim, kill or be killed. Where animosity exists, the globe itself is now hostage to a single ill-considered act, and in a timescale measured in minutes. Again, the nuclear sword of Damocles owes much to electronics and aviation.

Commercial and military aviation depends totally on increasingly complex automated functions that are routinely conducted with a speed, consistency and accuracy no human operator can approach. The development of control-law theory shows that, as complexity grows, people's response time and proneness to error increasingly limit the performance of a system if they function as a circuit element. Their most effective role in these circumstances is as monitors and decision-makers outside the system, with the ability to exert their will based on an appreciation of total situations that often cannot be rationalised simply as stored data in a computer. In this way people and machines best complement each other.

Technology v. unemployment

While this optimum relationship has been recognised for many years, the social environment has often been slow to implement it. The development of micro-circuitry and its application to data-processing can produce unemployment, not only among the unskilled and semi-skilled. On the flight-deck the radio operator began to disappear in the late 1950s, advances in automation since the 1970s have displaced the navigator, and the flight-engineer is now also an endangered species in the new generation of transports that will replace the older narrow-body airliners.

However, although some forms of surface transport operate automatically, with no human control agency aboard, no-one has seriously suggested that this policy should be extended to commercial and military aircraft (drones and remote-piloted vehicles are special cases, governed by other considerations). The capability of current electronic equipment is such that the designer's problem is not how to displace the pilot or flight crew but to work out how best to integrate their capabilities with the aircraft management and navigation systems. A classic case is the argument over collision avoidance: should the regional air-traffic centre be responsible for determining conflict situations, or should the conduct of the flight lie entirely with the crew? The technology exists to do either, and the decision is based on factors such as safety, cost, reliability and crew workload rather than on what is technically possible.

Many of those in the vast avionics industry are first and foremost professional electronics engineers, concerned with their own trade and product, and perhaps having little aviation background. For these, and particularly for the younger engineers and technologists who have recently come into the industry, it was felt worthwhile to paint in some of the principal milestones in the development of specialised aviation electronics and to see how they reflected the needs of the times.

The first recorded application of electronics to aviation goes back almost to the dawn of powered flight, to 1910 in fact. On 27 August of that year Canadian designer and pilot J D A McCurdy transmitted and received radio signals aboard a Curtiss aeroplane over Sheepshead Bay, New York State, using an H M Horton wireless set. During the following month another pilot, Robert Loraine, transmitted the message 'enemy in sight' from his Bristol Boxkite biplane over the British Army ranges on Salisbury Plain to a receiver a quarter of a mile away. It was the first time in Britain that airborne equipment of this nature had been used in a powered aircraft, though a ground-to-air transmission to a free-flying balloon had been accomplished two years earlier at Farnborough, headquarters of the British Army Royal Engineers Balloon Section.

These early trials led to the routine use of wireless (how already evocative of far-off times that term is!) as a means of recording fall of shot for the armies of the First World War. Again, it was not unusual for pilots to lose themselves, and already during the 1914-18 period advances in radio-navigation were taking place. At Cranwell, for example, a direction-finding system based on a null-sensing loop aerial was developed to work at 500 to 1500 kHz. Receivers of high sensitivity were essential to respond to the very low transmitter power available. They were thus susceptible to interference from magneto-based ignition harnesses and protracted efforts to perfect a workable system constituted the first example of what in decades to come would be called, in the jargon of the day, 'systems integration'.

Wartime developments

The First World War was a time of considerable technical development, and aviation was transformed from what many saw as a dangerous and impracticable pastime for the lunatic fringe into a new dimension of human experience, with unguessable but exciting potential. Much of the technology that was to form the basis of all subsequent commercial and military aviation came into being at that time. For example, the strategic bomber offensives launched by Britain and Germany against one another towards the end of the war soon called for new equipment. Both sides conducted raids by night when the risk of being shot down was less. To assist their crews, the German Gotha biplane bombers were fitted with the first artificial horizons, and by the war's end both sides had some feeling for the difficulties and dangers involved in nocturnal operations. Blind or night flying, and navigation were recognised by Europeans (who had taken the lead in aviation) as the two great early challenges that might be amenable to new equipments and technology.

But it was Lawrence Sperry in the United States who provided the basis for all-weather flying. To the artificial horizon he added the combined turn-and-slip indicator and the directional gyro. While horizons were regarded by some as an unnecessary indulgence, the turn-and-slip was soon recognised as essential in any situation lasting more than a few seconds in which natural orientation by observation of the ground was not possible. The directional gyro's value lay in helping the pilot to come out of a turn on to a pre-determined heading with precision, an ability not possible with the magnetic compass on account of the swinging that takes place.

Despite these developments there was considerable resistance from pilots to using more than the simplest aids, on the grounds that undue reliance on them might cause a loss of flying skills. A dashboard mounting, an engine rpm indicator and oil-pressure gauge, air-speed indicator and altimeter, and perhaps compass, was considered handsomely equipped; indeed, most (and with good reason) regarded only the engine instruments as essential. There were no air-traffic procedures to follow, and so pilots could choose their own cruising altitudes. Most spent their time below 5000 feet, finding their way by course of road and railway, and their judgement of height soon became accurate, the more so the lower they flew. Likewise pilots quickly became adept at controlling their speed from the attitude and 'feel' of the aircraft, and from the throttle setting. The speed range for most types was small, governed at the upper end by lack of power, and at the lower by the stall. None of these instruments was electronic, or even electrical, in nature, but they were later to become the basis for much more sophisticated types.

The period between 1920 and 1930 was one of consolidation. New aircraft were quick to build and relatively cheap, and their performance and reliability improved rapidly under the stimulus of the many pioneering flights that were being undertaken. Exploits such as the long-distance missions by European and American flyers, were largely setting the scene for closer links with dependencies, allies, or (in Britain's case) the Empire.

In parallel with or following up these often spectacular aerial voyages were the less sensational but in other ways equally demanding operations of the first commercial carriers. In Europe small airlines were set up to provide quite local services, often just to link two cities, for example London and Paris. In the USA passenger air transport was slower in establishing itself but lines were set up to service the US Mail, a commitment that made stringent reliability demands on its operators. All these activities revealed a pressing need for better and more comprehensive radio and navigation aids. The potential of radio as a navigational facility as well as for communication had been realised during the First World War, when it was used to fix the position of Zeppelin airships on their way to bomb London. The 'ships transmitted signals to radio stations in Germany, and the measured bearings were co-ordinated by a master station to provide position, which was then transmitted to the raider. Position-fixing by triangulation of radio transmissions was also the basis of a remarkable 3000-mile, non-stop flight by a Zeppelin which in 1917 was sent from Bulgaria to the relief of a military force in German East Africa.

Radio communication was becoming recognised as an essential adjunct to the safe conduct of any routine air-transport activity. By the end of the 1920s it was extensively used to keep commercial aircraft in touch with airline operations departments, and to provide pilots with weather reports and surface-pressure settings for altimeters. It was not much used for terminal operations, ie at airfields, where signalling by red and green lights was still *de rigueur,* but could and did provide verbal guidance to aircraft overhead that could not see the field for fog or mist.

For long-range flying, Morse code was employed in preference to speech since the limited transmitter power available could be channelled into the much smaller bandwidth needs of dots and dashes, when Morse signals could 'punch' their way across long distances and be intelligent in a way that speech transmissions often could not.

Radio navigation

But more than anything else it was the radio range that really set the infant airline industry on its navigational feet. This was an American invention of the late 1920s, in which the direction of four radio beams defining the approach to an airport was provided by transmitting, in Morse code, the letters A and N. The pilot could find the centre-line of the beam by listening, in headphones, until the letters came through with equal strength. To help the pilot fix his position along the beam away from the intersection of the four beams that comprised the system, so-called fan-markers were installed that broadcast a radio beam vertically upwards. Radio ranges, representing the beginnings of today's complex system of airways, could however be tricky to fly, operating in the hf band and so susceptible to interference by weather. They could also be 'bent' or deflected by certain obstructions, and knowledge of these peculiarities and how to compensate for them constituted a large part of a pilot's stock-in-trade of knowledge.

By the early 1930s the airlines were becoming busier, and their aircraft were becoming more sophisticated and capable of longer journeys. They were by then routinely flying by night and in bad weather, putting increasing demands on the flight crew — usually two (captain and first officer) by this time. During long flights, often in turbulence, the new and bigger aeroplanes were becoming tiring to fly, impairing the crew's ability to carry out other tasks. So it was that in 1933 Eastern Airlines (then known as Eastern Air Transport) began taking delivery of its new fleet of Curtiss-Wright Condor biplane transports fitted with the world's first commercial autopilots. These Sperry A1 electro-hydraulic systems did much to alleviate the difficulties of operating the New York-Miami route, with its notoriously bad East Coast weather.

The principles of automatic flight control had been known long before that, however, in fact since about 1873, 30 years before the Wright brothers demonstrated powered flight for the first time. The aeroplanes in the first decade of sustained flight, ie up to the beginning of the First World War, were frequently difficult to fly, being unstable and lacking control power to restore quickly level flight when they were upset by turbulence or cross-winds. From about 1909 onwards there were two schools of thought as to how these difficulties might be alleviated or removed. The first held that designers should give more attention to 'building in' natural stability so that aeroplanes were less easily upset. By implication, this would make them less manoeuvrable, calling for more effort on the part of the pilot in anything other than straight and level flight. The second school challenged this view, saying that a more acceptable solution would be to modify the handling characteristics of a design with relatively low stability by means of a control system which would provide artificially the desired characteristics.

The argument was resolved in June 1914 when Lawrence Sperry demonstrated from the River Seine at Bezons the impressive performance of his Curtiss flying boat stabilised by an elegant, 40 lb mechanical/pneumatic gyroscopic system that automatically maintained level flight. The aircraft was entered for the aeroplane safety competition which the Aero Club of France was conducting on behalf of the French War Department, and won the $10 000 top prize, handsomely covering the $8000 cost of development. Certainly the device, constituting what would now be termed an auto-stability system, had many drawbacks: a lengthy procedure was needed to set it up for each flight and constant adjustments were required. The gyros, for example, had a high drift rate, and so had to be made pendulous, which in turn made them sensitive to disturbance and protracted acceleration, for instance, during turns. Friction in the gimbal bearings and deadspace and flexibility in the control runs and the airframe itself added to the difficulties. Proportional amplifiers and signal-mixing devices were not then available, so that the system could not in any way have been optimised.

Autopilot technology

Despite these difficulties the Sperry stabilising system showed what could be done, and from it — with the addition of automatic steering, developed during the First World War — came the automatic pilot, or autopilot. By the mid-1930s autopilots were in general airline use, and their capabilities were dramatically demonstrated in a number of long-distance flights; Wiley Post's Lockheed Vega *Winnie Mae* for example had a Sperry A2 autopilot to help him in his solo, round-the-world flight of July 1933, the year (as noted earlier) that saw the commercial introduction of these systems.

In the late 1930s and early 1940s the first four-engined heavy bombers were making their appearance. The performance of Britain's Lancasters and Halifaxes, and America's B-17 Fortresses and B-24 Liberators, with twice the speed and three times the weight of the contemporary Boeing 747 and Douglas DC-3 commercial transports, were by then such as to call for autopilots with more facilities and better performance to offload crews on flights up to 12 hours long. In the UK pneumatic systems continued in use, developed by the Royal Aircraft Establishment, but in the USA more refined electrical systems were taking the air. The Minneapolis-Honeywell C.1 was just such a system; in addition to the usual attitude- and turn-control modes, it could be slaved to the new, highly secret Norden bombsight. Considerable technical sophistication was needed to match the performance of this and other contemporary bomb-sights, and it is hardly surprising to recollect that the Sperry A.5 autopilots fitted to Fortresses and Liberators weighed some 250 lb apiece. But despite their complexity, autopilots were still 'add-on' items — they could be designed in isolation from the aircraft in which they were to be fitted. The word 'integration' had still not yet entered the designers' vocabulary.

The 1930s were times of rapid technical growth in aviation, developments in structures, propulsion and electronics benefiting not only those areas but cross-fertilising others. The implications of being able to cross national boundaries at will with heavy warloads to put at risk national industrial resources and terrify civilian populations in a way that battleships and armies could not was giving urgency to fresh considerations on defence.

Accordingly the major powers began to seek ways and means of detecting still-distant hostile bomber streams in order to gain as much time as possible to launch defending fighters and get them into position. In the UK and elsewhere work was done on measuring the infra-red emission from engines as a possible way of detecting aircraft while still far off. However the detectors of the time were nowhere near sensitive enough to measure the small amount of radiation from aircraft exhausts at the distances called for. But simultaneously work going on into the properties of radio waves was laying the foundation for a new tool of extraordinary power and utility.

Almost simultaneous research by the UK, Germany, the Soviet Union, and the USA in the mid-1930s showed that aircraft at quite long distances illuminated by radio beams of around 50 metres wavelength could reflect sufficient electro-magnetic energy to be detected back at the transmitter. Under the threat of German rearmament the British Government and industry worked urgently to perfect the technique of radiolocation, as it was called in Britain, and by 1940 had a chain of early-warning stations along the south-eastern coastline. They were to be the means whereby the RAF's scanty fleet of Spitfires and Hurricanes were husbanded and efficiently directed against vastly greater enemy bomber forces.

Foiled by a meagre force of intercepters during the daytime raids of the Battle of Britain, the Luftwaffe, switching tactics, turned to night bombing. To counter this new threat, a GCI (ground-controlled interception) chain, a variation of the daytime early-warning system, was established around south-east England to guide, or 'vector', defending fighters onto individual enemy raiders. The ground controller employed long-range surveillance and detection equipment to bring the fighter into position at a closing range of a few thousand yards. Visual contact with the enemy was then established by means of detection equipment carried in the aircraft, operating at shorter wavelengths and lower power, to bridge the remaining separation. (The British term 'radiolocation' was eventually replaced by the more precise American word 'radar (radio detection and ranging) as a collaborative programme of research and development got under way between the two countries.) These sets, flown aboard Blenheim 1F night-fighters during the winter of 1940-41, were the world's first AI (airborne interception) radars. Their operation called for great skill and interpretative ability, and mandated a specially trained pilot and radar operator, the two operating as a close-knit team. It was the lack of single-engine, two-seat aircraft with the performance to match that of their fast and manoeuvrable adversaries, that obliged the RAF to adapt twin-engined types; the Blenheims were only stopgap night-fighters, soon being replaced by Beaufighters and Mosquitos with their much greater speed and firepower.

The pacing factor in these early radar developments was the need to increase radiated power and reduce wavelength to provide longer range and better resolution. Initial work with ground-based detection transmitters at around 50 metres soon produced systems in Britain and Germany operating at 1½ metres, adequate for early warning at that time. The need for better resolution and accuracy in the necessarily small interception radars, with their diminutive aerials, brought about the 'centimetric revolution', with wavelength of around 10 cm, owing much to the UK invention of the resonant-cavity magnetron as generator, the wave-guide as a low-loss feeder between magnetron and aerial, and the shaped dish aerial. In particular, the adoption of shorter wavelengths permitted the use of dish aerials small enough to be carried internally (thus without incurring speed-reducing aerodynamic drag) but still providing satisfactory resolution.

Signal processing

Along with developments in radar came the first attempts in its twin technology: signal processing. In the UK this was first applied to a radar display for single-seat aircraft intended for night-interception, and in which radar management was only one of the pilot's tasks. Arguments as to whether a second member is really essential to operate an AI radar continue to this day; the appearance of such outstanding single-seat fighters as the F-15, F-16, and F-18 seem to show that with the current generation of radars and their sophisticated information-processing capabilities, a second crew member may now be finally considered redundant.

As the tide of war for the Allies turned from defence to attack, so the emphasis on radar development in the UK turned from airborne interception to ground target discrimination. At the beginning of 1943 the RAF stood ready to launch its long-awaited

strategic bomber offensive. This failed to achieve the results expected by many (though not the chief of Bomber Command) because of crews' inability to navigate to their aiming-points in the dark. To help it find targets by night and in bad weather was a new radar device known as H2S, which painted a radar 'map' of the target and surrounding area on a cathoderay tube so that bomb-aimers could identify the precise aiming-point.

The same year was to see the steady throttling of Britain's crucially important sea-supply lanes by German submarines, and so the first ASV (air-to-surface vessel) radars were hurriedly developed and rushed into service aboard the Sunderland maritime-patrol flying boats assigned to convoy patrol. These proved an immediate success, as the U-boats were unable for a long time to detect the centimetric ASV transmissions (Germany not yet having developed transmitters or receivers to handle these short wavelengths) and many 'boats, caught on the surface, were sunk because they were unable to dive in time.

In the early 1930s Germany's Lorenz company developed a blind-landing system to help commercial aircraft find their destinations. It was widely used, not only by the airlines, but also by many air forces, including the RAF and the Luftwaffe. In essence it employed two radio beams transmitted along the desired flight path, and making a small angle with one another in the horizontal plane. One was coded with Morse dots, the other with dashes, both with the same strength. Where they overlapped a radio could pick up both dots and dashes. At the centre of the beam a continuous note was heard (the dots exactly occupying the interval between the dashes) and so the pilot manoeuvred his aircraft until he reached this area, then held the aircraft on course until the destination airfield came into view.

Blind bombing

The Lorenz beam was the basis of an ingenious scheme by German scientists to provide an accurate bombing system known as X-Gerat. A single beam defining the course to reach the target was intersected at quite large angles by three more beams, the first two of which notified the navigator that he was within a certain range of the target, the third intersection triggering an automatic release. Another system, lacking the refinements of X-Gerat but much simpler and based exactly on the Lorenz beam frequencies, was Knickebein, developed by Telefunken. It was put into operation in September 1940, just as the Luftwaffe was turning its attention from day to night bombing. The receiver, the avionics part of the system, was a slightly modified radio.

The usefulness of Knickebein was sharply reduced by a new electronic technique known as jamming, in which its signals were 'doctored' by transmitting stations in England so that instead of the expected 'continuous note' zone, aircrews could hear only a mixture of dots and dashes. They could not therefore locate the centre of the beam, and thus the science of electronic warfare was launched on its exotic, and for the most part highly classified, career.

Germany's strategists placed great emphasis on the development of electronics to help the war effort, and in the context of Luftwaffe operations set up a special unit staffed by highly skilled crews, known as Kampfgruppe 100, to exploit the new beam-flying techniques and secure squadron experience with equipment.

Britain's night-bomber offensive began early in 1943, but its efforts were largely ineffective owing to the difficulty (noted earlier) of locating targets by night. Eventually the RAF hit on the plan of establishing a force of specially trained 'Pathfinder' crews. The Pathfinder squadrons, sent out ahead of the main force to identify the target with a sophisticated system of flares or 'markers', used a system called Oboe to give them a fix on the target. Like Knickebein, it employed a radio triangulation technique, but the method was different. Instead of flying along a beam, the Pathfinder flew an arc of a circle at constant radius from a UK-based transmitter, the arc being arranged to pass over the target. Range was measured by interrogating a transponder on the aircraft by means of transmissions from the station and deriving the time taken for pulses to reach the aircraft and return. A Morse-coded signal of dots and dashes was then transmitted to the pilot, who flew a course so that they blended to produce a continuous, oboe-like note in the headphones. Approach to the target was announced by a second station, which provided a radio fix as the aircraft passed over it. The use of transponders to provide range was a development of their earlier application to code radar echoes in order to differentiate hostile from friendly aircraft in a technique called IFF (identification: friend or foe). The two applications were the first in a technology that was later to become known as secondary radar.

The limitations of these British and German devices was that an accurate fix (indeed, any kind of fix) was possible only at the intersections of beams. Risk of collision therefore prevented their use by more than one aircraft at a time. Moreover, bombers could make their approach to the target by only one route, which could be changed only by moving the ground stations. Even then, only slight variations were possible. The method was good for Pathfinder aircraft, though limited to 'line of sight' range, about 250 miles. It was also reliant on the effectiveness of just one or two aircraft. If these were shot down or lost en route, the raid failed disastrously.

Clearly, what the main bomber force needed was an 'area' navigation system, in which aircrews could accurately locate themselves by night or in bad weather, over very large regions. With the large-scale bomber offensive against Germany in prospect, scientists in Britain between 1938 and 1941 developed such a system. Called GEE, it comprised three widely separated radio stations along the east coast, radiating omni-directionally and in synchronism with one another pulsed trains of vhf energy. Special receivers measured the time differences between the arrival of pulses, and by comparison with purpose-made charts the navigator could determine the aircraft's position. In effect, the GEE transmitters overlaid Germany with an invisible and stationary 'grid', and thus was born hyperbolic navigation, the first rnav system.

GEE proved a most valuable navaid, and was kept in service till the late 1960s. Although different in operation, it laid the credibility ground-work for a succession of 'grid navigation' systems developed by Decca using low-frequency, continuous-wave radiation that set up hyperbolic phase difference patterns that were analysed by means of special phase-measuring instruments and charts. The Decca Navigator system and its variations proved highly successful, and remains in large-scale service, though the principal users are the shipping companies.

Component reliability

After 1936, when it became inevitable that war would begin, companies in Britain and Germany made great efforts to improve standards and reliability in the components industry. Domestic radio was by far the biggest market — by 1939 there were no fewer than ten million domestic radio sets in Britain — and set the standards for industry to follow. Anything else was virtually 'special order'. It was against a background of increasingly urgent rearmament that the Royal Aircraft Establishment at Farnborough in 1938 began testing a wide range of components to a Wireless Telegraph Board specification calling for improved performance and reliability. Similar work was under way in Germany. In the normal course of events there would have been little incentive for manufacturers to make special efforts to improve component quality for a market a thousand times smaller than that for domestic radio. As it was, industry responded with urgency to the worsening political situation under the direction of government bodies. Manufacturing output was stepped up (in Britain, to three times the normal peace-time volume), reliability was improved, and new devices developed. In particular, the need for compact, mobile equipment was emphasised, and resulted in the development of miniature, and then sub-miniature, electronic valves.

In the early post-war period the initiative in electronics technology passed from Britain and Germany to the United States, which had by then built up a huge war machine. It was an American company, Bell Telephone Laboratories, that in 1948 invented a component — the transistor — which was to launch the electronics revolution. The invention of the germanium transistor overshadowed the equally important but somewhat later development of silicon technology. In fact germanium as a semiconductor material was to have a relatively short life; already, by 1960, its disadvantages were becoming apparent by comparison with the performance of silicon.

Printed circuits

Along with transistors came a new concept of mounting components, the printed circuit (strictly speaking, it was already in use with subminiature valves, but its full potential was to await the transistor). This was to provide rewards in the shape of reduced construction and maintenance costs, because pcbs could be built

and tested as individual units. Meanwhile in the laboratories of another but this time little-known US company, Texas Instruments, a further silicon-technology development was taking shape, this time the integrated circuit. This device symbolised the transformation from three-dimensional components to two-dimensional circuits, with the accompanying benefits of greatly increased packing densities.

From the aircraft designers' point of view these developments added up to smaller and lighter black boxes, less power consumption, less demand for expensive cooling air, greater resistance to shock and vibration, and brought nearer the possibility of airborne computing power.

The war in the Pacific, a largely but not entirely American battle against the Japanese, had shown up the need for accurate, long-range and 'secure' navigation equipment, preferably independent of ground-based aids. It was, again, US industry that got to grips with the problem of providing a self-contained navigation aid, and arranged a marriage between precision gyros and highly accurate linear accelerometers to produce inertial navigation systems. The first such system, by Sperry, was flown experimentally aboard a DC-3 airliner in 1950. It was very expensive, took several hours to warm up, erect, and align before take-off, and was subject to severe drift.

Another self-contained navigation aid designed for long-range aircraft which spend much of their time outside the coverage of ground beacons is Doppler, developed in the USA during the 1950s. This is a radar system in which the echoes from ground or sea returned from beams transmitted ahead and behind the aircraft are compared with one another, their frequency differences being used to derive groundspeed and drift angle.

Whereas inertial navigation has great accuracy over the short term, but is subject to drift, which gives appreciable errors over longer periods, Doppler-derived speeds are less accurate over the short term, but do not drift. Both systems are used extensively by the airlines, providing adequate accuracy, but a combination of the two (known as Doppler/inertial mixing) provides the very much greater accuracy needed by a bomber, flying at 600 knots and 200 feet, to creep up on a target and destroy it with one pass, and is standard on many advanced combat aircraft; for example the multi-national Tornado.

Meanwhile for medium- and short-range aircraft, flying mostly over land, things were also changing. Vhf technology developed in the late 1930s, largely by GEC, was immediately incorporated into the radios of the Battle of Britain Hurricanes and Spitfires to bring about a new level of clarity in voice communication. Vhf was also to revolutionise radio-beacon navigation. The medium-frequency Radio Range system, so widely used by 1939, was by the early 1950s giving way to vor, or vhf omni-directional radio, free from the fading and distortion of the former. Vor was adopted as the world-wide standard short-range navaid to define the new system of airways that were established by international agreement in the 1950s to expedite the flow of air traffic.

Vhf technology was the basis for ILS, the instrument landing system introduced in America during 1946, and which has become another world standard. Along with developments in autopilot technology, it provides the foundation for all-weather automatic landing capability, closing the last gap in the establishment of fully automatic flight from taxi-out to switch-off.

VLF technology

Although inertial navigation (IN) was introduced into airline service in 1968 after a lengthy gestation period, and has become a standard fit in the wide-body transports, it is already being challenged by ground-based systems using vlf (very low frequency) technology, originally developed by the United States Navy for communicating with submerged submarines. Whereas three-channel IN was popular during the early 1970s when the wide-bodies were introduced (a triple-redundant system at $100,000 per channel was considered necessary to assure the desired reliability), twin-IN plus vlf now seem to provide the same performance and reliability at a much reduced cost. The financial benefit stems from the eight vlf stations established around the world being maintained by government agencies for military use, with no financial commitment on the part of the airline, while the receiving equipment itself does not contain the expensive gyros, accelerometers and precision mechanical elements that put up the cost of commercial IN.

The introduction of the transistor and, later, the integrated circuit, permitted the development for airborne use of a technology with potential even now only dimly perceived: digital data processing. The invention of the turbine engine and its development in the late 1940s, better aerodynamic prediction through the use of new wind tunnels and mathematical modelling, and the simultaneous arrival of the transistor gave promise of huge advances in aeronautical design. These developments between them lifted aviation to the status of a huge, multi-million dollar international industry. Continuing political instability everywhere, the three-year war in Korea, the Cold War, and the U-2 incident of 1960 all combined to loosen military purse strings as the USA sought to maintain a lead in arms quality. As a result, laboratory ideas were hurried out into the wind-tunnel, test-chamber or electronics rig as quickly as possible, with less concern for financial considerations than is the case today, Greater strides were made between successive generations of aircraft than before or since.

At this time the biggest US aircraft company was North American Aviation, and in 1955 it began work for the US Navy on a carrier bomber, the A-3 Vigilante, that was to combine for the first time on any aeroplane a comprehensive navigation and attack system, to be controlled by a digital computer. The A-3 was the prototype for what has come to be known as a fully integrated weapons system. The heart of the aircraft was an IN unit for navigation, and an advanced forward-looking radar that was connected to the flight-control system so as to be able to command the aircraft to fly a path only 200 to 500 feet above the ground at high speed, so keeping under hostile radar coverage — a mode known as terrain-following. The radar also provided the 'eyes' needed to release the weapon accurately.

The Vigilante was not an operational success because of problems with its unique nuclear-stores ejection system, but it represented a technological watershed in terms of new equipment and integration which all aircraft of equivalent performance were afterwards to follow.

Integration of similar functions reduces size, cost, maintenance demands, and weight. For example: flight instruments comprising air-speed indicator, altimeter, rate-of-climb, and machmeter were for decades fed separately by static air pressure (decreasing with height), and dynamic air pressure (increasing with speed or Mach number). Each had its own barometric capsules and electro-mechanical computing mechanisms, so that there was considerable duplication. Eventually all these functions, depending on air density, pressure and temperature, were calculated in a single unit known as an air-data computer, which then supplied electrical analogues to the appropriate instruments. This has resulted in great savings, particularly in the reduction of bulky piping needed to feed pitot and static pressures from the sensing heads to the various instruments. Again, many instruments and systems need to know the pitch and roll attitude of the aircraft. In earlier times, each system had its own set of gyros, which was uneconomic in terms of cost, power and space. The vertical gyro unit now provides a central facility for generating electrical attitude signals and feeding them wherever they are needed in the aircraft.

Television display

Modern aircraft derive and use vast amounts of information, and the problem for the designers is to decide how much is needed by the crew. The flight deck of a medium or large airliner or the cockpit of a multi-role fighter has over the years become increasingly congested with controls, dials and gauges. The crew needs to monitor or control engines, fuel, hydraulics, air-conditioning, electrics, pneumatics, auxiliary power, and several different types of radio and navigation equipment, besides controlling the aircraft through a diversity of attitude and direction indicators. The assumption was that the flight crew needed to be 'in the loop' of each system all the time. Over the past few years, developments in display technology based on cathode ray tube and other methods, along with computer storage, sequencing and 'discernment' of data (particularly of dangerous situations), now have made it possible to choose and display information as required, so dispensing with large numbers of heavy, expensive and space-consuming indicators. A central warning system, similar to but more comprehensive than those that have been used for so many years, alerts the crew to the onset of emergencies (fire in an engine, for example) and indicates the computer 'page' that can be displayed to provide the crew immediately with the appropriate emergency procedures.

The move towards replacing many traditional, permanent instruments by displays is well under way in the new and 'derivative' airliners. It began in the late 1970s with the McDonnell Douglas DC-9-80, gained momentum with the Airbus Industrie A310 and Boeing 757 and 767, and is now even penetrating to the business-aircraft level. These displays are mounted on the pilots' instrument panels, and in conjunction with extended automation of navigation, flight-control, engine and systems functions now permit aircraft to be certificated for operation by crews of two. Thus (as mentioned earlier) flight engineers now see themselves as an endangered species and just as likely to vanish as navigators did when IN and other automated, pilot-monitored navaids came along. Airline finances have taken heavy blows in recent years: the 1973 oil crisis obliged them to cut back on services without being able to reduce costs, and then, just as they were beginning to recover, the huge price rises imposed by the oil-producers in concert with international recession in 1980-81, dealt them another swingeing blow. They now seek desperately to implement all possible money-saving measures, and see the elimination of a superfluous but expensive crew member as a major step to financial solvency.

The importance of digital technology can hardly be overstressed, but there are two particularly significant areas of application: propulsion and flight control.

First, propulsion. Engine and airframe manufacturers are striving to ensure that the propulsion system (nowadays not just the engine itself, but a number of variable-geometry devices within it, and in some military aircraft and supersonic transports the highly refined movable intake and afterburner sections) will behave itself and work efficiently over the great range of speeds and altitudes in which modern aircraft operate. Military engines can be easily upset by sudden changes in airflow, caused for example by turbulence resulting from gun-firing or missile launch, while commercial engines need to operate at their most economical settings to keep down fuel costs.

Traditionally engines have been controlled by hydro-mechanical systems, which have now been brought to a fine degree of reliability and optimisation. Nevertheless they are relatively slow-acting, and are less and less able to control adequately the new civil and military engines.

The introduction of fast-acting digital electronics, at present only in supervisory capacity but later, as experience grows, with full authority, will eventually provide all but the simplest turbine engines with efficient control systems. Since engine removals and overhauls are such a prominent part of an operator's running costs, it is likely that control and data signals used in these systems will also form the basis of more sophisticated health-monitoring equipment that will permit problems to be identified at an early stage, before they cause damage.

Secondly, flight control. Digital technology permits three successively more beneficial modes of control to be applied to an aircraft. The simplest is known as electrical signalling, whereby pilot demands to change the aeroplane's attitude to initiate a manoeuvre are transmitted from the control column to the control surfaces (ailerons, rudders, and elevators) by means of electrical signals commanding movement of hydraulic jacks which in turn move the flying-control and other aerodynamic surfaces. By contrast with the mechanical tubes or rods that traditionally connect control column and rudder pedals to these surfaces, electrical signalling is lighter — at least, in large aircraft — and is easier to accommodate in the airframe, where space for moving rods is always difficult to find. It also reduces spurious control demands due to flexure or distortion in the airframe, and cuts down mechanical friction.

Fly by wire

However, the considerable development risks involved do not justify electrical signalling for such relatively small benefits. The real justification for electrical signalling is that it makes possible, with relatively minor additional risk, a mode of control known as fly-by-wire. While electrical signalling merely transmits control column or rudder pedal movements to the flying control surfaces indistinguishably from mechanical links, fly-by-wire (or FBW) 'shapes' or modulates them so as to provide the ideal aircraft response. In this case sensors measure exactly how the aircraft responds to pilot commands, and superimpose their own demands on the flying controls in order to suppress undesired responses such as overshooting a commanded attitude or cross-coupling between roll and yaw. In other words, the aeroplane does exactly

INTRODUCTION

what the pilot tells it to do, the system as it were surreptitiously removing any false motion or undesirable behaviour of the aircraft.

Such tailoring of aircraft response is known as manoeuvre demand, and vastly improves flying precision. It is particularly important in combat flying, where large rate changes of attitude have to be initiated instantly, and be 'killed' without overshoot, in order that the pilot may line up the guns quickly and accurately. FBW also confers another very great benefit: it permits safety limits to be built into the control system. Under the stress of air combat the pilot has little time for the niceties of flying, and it is quite easy to inadvertently take the aircraft outside what is called the manoeuvring envelope, the combination of speed, height, and g-loading permitted by the designer. If this happens the aerodynamic buffeting may be so severe that the pilot is unable to bring guns to bear, or the aircraft may stall and spin, or catastrophic structural damage may occur. By 'fencing-in' the permissible performance boundaries, the autopilot (now more properly known as an AFCS, an automatic flight control system) can protect the aircraft at a time when the pilot is concerned only with what is happening outside the cockpit.

The third benefit that follows on from electrical signalling and fly-by-wire stems from the application of what is called ccv (control configured vehicle) technology. Ccv is the application of electronic command equipment to the flying control surfaces or lift-augmentation devices (flaps and leading-edge slats) in such a way as to reduce structural weight and increase performance.

By shifting the centre of gravity towards the rear of the aircraft the wing can be made smaller as the tailplane is made to support a part of the aircraft weight in flight. The wing typically accounts for some 60 per cent of the structure weight of an empty aircraft, and so savings in this area can be exchanged for payload. This technique is not possible without ccv technology because movement of the centre of gravity rearwards makes an aeroplane unstable, eventually so much so that it becomes manually unflyable.

Other ccv applications permit the bending stresses in wings to be kept within limits by using the ailerons together in such a way as to vary the lift as the aircraft flies through a gust. Further away in time a development of this application may be used to prevent or delay the onset of flutter — a catastrophic torsional resonance of wings or tailplane at a particular speed. The normal way of preventing flutter is to make the wing rigid enough to withstand the natural tendency within the flight envelope. But with ccv controlling special aerodynamic, 'anti-flutter' surfaces on the wing, it would be possible to make a much lighter structure and constrain its destructive tendencies artificially.

There are other possible applications of ccv, but these examples serve to show that the technique can be used to secure substantial benefits in weight-saving, or performance, or a combination of the two as the designer wants.

From what has been said, it is clear that avionics equipment can have considerable authority to operate 'out of sight' of the pilots or flight crew, and can have an impact on safety as well as performance. Broadly, it can be categorised under three headings depending on permitted behaviour of the aircraft in the event of a fault: non-essential (failure of equipment to operate does not compromise safety and may not significantly degrade the mission, for example, loss of a radio); essential to accomplish mission, but not compromising safety (for example, failure of weapon-aiming system); and essential at all times, such as a full-authority flight-control system.

Reliability

Reliability is therefore a key factor, and airframe or engine designers and avionics engineers have to work closely together to decide how reliable a given system or piece of equipment must be to fulfil its function effectively and without jeopardising the safety of the aircraft and its crew, but at the same time giving as much thought to weight and cost.

There is a good analogy here with the structure of the aircraft. Mechanical reliability in the early days of flying was based on 'single channel' philosophy; the wing had one mainspar, strong enough to take the predicted loads, and was assumed never to

break provided the aircraft was operated within the flight envelope (the combination of speed and g-loading permitted by the designer). Sometimes it did, and the results were always catastrophic (in aviation the term 'catastrophic' implies a failure resulting in the total loss of the aircraft, and is usually fatal for the crew and other occupants).

Much later it became recognised that catastrophic failures of the metal structure could occur under conditions of continued stress well below the design break point. Accordingly structures are now designed so that the loads are distributed over a number of paths. If one fails, the other paths support the increased loads until the failure is detected and repaired. Structural components are so built that when failures occur, they propagate slowly, giving time for detection before degradation has progressed too far. These are called 'soft' failures.

So, too, with avionics. Advisory or non-essential systems can be built on the single-channel principle: a failure can be accepted in flight because it is not critically important to the progress of the flight or because their functions or the information they provide can to an extent be duplicated by or inferred from other equipment. On some aircraft, for example, yaw dampers (devices for countering sometimes unpleasant yawing motion in swept-wing aeroplanes) can fail without causing undue concern. Again, if an engine turbine temperature indicator fails, the powerplant can be kept running, if necessary at a lower speed, by reference to other instruments such as the rate of fuel flow and engine rpm or pressure ratio.

For more complex systems two or three channels may be required. Complex mathematical analysis backed by fine judgement is usually called for, because other, associated, systems may be involved and may need upgrading to secure the overall performance or reliability demanded. The relationship between autopilots and automatic landing systems provides an example of how carefully the customer needs to specify requirements. Two-channel autopilots are perfectly adequate to fly any aeroplane anywhere in the world. Each channel continuously compares its performance with the other, and signals the pilot or flight crew if discrepancies occur, so that the system can be switched off. However, if the operator then specifies automatic landing capability, the statistical safety arguments based on the known failure rates of components may dictate three channels so that a faulty channel may be outvoted. Since the autopilot commands the aeroplane to follow a given flight path via a number of other systems (eg ILS receiver) those systems may also need upgrading. All this involves a greater burden in terms of first cost and maintenance, and in addition there is a crew training cost associated with equipment and procedures. It may be that after consideration, the operator will decide that the particular category of automatic landing performance desired initially is not worth the extra cost in relation to the number of times during the year that it can expect to be used, and the amount of extra crew-training needed to maintain the required level of proficiency.

It is hardly possible to overstate the importance of the digital revolution that has taken place during the last 15 to 20 years. What has been said above about digital flight and engine control is little more than the tip of the technology iceberg that is bringing a new dimension to aviation, the end products of which will be cheaper business travel and package holidays, more efficient and economic national defence, and a continued trend to safer flying.

The capability of electronic systems is now such that virtually any technical challenge can be met, and probably with 'off the shelf' equipment at that. The question is no longer, 'when will that technology be available?' but, 'what is the risk involved and how much will it cost?' The real world of commercial judgement and military effectiveness demands a cautious, evolutionary approach in deciding the level of technology to be incorporated in a new aircraft. A wrong decision, such as the premature introduction of ccv to save airframe weight, could run an airframe manufacturing company out of business or at least deprive it of a runner in that generation of aircraft. On the other hand design teams starting out with clean sheets of paper must not be so conservative that they penalise the commercial prospects of a project that will probably be in the air for the next 30 to 40 years. To an increasing extent, electronics calls the tune.

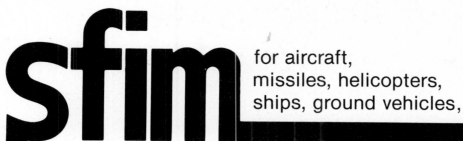

Glossary and abbreviations

ac	Alternating current		ga	General aviation
acdp	Armament control and display panel		GaAs	Gallium arsenide
a-d	Analogue-digital (conversion)		gmti	Ground moving-target indicator
adau	Auxiliary data acquisition unit		gpws	Ground-proximity warning system
adf	Automatic direction finder			
adi	Attitude director indicator			
AFB	Air force base		hf	High frequency
afc	Automatic flight control		hsi	Horizontal situation indicator
afcs	Automatic flight control system			
ahrs	Attitude and heading reference system			
ai	Airborne interception		ias	Indicated airspeed
aids	Aircraft integrated data system		ic	Integrated circuit
alcm	Air-launched cruise missile		iff	Identification, friend or foe
am	Amplitude modulated		ifr	Instrument flight rules
amti	Airborne moving target indicator		ils	Instrument landing system
apms	Automatic performance management system		imc	Instrument meteorological conditions
apr	Automatic power reserve		in	Inertial navigation
apu	Auxiliary power unit		ins	Inertial navigation system
ARINC	Aeronautical Radio Incorporated		inu	Inertial navigation unit
asw	Anti-submarine warfare		irs	Inertial reference system
atc	Air traffic control		ITU	International Telecommunications Union
Atlis	Automatic tracking and laser illumination system		ir	Infra-red
bisjet	Business jet aircraft		JTIDS	Joint Tactical Information Distribution System
bit	Built-in test			
bite	Built-in test equipment			
			kt	Knots
cdi	Compass direction indicator			
cdu	Control and display unit		lcd	Liquid-crystal display
cep	Circular error probability		led	Light-emitting diode
cg	Centre of gravity		lltv	Low-light television
cmos	C-type metal-oxide silicon		lru	Line-replaceable unit
cpt	Cockpit procedures trainer		lsb	Lower sideband
cpu	Central processor unit		lsi	Large-scale integration
crt	Cathode ray tube			
css	Cockpit systems simulator			
cw	Continuous wave		mad	Magnetic anomaly detector
			mcu	Management control unit
			MCU	Modular concept unit (⅛ ATR = 1 MCU)
dc	Direct current		MIL-STD	Military Standard
df	Direction finder		mls	Microwave landing system
dfdau	Digital flight data acquisition unit		mraam	Medium-range air-to-air missile
dicass	Directional command active sonobuoy system		mtbf	Mean time between failures
dits	Digital information transport standard		mtbo	Mean time between overhauls
dme	Distance measuring equipment		mti	Moving-target indication
dmet	Distance-measuring equipment with respect to time		mttr	Mean time to repair
dmu	Data management unit			
dr	Dead-reckoning		NASA	National Aeronautics and Space Administration
drlm	Digital radar landmass simulation		NBAA	National Business Aircraft Association
			ncdu	Navigation control and display unit
eadi	Electronic attitude director indicator		Nmos	N-type metal-oxide silicon
earom	Electrically alterable read-only memory			
ecm	Electronic countermeasures			
eccm	Electronic counter-countermeasures		oem	Original equipment manufacturer
efis	Electronic flight instrument system		on	Omega navigation
egt	Exhaust-gas temperature			
ehsi	Electronic horizontal situation indicator			
eicas	Engine indication and crew-alerting system		pa	Public/passenger address
elint	Electronic intelligence		pas	Performance advisory system
emc	Electro-magnetic compatibility		pcb	Printed circuit board
epr	Engine pressure ratio		pdcs	Performance data computer system
esm	Electronic support measures		pds	Passive detection system
eta	Estimated time of arrival		pmc	Performance management computer
			pms	Performance management system
			pncs	Performance and navigation computer system
FAA	Federal Aviation Administration		ppi	Plan-position indicator
fac	Forward air control		pps	Pulses per second, pilot's performance system
Fadec	Full authority digital engine control		prf	Pulse repetition frequency
fbw	Fly-by-wire		prom	Programmable read-only memory
fdau	Flight data acquisition unit			
fdep	Flight data entry panel			
fet	Field effect transistor		R & D	Research and development
flir	Forward-looking infra-red		ram	Random-access memory
fm	Frequency modulated		rcs	Radar cross-section
fmc	Flight management computer		rf	Radio frequency
fmcs	Flight management computer system		rmi	Radio magnetic indicator
fms	Flight management system		rnav	Area navigation
FRG	Federal Republic of Germany		rom	Read-only memory
fsk	Frequency-shift keying		rpm	Revolutions per minute
ft	Foot		rpv	Remote-piloted vehicle

rvr	Runway visual range		**uhf**	Ultra-high frequency
rwr	Radar warning receiver		**USAF**	United States Air Force
			usb	Upper sideband
sar	Search and rescue		**UVEROM**	Ultra-violet eraseable read-only memory
selcal	Selective call			
sid	Standard instrument departure			
'single'	Single-engined light aircraft		**vasi**	Visual approach slope indicator
STAR	Standard terminal-arrival route		**vhf**	Very high frequency
swr	Standing-wave ratio		**vlf**	Very low frequency
			vlsi	Very large-scale integration
			vor	Very high frequency omni-directional radio
Tacan	Tactical air navigation		**vor/loc**	Very high frequency omni-directional radio and localiser ils
tas	True airspeed		**vortac**	Combined vor and Tacan
TDMA	Time-division multiple access		**vswr**	Voltage standing-wave ratio
Tercom	Terrain contour-matching			
Tiseo	Target identification system, electro-optical			
TSO	Technical Service Order			
tvor	Terminal vhf omni-directional radio		**wra**	Weapon replaceable assembly
'twin'	Twin-engined light aircraft			

Flight Simulation Systems

CANADA

CAE

CAE Electronics Ltd, 8585 Côte de Liesse Road, St Laurent, Montreal H4T 1G6

CAE Electronics designs, develops, and produces flight, radar, and weapons simulators and electronic equipment for Canadian defence programmes. CAE has introduced many innovations in simulation technology, such as the hydrostatic six-degrees-of-freedom motion system, the use of general-purpose computers, and instructor facilities based on CRT displays. Fixed-wing simulators have been made for the A 300, B-727, B-737, B-747, DC-8, DC-9, DC-10, L-1011, F-28, CL-600, and for the new-generation A 310, B-757 and B-767 airliners that are entering service in 1982/83. Among simulators for military aircraft the company has built systems to represent the Tornado for West Germany. The company has also built up an extensive helicopter simulation business. This has included systems for Agusta AB-205 and AB-212, Bell UH-1D, Boeing-Vertol CH-47, Sikorsky CH-53, and Westland Sea King Mk 41 supplied to the West German Navy.

A particular point of note is that CAE has provided five full flight simulators to United Airlines (two for B-727s, one for DC-10, and two for B-767s) and has a technology-sharing agreement with the airline for access to extended aircraft performance data. The Canadian company anticipates that the inclusion of such data in future designs, and for restrospective improvements to existing simulators, will enhance their value as training devices.

Full flight simulator for TWA Boeing 727, by CAE

CAE full flight simulator for British Airways Boeing 737

CZECHOSLOVAKIA

Letov

Rudi Letov np, 199 02 Praha 9-Letnany

TL-410 flight simulator for LET 410 transport

The TL-410 system simulates the characteristics of the LET 410 twin-turboprop light transport and feeder-liner, and is the first Czechoslovak system to employ a digital computer for control and for modelling the performance and behaviour of the aircraft represented.

The system comprises a cockpit mounted on a three-axis motion platform, a visual display system fixed ahead of the windscreen, an airport model with lighting, instructor's station, flight-track recording console, a KRS 4201 digital computer supplied by VEB Robotron of East Germany, and an hydraulics power package. The input analog/digital and output digital/analog signal conversions are accomplished on a Dasio 600 interface system manufactured by Metra Blansko.

The visual system is fed from a TV scanner at the airfield model which is built to a scale of 1:1000. Runway visibility can be reduced from 2 km down to zero, and the height from 400 m to zero.

The instructor's station is isolated from the flight-deck enclosure, communication being maintained by means of a simulated VHF radio link. The instructor can choose any combination of failures from 52 specific faults, and a TV monitor duplicates the pilot's view of the visual display.

STATUS: In production

FRANCE

Thomson-CSF

Thomson-CSF, Simulator Division, 3 avenue Albert Einstein, BP 116, 78192 Trappes Cedex

In 1966 Thomson-CSF built the first digital simulator in Europe, not for training but to assist in the Concorde development programme. Its activities now embrace all types of system, from cockpit procedures trainers up to full flight simulators, for military and commercial aircraft and helicopters, armoured vehicles, ships, and nuclear and other power stations.

Responding to the standards for new simulation equipment set by the FAA with the aim of reducing training time on actual aircraft to a minimum, Thomson-CSF's simulators embody a number of standard facilities. These are: six-degrees-of-freedom motion systems mounted on hydrostatic bearings; advanced digital-based control loading systems; computer-generated image displays; high-resolution alpha-numerics and graphics displays, together with preprogrammed lessons, enabling greater consistency of teaching at the instructor's console; real-time computers specially adapted to simulator requirements, permitting optimisation of performance and flexibility;

Thomson-CSF full flight simulator for Alpha Jet strike/trainer

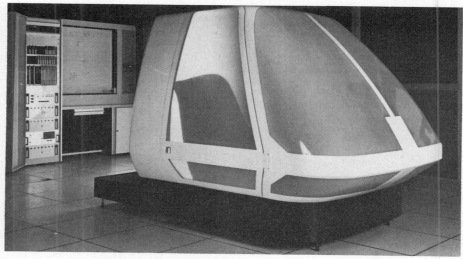

150H helicopter IFR and radio-navigation flight trainer by Thomson-CSF

programming in high-level or assembler language; and built-in test equipment that continuously verifies the fidelity of simulation.

For military users Thomson-CSF offers two simulator configurations: a trailer installation for customers needing mobility, and a fixed version where the training site is permanently located. These military systems embody motion platforms with four or six degrees of freedom, *g*-trousers to provide initial acceleration cues, *g*-seat to reproduce sustained acceleration, and computer-generated image displays representing airfield or aircraft carrier layouts, air/ground attack situations, air/air combat, formation flight, and air-refuelling.

For simulators appropriate to the very latest combat aircraft such as the Mirage 2000 and 4000, with their advanced radar and weapon-delivery systems, enhanced facilities include devices to represent the operational environment more realistically. These include digital simulation of the radar landmass, air/air radar,

ECM, digital displays for ground-attack, reconnaissance and penetration, and the provision of special systems to develop proficiency in air combat.

In addition to simulators representing specific aircraft types, Thomson-CSF provides families of procedures trainers under the designations LMT350 (fixed-wing transport and executive aircraft), LMT531 (combat aircraft and trainers) and LMT150H (helicopters) for training in IFR flight and radio-navigation. Aircraft whose equipment is simulated for this purpose include Nord 262, Falcon 20, Beech C90 King Air, Fokker F27, BR 1150 Atlantic, Aerospatiale Corvette, MS 760 Paris, Caravelle III, A.300, Boeing 727, and Aerospatiale SA.341 Gazelle.

Military aircraft simulators include Mirage, Alpha Jet, Jaguar, Transall, Etendard IV, Super Etendard, and SA 330 Puma helicopter.

Thomson-CSF full flight simulator for Iberia A 300

UNITED KINGDOM

Flytsim

Flytsim Ltd, Bell Street, Maidenhead, Berkshire SL6 1BU

Flytsim designs and builds high-quality simulators and training aids tailored to specific customer requirements. Target markets comprise flying schools, air-taxi and charter operators, aeronautical colleges or universities with appropriate departments, research establishments, feeder and international airlines, and military forces.

The company specialises in single- and twin-engined simulators, procedures trainers, and other instructional systems for piston-engined and jet aircraft, and most of them are built around five standard units, tailored as necessary to particular needs. These are:

Digital computer: Designed and built by Flytsim for radio aids computation, permitting any combination of ground stations from 10 to

140 to be incorporated into the equipment. These stations can represent any standard navigation or communications facility, eg ADF, VOR, ILS, DME, VOR/DME, or Vortac. Other facilities include digital QDM selection, digital station idents, and automatic landing markers.

Solid-state flight computer: This can be programmed to represent the full flight envelope of any type of aircraft from light singles to wide-bodies.

Two-axis hydraulic motion system: Permits excursions up to ±10° in pitch and roll. The system also includes effects to simulate take-off and landing roll, application of brakes, and runway irregularities.

Simulator: A glass-fibre cockpit, cabin, or flight-deck section.

Plotters: An on-board plotter records the approach pattern of the simulator, and a second 36 by 36 inch unit external to the

Flytsim representation of Boeing 707-369 for Kuwait Airways

simulator displays en route or approach patterns.

STATUS: In production

Link-Miles

Link-Miles Division, The Singer Company (UK) Ltd, 27 Churchill Industrial Estate, Lancing, Sussex

Hawk flight simulator

Delivery last year to Finland of a flight simulator to represent the British Aerospace Hawk light strike/trainers operated by that company, symbolised the first of a family of Link-Miles training systems in which mainframe computers have been replaced by fully distributed multiprocessing computing systems using microprocessors. It is claimed to represent a

significant advance in technology. The principal advantages to operators of light jet aircraft, both civil and military, are that it provides an economic solution not previously available to the simulation of this class of aircraft. The distributed processing system is capable of expansion to meet additional demands.

The company says that the new computing method greatly simplifies initial programming, the mechanical system itself, and fault-diagnosis and maintenance. The technology

Link-Miles simulator modelling front cockpit of BAe Hawk light-strike trainer

has so far been developed only for light turbine aircraft; the principles have not yet been applied to high-performance military or large commercial aircraft.

The simulator employs six Intel 8086 16-bit microprocessors and their input/output boards, providing 64K memory. Two are devoted to characterising aircraft performance, one to radio and navigational equipment, one to engine functions, and a fifth to auxiliary systems such as fuel management, electrical and hydraulic and other systems. They are accommodated around the cockpit on a motion platform, together with the analogue/digital and digital/analogue conversion equipment. The sixth microprocessor and board is housed in the instructor's console.

The single-window visual display is a CGI system driven by a PDP-11/55 minicomputer.

Instructor's console for Hawk simulator showing plotters, set of repeater flight instruments, and visual display

Rediffusion Simulation

Rediffusion Simulation Ltd, Gatwick Road, Crawley, Sussex RH10 2RL

This company, which in March 1981 changed its name from Redifon Simulation Ltd, has sold some 300 commercial and military full-flight simulators since building Europe's first electronic simulator in 1950. It builds all types of simulator, from full-flight and fixed-base systems with displays to cockpit and flight-deck procedures trainers, and is now using its aviation background to bring advanced simulation technology into industrial activities such as oil-rig management.

In January 1981 the company announced a contract with Boeing for ten simulators to be delivered during 1982. The order comprises full-flight simulators for Boeing 737 and 727 transports and for the new 757 and 767 airliners, and fixed-base and maintenance simulators for the 757 and 767. Substantial fuel savings, a changing air-traffic situation, demands for lower noise levels, increased training efficiency, and the promise of generally better economy were cited by the US company for increased dependence on simulators. It foresees a growing proportion of crew-conversion training being conducted on simulators, ultimately with all such training being accomplished on the ground.

During February 1981, in what it sees as the most important breakthrough in commercial flight simulators during recent years, Rediffusion Simulation announced that its new-technology simulator at Braniff International's training school in Dallas had become the first system to be approved by the FAA under Phase II of its Advanced Simulation Plan. As a

First full-flight simulator for 767 transport, destined for Boeing training centre, under construction at Rediffusion

Rediffusion's simulator for Sikorsky S-61N helicopters of British Airways installed at Aberdeen

result the airline is cleared to convert pilots from one type to another, and to upgrade first officers to captain, entirely by simulator. Phase II approval means that, for the first time,

First of Rediffusion's flight simulators to incorporate its New Concept technology is this Boeing 747 system for Japan Air Lines. It has been certificated by the Japanese Civil Aviation Bureau and is installed at JAL's training centre at Haneda Airport, Tokyo

complete flight-training programmes can be conducted without having to take aircraft out of service.

Then in April of that year the Crawley company, in supplying an advanced-technology Boeing 737-200 simulator to Air Florida, undertook to meet the stringent criteria specified under Phase III of the FAA's Advanced Simulation Plan, which permits airlines to qualify their pilots initially in these devices as well as have them convert from one type to another. The system was due to be commissioned in autumn 1982.

CGI (computer generated image) visual systems are built by Rediffusion Simulation Inc, of Arlington, Virginia under the name

Novoview (see below) for use with these and other manufacturers' simulators. By May 1981 some 225 had been sold, and the company says that it provided more than 80 per cent of all such systems ordered for commercial simulators during 1979.

Rediffusion Simulation is equally active in the military field, where the most recent contracts have included Tornado simulators (in conjunction with Link-Miles) for Britain's Ministry of Defence, and a B-52 refuelling trainer. It also has an agreement with British Aerospace to manufacture and market equipment based on BAe's air-combat simulation system (see below) originally designed as a research tool to assist in the development of new fighter projects.

Novoview SP visual simulation

The Novoview SP series of CGI visual simulation systems covers the full range of training needs, from bright day to darkest night. All weather, cloud and visibility conditions can be selected by the instructor.

Novoview SP1 uses a relatively low brightness beam-penetration CRT to display night or dusk scenes. It has 6000 lights, 200 to 400 surfaces, 256 edges, five colours and 64 grey shades, and a field-refresh rate of 30 Hz.

Novoview SP2 uses a high-resolution, high brightness shadow-mask CRT to display scenes in night, dusk, or bright daylight conditions. SP2 facilities can be retrofitted to SP1 systems 'in the field'. It has 1000 lights (6000 in night/dusk), 450 to 1000 surfaces (225 to 450 in night/dusk), 512 edges (256 in night/dusk), eight colours, 128 combinations of red, green and blue for surfaces, and a field-refresh rate of 40 Hz (30 Hz in night/dusk).

STATUS: In production

Novoview SP1 computer generated picture of an oil rig helipad

Rediffusion simulator representing the Boeing E-3A AWACS (airborne warning and control system) built for Boeing to train USAF aircrews in air-refuelling techniques

Air combat simulator under development by Rediffusion

Air combat simulation

Under licence agreement with British Aerospace, Rediffusion Simulation will undertake the commercial exploitation of the air combat simulation system devised by BAe. The system is designed to provide training in the key roles of combat-techniques familiarisation, combat between different types of aircraft, energy management, and tactical evaluation of new and existing aircraft. In its standard, single-dome configuration, the system provides a computer-controlled target image for the pilot to fly against. In the dual-dome configuration, it allows pilot-pilot combat interfaced through a central computer system, each pilot 'flying' his own cockpit in the dome, the target motion resulting from a combination of his own manoeuvres and those of the pilot in the other

dome. The system comprises three types of unit: computer, visual display, and cockpit.

Computer: Manufactured by Systems. Has 32-bit architecture with powerful instruction format

Visual display: 30 ft (9·15 m) diameter inflatable display spherical dome with sky/earth projector for all-round horizon, and solid-state target projector

Cockpit: Representative of typical single-seat fighter, with attachments for g-suit and helmet interface connections

STATUS: In development

Laser visual simulation

Claimed to be the world's first wide-angle laser visual system, the venture is being undertaken jointly by Rediffusion and American Airlines for the United States armed forces. Full panoramic

views are possible with very high levels of surface detail. Scenes may be produced by either laser-generated or computer-generated image techniques. The high bandwidth video system of the laser camera produces 20 times more picture elements than conventional television systems. The laser image display is projected onto a hemispherical screen with a 180 × 60° field of view at very high resolution. It is claimed to be the only military training system currently in development likely to provide a scene quality essential for tactical mission simulation. The system is applicable to fixed and rotary wing aircraft, and also to many land and seaborne weapon systems. The development system operates in monochrome (argon green), but production equipment will provide red, green, and blue for more realistic colour rendering.

STATUS: In development

UNITED STATES OF AMERICA

ATC

ATC Analog Training Computers, 185 Monmouth Parkway, West Long Branch, New Jersey 07764

ATC-610 flight simulator/procedures trainer

This is a 'personal' flight simulator representing the general configuration of light, single-engined aircraft, for private ownership, flying schools and clubs. The principal unit is a portable panel with an array of typical flight and navigation instruments, panel-mounted yoke providing pitch and roll inputs, and power quadrant with propeller, power, and mixture controls. A separate rudder-pedal assembly permits co-ordination of all control axes.

Optional equipment includes a monitor communications module, marker-beacon audio unit, flight-plotter, and more extensive navigation facilities. The monitor unit permits instructor briefings and pupil/instructor conversations to be recorded for post-exercise analysis. Use of the approved plotter permits the ATC-610 to be used for the entire period of simulation training time allowed by FAR Part 141 to count towards a pupil's total training time.

Simulator: 29·1 × 21·3 × 17·1 in (739 × 541 × 434 mm)
Rudder pedal unit: 18 × 7·3 × 19·8 in (457 × 185 × 503 mm)
Flight plotter: 19 × 6 × 14 in (483 × 152 × 356 mm)

STATUS: In production

ATC-710 flight simulator/procedures trainer

A development of the ATC-610 (described above) for professional pilots, the ATC-710 has an enclosure that, in conjunction with an approved flight-plotter, qualifies it to meet FAA requirements for private, commercial, and instrument ratings. Customers include flight schools, fixed-based operators, and corporate training departments. A model aircraft linked to the simulator and mounted above the glare shield moves against a land/sky background to provide realistic responses to pilot inputs for those needing an FAA-approved, moving visual reference model.

An instructor's console is an option where ATC-710 simulators are used singly, but is standard where a number of units are to be operated simultaneously. It provides two-way communication with any or all stations, master tape player, intercom, and headsets.

Simulator: 41 × 60 × 66 in (1041 × 1524 × 1676 mm)
Instructor's console: 24 × 36 × 40 in (610 × 914 × 1016 mm)
Flight plotter: 19 × 6 × 14 in (483 × 152 × 356 mm)

STATUS: In production

ATC's Model 610

ATC-810 flight simulator/procedures trainer

The ATC-810 is a microprocessor-based procedures and IFR simulator that represents the dashboard layout and performance of 6500 to 8000 lb turbo-charged cabin-class twin-engined aircraft such as Piper Navajo/Chieftain, Beech Baron/Duke, or the Cessna 400 series. It is designed for air-taxi and Part 135 commuter operators, flight schools, airlines, and corporate and other multi-engined users.

The simulator has a comprehensive set of flight, systems, and navigation instruments and controls, and failures can be commanded by means of a fault console, operated by an instructor, that plugs into the simulator. The system's PROM creates a 150 × 150 nm navigational region representing an ATC region, with 65 airports, 36 VORs, 18 ADFs, 38 ILS approaches, four Cat II approaches, and 153 instrument approaches. Options include a visual display system, with take-off and landing capability, flight director, and RNAV.

Simulator: 44 × 67·5 × 60 in (1118 × 1714 × 1524 mm)

STATUS: In production

ATC-112H helicopter IFR flight simulator

Representing the general characteristics of light to medium turbine-powered helicopters, the ATC-112H is an IFR flight simulator for flying schools, corporate or charter fleet organisations, fixed-base operators, educational institutions, military establishments, and private owners. Performance and handling very closely follow actual aircraft characteristics, and rotor torque behaviour can be switched to represent clockwise or anti-clockwise rotation. Realistic response to control inputs are provided and correlated by hybrid computer circuitry. Dynamically balanced cyclic and collective-pitch controls can be adjusted to provide the desired control stabilisation. The computer is programmed for a speed range of 40 to 160 knots. Winds up to 50 knots can be introduced from 12 directions, and five degrees of turbulence set up to affect pitch, roll, and altitude.

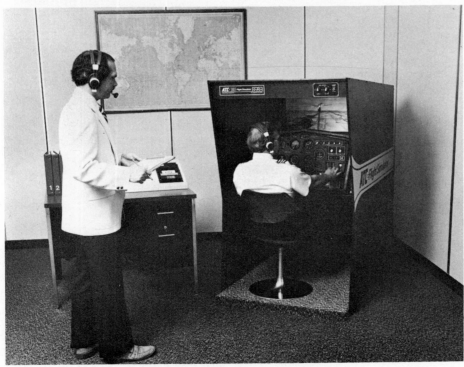

ATC-710 flight simulator/procedures trainer

Accessories include x-y flight plotter, monitor communications system, headset, and an enclosure for IFR training in accordance with FAA FAR Part 141. The flight plotter can be used with standard low-altitude charts to provide records for post-exercise study.

Simulator: 34 × 46 × 64 (864 × 1168 × 1626 mm)
Flight plotter: 19 × 6 × 14 in (483 × 152 × 356 mm)

STATUS: In production

Helicopter IFR simulator showing collective-pitch lever mounted on left-hand side of seat

Aviation Simulation Technology

Aviation Simulation Technology Inc, Hanscom Field-East, Bedford, Massachusetts 01730

The company builds simulators representing single- and multi-engined aircraft in the general-aviation sector. The principal market is for instrument and procedures training, though increasingly they are providing an introduction to the airborne environment and aircraft handling.

All units employ four independent computers to familiarise the student with the four separate aspects of flying: handling, instruments and systems, avionics' and navigation. They are supplied with computer-generated visual displays, claimed to be unique among general aviation simulators, that present sky, horizon, ground, and runway symbology. PROMS, either standard or to special order, and containing the appropriate data, permit students to fly routes and approaches into any airport in the world. Flights can be started at any desired location, and instantly repositioned, eg to return to an initial approach position so that consecutive approaches can be flown in quick succession. Digital design and modular construction together allow new facilities or technology to be incorporated as it becomes available, so that the equipment can be updated rapidly and economically.

A number of accessories are offered, including x-y plotters, digitally adjustable from 0·1 to 99·9 nm/inch, that can be used with any standard aeronautical chart or blank paper. A portable console permits the instructor to fail instruments and engines from outside the cockpit. Conversion kits permit customers with multi-engine simulators to convert them to single-engine units.

Simulators and displays have full FAA approval, and so time on them can be counted

Aviation Simulation Technology's multi-engine flight simulator

towards the total flight and ground time needed for a student to qualify for his licence. Four models are available.

Model 201: Basic representative single-engine simulator
Model 300: Basic representative of cabin-class twin

As supplied with the full range of optional extras, they become, respectively:

Mooney 200X: Fully comprehensive simulator, with layout resembling that of the Mooney light single
Model 300X: Fully comprehensive simulator representing cabin-class twin

STATUS: All in production

Burtek

Burtek Inc, PO Box 1677, 7041 East 15th Street, Tulsa, Oklahoma 74101
(In 1979 Burtek Inc was purchased by the Thomson Corporation of America, a US holding company of the Thomson-CSF group of France)

Some 29 military services, commercial airlines, and aircraft manufacturers use the cockpit systems simulators and cockpit procedures trainers produced by Burtek. This company designs, develops and produces custom-built simulators and training equipment. Burtek claims to be the world's leader in the production of cockpit procedures trainers, cockpit systems simulation, and other types of aircraft simulation equipment.

Performance of equipment is tailored to customers' individual needs. At the lower end of the product line, CPTs are designed to facilitate crew conversion to new aircraft types, and

Production area at Burtek, with CPT for Boeing transport and a line of systems trainers

feature the dynamic operation of all systems controls in both normal and emergency conditions. Limited flight capability can also be incorporated. In CSS equipment the flight and navigation capability is increased so that it becomes virtually a fixed base simulator. CSS devices are used typically to supplement training in full flight simulators, and so reduce simulator operating costs.

The Lockheed C-141 cockpit systems simulator is an example of how the company's CSS equipment duplicates many of the capabilities of full flight simulators. It includes simulation of the inertial navigation system, and has full communication systems logic. Dual 32-bit central processors simulate aircraft systems and can program up to 600 system malfunctions.

Operational trainers for the Northrop F-5 fighter are a recent addition to the company's product line. Digital software controls all aircraft dynamics, and special effects are included such as stall, speed-brake buffet, and weapons-firing vibration. Some of the features used to add realism include engine noise, gunfire, and effect of stores release. Full mission capability, with a fire-control system incorporating tactical radar and a lead-computing optical sight, permits tactical training throughout the F-5's operational envelope.

STATUS: The company has supplied or is building CSS and CPT equipment representing

Burtek procedures trainer for Boeing 737 twin-jet transport, with partial view at left of instructor's console

TH-57 and OH-58 helicopters, and F-4, F-5, F-15, S-3A, C-141, DC-8, DC-9, DC-10, B-707, B-727, B-737, B-747 and L-1011 fixed-wing aircraft.

Display Workshops

Display Workshops Inc, 150 Huyshope Avenue, Hartford, Connecticut 06106

This company, formed in 1951, specialises in cockpit and flight-deck procedures trainers built to customer specifications. A major line is the production of Boeing 727 simulators, but other CPTs have been built for operators of corporate and commuter aircraft. The B-727 units are claimed to be an inexpensive alternative to full-scale simulators, and are FAA-approved.

Cockpit procedures trainer by Display Workshops representing principal flight-deck control and instrument panels of Boeing's Advanced 727-200 transport

Frasca

Frasca International Inc, 606 South Neil Street, Champaign, Illinois 61820

Model 121 flight simulator

The Frasca 121 flight simulator represents the layout and performance of a typical single-engined light aircraft. All handling characteristics are accurately duplicated, from engine start to shutdown: stall speed varies according to bank angle and flap extension, control response is appropriate to speed, and behaviour on the back side of the drag curve is realistic. Changes in gross weight and centre of gravity, manipulated from the instructor's console, affect flight response and stability, and true airspeed and total performance are sensitive to altitude. In particular, the system computes total thrust based on manifold pressure and rpm taken together. Also included is a system, developed by Frasca, for representing phugoid oscillation after pitch upsets, to demonstrate fore-and-aft stability.

The system comprises two principal units: instrument panel and instructor's console. The instrument panel and surrounding structure accurately reproduce the typical cockpit environment of a single-engined aircraft. The instructor's console is in four sections: a radio-station control panel, audio panel, systems panel, and an engine panel. The radio panel permits the instructor to duplicate the coverage of radio navigation areas for any part of the world. The audio panel controls communication between student and instructor. The systems panel can generate navigational or aircraft systems failures, and can manipulate environmental factors such as wind speed and direction, and barometric pressure. The engine panel permits powerplant-related failures to be simulated. An x-y plotter with initial position-setting facility and instructor's controls tracks the course followed during the exercise over an area of 60 × 85 nm at a scale of 1 inch to 2·5 nm.

STATUS: In production

Frasca Model 122 simulator representing a typical light twin. Model on coaming moves in response to pilot demands

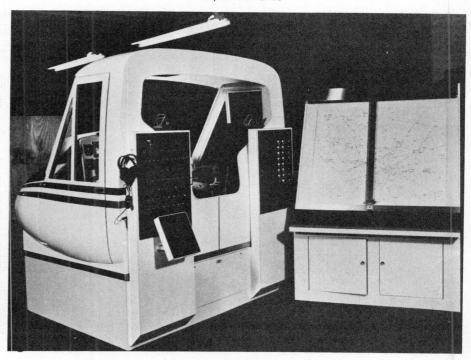

Simplest Frasca simulator, the Model 121

Frasca Model 300H/205 representing Bell 205/UH-1H turbine helicopter, with instructor's console and plotter board

Model 122 flight simulator

The Model 122 represents the layout and performance of a typical twin-engined general-aviation aircraft. It follows the design philosophy of the Model 121 described above, but is particularly equipped to teach comprehensive IFR flight and twin-engine handling. The instructor can 'kill' an engine without notice, or gradually take it out, eg by slowly raising the cylinder-head temperature or lowering the oil pressure. Opposite rudder then has to be applied, along with aileron, to maintain attitude and heading. The difference in drag between feathered and windmilling propellers can be demonstrated.

STATUS: In production

Model 125H flight simulator

The Model 125H is a low-cost simulator duplicating the features of typical light turbine helicopters, but can be programmed to characterise particular types of aircraft in that category as specified by customers. It can represent all manoeuvres encountered in normal instrument flight, and special additional capabilities include translation and autorotation.

The system comprises two principal units: an instrument panel and surround that represents the cockpit environment, and an instructor's console. An x-y plotter with initial position-setting facility and instructor's controls tracks the course followed during the exercise over an area of 60 × 85 nm at a scale of 1 inch to 2·5 nm.

STATUS: In production

Models 300H/205 and 300H/206 flight simulators

These two simulators represent, respectively, the Bell 205/UH-1H/212/Agusta A109 helicopters, and the Bell 206 and equivalent helicopters. Both comprise two units: the instrument panel and surround, which together reproduce the cockpit environment, and an instructor's console. A special feature of the 300H/205 is the portable instructor's panel that allows him to control most functions while seated next to the student. Plotters based on x-y co-ordinates permit courses to be tracked during exercises over areas of 97·5 × 135 nm (Model 300H/205) and 60 × 85 nm (Model 300H/206), both at scales of 1 inch to 2·5 nm.

STATUS: In production

Gould

Gould Government Systems, Simulation Systems Division, 50 Marcus Drive, Melville, New York 11747

This company produces a range of flight simulators and procedure trainers for specific aircraft types.

Lockheed C-5A flight deck procedures trainer

This fixed-base system provides controlled training for flight crews (each comprising a pilot, co-pilot, and flight engineer) in team skills and the procedural knowledge needed to operate the USAF's largest transport aircraft under normal and emergency conditions. Two instructor stations are provided, one for the pilots, the other for the flight engineer. The system comprises a computer complex and four major flight-deck units: the forward-facing pilots' station and panels, the flight engineer's station, and the two instructor consoles.

Size: 78 × 108 × 180 in (1981 × 2743 × 4572 mm)
Weight: 11 924 lb (5409 kg) including computer complex

Beech T-44 operational flight trainer

This is a facility for the military version of the Beech King Air cabin-class twin that fully trains pilots and co-pilots in all aspects of operating, ground manoeuvring, take-off, instrument flight, navigation, approach and landing. Three degrees of motion, aural cues, artificial feel on yoke and pedals and a fully 'live' panel of instruments provide a high degree of realism, all driven by a digital computer system. A computer-generated image system is also available. An instructor seated behind the aircrew on the motion platform controls the situation by means of a console containing graphics CRT displays and an alpha-numeric keyboard. These facilities permit automatic fault insertion, procedure monitoring, repeat instrument displays, and cross-country map generation.

Flight deck
Dimensions: 94 × 60 × 60 in (2388 × 1524 × 1524 mm)
Weight: 1500 lb (679·5 kg)
Instructor station
Dimensions: 74 × 88 × 72 in (1880 × 2235 × 1829 mm)
Weight: 2000 lb (906 kg)
Computer system
Dimensions: 60 × 40 × 71 in (1524 × 1016 × 1083 mm)
Weight: 1000 lb (453 kg)
Hydraulic power pack
Dimensions: 36 × 96 × 60 in (914 × 2438 × 1524 mm)
Weight: 2000 lb (906 kg)

Beech T-34 cockpit procedures trainer

This CPT uses digital simulation to train pilots in the normal and emergency procedures appropriate to the Beech T-34C Mentor turbine-powered primary trainer. It comprises two sections: a trainee station and an instructor station. The first is a replica of the T-34C forward cockpit with motion system, the second an assembly of control and monitoring facilities including a computer terminal with video display and a hydraulic power pack.

Cockpit with motion system
Dimensions: 98 × 56 × 105 in (2489 × 1422 × 2667 mm)
Weight: 4000 lb (1812 kg)
Instructor station
Dimensions: 66 × 66 × 82 in (1676 × 1676 × 2083 mm)
Weight: 1200 lb (543·6 kg)
Computer system
Dimensions: 60 × 40 × 71 in (1524 × 1016 × 1803 mm)
Weight: 1000 lb (453 kg)
Hydraulic power pack
Dimensions: 76 × 46 × 86 in (1930 × 1168 × 2184 mm)
Weight: 2000 lb (906 kg)

Link Flight Simulation

Link Flight Simulation Division, The Singer Company, Binghampton, New York 13902

In parallel with the inception and development of the new-generation transport aircraft exemplified by the Boeing 757 and 767, Link has been refining a package of extensive and fundamental technical improvements for its flight-simulators, and these are prefixed by the descriptive initials AST (advanced simulation technology). Apart from this new development, the company has for many years been supplying simulators and training devices across the entire aviation spectrum, as well as for nautical applications, armoured vehicles, spacecraft, and powerplants.

AST technology

The initials AST describe systems incorporating a wide range of fundamentally new improvements, resulting basically from modern electronic packaging techniques. The '100 per cent simulation' goal, around which AST technology was developed, was defined in the USA during 1980 by FAR 121-14C requirements, which specify simulator characteristics appropriate to three phases of training capability having to do with crew proficiency, conversion, and upgrading.

A new instructor's station within the simulated flight compartment can seat the two instructors serving the two pilots and flight engineer, and an observer. Twin, interchangeable CRT systems provide alpha-numeric or graphics information, and the control panels are also identical. Control panels and the new interface equipment operate independently of the displays, connected only by software. The programmable nature of the displays permits changes in the type or arrangement of information without electrical or mechanical changes.

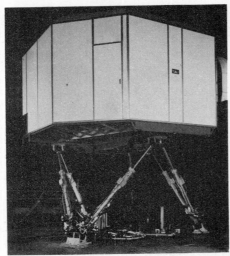

Link's new family of AST simulators, housed in octagonal enclosures containing flight decks and electronics and mounted closer to ground

Flight deck of Link simulator for B-52 bomber

The most significant development in the AST motion system is the new actuator assembly, using much smoother, hydrostatic bearings. A new ultrasonic linear displacement transducer eliminates all mechanical connection between actuator and position sensor, resulting in a very clean feedback signal. More than two-thirds of the plumbing of previous designs has been eliminated and platform, joint, and electronic assemblies have been simplified, reducing the likelihood of oil leaks and simplifying maintenance and improving accessibility and reliability. Again, the interconnection between the simulator's flight compartment and the computer complex has been greatly simplified by the substitution of a digital bus, using serialised data transmission. This permits simpler cable runs, less documentation, and easier trouble-shooting.

AST simulators are packaged into an octagonal enclosure that houses both simulator flight deck and electronics cabinets, reducing by as much as 80 per cent the space previously needed for these systems. Proximity of simulator flight deck and electronics reduces the lengths of interconnecting cables, so cutting down the risk of electrical interference.

Boeing B-52 weapons system trainer

One of the company's most significant programmes was the production of 16 weapons system trainers for the USAF's Strategic Air Command fleet of B-52 bombers. In the first-ever 'fly-off' competition to determine the winner of a simulator programme, Link was awarded the largest contract in history.

Each simulator has three separate training stations: the flight-deck, with a six-degrees-of-freedom motion system, an offensive station with limited motion, and a defensive station without motion. The simulator is designed to train the six-man crews for entire missions, including penetration into hostile airspace. For the first time, a simulator provides completely integrated representations of outside world, radar landmass, low-light television, and forward-looking infra-red imagery. All sensor data is correlated with itself and with outside-world information, and the simulation includes rendezvous and coupling with a KC-135 tanker.

The defensive station simulates, with great fidelity, friendly and hostile environments for the electronic warfare officer and gunner. The EWO learns to recognise electronic 'threat' signals and how to counter them effectively with sophisticated jamming techniques. If jamming is successful, the 'threat' goes off the air; if unsuccessful, appropriate damage is simulated.

F-16 tactical flight simulator

The huge F-16 lightweight fighter programme that has been under way for several years in the USA and Europe has given rise to one of the most extensive projects ever undertaken by a simulator manufacturer. Link Flight Simulation Division is collaborating with companies in Norway and Denmark to provide 18 flight simulators for the United States, Belgium, Denmark, the Netherlands, Norway, and other countries that also operate the General Dynamics/USAF fighter. Each system comprises a fixed-base, fully representative cockpit section, instructor's station, digital computer and interface equipment, with optional motion simulation, electronic warfare, and radar ground-mapping simulation.

The system provides take-off and landing, instrument flight, and navigation simulation, instruments and systems familiarisation, air-to-air intercepts, air-to-ground weapon delivery, and tactical mission operations. Simulation of the F-16's navigation/weapon aiming system makes use of the computer tapes prepared for operational flight and combat, and considerable effort has gone into optimising the integration of the aircraft computer and the simulation, or host, computer.

Simulation of what is termed the 'tactical environment' includes a facility for scoring, or assessing the pilot's accuracy in delivering weapons to the target, and another for representing the rendezvous and contact procedures with a Boeing KC-135 tanker. Combat realism is extended by the inclusion of a mechanoreceptor cueing system, comprising a *g*-seat, anti-*g* suit, and vibration and buffet inducing system. The purpose of the facility is to provide realistic cues, or indications, to the onset of aircraft rotation about the pitch axis. During high-*g* manoeuvres, pneumatic pads in the seat inflate or deflate to give a sensation of 'hardness', while a drive attached to the underside of the seat simulates aerodynamic buffet and vibration as the 'aircraft' reaches its limiting angle of attack.

The computational system comprises four 32-bit Nord 50 processors controlled by a Nord 10S processor. Among the special interface units are the actual Delco Magic 36F fire-control computer fitted to the F-16, a Sanders display driving three CRTs and a keyboard at the instructor's console, and stores management and radar processing sub-systems.

STATUS: In production

Visulink laser image generator

Link's LIG system provides high resolution and scene detail at all altitudes, but is particularly suitable for simulating low-altitude flight. It offers marked improvement in performance and lower maintenance costs while retaining many of the components of the high-resolution systems based on television cameras. Thus, says Link, it is a natural step in the evolution of advanced visual display technology.

The system employs a multi-coloured laser beam to scan a high-detail model board within the pilot's field of view. The reflected light from the board is detected by a bank of photomultipliers and processed to generate full-colour signals to the simulator display, which may be a CRT or projector. The principal differences from the television-based system lie in the replacement of the TV camera by a bank of light-sensitive photomultipliers and the substitution of the lights illuminating the model by a scanning laser beam. Benefits claimed for the system are: no degradation of resolution due to image lag; simpler and more stable colour registration and alignment; improved signal/noise ratio; improved resolution due to more efficient use of projection aperture; elimination of power demands to drive the lighting system; better simulation of night and dusk as well as daylight; greater depth of field; and wider field of view.

The wide field of view and picture quality associated with the LIG system are particularly appropriate to nap-of-the-earth attack helicopter missions and V/STOL operations.

STATUS: In development

McDonnell Douglas Electronics

McDonnell Douglas Electronics Company, Box 426, St Charles, Missouri 63301

Vital IV visual simulation system

Vital IV is the most recent in a family of CGI (computer generated image) visual simulation systems that go back to the Vital I laboratory prototype of 1969. Vital stands for virtual image take-off and landing. Vital II became in 1972 the first CGI visual system to be approved by the FAA. This system and its developments have since been installed on many flight simulators around the world. Vital IV builds on the successful base of CGI technology to provide finer and sharper images with greater detail than previously, with the option of panoramic views across all flight-deck windows, and presents to the flight crew images in realistic perspective and virtually without distortion. The computer-generated scenes made up of light-points and surfaces change in response to flying control inputs.

Types of scene: Airports, surrounding districts, aircraft carriers, general terrain
Scene conditions: IFR, VFR, twilight, night, dull day, above, in, or below cloud

La Guardia Airport as represented by McDonnell Douglas Vital IV computer-generated image system

Captain's-eye view of runway as seen by McDonnell Douglas Vital IV simulation system representing Douglas DC-10 of French airline UTA

Data-base coverage: Normally 170 × 170 nm or, with combined data-base techniques, unlimited
Optical resolution: <3 minutes
Lightpoints: 400 000+ carried on-line, with 8000+ displayed at any one time. Used for airport lighting, roads, moving traffic, beacons, and other illuminated representations
Surfaces: 15 000+ carried on-line, with 300+ displayed at any one time. Used to represent terrain, buildings, moving vehicles, runways, and other hard surfaces
Scene access: On-line access to 50 separately stored scenes, each loaded and brought into view in a few seconds. Alternatively, large contiguous areas represented by automatically accessed sub-scenes

Weather: Variable visibility, cloud, fog, haze, and thunderstorms
Colours: 10+ separate colours, ranging from red to green
Dynamics: Smooth, flicker-free bright image with motion in direct response to simulated aircraft manoeuvres, without smearing, comet-tailing, or break-up. 30 completely new pictures presented each second. All practicable manoeuvres and speeds.
Occultation: Complex buildings, mountains, and vehicles programmed to have a solid appearance, occulting lights and solid objects behind them from the pilot's point of view
The Vital IV system comprises three standard equipment modules packaged as needed to match any simulator flight-deck:

Image generator: Single cabinet containing a general-purpose computer that shares the job of equation-solving with special processing circuitry to generate the real-time scenes. Programs and extra scenes are stored on-line in a high-speed disc unit.
Display unit: Combination of electronic CRT and reflective virtual-image optics mounted on the simulator flight-deck.
Instructor's control panel: A small facility that may be incorporated into the simulator's instructor system, giving him control over such characteristics of the visual scene as runway light visibility and weather.

STATUS: In production for commercial and military customers

Pacer

Pacer Systems Inc, 87 Second Avenue, Northwest Industrial Park, Burlington, Massachusetts 01803

Pacer is a small company engaged in a number of diverse activities – management, systems synthesis, applied aerodynamics, cartography and photogrammetry, the development of air-data systems, information handling, and training and simulation.

The company has developed a device called CPIFT (cockpit procedures and instrument flight trainer) in which the capabilities of a conventional CPT are expanded so that it can be 'flown' and navigated, in addition to its use as a ground procedures trainer. The microprocessor-based system combines all the functions of a CPT with most of those of a full simulator at a cost claimed to be lower than that of any other system with this level of capability.

Simulators are built to customers' specifications around six sub-systems: cockpit or flight-deck simulation and enclosure, power plant, aircraft systems, radio/navigation, flight dynamics, and instructor station. Moving-base capability and visual displays are not provided except to special order. Costs are between 25 and 50 per cent of those appropriate to the aircraft they represent.

Instructor's station for Pacer's de Havilland Canada DHC-6 STOL transport simulator

Pacer CPIFT simulator system for Short 330 commuter airliner

Reflectone

Reflectone Inc, 5125 Tampa West Boulevard, Tampa, Florida 33614

A subsidiary of Dunlap and Associates Inc, Reflectone is a specialist simulation and training-device company, with activities in aviation and other fields. The following systems represent some of the more important aircraft simulation projects:

Fairchild A-10 operational flight trainer

Built for the USAF to represent the anti-tank battlefield bomber/gunship, the system permits proficiency development in all phases of instrument flight from pre-flight and start-up through navigation and combat, visual and GCA approach, to landing and post-flight debriefing. The weapon system can also simulate the electronic warfare environment and, while the simulator is mounted on a fixed base, a *g*-seat with inflatable air bladders provides cues representing the onset of acceleration and deceleration.

Day/night capability is enhanced by a McDonnell Douglas Vital visual display mounted on the windscreen. The entire system is based on an SEL 32/55 computer with 128K words of core memory.

Boeing-Vertol CH-46E operational flight trainer

Developed for the US Marine Corps, this system represents the twin-turbine tandem-rotor assault helicopter and is intended to build up and maintain pilot proficiency in all aspects of operation. The system is mounted on a six-degrees-of-freedom motion base and is controlled by a Harris Slash-4 computer with 82K words of core memory and more than 40 megabytes of disc storage. The full-daylight CGI system is provided by Reffusion Simulation in collaboration with Evans and Sutherland Computer Corporation. The visual display, added in late 1980, permits a number of new training missions to be flown: landing and take-offs in confined areas, operations aboard LST-class ships, IFR and VFR in poor weather in daylight, dusk, or darkness, flight with a slung load, and formation flying.

The system employs what is claimed to be a unique feature, a remote trainer control panel at the centre console of the flight deck, permitting the instructor to fly as a pilot or co-pilot and so exercise limited control of the training session. In addition it allows the student to conduct self-training, without instructor assistance, and enables playback of the most recent five minutes of flight.

Sikorsky S-76 instrument flight simulator

This six-degrees-of-freedom motion simulator was designed for American Airlines Training Corporation to duplicate the characteristics of the Sikorsky S-76 twin-turbine executive helicopter that is now being used by large corporations, air taxi operators, the offshore oil industry, and other large industrial ventures. It develops crew proficiency in the operation of controls, interpretation of instruments, management of navigation and communication systems, and in coping with a variety of emergency situations through instructor-induced faults.

The simulator comprises a fully representative cockpit, motion and hydraulic system, instructor's station mounted on the motion base behind the cockpit, and a Harris Corporation Slash-6 computer with 48K words of core memory. The system characteristics are based on flight-test data obtained by American Airlines specifically for this application, and include duplication of the vibration spectrum generated by the main rotor and transmission.

Sperry

Sperry Flight Systems, Defense and Space Systems, 21111 North 19th Avenue, Phoenix, Arizona 85027

Sperry's simulation activities complement its large-scale and historical involvement in auto-pilots, fight-control systems, and navigation and flight instruments, utilising in particular its extensive experience in mathematical modelling and real-time data processing. Apart from aircraft, Sperry builds simulators and training aids for military ships and vehicles.

Grumman EA-6B weapons systems trainer

The US Navy's most sophisticated EW aircraft is the carrier-based EA-6B Prowler. It uses high-power jammers and other electronic countermeasures to disrupt enemy radar and communications, so screening friendly strike aircraft from surface/air missiles. It is also used to protect surface ships from radar detection by enemy aircraft and from cruise missiles.

Sperry's trainer is the US Navy's largest and most technically advanced flight simulation device. It accurately reproduces the EA-6B's performance and operational characteristics, the environment within which it operates, and the inter-relationship of performance and environment. High fidelity is achieved by extensive simulation of radars, visual scenes, and radio communications. The system incorporates the most recent features of digital flight simulation technology, permitting it to cover the entire flight envelope and every variation of flying qualities, including stalls and spins.

For realism in its displays, the system includes a digital radar land-mass simulation of search-radar returns, with a storage capacity of 1.6 million square miles. Terrain and cultural features that can be stored for display include shadows, refraction and earth curvature, far-shore brightening, range attenuation, moving targets, and emitter occulting. Electronic countermeasures simulation covers tactical and defensive jamming, communications jamming, and the effects of chaff dispensation.

The four-seat cockpit, fully representative of the EA-6B, is mounted on a six-degrees-of-freedom motion platform, and a computer-generated visual-image system displays outside-world scenes for both shore- and carrier-based operations. As many as three instructors can co-ordinate and control a training mission, integrating the operation of the aircraft with the military environment. The instructor station has five CRT graphics displays with interactive terminals, and repeat displays there permit the instructors to monitor the out-of-window scenes, radar, and other displays.

Sperry weapons-system trainer for the US Navy/Grumman EA-6B

F/A-18 operational flight trainer

The F/A-18 is the US Navy's newest fighter/attack aircraft, with a capability greater than that of any other carrier-based aircraft in its class. The Sperry simulator represents the systems and performance of the fighter, the environment in which it operates, and the relationship between the two. Specifically, it represents the environment of the USS *Nimitz*, including approach, arrested landing, wave-off, bolter, touch-and-go, barricade arrestment, deck taxi, and catapult tensioning and launch.

The simulator comprises a representative

Instructor's station for EA-6B simulator

F/A-18 cockpit, a three-window, three-channel visual system, and a facility for transmitting seat and buffet loads and shocks to the cockpit floor. A g-seat and g-suit system combined with seat-buffet generator provides acceleration cues. The instructor station can be operated either individually or jointly by an instructor and a device operator, and can also accommodate an observer. Faults can be inserted or programmed from the CRT keyboard, and the design can be modified and expanded in accordance with aircraft development and new training requirements.

Sikorsky CH-53 operational flight trainer

Sperry's experience in helicopter flight simulation includes the design, development, and construction of training equipment for the CH-53, HH-3F, HH-53C, HH-52A, UH-1E, TH-1L, and CH-47 helicopters. This work has resulted in the development of new mathematical modelling techniques for helicopter aerodynamics and engine operation. These mathematical models now permit complete and accurate reproduction of performance and flight characteristics throughout entire flight envelopes; they include buffet and rotor stall effects, autorotation, power settling, ground effect, ground resonance, and variable turbulence.

The simulators for the US Marine Corps' CH-53D and -E assault helicopters comprise the complete cockpit mounted on a six-degrees-of-freedom motion platform, instructor station, and digital computation system. The cockpit module has a six-window, six-channel CGI visual system with 196° horizontal and 60° vertical field of view.

Full flight simulator for USMC/Sikorsky CH-53 helicopter under construction at Sperry

Radar

Attack Radar

FRANCE

Electronique Serge Dassault

Electronique Serge Dassault, 55 quai Carnot, 92214 St Cloud
Formerly Electronique Marcel Dassault, this company changed its name in February 1982.

Antilope V TC

The Antilope V TC radar has been designed for the penetration version of the Mirage 2000N. Its basic functions are terrain following, air/air, air/sea, air/ground and navigation with ground mapping and navigation updating. It employs a travelling-wave tube J-band coherent transmitter and a flat slotted-plate antenna. Radar information is displayed on a head-up display and on a three colour multi-mode cathode ray tube head-down display.

STATUS: Electronique Serge Dassault is the design leader on this programme but shares it with Thomson-CSF. The Mirage 2000N radar is planned for service in 1986.

Aida II

This miniature lightweight system is designed for light interceptors and aircraft with restricted accommodation in the nose. It can also be installed in pods for under-wing mounting and in some versions of the Mirage V this is the method adopted. The system automatically searches for, acquires and provides ranges of air or sea targets within a cone of 18 degrees. The antenna is fixed and the pilot has to point the aircraft more or less in the direction of the target. Used in conjunction with a gyroscopic gunsight it supplies all the information necessary for interception and attack with guns, rockets, bombs or missiles. The operating frequency lies in the I/J-band and the transmitter power is between 80 and 100 kW.

Antilope V TC terrain following and navigation radar by ESD

ESD's Aida II automatic fire-control radar for guns and IR missiles

Thomson-CSF

Division des Equipements Avioniques, Thomson-CSF, 178 boulevard Gabriel Péri, 92240 Malakoff

Cyrano IV family

The T-CSF Cyrano IV was introduced in 1972 and became the first of a family of air/air and air/ground radars. The following models are available or planned: Cyrano IV-0, IV-1, IV-2, IV-3, IV-M, and IV-M3.

Cyrano IV-0

The first Cyrano radar, a monopulse system, was designed for air/air interception, searching for and tracking hostile aircraft and providing flight, firing, and break-off information to the pilot via a weapons sight. In the search mode the radar scans ± 60° in azimuth and ± 30° in elevation. When the pilot puts a marker on his display to designate a particular target, the radar moves into the track mode, measuring the range and relative velocity. In the final, inter-

ception mode, the system signals the earliest and optimum times to fire a designated weapon, and follows this with an indication of when the firing 'window' has ended. The system, which operates in the eight to ten GHz band (the generator being a coaxial magnetron), was originally designed for the French Air Force's Mirage F1 fleet and subsequently remains in production for other customers ordering these fighters.

Thomson-CSF Cyrano IV-0 AI radar

Thomson-CSF Cyrano IV-M multi-mode radar

Cyrano IV-1

By adding a fixed-target rejection filter the basic Cyrano IV radar can be upgraded to include MTI capability so that hostile aircraft can be tracked amid ground clutter in the 'lookdown' mode

STATUS: in production.

Cyrano IV-2

With the addition of beam-sharpening circuits and other modifications, the Cyrano IV-0 converts into the first multi-mode member of the family. Besides air interception, the radar can now be used for ground-mapping and low-altitude navigation. The latter capability includes contour mapping, terrain avoidance and blind penetration. For ground-attack the system can also provide range-to-target with an accuracy of about 150 feet (50 metres).

STATUS: in production.

Cyrano IV-3

This progressive development of the series brings together all the improvements incorporated in the -0, -1, and -2, radars.

STATUS: in production.

Cyrano IV-M

The first Cyrano system to be designed specifically for multi-mode operation (though still aimed basically at air interception), the -M can accomplish a wide variety of air/air and air/ground missions and has a high degree of resistance to electronic countermeasures. Radar video signals are put up on a head-down cathode ray tube with a B-type display for air/air operation and a plan-position indicator presentation for air/ground operation. The 57 cm diameter inverted Cassegrain antenna handles 200 kW of power from a coaxial magnetron. Built-in test circuits permit rapid fault diagnosis down to individual line-replaceable unit level.

STATUS: the French Air Force has been upgrading all its Cyrano IV equipment to IV-M standard and ordering new sets of this type for its Mirage F1s.

Cyrano IV-M3

A development of the -M, this radar is designed for multi-role versions of the Mirage 50 and for retrofitting on Mirage III or V with air/air and air/ground missions. Intended for export, or for older aircraft with an earlier generation of electronics, this system lacks the refinement of the Cyrano IV-M.

STATUS: in development.

RDI (radar Doppler à impulsions)

This is one of two new pulse-Doppler radars being developed for France's Dassault Mirage 2000 fighter. The RDI is intended for the all-altitude, air superiority and interception version, and is based on a travelling-wave tube, I/J-band transmitter radiating from a flat, slotted plate antenna. The range is said to be around 60 miles (100 km). The performance of the radar, and that of other systems on the aircraft, benefits from the digital signal handling and transmission of information by data-bus. Considerable electronic countermeasures resistance is built into the equipment, which can operate in air/air search, long-range tracking and missile guidance, and automatic short-range tracking and identification modes. Although designed for air/air operation, the system is understood to incorporate ground mapping, contour mapping, and air/ground ranging modes.

STATUS: in development. Originally scheduled for the production Mirage 2000s for the French Air Force, there have been delays in development, and so the first operational aircraft, to equip the Second Fighter Squadron at Dijon in

Thomson-CSF Cyrano IV-M3 radar

Thomson-CSF RDI interception radar

Thomson-CSF RDM multi-mode radar

Dassault Mirage 2000 advanced multi-purpose fighter for which Thomson-CSF's RDI radar is being developed

1984, will have the multi-role RDM radar (see below).

RDM (radar Doppler multifunction)

Along with the RDI, this radar is under development for France's new fighter, the Dassault Mirage 2000. Whereas the RDI is designed for interception and air combat, the coherent, multi-mode, digital, frequency-agile RDM is intended for the multi-role export version. It operates in three modes: air defence/air superiority, strike, and air/sea.

In the air/air role, the system can look up or down, range while searching, track while scanning, provide continuous tracking, illuminate targets for medium-range air/air missiles, generate aiming signals for air combat, and compute attack and firing envelopes. For the strike role it provides real-beam ground-mapping, navigation updating, contour-mapping, terrain-avoidance, blind let-down and air/ground ranging, and can detect moving targets on the ground above given threshold speeds and apparent sizes. In the maritime role it provides long-range search, track while scan, continuous tracking, and can designate targets for active missiles.

Options include a continuous-wave illuminator and GMTI and the potential for further growth includes Doppler beam-sharpening, identification friend or foe air/air identification and raid assessment.

The system is used in conjunction with a head-up display, a three-colour head-down display and a cathode ray tube counter-

Thomson-CSF RDM multi-mode radar

measures display. Self-test circuits localise faults to line-replaceable unit level and significant improvements in reliability have been made as a result of microprocessor technology, hybrid circuits, and microwave integrated circuits.

STATUS: equipment is understood to be planned for delivery in 1983. It has been chosen as a stop-gap for initial deliveries of the French Air Force Mirage 2000s in 1984 following delays in development of the RDI radar designed for the interceptor (see above).

ISRAEL

IAI

Elta Industries, Israel Aircraft Industries Ltd, Ben-Gurion International Airport

EL/M-2001B

This is a range-only radar for single-seat tactical aircraft operating in air/air and air/ground modes. The target is detected visually with acquisition and tracking being accomplished automatically by the radar. The system can operate in heavy ground clutter. Information from the radar can be displayed on the head-up display or fed into the weapon control computer for weapon delivery computation. The six line-replaceable units are based on solid-state technology with the exception of the travelling wave tube and have considerable reserves for future growth.

Base diameter: 450 mm
Length: 790 mm
Antenna diameter: 195 mm
Power: 115 V 3 phase 400 Hz

STATUS: in service with Mirage fighters of the Israeli Air Force.

EL/M-2021B

The EL/M-2021B is a second generation

Elta Industries' EL/M-2001B

coherent radar operating in air/air and air/surface modes, designed specially for single seat aircraft, with multi-mode operation and integrated control. The system operates either in the long range mode with a built-in track-while-scan function or in the short range mode with an automatic lock-on feature. After lock-on to the target, a track-while-scan mode provides the pilot with additional information on other enemy aircraft in the vicinity. In the air/surface mode the radar provides range-to-target as part of the aircraft weapon delivery system and it can also be operated in a ground mapping mode. The performance can be improved in this mode by selecting Doppler beam sharpening. It is also possible to operate the radar in the air/sea search mode, when boats can be detected in both high and low sea states. Navigation, especially at low altitudes and in poor weather, is facilitated by using terrain following and terrain avoidance features. There is a beacon display in both air/air and air/ground modes which further enhances navigation accuracy.

The system can operate either as an integral part of a full fire-control system or it can stand by itself. As part of a full weapon system, aircraft information and pilot commands to the radar are controlled by a central computer. The EL/M-2021B data is fed to the computer and information is displayed on the system's head-up and head-down displays. Communication with the system is mainly through a multiplexed bus. In the stand-alone configuration the radar communicates directly with the appropriate aircraft sub-systems. Information is presented to the pilot on a head-down display.

Weight: 120 kg
Power: 115 V 400 Hz 2·5 VA
Maintainability: BITE system locates fault to specific lru (of which there are 6), simultaneously alerting pilot
Reliability: designed for 100 h mtbf, complying with MIL-E-5400N

STATUS: in production for, and operational with, Kfir fighters of the Israeli Air Force.

Radome swung aside on Israeli F-4E Phantom revealing Elta Industries' EL/2021B attack radar

Production of EL/M-2001B radar sensing heads

SWEDEN

L M Ericsson
Telefonaktiebolaget L M Ericsson, MI Division, S-431 20 Mölndal 1

PS-37/A
This multi-mode I/J-band, monopulse radar was designed for the JA 37 attack version of the Viggen fighter, Sweden's principal combat aircraft. It stems from mid-1960s designs and is largely integrated with the navigation, display and digital computer-based data-processing sub-systems, and comprises two units: a scanner and a transmitter/receiver package, the latter being made up of 13 line-replaceable units.

Except for some of the high-frequency components, the PS-37/A is of solid-state design. Elaborate signal-processing provides a high degree of immunity from both natural interference and electronic countermeasures, and accuracy is enhanced by lobe-shaping, whereby the aperture (effectively the scanner diameter) is artificially increased to provide better resolution and reduce side-lobe effects. The radar is semi-automatic in operation to reduce the work-load on the single crew-member, and information is presented to him on both head-up and head-down displays.

Operating modes are search, target acquisition, air-target ranging, obstacle warning, beacon-homing and terrain mapping. By adding a further unit a terrain-following capability can be provided, the pilot flying the aircraft in response to head-up display demands.

STATUS: in service.

PS-46/A
This software-controlled, multi-mode, pulse-Doppler AI radar has been developed for the JA 37 interceptor versions of the Swedish Air Force's Viggen fighter. In view of the numerically small size of Sweden's defence force, great emphasis has been placed on operational availability and readiness, and all-weather capability and effectiveness in an electronic countermeasures environment are also premium requirements. Designed to cope with high-performance aircraft, transports, and helicopters, the system has wide-angle coverage, look-down capability, and can operate at all altitudes.

The multi-mode requirements of the PS-46A fall into two categories: air/air and air/ground. The latter are met by using conventional non-

PS-37/A radar by L M Ericsson

PS-46/A radar for Viggen interceptor

coherent pulse waveforms, but the former call for more sophisticated waveforms. The standard radar functions are controlled by a data-processor that extracts information from the raw radar and transfers it to other aircraft systems. A digital bus distributes all signals within the radar itself with minimum wiring. For the guidance of semi-active homing missiles an illuminator transmits a continuous-wave radio-frequency signal through the radar antenna.

Control of the system through suitable software, says LME, enables parameters to be changed or optimised according to the needs of flight development programmes without time-consuming changes to equipment; similar changes can be introduced during service according to changing military requirements, and radar signatures adopted for peace-time training and exercises can be easily changed during times of conflict to thwart enemy intelligence and countermeasures.

Type: medium prf pulse Doppler
Frequency: x-band, bandwidth >8% of spectrum
Performance: detection range 50 km in look-down mode
Weight: 300 kg
Power: (coherent transmitter) 50 kW (cw illuminator) 200 W
Antenna: 700 mm diameter
Processor: high-speed 16-bit word-length system with 32 K word program memory
Modes: search, acquisition (automatic via hud, semi-automatic via hdd), tracking (track-while-scan, continuous track), target illumination, air/ground ranging
High-resolution ground-mapping is optional
Reliability: mtbf 100 h

STATUS: in production.

Saab JA 37, Sweden's distinctive double-delta fighter, for which PS-37/A monopulse radar was designed

---

UNITED KINGDOM

MEL

MEL, Manor Royal, Crawley, Sussex RH10 2PZ

ARI 5955/5954

This radar for asv, asw, mcm, and sar roles was designed specifically for installation on helicopters. The main radar is the ARI 5955 intended for detecting air and surface targets. The ARI 5954 is the corresponding equipment with this system providing identification of friendly aircraft and surface craft.

The system operates in the I-band and a stabilised dish scanner is housed in a rotor located in the upper surface of the fuselage just behind the rotor pylon. Antenna tilt is adjustable, both above and below a horizontal datum, from the radar operator's position. The receiver gain is adjustable and variable swept-gain facilities are provided. A selection of range presentations up to 50 nautical miles is available and information is displayed on a projection cathode ray tube in conjunction with a Schmidt optical system. This provides unusually high brightness over a 430 mm square plotting surface. An illuminated parallel-line protractor is incorporated. The operator has the choice of three modes of presentation: conventional plan-position indicator, ground stabilised or ground stabilised with offset.

Royal Navy/Westland Sea King sub-hunter, which uses MEL's ARI 5955/5954 attack radar

STATUS: the programme was initiated in response to a UK Government contract constructed in collaboration with the old Ministry of Technology. The ARI 5955 is installed on RN Sea Kings, Wessex helicopters, and RAF Sea Kings. A number of other countries have also bought this equipment.

Ferranti

Ferranti Ltd, Electronic Systems Department, Ferry Road, Edinburgh EH5 2XS

Blue Fox interception radar

Blue Fox is a lightweight (less than 190 lb (86 kg)) high-performance pulse-modulated radar that combines airborne interception and air/surface search and strike for the Royal Navy Sea Harrier V/STOL fighter. The specific roles are:

Search, in which there is a choice of scan patterns in azimuth and elevation for use against air/air and air/surface targets.

Detection, in which the radar embodies special features to facilitate detection of small air and surface targets.

Lock-on track, to provide the necessary input to the weapon aiming computer for weapon release.

Radar lock-on visual acquisition, where the pilot flies the aiming mark on the head-up display onto either air or surface targets and then locks the radar onto the target by pressing the 'accept' button.

Air/surface ranging, where during an attack the radar boresight can be slaved continuously to the aiming mark to provide ranging irrespective of whether or not the surface target is radar discrete.

First set of Ferranti Blue Fox attack radar to be delivered for the Indian Navy's Sea Harrier programme

Five units of Ferranti's Seaspray surveillance radar designed for naval version of Lynx helicopter

Identification, in which transponder returns from friendly ships and aircraft are displayed for identification and location.

Visual identification, where the radar incorporates a short range scale to enable the Sea Harrier to close to within a short distance so that the pilot can identify visually any intruder aircraft.

Navigation, in which the radar can be used in a ground-mapping role as well as interfacing with the navigation computer.

The main radar unit is housed within the folding nose of the aircraft and comprises five line-replaceable units: the transmitter, receiver, processor, amplifier electronic control, and scanner. The remaining four units within the cockpit are also first-line replaceable. The display which makes up two of them is divided between the actual presentation and its associated drive unit. Of the other two units, the radar control set contains those controls which can be preset before an attack and the hand controller contains those that may need to be operated during an attack.

The radar has a flat aperture, slotted array antenna which provides better detection ranges than more conventional forms; it also reduces sidelobes, another benefit in an ECM environment. It is roll-stabilised to provide increased accuracy during air/surface attacks. Within the processor unit is a digital scan-converter which increases the brightness of the display in all conditions of ambient lighting, even when the sun is shining directly onto the face of the display; its storage and variable persistence can be used to advantage elsewhere. By selecting a 'freeze' facility the radar picture can be retained indefinitely after the transmitter has been switched off, a useful capability in an ECM environment when approaching a potential

Ferranti's Blue Fox radar showing flat-plate antenna

surface target for subsequent attack. Alternatively a 'tails' selection can be made which will cause fast moving targets to produce a comet-tail effect on the display and thus indicate relative track histories in an air/air encounter. The system is frequency-agile, that is within a wide frequency band each pulse is transmitted on a different frequency from the previous one, the rate of change of frequency being very rapid.

This technique results in several advantages:

Effects of sea and rain clutter are reduced and therefore the system's detection capability under these conditions increases.

It combats the effect of ECM since the enemy is forced to jam over a wider frequency band, decreasing its effectiveness.

It counters the effect of 'glint'. When the radar target with a complex reflecting surface such as a naval vessel carrying radar and radio antennas changes aspect relative to a radar transmitter, the effective radar centre of the target changes; it can even move to a point outside the physical outline of the target. The effect on the radar tracking target is to introduce low frequency noise into the tracking loop as the radar attempts to follow the rapidly shifting centre of reflection. Using the frequency-agile mode, this 'glinting' action takes place much more quickly so that in a short time it is possible to determine a sharply defined average position of the target. The elimination of glint is important for missile guidance systems.

Frequency agility eliminates the confusing effects of second trace returns, ie return echoes from large targets outside the range scale that appear on the display in random fashion. This advantage is particularly evident when operating near coast lines or other features having large radar echoes.

Reliability: >120 h mtbf

STATUS: in production for British Aerospace Sea Harrier V/STOL strike aircraft, which entered Royal Navy service in 1981, and were used in May 1982 in the Falkland Islands conflict.

Seaspray

Seaspray is a lightweight, high performance I-band radar specifically developed and produced for the Royal Navy to counter fast patrol boats, its initial application being the Westland Sea Lynx.

The system comprises five line-replaceable units: a scanning antenna, transmitter, receiver, control/display unit and processor. With these is associated a heat exchanger. The control/display unit is a bright, flicker-free television display. A Ferranti scan converter displays the radar pictures in a television raster format which would allow the display to be used for other sensors such as forward-looking infrared and low-light television. Target range and bearing is provided on the screen in alphanumeric form and the display persistence may be varied according to the operator's requirements.

There are eight operational modes:

Search: the targets are large and small surface vessels in bad weather and high seas, low flying aircraft, and periscope detection in the anti-submarine warfare role.

Weapon system integration (employing monopulse tracking techniques): guidance for the Sea Skua air/surface missile and accurate target range and bearing inputs (through a data link) for surface/surface missiles.

Identification: the system can be used in an interrogate mode for target identification.

Navigation: in normal operation, and by integration with a Decca TANS.

Search and rescue.

Station keeping.

Weather warning.

Self let-down and recovery to base in bad visibility.

The adoption of frequency-agility affords protection from ECM, rejection of clutter and glint, and the elimination of second trace returns.

Weight: 140 lb (64 kg)

Antenna dimensions and scan: 27 inches wide (686 mm), 10 inches deep (254 mm), ±90° azimuth about the fore and aft axis.

STATUS: Seaspray is in production and in service with Sea Lynx helicopters of the Royal Navy, Royal Netherlands Navy, Royal Danish Navy and the Brazilian and Argentinian navies.

Royal Navy/British Aerospace Sea Harrier FRS. Mk 1, carrying Ferranti Blue Fox radar

Seaspray radar display in Lynx cockpit

Royal Navy/Westland Lynx HAS. Mk 2 subhunter, which uses Ferranti's Seaspray attack radar

Blue Kestrel

Following a decision by the UK and Italian governments to proceed with a replacement for Sea King anti-submarine warfare helicopters, Ferranti has reached the detailed design stage of a new surveillance and attack radar known as Blue Kestrel, to be used on the UK version of a new collaborative helicopter to be known as the EH101.

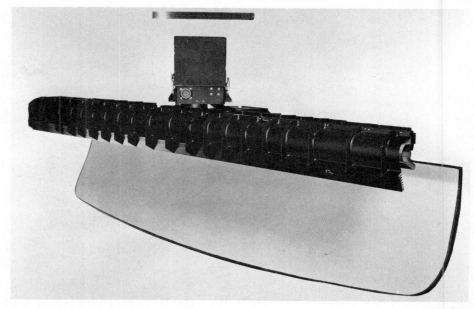

Antenna of Ferranti's Blue Kestrel radar for Westland/Agusta EH101 helicopter

Royal Navy Sea Harrier V/STOL attack-fighter which carries Blue Fox nav-attack system

Marconi

Marconi Avionics Ltd, Airport Works, Rochester, Kent ME1 2XX

Tornado F.2 Foxhunter interception radar

This all-British pulsed-Doppler AI radar has been developed for the air-defence version of the Panavia Tornado, 165 of which will re-equip two Phantom and seven Lightning squadrons in the RAF. Its existence was publicly revealed in October 1979 when the first development Tornado F.2 made its initial flight. Ferranti of Edinburgh is a sub-contractor to MAV in the programme.

The primary role of the system is the search for, detection, and tracking of airborne targets, but it will have an effective secondary capability in air/ground ranging and ground mapping. It also provides the target illumination for the

First prototype Tornado F.2 on test flight

Skyflash MRAAM semi-active-homing missiles. Significant size and weight reductions have resulted from the adoption of microprocessors and hybrid and thick film electronics. Designed to operate in and around Europe and the Atlantic approaches, the world's potentially most dense electronic countermeasures environment, the system relies heavily on sophisticated counter-countermeasures to perform its task and is designed for two-man operation. The navigator has two multi-function television tabular displays on which he can portray navigation, search, tactical evaluation, and approach-course information. The pilot has a head-down display that can repeat any of the information available to the navigator, and a head-up display for approach and attack guidance.

During the search phase the I-band (three-centimetre) radar picks up targets at a range of about 100 nautical miles and the navigator can build up a picture of the tactical situation by designating those that constitute potential threats and instructing the radar to continue automatic surveillance of the others. This ability to track targets while scanning for others is made possible by the radar's fmicw (frequency-modulated interrupted continuous wave) mode of operation. All selected targets are automatically tracked and information on their positions, heights, speeds and flight paths is continuously updated for display to the crew. Several targets can be designated for simultaneous attack, the radar computing the firing 'windows' for each one.

In fast-moving situations, eg close combat, the pilot can direct the scanner as he wishes. When targets appear in the designated area

Dielectric radome swung aside to reveal Marconi Avionics' Foxhunter AI radar fitted to Buccaneer aircraft for trials

lock-on follows immediately and is maintained during periods of high angular rates of motion, and medium- or short-range missiles or guns can be used as appropriate.

The Foxhunter system is designed to provide the Tornado with an autonomous combat capability, since it was reasoned that battle-damage could cancel assistance from ground or airborne early-warning radar. However, in line with more recent thinking, there is now a requirement for a JTIDS data-link so that information can be pooled.

UNITED STATES OF AMERICA

General Electric

General Electric Company, Aerospace Electronic Systems Department, 831 Broad Street, Mail Drop 508, Utica, New York 13503

AN/APQ-114 attack radar

A member of the AN/APQ-113 family, the APQ-114 was developed in the mid-1960s specifically for the FB-111 strategic bomber version of the General Dynamics F-111 tactical fighter. The FB-111A was planned to supplement the obsolescent Boeing B-52 bomber in some roles, particularly deep penetration at low level. The APQ-114 is basically similar to other members of the family, but in addition features a north-orientated display, beacon-homing mode, and automatic photo-recording.

STATUS: in service.

AN/APQ-144 attack radar

This is another member of the APQ-113 family of attack radars, and was developed in the late 1960s for the General Dynamics F-111F tactical fighter. In addition to the basic features of that family, the radar contains a number of improvements. Notable among these are the addition of a 0·2 microsecond pulse-width capability and a 2·5-mile (4-kilometre) display range.

Schemes for adding digital moving target indication and K-band transmission capabilities were successfully flight-tested, but have not been incorporated into the service aircraft. A modified APQ-144 radar was built in small numbers for the US Air Force/Rockwell B-1 strategic bomber programme, cancelled by President Carter in 1977 but revived by President Reagan in 1981. This system was designated AN/APQ-163 (see below).

STATUS: in service with US Air Force/General Dynamics F-111F tactical fighters.

AN/APQ-163 attack radar

A development of the AN/APQ-144, the APQ-163 was designed for the US Air Force/

General Electric AN/APQ-114 attack radar in the nose of a US Air Force/General Dynamics FB-111 strategic bomber

Rockwell B-1 strategic bomber. Three sets of equipment had been built and supplied for the five-prototype development programme by the time it was cancelled by President Carter in 1977, but the system continued to fly in the restricted flight programme involving a single aircraft that was subsequently authorised.

Among modifications to the APQ-144 to suit the new application were an increase in the scan angle from plus or minus 45 to 60 degrees and nuclear hardening of the electronics. Subsequently, beginning in September 1978, a Doppler beam-sharpening mode of synthetic aperture operation has been tested on the one B-1 permitted to fly. This technique is implemented by a coherent-on-receive modification to the APQ-163.

STATUS: unclear at present.

Multi-mode radar for Northrop F-5G Tigershark

In June 1981 General Electric announced a $38 million contract from Northrop to develop a multi-mode radar for that company's private-venture F-5G derivative fighter unveiled that month at the Paris Air Show. An additional $19 million was ear-marked for test equipment and a production option of 12 units was expected in late 1981. Successful sales of this new fighter, intended to narrow the difference between the F-5 family and the General Dynamics F-16 fighter for third-world countries, could result in business worth more than $200 million by 1986, according to the radar company.

Based on an extensive development of the GE MSR (modular survivable radar, see below) programme, the new radar provides long-range look-up and look-down detection of airborne targets. It also employs Doppler beam-sharpening for high-resolution ground-mapping, a feature that had already been tested on the APQ-163 radar for the US Air Force/Rockwell B-1 bomber. Air and sea modes are incorporated for maritime detection and tracking.

STATUS: in development.

Modular Survivable Radar demonstrator programme

GE's MSR (modular survivable radar) is a bench-top demonstrator programme to investigate the modular technology that the company believes will be needed in future airborne radars. The aim is to demonstrate a radar whose modules are so planned that changes or improvements can be incorporated simply by substituting new line-replaceable units with the minimum impact on weight and volume.

This modularity, says GE, is not just a sophisticated method of system partitioning, but a design discipline that extends down to shop-replaceable level. MSR is termed 'survivable' because of features that permit it to operate in a severe electronic countermeasures environment.

The term 'technological transparency' was applied to the system by GE to indicate that it is broken down for design purposes into a number of functionally specified modular elements that can be upgraded with new technology as it becomes available. Equivalent functions can therefore be performed by succeeding generations of MSR at a fraction of the cost of replacing the entire system as a result of newer technology in just a few areas.

This functional building-block approach to the programmable signal processor, which is based on large scale integrated circuits, permits it to be matched to whatever level of processing is required. The company has made a major commitment to very large scale integration technology and this is compatible with MSR in handling foreseen improvements during the 1980s.

STATUS: research tool, and one from which the combat radar for Northrop's new F-5G fighter design has been partly derived.

Hughes Aircraft
Radar Systems Group, Hughes Aircraft Company, El Segundo, California 90009

AN/AWG-9 interception radar
In the mid-1950s airborne radars still had one major weakness: they could not look down without being blinded by clutter. Target aircraft could therefore escape detection by flying close to the ground where their echoes were masked by the much stronger ground echoes. This was eliminated by the advent of coherent-pulse Doppler radar. One of the first such systems was the Hughes AN/ASG-18 designed for the North American F-108 interceptor. When that project was cancelled, work continued on the radar and the technology found its way onto the Mach 3 USAF Lockheed YF-12 interceptor, which became the first application.

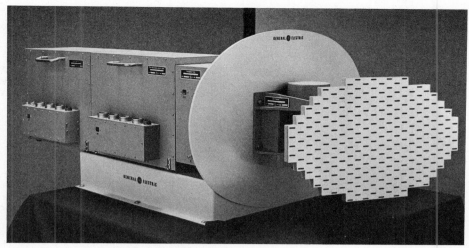
Mock-up of General Electric privately financed multi-mode radar for Northrop's F-5G Tiger Shark export fighter

Full-size mock-up of Northrop F-5G Tiger Shark derivative of long established F-5 family, principal combat sensor for which will be General Electric radar

Units of General Electric's modular survivable radar

Eagle fighter with nose radome swung aside to give access to Hughes AN/APG-63 radar

Developed still further in the late 1960s, the pulse-Doppler radar found its first volume-production application in the Hughes AN/AWG-9 radar (strictly a weapon control system) for the USN Grumman F-14 Tomcat air-superiority fighter. This radar is unique in several respects. To take advantage of the long-range detection capability conferred by the high pulse-repetition frequency waveform, it was designed in conjunction with a new missile (also by Hughes), the Phoenix, capable of launch ranges of more than 100 miles (160 km). The radar incorporates velocity-search, range-while-search and track-while-scan modes for target detection and tracking.

Range-while-search is a technique for coding the pulses in the high pulse-repetition frequency Doppler waveform so that target distance information can be recovered. Whereas range obtained in this way is not as accurate as that obtained using a low pulse-repetition frequency waveform, it can be obtained at longer distances, which is a valuable capability for fleet defence. This feature, coupled with what Hughes claims is the most powerful general purpose computer ever to be used with a fighter radar up to the time of its introduction, made it possible to implement a track-while-scan mode in which the system can track up to 24 targets simultaneously. Prior to this, tracking could be accomplished only by pointing the antenna at the target, restricting the fighter's capability to single-target engagement. The AWG-9 system can engage up to six targets simultaneously with six Phoenix AIM-54 missiles.

In the late 1970s the US Navy and Hughes together instituted an enhancement programme, a key element of which is the programmable signal processor. Apart from the much greater flexibility in adding or changing modes, the new system will improve the countermeasures capability, widen the missile launch zone, add medium pulse-repetition frequency, and provide surface attack ability.

STATUS: in service.

AN/APG-63 fire control radar

This radar is the principal sensor for the USAF/McDonnell Douglas F-15 Eagle which, in the hands of the Israeli Air Force, is already combat proven. The system is built by Hughes under contract to McDonnell Douglas and was one of the three major contracts (airframe, engine and radar) let in about 1970 when the US Air Force chose the St Louis firm to build the FX (fighter, experimental) design publicised at that time as the 'Foxbat Killer'.

Hughes Aircraft AWG-9 fire control radar for F-14 fighter 'opened up' for maintenance. Note iff array on front face of flat-plate antenna

New digital display for F-14's AN/AWG-9 system

Hughes' AN/APG-63 multi-mode radar in US Air Force/McDonnell Douglas F-15 Eagle air-superiority fighter

The system was designed around three main objectives; capability, reliabiity, and maintainability. Capability includes one-man operation (the F-15 is a single-seater) that can locate and track hostile aircraft at long range; close-in and look-down capability in situations that would blind earlier-generation radars; a clutter-free display with all appropriate target information; tracking and steering guidance information on the head-up display so that the pilot can keep his eyes continuously on the target; simplified controls and weapons co-ordination; and a certain air-to-ground capability.

The APG-63 meets possibly the highest reliability standards yet set for radar of its class. As a result, and despite its superior capabilities, maintenance time is reduced to about a quarter that of previous systems. High reliability and relatively simple maintenance schedules reduce the turn-round time, which is reflected in the smaller numbers of flight-line and maintenance personnel needed. This is largely due to the use of hybrid and integrated circuits to

reduce the number of components (parts count in American technology), and to the use of only four basic module sizes in the entire system.

Primary controls for the multi-mode, pulse-Doppler X-band radar are located on the control column, freeing the pilot to keep his head 'out of the cockpit' during fast-moving situations. Two special modes are provided for close-in combat, supersearch and boresight, enabling automatic acquisition of, and lock-on to, targets within ten miles. In the supersearch mode the radar locks on to the first target entering the head-up display field of view and displays its position. This alerts the pilot and greatly increases the range of visual detection. In the boresight mode the antenna is directed straight ahead and the radar locks on to the nearest target within the beam.

The system has a wide look-angle and its antenna is gimballed in all three axes to hold target lock-on during roll manoeuvres. The clutter-free head-down radar display gives a clear look-down view of target aircraft silhouetted against the ground, even in the presence of heavy 'noise' from ground returns. This look-down, shoot-down ability is achieved by using both high and medium pulse repetition frequencies (a technology perhaps unique at the time of the radar's appearance), by the digital processing of data, and by using Kalman filtering in the tracking loops. Also, the system's gridded travelling-wave tube permits variation of the waveform to suit the tactical situation.

False alarms are eliminated regardless of aircraft altitude and antenna look angle by a low-sidelobe antenna, a guard receiver, and frequency rejection of both ground clutter and vehicles moving on the ground so that only real targets are displayed.

When the FX fighter was being planned in the late 1960s, it was designed specifically for the air-to-air roles, and great efforts were made to preserve the integrity of this role to the exclusion of all other applications (the Air Force slogan 'not a pound for air-to-ground' called for all efforts to resist the weight compromises needed to accommodate other roles). However, since that time the F-15 has emerged as a potent ground-attack fighter, and in fact the APG-63 does have target-ranging for automatic bomb release for visual attacking, a mapping mode for navigation, and a velocity update for the IN system; a version of the F-15 known as the Strike Eagle acknowledges its ground attack capability and its radar reflects this new emphasis.

The APG-63 is compatible with both the AIM-7F Sparrow and AIM-9L Sidewinder air-to-air missiles. Antenna search patterns and radar display presentations are selected automatically for the type of weapon (medium-range radar-homing, or short-range infra-red missiles or M61 cannon) to be used; the pilot simply chooses the weapon by means of three-position switch on the throttle.

The ability to handle equally effectively both air-to-air and air-to-ground situations is the result of combining high and medium pulse repetition frequencies and the APG-63 is the first radar to do this. High pulse repetition frequencies are necessary to detect targets at long range but are not suitable for measuring range because there is insufficient time for pulses to return and be correlated before the next ones are being transmitted. On the other hand low pulse repetition frequencies that enable accurate range measurement and elimination of ground clutter do not have the power to give detection echoes at long range. But the APG-63 interleaves, for the first time in a production radar, high and low pulse-repetition

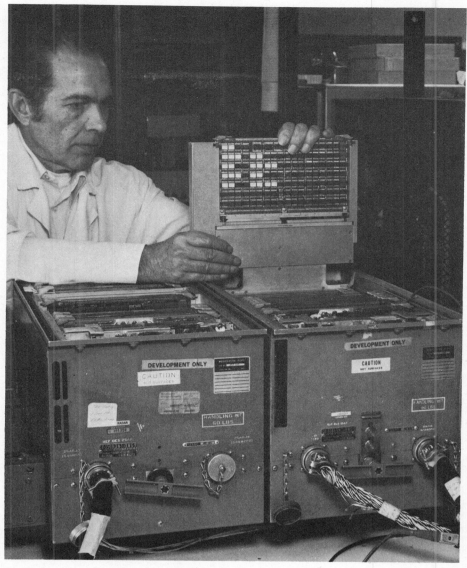

Programmable signal processor (right) alongside radar data processor for AN/AWG-9 update

frequency waveforms. The key to this technology was the substitution of heavy and bulky Doppler filters with a digital signal processor. By this means, incoming signals are sampled and their frequency content analysed by performing Fourier transforms on individual samples.

The APG-63 system comprises nine units occupying a total on nine cubic feet (0·25 cubic metre), weighing 486 lb (220 kg), and consuming 12·975 kW of power.

STATUS: in production and service with the McDonnell Douglas F-15 Eagle.

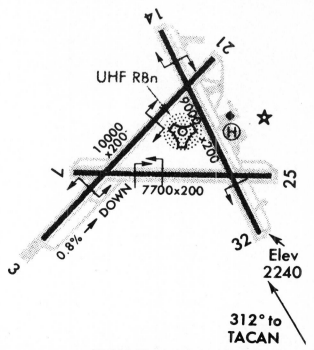

AN/APG-65 radar's Doppler beam-sharpening map (right) compared with actual aerodrome layout

AN/APG-63 with programmable signal processor

The use of synthetic-aperture radar for reconnaissance is well established. However search and rescue systems for this application typically look abeam of the aircraft only, and employ ground-based optical processing to secure usable images. But the advent of high-speed low-power large scale integration technology has made possible the development of a signal processor that can be programmed so that the performance of the radar system can be governed by software rather than by the 'hard-wired' logic currently built into the APG-63. This capability permits rapid switching among the diverse modes stored in a large off-line memory.

STATUS: in production. This enhancement was included first in F-15C Eagles operating in Europe.

AN/APG-65 multi-mission radar

Escalating procurement costs have underlined the need for multi-mission tactical aircraft, but until recently the necessary compromises in airframe and equipment caused by inadequate technology have prevented initial promises from being fully achieved. The problem has been that air/air systems need extremely fast data-processing rates (closing speeds can be up to a mile a second) and variable waveform flexibility to achieve all-aspect, all-altitude target-detection capability, while air/surface systems need very large data storage and processing capabilities for high-resolution mapping. These requirements have hitherto been incompatible with one another.

The digital breakthrough stemming from the early 1970s has gone a long way to changing the situation and the F/A-18 Hornet is perhaps the first combat aircraft to embody in one airframe and fire-control system an optimised air superiority and ground attack capability. For this reason the Hornet is the first US aircraft ever to have the dual designation F/A, indicating fighter/attack.

The key to combining the two into a single flexible unit is the programmable signal processor. While the AN/APG-63 in the F-15 fighter introduced digital signal processing to tactical fighters in the form of 'hard-wired' logic with a fixed repertoire of modes, the Doppler filter and range gate configurations that make up the programmable signal processor's capability in the APG-65 are defined by program coding; existing modes can be modified or new codes added by changes to software.

In the air/air role the APG-65 radar incorporates the complete range of search, track and combat mode variations, including several not previously available in an operational radar. Specifically these modes are head-up display acquisition (the system scans the head-up display field of view and locks on to the first target it sees within a given range); vertical acquisition (in which the radar scans a vertical 'slot' and automatically acquires the first target it sees, again in a given range); and boresight (the radar acquires the target after the pilot has pointed the aircraft at it).

In another, raid assessment, mode the pilot can expand the region around a single target that is being tracked, giving increased resolution around it and permitting separation of closely spaced targets.

As a gunsight the radar operates as a short-range tracking and lead-computation device using frequency-agility to reduce errors due to target 'scintillation'. All the pilot has to do is to put the gunsight 'pipper' on the target and press the gun-button.

Other modes are long-range velocity search (using high pulse-repetition frequency waveform to detect oncoming aircraft at high relative velocities); range-while-search (high and medium pulse-repetition frequency waveforms being interleaved to detect all-aspect targets, ie

High-resolution radar mapping (top) compared with conventional air photography show performance of AN/APG-63 radar is modified to include synthetic-aperture capability. Picture resolution is said to be ten times greater than previously obtainable with tactical radars

not only oncoming ones but those at any line-of-sight crossing angle); and track-while-scan, which can track simultaneously up to ten targets and display eight. When combined with future autonomous missiles such as AMRAAM, this mode will confer a 'launch and leave' capability as well as the simultaneous engagement of multiple targets.

In the F/A-18's air/ground role, the APG-65 has six modes: terrain-avoidance, for low-level penetration of hostile airspace; precision velocity update, when the radar provides Doppler speed signals to update the aircraft's IN system and if necessary to align it; the tracking of fixed and moving ground targets; surface-vessel detection, in which the system suppresses sea

clutter by a sampling technique; air/surface ranging on designated targets; and ground mapping. Two Doppler beam-sharpening modes are provided for these air/ground modes.

Number of units: 5

Antenna: low sidelobe planar array with fully balanced direct electric drive replacing hydraulics and mechanical locks of previous systems

Transmitter: liquid-cooled, contains software-programmable gridded travelling-wave tube amplifier

Receiver: contains A-D converter

Radar-data processor: general-purpose with 250 K 16-bit word bulk-storage disc memory

Signal processor: fully software programmable, runs at 7·2 million operations/s
Reliability: 106h mtbf. BITE detects 98% of faults and isolates them to single replaceable assemblies that can be changed in 12 minutes without adjustments or setting up
Weight: 340 lb (154·6 kg)
Volume: <4·5 ft³ (0·42 m³) excluding antenna

STATUS: in production and service.

Hughes Aircraft AN/APG-65 for F/A-18 being tested in modified T-39D Sabreliner

USN/McDonnell Douglas F-18 Hornet, which uses Hughes Aircraft's AN/APG-65 attack radar

Texas Instruments

Texas Instruments Inc, Equipment Group, 13510 North Central Expressway, Dallas, Texas 75266

Forward radar for Tornado IDS

As one of the principal specialist producers of terrain-following nav-attack radars in the West, Texas Instruments was chosen to design and develop the system for the IDS (interdictor/strike) version of the Tornado bomber being built for the air forces of Britain, West Germany, and Italy.

The all-weather radar comprises two essentially separate systems that share a common mounting, power supply and computer/processor. They are the TFR (terrain-following radar) and the GMR (ground-mapping radar). The first is used for automatic high-speed, low-level approach to the target and escape after an attack. The second is the primary attack sensor for the Tornado and operates in air/ground and air/air modes to provide high-resolution mapping for navigation updating, target identification and fire control.

The radar enables the crew to fix the aircraft's position by updating the Doppler-monitored INS, provides range and tracking information for offensive or defensive weapon delivery, and commands via the autopilot a contour-hugging flight profile that shelters the aircraft as far as possible from detection by hostile air-defence radars. It makes extensive use of electronic counter-countermeasures to provide relative immunity from interference in severe electronic countermeasure environments. The three units comprising the system are the radar sensor (transmitter/receiver package for TFR and GMR), a digital scan converter, and a radar display unit in which a moving map image can be superimposed on to a radar 'picture' for

Two of Texas Instruments' attack radar units for Tornado IDS

navigation updating and target identification purposes.

The GM radar operates in the Ku frequency band with nine modes: readiness, test, ground-mapping, bore-sight contour mapping, height-finding, air/ground ranging, air/air tracking, land/sea target lock-on, and beacon homing.

The TFR also operates at Ku frequencies and has three modes: readiness, test, and terrain-

following. In the last-named the pilot can elect either to have the aircraft flown automatically or to fly it himself through head-up display steering commands. He can also select ride comfort (for a given speed, the closer the allowable ground clearance, the less comfortable the ride owing to the greater g-levels needed to stay on the commanded flight profile).

Both systems have extensive built-in test features to ensure a high degree of fault isolation and comprehensive reversionary modes.

STATUS: in production by a consortium of companies under licence to Texas Instruments. The companies are Ferranti/Marconi Avionics (UK), AEG-Telefunken/Siemens (West Germany), and FIAR/Elletronica (Italy).

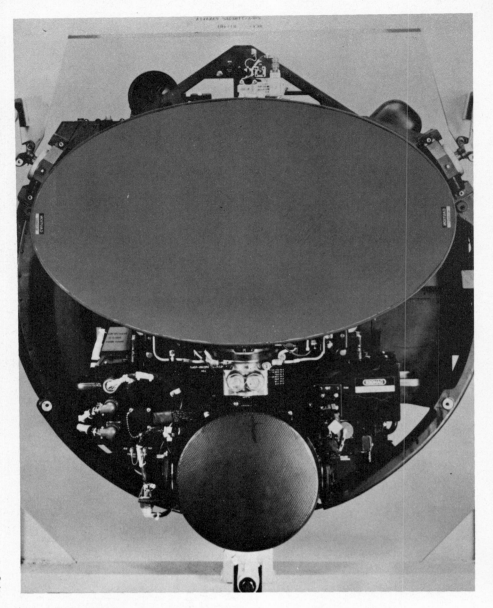

Texas Instruments' attack radar for Tornado IDS

Westinghouse

Aerospace Division, Westinghouse Electric Corporation, PO Box 746, Baltimore, Maryland 21203

Night/adverse weather attack radar

This compact, lightweight, millimetre-wave attack radar was designed for close-support aircraft operating at night or in bad weather. As a non-coherent sensor operating in the Ka band it offers four simultaneous air/surface modes:

Terrain-following/terrain-avoidance, in which the system displays on a head-up display contour lines indicating ground profiles up to three miles ahead of the aircraft. A steering 'box' is provided on the display, permitting the pilot to fly as low as 200 feet (60 metres) above ground level.

Emitter locator, in which by using the broadband characteristics of the reflector antenna, the radar can detect and isolate high-priority targets in the Ku/Ka bands (such as anti-aircraft missile radars) and provide azimuth steering information for fire-control.

GMTI, or ground moving target indication, in which the radar can discriminate vehicles, vessels, or other mobile targets moving at speeds as low as three mph (five km/h) using clutter-referenced rejection techniques. Helicopter- or tank-size targets can be detected at ranges up to ten miles (16 kilometres).

Ground-mapping, in which picture resolution is sufficient to permit navigation and the identification of tactical targets up to 15 miles (24 kilometres) distant.

Future modes envisaged for the system, which is based on the Westinghouse WX series

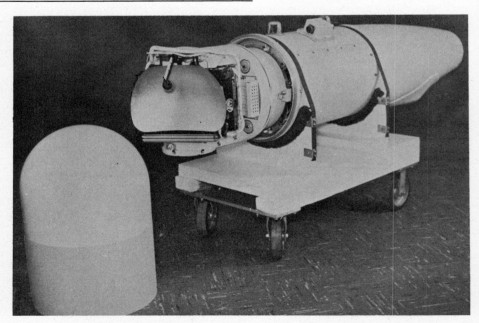

Prototype of Westinghouse night/adverse weather attack radar

that (claims the company) pioneered digital and modular technologies, include air/ground, ranging and beacon-homing.

Installation: external pod is alternative to nose-mounting
Volume: 3·4 ft³ (0·1 m³)
Weight: (nose-mounted) 145 lb (65·9 kg) (in pod) 290 lb (131·8 kg)

STATUS: private venture. An earlier version of the radar, designated WX-50, was flight-tested by the US Navy on TA-4J and OV-10 fixed-wing aircraft, and on a UH-1N helicopter. An evaluation set was later supplied to Fairchild to demonstrate the A-10's ability to operate as an all-weather battlefield bomber.

AN/APG-66

The F-16 Fighting Falcon air superiority and ground attack fighter is the subject of the largest international military co-production programme in history. The industries of five nations, United States, Belgium, Denmark, Netherlands, and Norway, are jointly producing components structures and equipment for well over a thousand F-16s on order by one North American, four European and two Middle Eastern air forces. The F-16 is a day fighter designed to augment the much larger F-15 Eagle air superiority fighter and in any European conflict would be involved in a high proportion of the air fighting.

This system was designed around the USAF/ General Dynamics F-16 Fighting Falcon light-weight fighter, launched out of the LWF prototype programme in January 1975. It has a look-down range, in ground clutter, of 20 to 30 nautical miles and a look-up range of 25 to 40 nautical miles. Great emphasis was placed on simplicity; the production system has no associated hydraulics, rate gyros or roll gyro, and there are only 9500 components. Despite this, there are ten operating modes, several of them associated with frequency agility to resist jamming.

The AN/APG-66 is of modular design so that foreseen improvements can be easily incorporated. A programmable signal processor permits new modes for air/air and air/ground to be added without increasing the space occupied by the system.

STATUS: in production for the USAF and the air forces of Belgium, Netherlands, Denmark, Norway, Israel and Egypt.

Improved AN/APG-66 fire-control radar

In August 1980 the US Air Force authorised Westinghouse to develop a substantially improved version of the AN/APG-66 fire-control radar for the F-16. It will improve the fighter's ability to deliver air/air and air/ground weapons in all weather conditions; in particular, it will permit F-16 pilots to launch from beyond visual range the new AMRAAM (advanced medium range air/air missile), and will improve the ground-mapping performance.

These important new capabilities will be gained by the addition of a programmable

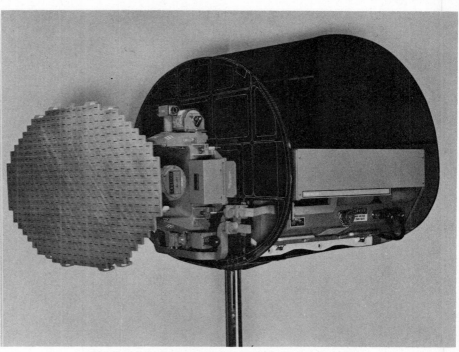

Westinghouse Improved AN/APG-66 radar for F-16 fighter

Modified Boeing 707s with rotordome housing Westinghouse AN/APY-1 surveillance scanner, and designated Boeing E-3A AWACS

signal-processor and a dual-mode transmitter. The former incorporates advanced modular processing architecture and a reliable high-density, solid-state memory. The latter will select the best waveform for each mode of operation, ranging from low prf in air/surface engagement to medium and high prf for long-range air interception.

In the air/air mode, the range at which on-coming targets are detected is increased by using a high-prf, velocity-search mode (ie the radar is searching for targets with such high relative velocities that they are inferred to be approaching nearly head-on). Once detected, a medium-prf range-while-search mode can be employed against targets with any aspect to gain additional range and angle information. In the track-while-scan mode, the radar can track simultaneously a number of targets, assess the degree of threat from each, and launch missiles as appropriate. By using high-resolution Doppler techniques, closely spaced targets can be distinguished and tracked separately.

In the air-combat mode, the radar scans selected airspace and automatically acquires the nearest target. In 'look-down' situations (where targets are seen against ground), land-based targets such as moving vehicles or vessels are ignored because the radar rejects returns with less than a specified threshold. This MTI technique ensures that only airborne targets are portrayed on the radar scope.

The air/ground performance of the new system has been improved by a 64-fold increase in the ground-mapping resolution and substantial spare computing power is available to cater for future growth, which could include, for example, automatic terrain-following or terrain-avoidance or very high resolution synthetic aperture mapping.

STATUS: in development. Prototype systems were delivered during 1981, and the first production sets will be supplied in late 1983.

AN/APY-1 surveillance radar for US Air Force/NATO Boeing E-3A Sentry

The principal sensor for the Boeing E-3A AWACS airborne warning and control system is the Westinghouse AN/APY-1 developed specifically as a surveillance and early warning radar. The scanner, which has a range of several hundred miles from the Sentry's operating altitude, is mounted back-to-back with a complementary identification friend or foe – secondary surveillance radar antenna and both are contained in a radome carried over the rear fuselage of the aircraft. The system works at X-band frequencies (ten centimetres) and has

General Dynamics F-16 Fighting Falcon, prime vehicle for Westinghouse AN/APG-66 and Improved APG-66

seven modes of operation. These modes are:
Pulse-Doppler non-elevation scan, enabling aircraft to be detected and attacked right down to ground level, though their height is not measured.
Pulse-Doppler elevation scan, in which target elevation is measured by electronic scanning of the beam in the vertical plane.
Beyond-the-horizon, using pulse radar without Doppler for extended range surveillance where ground clutter is in the horizon shadow.
Passive, in which the radar transmitter can be shut down in selected sectors of the scan while receivers continue to process ECM information coming from that direction. A single strobe line passing through the position of each jamming source is generated on the display console.
Maritime surveillance, involving the use of very short pulses to reduce the effect of sea clutter and enhance the detection capability of moving or stationary surface vessels.
Test maintenance, in which control of the system is delegated to the radar technician for maintenance.
Standby, in which the radar is kept in an operational condition, ready for immediate use, but the receivers are shut down.

STATUS: in production for the E-3A of the United States Air Force and for NATO. The full scale development effort on the E-3A did not begin until the radar was proved to be capable

of conducting the operational mission. Accordingly, Boeing received a US Air Force contract in July 1970 to develop and flight test two competing radar designs: one developed by Hughes and the other by Westinghouse. Flight tests on both systems were conducted in 1972, as a result of which the Westinghouse radar was selected by Boeing with the approval of the USAF. All of the test aircraft were refurbished and delivered as fully operational aircraft. The first AWACS was delivered to the USAF in March 1977, and the 552 AWACS Wing at Tinker Air Force Base became operational in the spring of 1978. By the end of 1979 20 aircraft, by now designated E-3A Sentry, had been delivered. The type began service with NORAD (North American Air Defence Command) in January 1979. Deployment in Europe, the Far East and Alaska was completed to test the interoperability of the system with existing air defence installations. In December 1978 the NATO Defence Planning Council, made up of defence ministers of member nations, gave approval for the purchase of a fleet of 18 AWACS for the NATO airborne early warning requirement. This is a $1800 million undertaking. USAF Sentries have been deployed on a number of occasions in trouble spots throughout the world. Most recently they were involved in monitoring the Iran/Iraq war.

UNION OF SOVIET SOCIALIST REPUBLICS

Jay Bird

Jay Bird is the NATO designation given to the air/air interception radar installed in the shock-cone intakes of the MiG-21 Fishbed-J version of the USSR's still widely used delta-tailed fighter. The system operates in the J-band, between 1280 and 1320 MHz, according to one source. Antenna size is limited probably to about 400 millimetres and three pulse-repetition frequencies have been quoted: 2042 to 2048, 1592 to 1792 and 2716 to 2724 pulses per second. Search and tracking ranges for this 'short-legged' fighter are perhaps 30 km and 20 km respectively with lobe-switching for tracking targets. Transmitter peak power may be around 100 kW. Acquisition range against typical targets is said to be less than 32 kilometres and there is believed to be no look-down capability. Fishbed-J may carry a radar-homing version of the Atoll air/air missile, in which case a target illumination capability would be necessary.

Look Two

This is the NATO designation for an I-band weapon-delivery and navigation radar. Western

MiG-21 Fishbed fighter which carries AI radar in centre-body shock-cone

electronic intelligence gives the operating band as 9245 to 9508 MHz in conjunction with frequency agility and four pulse-repetition frequencies: 320 to 336, 619 to 623, 1247 to 1253 and 1871 to 1879 pulses per second. This radar may be fitted in some versions of the Yak-28 Brewer, a 1950s-vintage swept-wing tactical bomber equivalent to the Canberra and now out of front-line service with the Soviet Air Force.

Scan Fix

This is the NATO code for the interception radar fitted in the pitot intakes of the MiG-17 and -19 interceptors. The intake configuration of these aircraft would not have permitted either a large antenna or a wide scan angle. Both I- and E/F-band versions are thought to have been produced. The former may have equipped MiG-19 Farmers, the latter going to MiG-17 Frescos. Both these fighters have long since been withdrawn from front-line service with the Soviet Air Force, though they serve elsewhere, eg in Warsaw Pact countries, and a version of the MiG-19 known as the Shenyang F6 has been built in China and equips more than 40 air regiments there. The radar, like the aircraft themselves, is obsolescent.

Scan Odd

This NATO code-name is for an interception radar fitted to the MiG-19 Farmer interceptor. An I-band system operating at 9300 to 9400 MHz, it is believed to be a later system than Scan Fix. If, as reported, it has an unusually complicated scan pattern, this is probably due to the limitations imposed by the installation in the pitot-type nose intake. Since the system is geared to the MiG-19, it again must be considered obsolescent.

Scan Three

An I-band radar operating in the 9300 to 9400 MHz range of frequencies, the interception radar NATO code-named Scan Three is fitted to the Yak-25 Flashlight two-seat all-weather fighter, a late-1940s design that came into service in 1955.

Short Horn

This weapons delivery and navigation radar appears from Elint information to be an example of relatively recent Soviet technology. It operates in the J-band around 1400 to 1500 MHz with frequency agility. Four pulse-repetition frequency/pulse-width combinations have been identified: 313-316/1-1·8, 496-504/0·5-1·4, 624-626/0·4-1·3 and 1249-1253/0·01-0·09 (the second set of figures in each case being in microseconds). Circular and sector scans have been recorded and the system may have ASV and maritime applications. Aircraft reported to have this equipment include the B, C, D and E versions of the 1960s-designed Yak-28 Brewer tactical twin-engined two-seat fighter-bomber, the Tu-105 Blinder medium bomber (in A, C and G models), and the Tu-16 Badger H strategic bomber. About 150 maritime patrol Blinder Cs are still in service, together with some 800 of the venerable Badgers.

This version of Tu-16 Badger is one application of Short Horn nav-attack radar

Larger radome of this late-version MiG-21 Fishbed may house Spin Scan interception radar

Skip Spin

This I-band interception radar was introduced in the mid-1960s and is fitted to Su-15 Flagon A and Yak-28P Firebar fighters. Estimated output power is 100 kW with a range of 25 miles (40 kilometres) operating at 8690 to 8995 MHz, with pulse-widths of about 0·5 microsecond and pulse-repetition frequencies of 2700 to 3000. The radar presumably provides the searching and tracking modes for the Anab air/air missiles which arm these fighters, as well as illumination for the radar-homing version of the Anab.

Spin Scan

This is the NATO designation for a family of interception radars that commonly equip the USSR's most well-known fighter, the MiG-21 Fishbed. Soviet designations are believed to be R1L and R2L, the latter for export versions, for example for India. Specific models of the Fishbed thus equipped are the D and F, and the Sukhoi Su-9 Fishpot B also has Spin Scan. The radome of the Fishbed D is larger, and presumably accommodates a larger antenna, than that of the Fishbed F.

Weather Radar

FRANCE

Omera-Seguid

Société d'Optique, de Mécanique, d'Electricité et de Radio Omera-Seguid, rue Ferdinand Berthoud, 95101 Argenteuil

ORB 37

This radar for fixed-wing aircraft and helicopters is the company's first venture into weather warning systems of this type but is based on a substantial avionics and radar background including the Heracles surveillance radar. As well as weather, the monochromatic ORB 37 can be used in a ground-mapping mode for navigation and there is an interrogation facility for beacon-homing.

The system comprises seven units: a slotted-array flat-plate antenna, separate transmitter and receiver, a plan position indicator-type, high-definition circular display for ground-mapping at a navigator's position, a rectangular weather display on the flight deck, and two control units, one for each station.

The antenna scans at a low rate for the weather mode and at a high rate for the ground-mapping mode to achieve the greatest efficiency in each mode. The corresponding pulse widths are 2·5 and 0·4 microseconds.

Principal components of Omera-Seguid's ORB 37 weather radar

STATUS: in production and service with C 160 Transall military transports and SA 330 Puma helicopters of the French Air Force and French Army respectively.

UNITED STATES OF AMERICA

Bendix

The Bendix Corporation, Air Transport Division, 2100 NW 62nd Street, Fort Lauderdale, Florida 33310

RDR-150 weather/multi-function radar

One of the simplest models in the Bendix range of weather radars, the RDR-150 is designed for light and medium twins and the smaller helicopters. Two versions are available: the less expensive black-and-white Weathervision system, or the three-colour Colorvision type, both employing the same transmitter/receiver and antenna. The former may be upgraded to the latter, if space permits, by substituting a slightly larger control/display unit. Again, there is a choice of antenna: normal parabolic, or slotted planar array for greater range with the same diameter.

Storm intensity is indicated by three grey tones in the Weathervision and by three colours in the Colorvision version, measured as a function of rainfall. Red indicates severe weather, with rainfall more than 12 millimetres per hour; yellow shows more moderate conditions, with rainfall in the four to twelve millimetres per hour range, and green points to rainfall up to four millimetres per hour. A weather-alert mode causes the red storm centre to blink on and off, drawing crew members' attention to the possibility of severe weather. The standard control/display unit has a track line that plots an accurate diversion around severe weather.

The system also provides a ground-mapping mode that shows up prominent surface features such as lakes, bays, rivers, channel markers and offshore oil rigs. This can be a

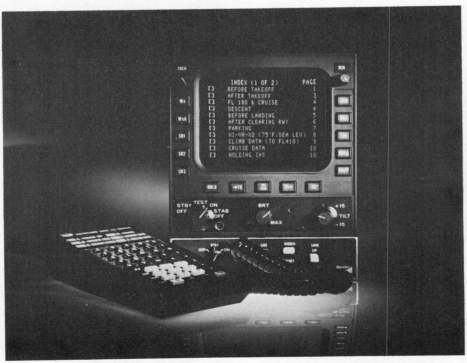

Bendix RDR-230 radar indicator with entry keyboard

valuable navigational cross-check when flying in poor visibility or above cloud.

The addition of a checklist control panel and a modified Hewlett-Packard HP-67 pocket calculator permits the system to store and display in alpha-numeric form up to 16 pages of normal and emergency flight or aircraft procedures such as standard instrument departures or approaches or engine fire drills. Another optional feature is a navigation mode. By the addition of information from Bendix BX-2000 communications, navigation, identification and horizontal situation indication systems, entire flight profiles can be displayed with waypoints, course deviation and track shown in different colours. In a third option, weather information can be overlaid with navigation information to form a moving-map display.

Characteristics for both Weathervision and Colorvision are given below, except where indicated.

Frequency: X-band (9375 MHz)
Rf power output: 8 kW/3·5 μs
Antenna scan angle and rate: 90°, 16 looks/minute
Antenna tilt: manually selectable to any angle between ±15°
Display storage: digital memory
Range: 160 n miles
Antenna stabilisation: none
Pressurisation: none needed
Antenna size and weight: 12 inch (305 mm) parabolic, or 10 or 12 inch (254 or 305 mm) flat-plate; parabolic 2·7 lb (1·2 kg), flat-plate 5·5 lb (2·5 kg)
Transmitter/receiver size and weight: ½ ATR Short, 10·5 lb (4·8 kg)
Cdu size and weight: (Weathervision) 6¼ × 4 × 9⅞ inches (159 × 102 × 251 mm), 5·5 lb (2·5 kg);
(Colorvision) 6¼ × 4·7 × 12·06 inches (159 × 119 × 306 mm), 10 lb (4·55 kg)
Qualification: TSO C-63B

STATUS: in production and service since 1976.

RDR-160

This low-cost weather radar was designed to suit the majority of general-aviation light twins and some singles. Recognising that space and economy are at a premium in this class of aircraft, Bendix has combined the transmitter/receiver and antenna into a single unit, obviating the need for waveguides and cable looms between the two, and resulting in a system weight of only 15·5 lb (6·9 kg) perhaps the lightest weather radar available anywhere. As with the RDR-150 from which it was developed, the -160 is available either as the monochromatic Weathervision or the three-colour Colorvision and provides basically the same facilities, namely weather and terrain mapping and weather alert. The main differences from the -150 are a reduction in transmitter power, from eight to six kilowatts, and the non-availability of a flat-plate, slotted-array antenna. The control/display unit is identical to the corresponding unit for the RDR-150 Colorvision.

STATUS: in production and service since 1978.

RDR-230HP

One of the most recent additions to the Bendix range of weather radars for light and medium twins, the RDR-230HP is essentially an upgraded version of the X-band RDR-160 with improved performance, mounting and packaging. Featuring a non-stabilised 12-inch (305 mm) flat-plate antenna and backed by a 5 kW peak-power transmitter, the all-colour system has a 240 nautical mile display range and a weather-avoidance range of 200 nautical miles. A special feature of the RDR-230HP is what Bendix calls Extended STC, which permits significant storm regions to show up as bright red even at great distances. Without this feature such weather developments might not appear in red until they grew stronger or the range was closed.

The system continues the Bendix-initiated ART idea which combines the antenna and transmitter/receiver into a single unit (again, a feature of the RDR-160). The 12-inch (305 mm) flat-plate antenna makes more effective use of the radar energy than does the parabolic dish by concentrating the beam into a narrower cone, providing better defined weather patterns and ground mapping. The antenna scans ±45° about the aircraft axis and tilts 15° up or down on pilot command.

Another feature is the track-line cursor. This is a line on the display that starts at the aircraft and can be rotated left or right as the pilot directs, the angular deviation from the aircraft

Bendix RDR-160 transmitter/receiver and antenna unit

axis being displayed in alpha-numerics on the radar screen so that the pilot can choose a convenient course to avoid the worst weather. The system also accommodates all the most widely used radar options including a 32-page programmable checklist with automatic high-priority page call-up for alerts, and the moving-map display option so that information from the most popular RNAV, Omega, IN, and Loran systems can be displayed.

STATUS: the RDR-230HP was introduced during late 1981.

RDR-1100

This is a lightweight, pitch and roll stabilised weather radar system with a small-face control/display unit for situations where space is limited, for medium turboprop and turbine

twins. The RDR-1100 is available in monochrome Weathervision or three-colour Colorvision alternatives, both using the same transmitter/receiver, antenna, and drive units. A version designated the RDR-1150 has a fourth colour (magenta) for weather. The systems also use the same aircraft wiring and are plug-to-plug compatible so that upgrading to colour from monochrome entails simply the substitution of a slightly larger control/display unit.

Standard features for the colour system, apart from the usual three-colour presentation of storm intensity, are a movable azimuth track line permitting the pilot to choose and read off for air traffic control purposes a new course to avoid bad weather; weather alert, whereby the red storm centres blink on and off to draw the crew's attention; and a ground-mapping mode.

Optional features available in the Colorvision

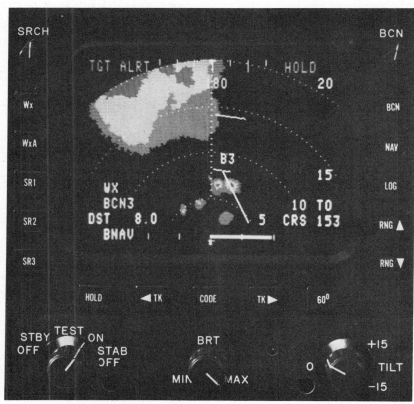
Bendix RDR-1400 control/display unit (see page 37)

version are checklist, navigation and weather/ navigation overlay modes. In the first of these, an additional unit, the pilot-programmable CC-2024B, C, or D unit provides 16 or 32 pages for the display of flight checklists or other information in alpha-numeric form. In the navigation mode, an appropriate interface unit permits the pilot to display waypoints, course deviation and planned changes of course. Bendix provides interface units for its own BX 2000 navigation system and for the most widely used RNAV, Omega and IN systems.

Characteristics are for both monochrome and colour except where stated.

Frequency: X-band (9375 MHz)
Rf power output: 8 kW
Antenna scan angle and rate: 120° (120°/60° colour), 24°/s (24°/s, 48°/s colour)
Display range: 200 n miles (240 n miles colour)
Antenna size and weight: 12-inch (305 mm) flat-plate, 8 lb (3·6 kg)
Transmitter/receiver size and weight: 1/2 ATR Short, 6 lb (2·7 kg), colour 10·5 lb (4·77 kg)
Cdu size and weight: 6 1/4 × 4 × 9 7/8 inches (159 × 102 × 251 mm)
Colour has same face dimensions but is 12·06 inches long (306 mm)
Stabilisation: ±30° combined pitch, roll and tilt
Pressurisation: none required
Power: 3·5 A at 28 V dc
Qualification: TSO C63b

STATUS: in production.

RDR-4A

Chosen by Boeing as standard equipment for the new, all-digital 767 and 757 narrow-body transports, the Bendix RDR-4A is claimed to be the most advanced multi-function colour radar currently available. Designed to meet the new ARINC 708 requirements, the X-band system features a solid-state transmitter and line-of-sight antenna with split-axis performance, and is compatible with the EFIS flightdecks of the 767, 757, Airbus Industrie A310 transports, and other projected designs. The range is 320 nautical miles.

STATUS: in production. The system was tested over a six-month period from July 1980 aboard a Pan Am 747, and is now in service with, among others, the A300s of TOA Domestic Airlines of Japan.

AN/APN-215

The AN/APN-215 colour radar is a weather, surface search and precision terrain-mapping system derived from the successful and widely used RDR-1300 commercial system. It is designed for heavy twins, turboprops and transport helicopters. Low weight and a 240 nautical mile range suit it to utility and reconnaissance aircraft, and it was chosen for the US Army's U-21s and RU-21s, military versions of the Beech King Air.

In conjunction with other equipment the APN-215 can display navigation pictorial information overlaid on the weather map, together with pilot-programmable pages of checklist information such as en route navigation data and emergency procedures.

Boeing 767 in foreground and 757 behind will have Bendix RDR-4A radars

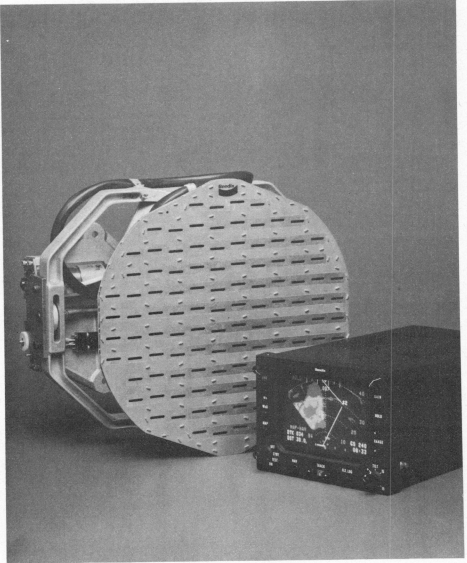

Bendix RDR-230 HP radar

The system comprises three units: a 12-inch (305 mm) pitch and roll stabilised antenna, transmitter/receiver and colour control/display unit.

STATUS: in production and service.

RDR-1FB (AN/APS-133)

This colour radar is a high-performance weather, beacon-homing and terrain-mapping system designed for commercial and military transports. The three-colour display can be

Bendix RDR-4A radar

Bendix AN/APS-133 four-unit weather radar

Moisture content of cloud ahead represented as three colours on AN/APS-133 system

used in conjunction with other optional equipment to show programmable checklists or to superimpose navigation or other information on the weather map. The system employs digital processing and microcomputer techniques, radio frequency semiconductors, and a solid-state modulator.

Significant landmarks and continental shorelines up to 300 nautical miles away can be portrayed in the ground-mapping mode by using the high power output concentrated into a pencil beam. At the same time discrete details such as lakes, rivers, bridges, runways and runway approach reflectors, readily show up on the colour display. To improve range resolution at short ranges the system operates with 0·5 microsecond pulses in contrast to the five microsecond pulses used for long-range ground and weather mapping.

In the air/air mapping mode, the RDR-1FB detects and tracks other aircraft during rendezvous, formation and air-refuelling. Aircraft of C-130/C-141 size can be tracked to 12 and 20 miles (19 and 32 km), but may still be resolvable at ranges as little as 600 yards (550 metres) depending on relative bearing, aspect and altitude.

To provide long-range homing to remote ground destinations or tanker aircraft, the RDR-1FB operates at X-band frequencies (9375 MHz) for beacon interrogation and reception. The identification of closely spaced pulse reply codes at long ranges is made possible by the marker and delay modes of the radar indicator. In the marker mode a variable marker is positioned on the screen just in front of the beacon reply. When switched to the delay mode the display presentation starts at the marker range. The range switch can then be moved to select a shorter range scale yielding an expanded view of the area containing the beacon reply.

Derived from the Bendix RDR-1F used on about 70 per cent of current wide-body transports, the APS-133 comprises four line replaceable units: a 30-inch (762 mm) fully stabilised split-axis parabolic antenna that provides specially shaped search or fan beams for terrain-mapping, a transmitter/receiver, a colour display and a control unit.

Frequency: X-band (9375 MHz)
Power output: 65 kW
Prf: 200 pulses/s
Pulse widths: (weather) 5 μs
(beacon) 2·35 μs
(mapping) 0·5 μs

Weights: (antenna) 37 lb (15·8 kg)
(transmitter/receiver) 55 lb (24·9 kg)
(cdu) 13·8 lb (6·3 kg)
(control unit) 2 lb (0·9 kg)

STATUS: in service with US Air Force transport aircraft, notably C-130s, KC-10 Extenders and C-5A Galaxies.

RDR-1200

This weather radar pioneered digital memory display weather radars for general aviation and is now recognised as one of the industry standards; it came on the market in 1974. The system is aimed at the heavy turboprops and jets that largely make up the corporate aircraft market, and is available in Weathervision (monochrome) or Colorvision versions. Special storage circuitry permits a continuous read-out of radar video and the system can be frozen for extended periods by a hold mode so that growth and movement in storm cells can be seen by switching back to the normal scan mode.

The system, which weighs 34 lb (15·5 kg) in its colour version, feeds 10 kW output power into a pitch and roll stabilised 12- or 18-inch (305 or 447 mm) flat-plate antenna.

STATUS: in production and service.

RDR-1400 multi-function weather radar

This new radar is designed for commercial helicopters, particularly those associated with the large international offshore oil and gas industry. It differs from almost all other Bendix weather radars by having a beacon interrogator that exploits the increasing use of portable radar beacons in these industries.

The original monochromatic RDR-1400 has now been joined by a colour-radar version with greater performance. The following operational modes are available:
Beacon navigation
The growing popularity of portable beacons is supported by several special RDR-1400 capabilities. Beacon signatures are denoted by short lines or obliques on the display, the actual location of the device being determined by the middle of the line and the pilot can overlay beacon returns on the weather map. The beacon's discrete code can be displayed for positive identification, an important factor when the pilot is trying to locate a specific rig in a drilling farm where numerous rigs may be

transponder-equipped. Beacon-detection range is 160 line-of-sight miles (257 kilometres) depending on altitude.
Beacon Trac
This mode, peculiar to Bendix, generates and displays on the weather radar screen an inbound course to the discrete beacon. This course line can be rotated 360° about the beacon by rotating the horizontal situation indicator course selector, thus allowing the pilot to choose a convenient course to the beacon. A number of programmes are under way to assess the role of these beacons in published approaches, for example as a final fix or for establishing a course.
OBS Trac
This mode provides another course-following option. When in a weather or search mode, a track line or course-bearing cursor can be generated from the aircraft position and controlled by the horizontal situation indicator course selector to provide a course line to the chosen target. The OBS Trac heading is displayed digitally in the lower right-hand corner of the indicator. Left/right deviations can be determined by comparing heading information to this number and by observing the movement of the track line in relation to background targets.
Search
Three search modes are available. Search 1 has special sea-clutter rejection circuitry in order to detect small boats or buoys, for example, down to the minimum range. Search 2 is for precision ground-mapping in situations where high target resolution is important. Search 3 includes maximum return clutter and can detect and track oil slicks.

With these capabilities the system is suitable for search and rescue, surveillance, aerial survey work and law enforcement applications, besides the aforementioned rig-servicing.

Frequency: X-band
Rf power output: 10 kW
Antenna size and scan angle: 12 or 18 inches (305 or 457 mm) flat-plate, 120° or 60°
Display size: 4·34 × 3·33 inches (110 × 85 mm)
Dimensions: (transmitter/receiver) ½ ATR Short 5 × 6¼ × 13⅞ inches (127 × 159 × 352 mm) (cdu) 6¼ × 6¼ × 10⅞ inches (159 × 159 × 276 mm)
System weight: 34·1 lb (15·47 kg) (12-inch antenna)
Qualification: TSO 63b

STATUS: in production and service.
(see page 35 for illustration)

Collins

Collins Avionics Divisions, Rockwell International, 400 Collins Road NE, Cedar Rapids, Iowa 52406

Micro Line WXR-200A

Designed for light twins and some singles, the monochromatic WXR-200A is the smallest system in the Collins range of weather radars. A high-resolution picture, the result of a memory-enhancement technique to provide the equivalent of 128 000 bits of memory, smooths out block-edged storm outlines and minimises target shift and smearing. The result is that fine detail in the picture, such as hooks, anvils and scallops, can be more clearly seen. For typical weather-mapping the system operates at 5·5 microsecond pulse widths, but for close-in weather and for ground-mapping, it automatically shifts to one microsecond for better picture clarity and definition.

The system comprises three units: a 12-inch (305 mm) slotted-array flat-plate antenna (optional alternatives are a 10-inch (254 mm) flat-plate of 12-inch (305 mm) parabolic dish) with pitch stabilisation, transmitter/receiver and control/display unit. The bright display is readable under high ambient lighting conditions and is presented in four levels. Maximum cell activity is shown as a black patch or hole surrounded by areas of lesser rainfall. Pilot-selectable receiver gain has four levels to help in some ground-mapping situations and to assist in weather analysis.

Frequency: X-band (9345 MHz)
Power output: 5 kW, 5·5 μs for long range, 1 μs for short range
Range: 180 n miles
Antenna scan angle: ±45° about fore-and-aft axis
Dimensions: (cdu) 6¼ × 4 × 10¹⁄₁₀ inches (159 × 102 × 257 mm)
(transmitter/receiver) 5 × 5 × 12½ inches (127 × 127 × 315·5 mm)
Weights: (antenna) 6·1 lb (2·77 kg)
(transmitter/receiver) 10·7 lb (4·86 kg)
(cdu) 5·9 lb (2·67 kg)
Qualification: FAA C63b

STATUS: in production and service.

Pro Line WXR-250A

For the larger twins, corporate jets and small/medium helicopters, the monochromatic WXR-250A incorporates many of the features of the smaller Micro Line WXR-200A described above. For long range mapping the system automatically generates 5·5 microsecond pulses, providing the power needed to scan the weather within a range of 240 nautical miles. When the pilot selects the mapping mode the system switches itself to one microsecond pulse-width for better resolution.

Seven modes are provided: WX gives cyclic contouring on alternate antenna scans; WX HOLD freezes the display; WX ID (weather ident) displays only the contoured areas; NORM displays all levels (no contour or cyclic contour); MAP gives maximum gain without cyclic contouring (there are four lower gain levels); TEST energises a test pattern to verify operation of the control/display unit; and STBY energises the system but without transmission.

At the time of its introduction in the mid-1970s the WRX-250A was perhaps the most economically packaged radar in its category. The transmitter/receiver package is still one of the smallest, enabling it to be mounted close to the antenna so that the wave-guide can be as short as possible, thus increasing performance.

Power output: 5 kW from magnetron, 5·5 μs long range, 1 μs short range
Frequency: X-band (9345 MHz)
Range: 240 n miles
Antenna: 12-inch (305 mm) flat-plate fully stabilised, scan angle ±60°

STATUS: in production and service.

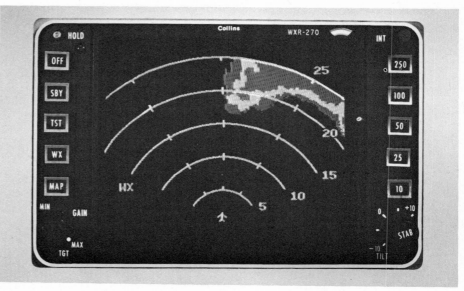

Presentation for new Rockwell Collins WXR-270 weather radar planned to replace WXR-250A

Simple presentation of Rockwell-Collins weather radar indicator, showing cloud 20 and 40 miles ahead

Centre-panel presentation of Rockwell Collins RNS-300 combined weather-radar and navigation display

Pro Line WXR-300

Designed for the larger twins and turboprop and jet corporate aircraft, the WXR-300 was the first of Collins' second-generation colour radars and the first such system to incorporate push-button control of mode and range. Good picture resolution is obtained by using 250 000 bits of display. One of the advantages of the radar is its capability as an alpha-numeric and graphics information display when connected to an appropriate computer. Information such as normal and emergency operating procedures or navigation data can be entered by the pilot through an optional keyboard entry device.

Frequency: X-band (9345 MHz)
Power output: 5 kW, 5·5 μs long range, 1 μs short range
Range: (10-inch antenna) 217 n miles
(12-inch antenna) 256 n miles
(18-inch antenna) 353 n miles
Weight: (system with 18-inch antenna) 28·6 lb (13 kg)
Qualification: FAA TSO C63b

STATUS: in production and service.

RNS-300 weather radar navigation system

The RNS-300 is essentially the Collins WXR-300 weather radar with an additional processing unit that makes it possible to superimpose, on the weather map, aircraft position with respect to ground stations, waypoints, and course and heading lines. En route waypoints or navigation stations appear to move from the top of the screen to the bottom, a major advantage being that, with course and heading lines being displayed simultaneously on the same CRT, any drift shows up immediately. This indication is particularly helpful when flying in the vicinity of storms, when considerable and rapid changes in wind speed and direction may occur due to local and random pressure changes.

The radar/navigation functions of the system make it possible to plot course lines to RNAV waypoints, Vortacs, or the intersections defined by heading and course lines, or omega tracks. Two course lines and a heading line can be plotted simultaneously to form several legs in a navigation segment such as a SID or a STAR. Clear presentation of such information facilitates the choice time and fuel-saving routes.

Collins claimed that, at the time of its introduction, the RNS-300 was the only system on the market capable of plotting intersecting course lines using navigation information from dual navigation radios or RNAVs.

STATUS: the system was introduced during 1980 and has been adopted by Gates Learjet, Beech, and the Sabreliner Division of Rockwell International. The system has also been fitted to British Aerospace HS125-700s.

WXR-700C/WXR-700X

The Collins WXR-700 series digital colour radar is one of the most advanced and most recent systems to appear, and in fact is the first airline-category digital colour weather radar to fly. Geared to ARINC 708, the system replaces the magnetron of earlier radars with solid-state power generation as the basis for alternative X-

Four-unit Rockwell-Collins WXR-700 weather radar showing two versions of control unit

or C-band radars. Along with electronic advances, the design team worked closely with the US National Severe Storms Laboratory to obtain the most recent understanding of how cells form and grow in storms.

The WXR-700 comprises four units: a slotted-array flat-plate antenna (for good side-lobe reduction), a microprocessor-controlled transmitter/receiver, a display unit, and a control unit. The microprocessor control system in the transmitter/receiver supervises all control and data transfers, programmes and controls the RF processes such as pulse-width, band width and pulse-repetition frequency selection, and likewise directs antenna scan and stabilisation. The unit also contains circuits to reduce ground-clutter suppression when operating in the weather mode. Another, optional, feature is what Collins describes as pulse pair Doppler processing in which, by the addition of a single circuit board to the transmitter/receiver, the horizontal velocity of rainfall can be sampled. This technique, thought to be used only by Collins, is recommended by the NSSL as being particularly suitable for the analysis of storm cells.

The cathode-ray tube indicator uses a high-resolution shadow-mask tube with multi-colour display scheme. The cathode-ray tube provides alpha-numeric identification of radar modes and incorporates annunciators and controls. The receiver has a sufficiently wide dynamic range to 'see' the Z-5 and Z-6 levels of rainfall that indicate a high probability of hail.

Apart from the traditional single channel configurations based on one set of equipment with perhaps a second indicator, the flight-deck standards of the new-generation transports such as the A310, Boeing 757 and 767 offer alternative possibilities. For example, dual transmitter/receivers fed from a single antenna could feed multi-purpose EFIS (electronic flight instrument system) indicators instead of dedicated weather cathode ray tubes (though

Air Canada Lockheed L-1011-500, one of first applications of Rockwell-Collins WXR-700 weather radar

these themselves can already display operating procedures or flight information).

Frequency: (C-band) 5440 MHz, (X-band) 9330 MHz
Power output: (C-band) 200 W, prf 180 – 1440, pulse-widths 2 – 20 μs
(X-band) 100 W, prf 180 – 1440, pulse-widths 1 – 20 μs
Range: (C-band) 240 n miles, (X-band) 320 n miles
Dimensions: (identical for C- and X-band) (antenna) ARINC 708
(transmitter/receiver) 8 MCU
(indicator) ARINC 708 Mk II
(control unit) 5·75 × 2·625 × 6 inches (146 × 67 × 152 mm)
Weights: (antenna) 27 lb (12·25 kg)
(transmitter/receiver) 27 lb (12·25 kg)
(indicator) 18 lb (8·16 kg)
(control unit) 2·3 lb (1·04 kg)

STATUS: in production and service, and chosen as standard on the Airbus Industrie A310.

King

King Radio Corporation, 400 N. Rogers Road, Olathe, Kansas 66061

KWX 56 digital colour radar

Introduced at the National Business Aircraft Association's annual meeting in 1981, this new system is the result of a two-year development programme to build an inexpensive pitch/roll stabilised colour radar.

The system comprises two units: a five-inch (127 mm) diagonal KI 244 panel-mounted high-contrast black matrix display and a KA 126 combined antenna/transmitter/receiver. The latter can be stabilised using a flight-director or vertical gyro; the addition of roll-stabilisation eliminates screen blanking caused by ground returns during medium or steep turns. The flat-plate antenna has a diameter of either ten or twelve inches (254 or 305 mm), and can be supplied either pressurised or unpressurised, suitable for altitudes of up to 20 000 and 50 000 feet (6100 and 15 200 metres) respectively.

Power: nominal peak pulse 7·5 kW, 3·5 µs pulse width
Ranges: 10, 20, 40, 80, and 160 n miles by rotary switch
Display: Conventional colours. In mapping mode:
red becomes magenta
green becomes blue
yellow remains unchanged
Stabilisation: From KI 256 flight director and ARINC-standard vertical gyros associated with most autopilot/flight-director systems, the KWX 56 can be used with Century 41, Century

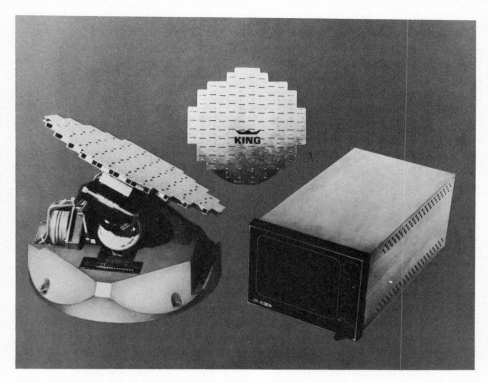

King's KWX 56 two-unit colour radar system including front-view of antenna

IV, Cessna ARC 400, 800 and 1000 series autopilots

STATUS: in production and service

Ryan Stormscope

Ryan Stormscope, 4800 Evanswood Drive, Columbus, Ohio 43229

WX-8, WX-9 and WX-10 Stormscope

These are not radar systems as they work on an entirely different principle, but are included in this section because their functions and presentations are similar to those of weather radars.

The purpose of these patented systems is to indicate the presence and direction of dangerous turbulence associated with thunderstorm activity. They detect the direction, relative to the aircraft, of the electrical discharge activity invariably associated with dangerous convective shear motion between vertical air currents in storm clouds, and present the information in the form of a plan-position indicator display. Electrical discharge activity generated by atmospheric convection directly corresponds to the same factor that produces gust loads on an aircraft, ie convective shear. The system detects the presence and intensity of storms by picking up radio frequency noise from the lightning discharges that accompany intense convection, processing it and displaying it on a 360° azimuth indicator.

The system (WX-9) comprises four units: receiving antenna, receiver, computer/processor and display. The receiver and display are mounted adjacent to one another on the dash panel. Each electrical discharge is analysed to provide azimuth and range and is plotted as a bright green dot in the appropriate position on the display. Azimuth is measured in the same way as for a radio beacon, but range is determined to an accuracy of 10 per cent by 'finger-printing' the returns in accordance with an understanding of the behaviour of lightning discharges. Since the discharges are momentary, the image is held in the system's memory for continuous display. As repetitive discharges occur, clusters of such dots form, indicating the extent of storm activity and the location of particularly dangerous cells. The rapidity with which they appear also indicates the activity of the storm. Range settings can be varied to show activity within 40, 100, and 200 nautical miles.

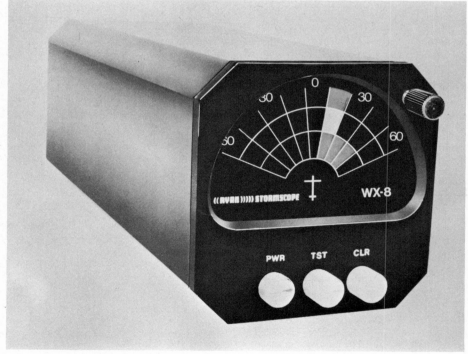

Ryan Stormscope's WX-8 weather mapping system showing colour sector scan

The system can memorise 128 discharges, after which the earliest recorded images are erased to make way for new information. With a medium to severe storm, sufficient discharges occur to update the display every 25 seconds. The WX-10 system is omni-directional (unlike a conventional weather radar) and this capability can be useful if the crew wishes to keep track of possible rearward 'escape routes'. Alternatively the pilot or crew may elect to dedicate all the memory to the forward 180°.

WX-8

This is a two-unit system comprising antenna and three-colour, liquid-crystal display unit that also contains a micro-computer to analyse and process storm activity.

WX-9

A more advanced version of the WX-8 with 180° scan in place of the former's 135° and separate processor.

WX-10

The most advanced Stormscope model, with maximum range doubled from 110 to 220 nautical miles, memory increased from 128 dots to 256, controls grouped in the cathode ray

New standard antenna used on all Ryan Stormscope systems

Ryan Stormscope WX-10 weather system

tube display instead of occupying a separate receiver unit, and 'omni'-presentation.

Display size: 2⅝ inches diameter
Power: 28 W at 28 V dc
Operating ranges: switchable, 25, 50, 100 and 200 n miles

Dimensions: antenna 4 × 8·16 × 0·85 inches (102 × 207 × 22 mm)
computer/processor 7·63 × 12·9 × 2·25 inches (194 × 5250 × 57 mm)
cdu 3·19 × 3·19 × 9·22 inches (81 × 81 × 234 mm)

Weights: computer/processor 4·3 lb (2 kg)
antenna 2 lb (0·9 kg)
cdu 3·4 lb (1·5 kg)
Qualification: FAR Part 135 section 173 approved for thunderstorm and severe weather avoidance, but not penetration

STATUS: in production and service.

Sperry

Avionics Division/Van Nuys, Sperry Flight Systems, 8500 Balboa Boulevard, PO Box 9028, Van Nuys, California 91409

Sperry's entry into the weather-radar market was the result of its purchase in January 1981 of RCA Avionics Systems, a company that had long been a leader in this field, and which in 1977 introduced colour radar to the aviation world.

WeatherScout II

This two-unit system is intended for single-engined aircraft (pod-mounted under a wing), and light twins. Weather up to 120 nautical miles away can be displayed at three levels of monochrome, using x-y scanning so that the screen area is completely filled.

Qualification: TSO C63b (DO – 160) Class 6
Power: 2 A at 28 V dc
Frequency: x-band, 9·345 GHz

Cdu
Dimensions: 6·25 × 4 × 9·8 inches (158·7 × 101·6 × 248·9 mm)
Display size and type: 5 inches (152·4 mm), 90° sector scan
Tilt control: ± 12°
Weight: 4·8 lb (2·18 kg)

Antenna/transmitter/receiver
Dish size: 10 or 12 inches (254 or 304·8 mm)
Average power: 2·28 W
Prf: 228 pps
Beam size: 10° at 10 inches, 8° at 12 inches
Scan rate: 17·8 looks/minute
Pulse-width: 4 μs for 12 and 30 mile ranges 10 μs for 60 and 120 mile ranges

STATUS: in production.

Primus 100 ColoRadar

This is a colour version of the WeatherScout II having a 200-mile range, target-alert and freeze mode. Weather is displayed in three colours: green (light rainfall), yellow (medium), and red (heavy precipitation). The radar can also be

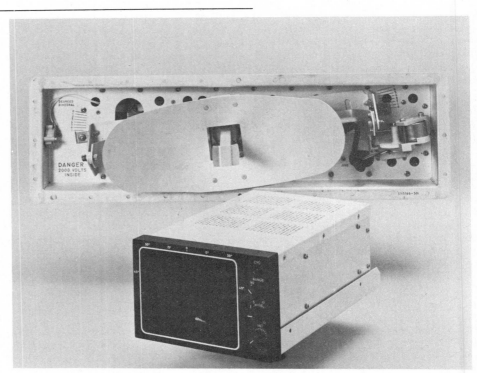

Sperry Weather Scout II for installation in wing leading-edge

used for mapping, as a navigational aid, but to avoid confusion, maps are shown in magenta, yellow and blue according to the strength of returns. Least expensive of the company's colour radars, the system can be installed in a pod for under-wing mounting on single-engined types, or conventionally in the nose for twins.

Characteristics
As for Primus 100 monochromatic, except that the control/display unit is housed in a 12·38 inch (314 mm) deep case.

STATUS: in production.

Primus 200 ColoRadar

This system represented a step forward in design by combining a compact, 4⅜ × 6¼ inch (111 × 159 mm) control/display unit with a lightweight transmitter/receiver and flat-plate antenna for use with light twins. Weather is displayed in red, yellow, and green, with maps showing in magenta, yellow and blue.

Primus 300SL ColoRadar

This second-generation colour radar is packaged into three units: 10, 12 or 18 inch (254, 314·8 or 457·2 mm) antenna, transmitter/

Sperry Primus 200 antenna and display

Sperry Primus 300SL antenna, transmitter and cdu

receiver, and control/display unit, and with a 300 nautical miles range appropriate to the faster turboprop twins or jets.

Qualification: TSO C63b
Power: 4·5 A at 28 V dc
Frequency: x-band, 9·375 GHz

Cdu
Dimensions: 6·25 × 4·37 × 12·38 inches (158 × 111 × 314 mm)
Display size and type: 5 inches (152·4 mm), x-y scan
Tilt control: ± 12°
Weight: 10·9 lb (4·95 kg)

Transmitter/receiver
Dimensions: ½ ATR Short 5·06 × 7·68 × 12·65 inches (128·5 × 195·1 × 321·3 mm)
Weight: 13 lb (5·9 kg)

Antenna
Size: 10, 12, or 18-inch flat-plate (254, 304·8 or 457·2 mm)
Weight: 8·4 lb (3·8 kg) with 12-inch antenna 12·8 lb (5·8 kg) with 18-inch antenna
Stabilisation: line of sight ± 30° pitch and roll
Scan rate: 12·5 looks/minute

STATUS: in production.

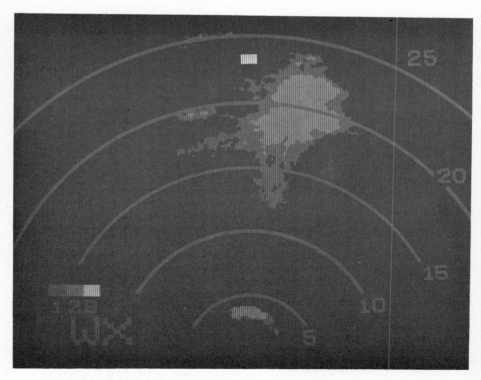

Sperry Primus 500 weather radar system

Primus 500 ColoRadar

This is claimed to be the first colour radar to combine simultaneously beacon position and weather mapping. Intended for helicopters and fixed-wing aircraft, the three-unit system can interrogate a radar beacon and show its position on the control/display unit in relation to weather and topography; the beacon-homing mode is particularly useful in parts of the world off normal routes, or for special-purpose operations such as servicing the off-shore oil and gas industry.

When combined with the RCA-designed Sperry Data Nav display systems the Primus 500 can show up to 120 pages of normal and emergency checks, eg take-off drills and engine-fire procedures, performance tables, and other information that can be entered by operators. The system also interfaces with a number of long-range navigation systems so that flight plans can display on the Primus control/display unit, the waypoints pictorially associated with the range markers and weather.

Qualification: TSO C63b (DO – 138) Class 7
Power: 5·6 A at 28 V dc
Range: 200 n miles
Frequency: 9·375 GHz

Cdu
Dimensions: 6·36 × 6·36 × 12·5 inches (161·5 × 161·5 × 317·5 mm)
Weight: 12·7 lb (5·77 kg)
Scan angle: 60° and 120°

Typical Primus 500 presentation showing cloud and radar beacon at 23 miles range

Tilt control: ± 15°
Display modes: radar only, beacon and radar simultaneously, beacon only

Transmitter/receiver
Dimensions: ³/₄ ATR Short
Weight: 17·5 lb (8 kg)
Peak power: 10 kW
Prf: 120 MHz

Antenna
Dish size: 12 or 18 inches (304·8 or 457 mm)
Weight: 8·8 lb (4 kg)
Scan rate: 14 looks/minute at 120° or 28 at 60°
Stabilisation: ± 30° pitch and roll

Primus 708

This third-generation weather radar is designed around ARINC 708 digital interface and installation requirements for flight-decks with the new EFIS (electronic flight instrument system) displays. In its primary mode the system provides a seven-colour display of weather within a range of 300 to 320 nautical miles. Colours are shown with equal intensity against a black background for greatest visibility in bright ambient light. Red, yellow and green show heavy, medium and light precipitation and the other colours can be used for ground-mapping, turbulence detection or for other multi-function display modes.

The Primus 708 has a solid-state impact diode transmitter which, according to Sperry, has less circuitry and is more reliable and easier to maintain in the field than varactor multiple chains. Two sets of pulse widths are employed, one for weather mapping, the other for long and short range ground-mapping. Receiver bandwidths are matched to the transmitter pulse widths and are automatically designated by microprocessor circuits when range and mode are selected. Antenna stabilisation is also microprocessor controlled. Optimum ground clutter removal is obtained by dedicated active circuitry used in conjunction with a flat-plate, slotted array antenna with minimum sidelobe performance. Another feature is what Sperry calls REACT (radar echo attenuation compensation technique), which maintains the radar signal at the correct level in the presence of intervening precipitation.

STATUS: in production and service.

Data nav display systems

During the early 1970s, when the aviation industry was beginning to investigate the

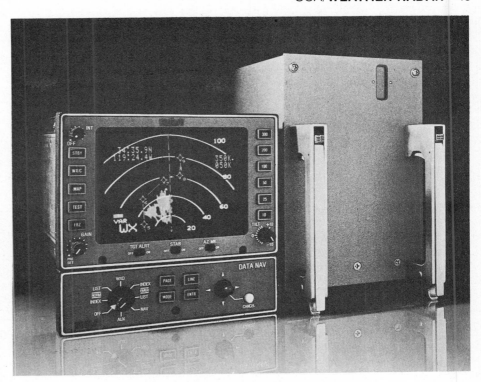

Sperry's Data Nav system in conjunction with Primus 300SL

benefits of electronic displays for centralised aircraft management, RCA recognised the potential of cathode-ray tubes to show more than just radar images of weather, and development of a more general graphic display capability was begun in early 1974. Working in conjunction with consultants in the industry, and with leading suppliers of navigation equipment, RCA has produced a range of interface units (essentially signal converters) that can process information into a form suitable for display on its weather-radar control/display units.

The information displayed falls into two categories: checklist data (eg for take-off and descent and emergency procedures) in alphanumeric form, and pictorial data displayed as conventional weather plots on which are superimposed alpha-numeric or symbolic information such as flight-plan waypoints. In the former category the systems can store for display up to 120 pages of checklist, procedures or other information. For check purposes each function can be 'ticked off' by means of a cursor which changes the colour of the list – green (unchecked), yellow (being checked) and blue (check completed).

Three systems are currently offered: Data Nav I electronic checklist, Data Nav II en route navigation display for JET DAC-7000 navigation system, and Data Nav III en route navigation display for use with Delco Carousel IVa IN, Garrett AiRNAV 300, Global GNS 500A omega/vlf, Litton LTN 211 omega/vlf, and Litton LTN 72R IN. These Data Nav systems are compatible with all the RCA (now Sperry) ColoRadar systems.

Apart from the Primus radar, the system comprises four units: an interface computer, control unit, control panel and pilot-entry keyboard (effectively a pocket calculator whose display is the radar screen).

STATUS: in production.

Surveillance Radar

FRANCE

Omera-Seguid

Société d'Optique, de Mécanique, d'Electricité et de Radio Omera-Seguid, 49 rue Ferdinand Berthoud, 95101 Argenteuil

Heracles II/ORB 32

The appearance of a new generation of maritime reconnaissance aircraft for patrolling waters within 200 nautical miles of coast-lines and the continuing development of sea-patrol helicopters call for surveillance equipment capable of high performance under harsh operational conditions, often with jamming present. Omera's answer to this challenge is the Heracles II/ORB 32 family of radars developed from the earlier Heracles I/ORB 31. Omera says that the Heracles I was the first radar to permit both direct firing of AM 39 type active air/ surface missiles and the first real over-the-horizon firing of long-range, sea-launched anti-ship missiles (Exocet MM 40 and Otomat).

The Heracles II is a modular system whose individual sub-units can be tailored to optimise specific tasks. The radar fulfils a wide range of missions under the headings of maritime patrol, anti-submarine warfare, active missile fire control, search and rescue, radar navigation and weather warning. Units common to all versions of the system are the antenna drive mechanism, junction box and control unit, while other parts of it, notably the antenna itself, transmitter/ receivers and indicators and control/display consoles, are specified by the user. There are basically two types of display: a five-inch (127 mm) high-brightness digital system for mounting on the flight panels and a 9- or 16-inch (229 or 406 mm) tactical unit for an observer or weapons-system operator. The antenna drive mechanism permits 60°, 120°, or

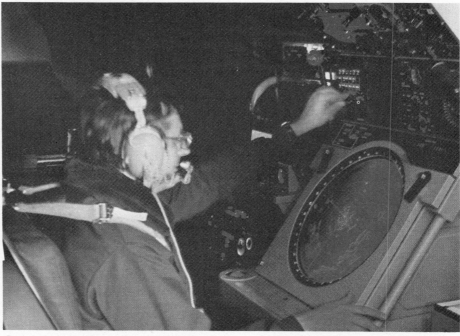

Radar display of Omera-Seguid ORB 32 aboard Aerospatiale SA.321 Super Frelon helicopter

180° sector scan, 12 and 24 rpm rotation speeds, ±15° beam elevation adjustment, antenna line-of-sight stabilisation up to ±20° and roll stabilisation up to ±35°. The system can drive wide-band antennas of various sizes and which consist of a parabolic reflector with rear illumination.

Detection ranges claimed are 50 nautical miles for a snorkel, 75 nautical miles for a

trawler, 100 nautical miles for a destroyer and 200 nautical miles in the weather mapping mode.

STATUS: in production and installed on SA 321 Super Frelon, SA 332 Super Puma, SA 365N Dauphin and Beech Super King Air 200, all in service with the French armed forces.

Thomson-CSF

Division des Equipements Avioniques, Thomson-CSF, 178 boulevard Gabriel Péri, 92240 Malakoff

Iguane/Agrion/VARAN surveillance/strike radar

Under French government contract, Thomson-CSF has been developing a family of new airborne radars for sea surface surveillance and maritime warfare applications. Three versions have so far been named. They are:

Iguane: this system is in production to replace the DRAA2A air-to-surface vessel radar fitted in the Breguet Alizé carrier-borne and maritime anti-submarine warfare aircraft, and has been chosen for the Breguet Atlantic ANG (Atlantique Nouvelle Generation) long range maritime patrol aircraft.

VARAN: this is essentially an Iguane radar with a smaller antenna which makes it suitable for virtually all the present and planned lightweight and ultra-light maritime control aircraft, for example the BAe Coastguarder, HS.125, and Embraer Bandeirante.

Agrion: this radar exists in several versions, and is primarily used for helicopters or light aircraft forming a part of task forces employed for support at sea or coastal protection. It also provides range and bearing information for the Aerospatiale AS.15TT air/surface missile.

Agave attack radar

This lightweight, multi-role radar, initially designed for the Super Etendard carrier-based fighter of the French Navy and more recently chosen for the Jaguar International strike aircraft, is designed for naval use. However the system has considerable air interception/ ground attack capabilities.

Basic functions are search (air/surface and air/air), target designation for the homing head of long range active missiles or a head-up display, automatic tracking (air/surface and air/air), ranging (air/air and air/surface) and mapping.

Since the system was designed for operation by a single crew member the system is automatic as far as possible; for example, the system has an instantaneous automatic gang control whereby ground clutter or other unwanted echos are considerably reduced in comparison with the pin-point reflections of target aircraft and surface vessels. In another case, when the air/ground mode for ground mapping is chosen, the best possible tracking elevation for maximum ground coverage ahead of the aircraft is automatically set up. However the pilot can still over-ride the computed elevation and have the option of final adjustment when short range, high power echos are present.

In the system developed for the Jaguar International all radar information can be

displayed on a raster head-up display of a scan-converter fitted to the combined map/reader head-down display.

TMV 118-B VARAN maritime surveillance radar

This system has been designed for coastal and maritime patrol aircraft with surveillance/ protection, pollution monitoring and anti-submarine warfare commitments, and can be used for navigation and weather-avoidance. Key factors are I-band operation, pulse compression over several pulse-widths, frequency agility for electronic counter-countermeasures, and beacon detection. The unspecified but low peak power level, associated with high receiver sensitivity, increases the difficulty of detection by hostile radars. Typical detection ranges, in sea state 3/4, are: snorkel 30 nautical miles, fast patrol boat 60 nautical miles, and freighter 130 nautical miles. Total weight of the six units comprising the system is 244 lb (111 kg).

STATUS: airborne trials are believed to have been conducted aboard an Aerospatiale Dauphin 2 helicopter.

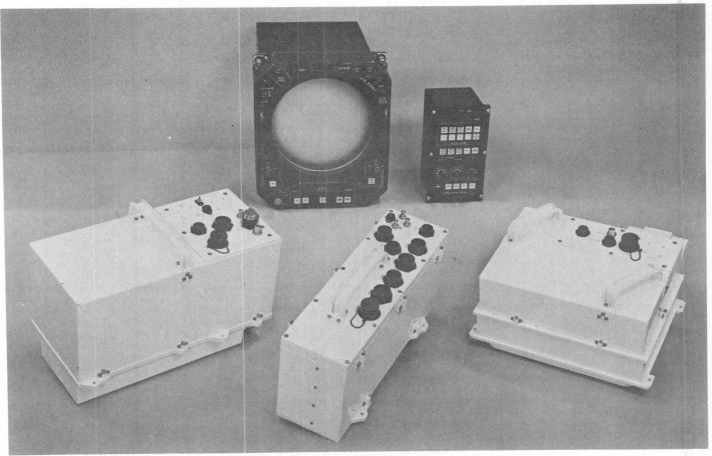

Thomson-CSF VARAN surveillance radar

ITALY

SMA

SMA-Segnalamento Marittimo Ed Aereo, PO Box 200, via del Feronne-Soffiano, 50100 Florence

MM/APS-705

This search and navigation radar was designed specially for naval helicopters, in particular the Agusta AB.212 and SH-3D types, built under licence from Bell and Sikorsky respectively. The antenna system is tailored to the space and location on the aircraft. For example, on the SH-3D the antenna is placed in the dorsal position, on top of the fuselage. Line-of-sight stabilisation is provided and there are alternative selectable antenna rotation rates, 20 or 40 rpm. Manually controlled antenna tilt provides plus or minus 20 degrees of movement. An alternative back-to-back antenna is another option for use where higher data rates are needed.

The display unit incorporates a nine-inch diameter (230 mm) cathode ray tube plan-position indicator presentation with electronic and mechanical cursors and markers, and complemented by a separate digital readout x-y reference display.

There is an expanded micro-B display for chaff or multi-target detection, and the system provides outputs for other displays and extractor units. Other facilities include sector transmissions and blanking, interfaces for beacon receiver, identification friend or foe, anti-submarine warfare and electronic countermeasure systems, built-in test, data link, track-while-scan and dense environment tracker.

Transmitter/receiver: dual-mode, I-band, transmitting 25 kW, with option for 75 kW transmitter/receiver with frequency agility.
Pulse width/prf: $0 \cdot 05 \mu s$/1600 Hz – $1 \cdot 5 \mu s$/650 Hz
Range settings: $0 \cdot 5$, 1, 2, 5, 10, 20, 40, 80 n miles
System weight: 87 kg

STATUS: in service with AB.212 and SH-3D helicopters of the Italian Navy and a number of other navies.

MM/APS-707

This system exploits the basic circuits of the -705 radar described above but uses a single 20 kW transmitter/receiver for fixed frequency operation in the I-band. It is intended for application where low weight, power consumption and cost call for less sophisticated equipment. In most other respects the two systems are identical and the -707 satisfies the helicopter requirements associated with navigation, surface search, tracking, mapping, search and rescue and anti-submarine warfare.

SWEDEN

L M Ericsson

Telefonaktiebolaget L M Ericsson, M I Division,
S-341 20 Mölndal 1

Sideways looking radar

This inexpensive, real aperture, pulse Doppler sideways looking radar is designed exclusively for maritime patrol and surveillance (eg for fishery protection, oil pollution monitoring, and search and rescue), with emphasis on simplicity and ease of operation. It comprises five line-replaceable units: antenna, transceiver, digital signal processor, control unit and television display. The radar images can be recorded on standard video recorders. Digital radar video can also be recorded for subsequent computer processing. With a radio link between aircraft and ground station or ship, images can be transmitted in real time with no degradation of quality.

The digital signal processor is designed around a 1.6 Mbit TV memory display. Up to 2000 range cells can be processed to an accuracy of 6 bits (giving 64 grey tone levels). The presentation can be varied to suit customers' needs, and other features include a reference grey scale, level mapping, positive or negative picture representation and automatic

L M Ericsson's lightweight surveillance radar in under-slung pod

positioning of targets. Monochromatic TV images recorded in the air can be converted to colour on the ground.

Installation: for double-sided coverage, glass-fibre pod containing two antennas carried under fuselage, or antennas carried in individual pods on each side of aircraft

Beamwidth: horizontal 0·5°
vertical 33°
Frequency: 9·4 GHz
Peak output power: 10 kW
Prf: 1 kHz
System weight: 70 kg

STATUS: in production.

UNITED KINGDOM

EMI

EMI Electronics Ltd, 135 Blyth Road, Hayes, Middlesex UB3 1BP

Searchwater

The collaborative efforts of EMI and the Royal Signals and Radar Establishment at Malvern have been directed over many years at identifying and evaluating the most appropriate techniques for long range airborne surveillance and classification of marine targets. As a result the company has become the sole supplier of such equipment to the UK's Ministry of Defence. The most recent system of its type is known as Searchwater, and has become standard equipment on the RAF's Nimrod Mark 2 maritime reconnaissance aircraft.

The system comprises a frequency-agile radar which uses pulse compression techniques and a pitch and roll stabilised scanning antenna with controllable tilt and automatic sector scan. Iff systems are included to interrogate surface vessels and helicopters.

A signal processor enhances the detection of surface targets (including submarine periscopes) in high seas and an integrating digital scan converter permits plan-corrected presentation and classification of target and transponder returns. The single radar observer in this aircraft is presented with a selection of bright, flicker-free television-type plan-position indicator B-scope and A-scope displays in a variety of interactive operating modes. Weather radar and navigation facilities are provided within the system. A real-time dedicated digital computer relieves the radar operator of many routine tasks, while continuously and automatically tracking, storing, and analysing data on many targets at the same time.

The facilities offered by Searchwater reduce the vulnerability of control aircraft by permitting them to operate effectively in an entirely 'stand-off' mode and also relieving them of the need to fly over the target to confirm its type by eye. The system is entirely modular with the interfaces and mechanical construction designed for ease of fault location and replacement. Major units are designed to be functionally self-contained as far as possible with a minimum of interconnections to other parts of the system. Extensive use is made of hybrid and integrated circuit techniques. The transmitter uses solid-state frequency generators and

EMI's Searchwater radar display in RAF Nimrod

mixers, followed by two cascaded travelling-wave tubes. The fluorocarbon liquid cooling system is employed with the scanner, which both transmits and receives radar and iff signals and uses a reflector of lightweight construction based on resin bonded carbon fibre.

STATUS: Searchwater was designed to replace the rather elderly ASV Mark 21 which was bequeathed to the Nimrods by the Shackletons they replaced. The system forms part of a major mid-life refit and the first squadron to get the new aircraft, No 206 at RAF Kinloss, was due to have completed its refit by mid-1982. Thirty-five Mark 2 Nimrods are eventually due for this modification.

RAF Nimrod MR Mk 2 submarine hunter/killers for which EMI's Searchwater radar is one of the principal sensors

Marconi

Marconi Avionics Ltd, Airport Works, Rochester, Kent ME1 2XX

Airborne early warning radar, Nimrod AEW 3

Impatient of the continuing delays by its NATO partners in reaching a decision for an integrated airborne early warning system for Europe, Britain decided to go ahead with a 'home grown' version based on its own Comet airliner in March 1977. This was the Nimrod airborne early warning version to replace the earlier piston engined Shackletons with their 1950s-designed APS-20 radars of US origin. The UK is building 11 AEW 3 Nimrods with airframes converted from standard reconnaissance Nimrods.

Marconi Avionics is responsible for the mission system avionics, which include the pulse Doppler radar made by that company; the integral iff is by Cossor and ESM, and data handling and display system by the US company Loral. Unlike the US Air Force and NATO AWACS early warning aircraft, the Nimrod AEW 3 has two fore and aft antennas, fed sequentially from a single transmitter, and placed in the nose and the tail of the aircraft where they scan the forward and rear sectors.

The radar has basically two modes of operation, used simultaneously. A high pulse-repetition frequency tracks fast moving low flying aircraft while a low pulse-repetition frequency is used for periodic updating of ship positions. The two modes are interleaved so that both fast and slow moving targets can be surveyed continuously.

The radar is a pulse-Doppler system operating at E/F-band frequencies and pulsing the transmission for range measurement while Doppler filtering is used for MTI and clutter rejection. The receiver analyses groups of returned pulses to detect any Doppler shift. The antennas have been specially designed to minimise sidelobes, and both are of identical shape. Their 180° scans are synchronised and they are automatically roll- and pitch-stabilised by a pair of gyro platforms which compensate for structural flexing and overcome the cyclic errors present in other systems.

STATUS: the first radar development aircraft, a modified Comet with a full system, was completed in March 1977. Flight trials began later that year, and three Nimrod AEW development aircraft had also been allocated to the programme. The first of the 11 aircraft was expected to become operational early in 1982 at RAF Waddington in Lincolnshire and it is anticipated that all will be in service by 1985.

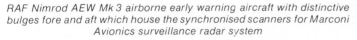

RAF Nimrod AEW Mk 3 airborne early warning aircraft with distinctive bulges fore and aft which house the synchronised scanners for Marconi Avionics surveillance radar system

Part of Marconi Avionics' mission system avionics suite in Nimrod AEW Mk 3

MEL

MEL, Manor Royal, Crawley, Sussex RH10 2PZ

MAREC II

The MAREC II (Maritime Reconnaissance Radar) is based on the company's helicopter radar system. Suitable for fixed-wing aircraft and helicopters, it has a true-motion plotting table display covering 360° in azimuth and a range of 250 nautical miles, together with a pilot's display. As well as general surveillance, operational roles include search and rescue, fishery and oil-field control, asv, asw and weapons control.

Plotting table display ranges: 17 – 219 n miles
Plotting table size: 430 × 430 mm
Pilot's indicator ranges: 10 – 250 n miles
Frequency: I-band 9345 MHz
Power output: 80 kW
Pulse width: 0·4 and 2·5 μs
Prf: 200 and 400 Hz

ARI 5991 Seasearcher

Seasearcher has a primary and secondary capability for asv, asw, sar and other maritime roles. It was designed principally for a Royal Navy Sea King update and is also applicable to other helicopters and fixed-wing aircraft.

ARI 5991 is a high power I-band (three-centimetre system) with selectable pulse width, 360° scanning and sector scan and multiple track-while-scan capabilities and digital read-out plot extraction. The indicator is a 17-inch (430 mm) diameter plotting table with a stabilised true-motion tactical display. It presents raw radar, secondary radar and numerous markers, and generates and displays multiple tactical symbols in response to digital position and identification information from a computer store. Optional features include: I-band identification friend or foe compatibility, pilot's repeater display and a weather radar mode. The effective range of Seasearcher is said to be about double that of previous equipment.

STATUS: in production and early service.

MEL's MAREC II radar-plot display

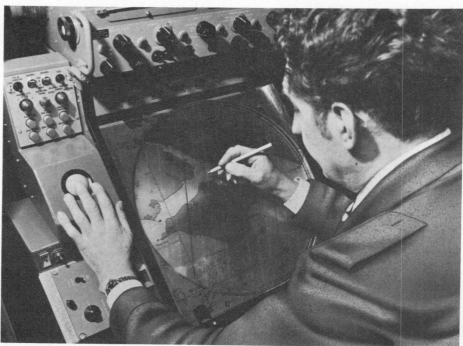

Display and control unit of MEL's new Sea-searcher radar

UNITED STATES OF AMERICA

AIL

Eaton Corporation, AIL Division, Deer Park, New York 11729

AN/APS-128 surveillance radar

The AN/APS-128 is perhaps one of the most widely used maritime patrol radars in the world, and claimed by AIL to be the only system of its class with frequency agility, sensitivity time control and a constant false-alarm rate. It has come into its own largely as a result of the recently instituted international 200 nautical mile fishing boundaries which need constant patrolling, but by low-cost aircraft to be economically effective.

The system comprises a rectangular flat-plate antenna and pedestal, transmitter/receiver, radar control module, range and bearing control module, and azimuth/range indicator.

Currently the AN/APS-128 is installed aboard all the Beech 200T maritime-patrol aircraft operated by the Japanese Maritime Safety Agency, and more recently has become operational with the Uruguayan Navy. The

Brazilian Air Force bought 16 sets of equipment for its EMB-111 Bandeirante MR aircraft, and Gabon has also chosen the system for its EMB-111s. The Royal Malaysian Air Force has the AN/APS-128 on its squadron of multi-purpose C-130 Hercules, and it is also offered as standard equipment for all MR versions of the Rockwell Sabreliner and Cessna's Conquest.

Claimed target detection ranges are:
fishing vessel (assumed six metres long, ten square-metres cross-section, in sea state three), 25 nautical miles
trawler (160 metres long, 150 square-metres cross-section, sea state five), 50 nautical miles
freighter (360 metres long, 500 square-metres cross-section, sea state five), 80 nautical miles
tanker (600 metres long, 1000 square-metres cross-section, sea state five), 100 nautical miles.
The system also functions as a weather radar with a range of 200 nautical miles.

Frequency: 9375 MHz
Frequency agility: 75 MHz peak/peak
Power output, pulse width, and prf: 100 kW peak, 2·4 and 0·5 µs, and 267, 400, 1200, and 1600 Hz
Antenna rotation rate: 15 and 60 rpm
Antenna stabilisation: automatic compensation for pitch and roll up to ±20°, with tilt to ±15°.
Azimuth/range indicator: ppi, P-7 phosphor, 178 mm crt, north or aircraft heading orientated, range scales 25, 50, 125 n miles
System weight: 174 lb (79·1 kg)

STATUS: in production and service.

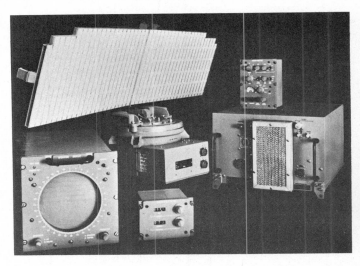

Component modules of AIL's AN/APS-128 surveillance radar, with flat-plate rectangular scanning array, operator's display, control units, and processor

Maritime-patrol Beech Super King Air with radome for AIL's AN/APS-128 surveillance radar visible on underside of fuselage

General Electric

General Electric Company, Aerospace Electronic Systems Department, 831 Broad Street, Mail Drop 508, Utica, New York 13503

AN/APS-120 surveillance radar

This surveillance radar is the principal sensor in the first 33 Grumman Hawkeye E-2C carrier-based early-warning aircraft to be built for the US Navy, and operates in conjunction with the same company's OL-93/AP radar detector processor, and an APA-171 antenna group. The purpose of the system is to detect distant aircraft in the presence of heavy sea or land clutter, but it also has a surface target detection capability. The system uses many of the techniques that were verified in the APS-111 (XN-1) flight-test programme to detect targets at longer ranges and with greater clutter than the earlier AN/APS-96.

Several features combine to give the APS-120 its long-range AMTI performance in high clutter levels, among them stability, coherence, pulse compression, and motion compensation. At the same time the system is no heavier or bulkier than its predecessor the APS-96. General Electric's own reliability programmes have found it to have a mean time between failure 50 per cent greater than contractually specified.

A feature of the APS-120 is the advanced radar processing system that combines increased sensitivity in noise and clutter with a sophisticated spurious-warning control and major electronic counter-countermeasures advances to provide the E-2C with an automated overland performance. This development has resulted in a new radar, the AN/APS-125 (see below).

STATUS: the last production models of the APS-120 and OL-93/AP were delivered in mid-1976, and subsequent production turned to the APS-125. Meanwhile a programme to modify the APS-120 up to -125 standard was begun in 1977.

AN/APS-125 surveillance radar

In order to improve the overland performance of the AN/APS-120 radar to include an automatic detection capability, the US Navy in 1972 placed contracts with both Grumman and General Electric to develop an advanced radar processing system that would combine greater sensitivity in noise and clutter with a greater resistance to false alarms. It would also have major electronic counter-countermeasures features new to airborne applications. Another improvement was the substitution of a digital AMTI system for the analogue unit of the earlier APS-120.

General Electric AN/APS-120 surveillance radar as fitted to some early US Navy/Grumman E-2C Hawkeye airborne early warning aircraft

Bulk and weight of earlier-generation electronics by comparison with present avionics demonstrated by General Electric's AN/APS-120 radar in laboratory test-rig

Initial deliveries for production aircraft began in mid-1976 and were scheduled to continue into the mid-1980s. A refurbishment programme to upgrade APS-120s to -125 standards began in 1977.

In 1981 Lockheed announced that it was studying a projected version of the Hercules designated EC-130ARE (the last letters standing for airborne radar extension) that would mount the 24-foot (7·3 metre) diameter APS-125 radar dish on the transport's fin. In this position, says Lockheed, the uhf system and its associated identification, friend or foe interrogation system could simultaneously track up to 300 targets, information being processed aboard the aircraft for action by the mission operators or transmitted by data-link to an air-defence centre on the ground. The *raison d'être* for the EC-130ARE, it is believed, was originally to provide an acceptable alternative to the Boeing AWACS for Saudi Arabia, the proposed sale of which was coming under heavy congressional criticism in 1981. In the event the Saudi deal went through, but Lockheed and General Electric continue to hawk the Hercules proposal around the world, particularly, it is thought, in the Middle East.

STATUS: privately sponsored feasibility study by the Lockheed Aircraft Service Company and General Electric.

Lockheed and General Electric are proposing a version of the C-130 Hercules transport with AN/APS-125 radar as a relatively inexpensive AEW aircraft

Sperry

Sperry Marine Systems, Charlottesville, Virginia 22906

AN/APN-59E(V) search radar

The APN-59E(V) was developed in the late 1970s to replace the earlier AN/APN-59B with greater performance and, particularly, reliability, in a great variety of retrofit situations. Sperry considers the new X-band radar a very mature design; it was tailored to a closely defined mission profile with unusually stringent quality assurance demands. For example, component selection was made on the basis of a number of engineering and data bank recommendations such as the Government/Industry Data Exchange Program, and the resulting system has been verified by rigorous testing to AGREE (Advisory Group on Reliability of Electronic Equipment) type testing. Mean time between failure is given as 219 hours.

The principal modes are search, navigaton, weather mapping and beacon homing, and to accommodate all these the operator can choose pencil or fan beam with a variety of pulse lengths and repetition rates, and the system can be set up for angle sector or 360° scan.

All line-replaceable units are interchangeable with those of the earlier system so that separate stocks of spares are not needed for flight line support, and gradual upgrading of a system can be accomplilshed over a period of time and without standing aircraft down.

Many alternative configurations are possible, ranging from single azimuth/range displays driven by the radar as a single and independent system, to more complex installations with up to three displays. Where requirements are particularly critical the system can be connected to a compass and dead-reckoning computer for the most accurate navigation fixes. The weight of a typical configuraton is about 185 lb (84 kg).

STATUS: in production under US Air Force contract to retrofit its C-130, C/KC-135 and

USAF Lockheed C-130 Hercules tactical transport, the fleet of which is being re-equipped with Sperry Marine's APN-59E(V)

RC-135 fleets, and the C-130 fleets of certain other air forces.

Texas Instruments

Texas Instruments Inc, 13510 North Central Expressway, Dallas, Texas 75266

AN/APS-124 search radar

The AN/APS-124 search radar was specially designed to be part of the comprehensive avionics suite for the US Navy/Sikorsky SH-60B Seahawk anti-submarine warfare helicopter built to satisfy the LAMPS Mk III requirement. One of the problems associated with the operation of these medium-size helicopters from the 'Spruance'-class destroyers on which they will serve is that of stowage, particularly in height limitation. The APS-124 is therefore designed around a low-profile antenna and radome.

Optimum detection of surface targets in rough sea is accomplished by the adoption of several unique features, including a fast-scan antenna and an interface with the companion OU-103/A digital scan converter to achieve scan-to-scan integration. The system is associated with a multi-purpose display and with the LAMPS data-link so that radar video signals generated aboard the aircraft can be displayed on LAMPS-equipped ships.

The system operates in three modes covering long and medium range search and navigation and fast scan surveillance. Mode 1,

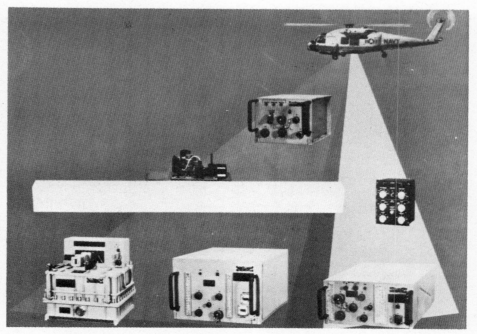

Principal detection sub-system of US Navy/Sikorsky SH-60B Seahawk sub-hunter helicopter Texas Instruments' AN/APS-124 radar

long range search, is characterised by long pulse length, low pulse repetition frequency, and slow scan, actual values being two microseconds, 470 pulses per second, and six rpm. Display ranges are selectable out to 160 nautical miles. In the medium range Mode II, these values change to one microsecond, 940 pulses per second, and 12 rpm respectively. For Mode III they become 0·5 microsecond, 1880 pulses per second, and 120 rpm, the display ranges are selectable up to 40 nautical miles and the false-alarm rate is adjustable to suit conditions.

The system is designed around the MIL-STD-1553 digital data bus to communicate with other aircraft equipment, and the modular design facilitates installation on other aircraft. The under-fuselage antenna provides 360 degrees coverage, and the entire six-unit APS-124 weighs 210 pounds (95 kg).

STATUS: in initial production, with service deployment of the Seahawk due to commence in 1983.

AN/APS-134 (V) periscope detection radar

The APS-134 (V) anti-submarine warfare and maritime surveillance radar is what Texas Instruments calls the 'international successor' to the US Navy's AN/APS-116 periscope-detection radar. Texas Instruments says that the APS-116 is the world's only radar specifically designed to detect submarine periscopes under high sea-state conditions and that its performance makes it the main-stay of the US Navy's Lockheed S-3A Viking anti-submarine warfare fixed-wing aircraft fleet. The APS-134 (V) incorporates all the features of the former system while improving performance and adding new capabilities, including a new surveillance mode.

The heart of the new radar is a fast-scan antenna and associated digital signal process-

AN/APS-134(V)1 surveillance radar which carries US Navy/Lockheed S-3A Viking carrier-based sub-hunter

ing which, says Texas Instruments, form the only proven and effective means of eliminating sea clutter. This technique is used in two of the three operating modes, the third being a conventional slow scan for long-range mapping and navigation.

In Mode I, periscope detection in sea clutter, high resolution pulse compression is employed with a high pulse repetition frequency and a false-scan radar, actual values being 1·5 feet (0·46 metre), 2000 pulses per second and 300 rpm. Display ranges are selectable to 32 nautical miles. There is no adjustable false-alarm rate to set the prevailing sea conditions and scan-to-scan processing is employed.

Mode II, long range search and navigation, operates at medium resolution and with a low pulse repetition frequency, low scan and display ranges selectable to 150 nautical miles. Actual values are 500 pulses per second and six rpm.

Mode III operates, again at high resolution, for maritime surveillance. A low pulse repetition frequency and 500 pulses per second are used

in conjunction with an intermediate scan speed of 40 rpm.

Display ranges are selectable to 150 nautical miles and an adjustable false-alarm rate is used together with scan-to-scan processing. The system interfaces with its own control and display equipment and with other aircraft systems by means of a MIL-STD-1553 digital bus, and a digital scan converter provides multiple display configurations using a 10 × 10 inch (254 × 254 mm) cathode ray tube. The weight of the entire APS-134(V), including the wave-guide pressurisation unit, is 527 lb (237 kg). The equipment is compatible with the inverse synthetic aperture radar techniques developed by Texas Instruments for long range ship classification.

STATUS: in service with the US Navy aboard its Lockheed S-3A Viking fleet and with Breguet Atlantic shore-based, fixed-wing, anti-submarine warfare aircraft of the West German Navy.

UNION OF SOVIET SOCIALIST REPUBLICS

Puff Ball

This NATO designation has been applied to a radar which has been stated to equip the Myasishchev M-4 Bison that still operates in small numbers with the Soviet Long-Range Aviation Force. It is probably also in some versions of the Tu-16 Badger and the Tu-142 Bear, both of which have broadly similar strategic roles.

Puff Ball is an I-band surveillance radar for large-area surface-vessel detection and it may perhaps be employed in the guidance of such air/surface missiles as Kangaroo, Kennel, Kelt and Kipper. American reports state that the system can provide friendly surface/surface missile batteries with target co-ordinates for missile guidance, the information being transmitted to the missile site by means of a data-link.

Tu-142 Bear-D, thought to carry Puff Ball surveillance radar

Secondary Radar

FRANCE

TRT

Télécommunications Radioélectriques & Téléphoniques, 5 avenue Réaumur, 92350 le Plessis-Robinson

TSR-718 air traffic control transponder

TRT's TSR-718 transponder is designed primarily for air-transport aircraft and provides identification on mode A with altitude-reporting on mode C. Transmitted power is 400 watts, plus or minus 100 watts on 1090 MHz.

The system employs fully solid-state construction, is microprocessor controlled, and uses software analysis of the input data to generate the uhf output. Internal desensitisation techniques provide echo protection. Other features include pulse-width verification, sidelobe suppression, and a suppression pulse which inhibits other pulse equipment in the aircraft each time a reply group is transmitted.

Self-test circuits monitor all TSR-718 functions, including uhf power amplifier output and frequency stability. Automatic fault identification is provided as standard, and a non-volatile fault-memorisation facility is available optionally. The unit's case is hinged at each side to provide direct access to all components for maintenance and repair. A mean-time between failures of more than 7000 hours is claimed. Antenna mismatch protection is incorporated and the transponder is not damaged by antenna short-circuit or open-circuit operation.

Dimensions: complies with 4 MCU (ARINC 600) form-factor
Weight: 10 lb (4·2 kg)

STATUS: in production and service.

UNITED KINGDOM

Cossor

Cossor Electronics Ltd, The Pinnacles, Elizabeth Way, Harlow, Essex

IFF 2720 transponder

This is a complete micro-miniature Mk 10A identification friend or foe/secondary surveillance radar transponder for use on all types of military aircraft and helicopters. On modes 1, 2, 3/A and B the full 4096 codes are available, and 2048 codes are available on mode C for altitude reporting. In addition there are circuits for the identification facility (SPI or I/P) and military emergency. The system comes in two units: a transmitter/receiver and controller containing easily accessible circuit boards.

Electronic warfare provisions include resistance to continuous wave, modulated continuous wave and pulse jamming, sidelobe rate limiting, short pulse and spurious interference protection, single pulse rejection, and long pulse discrimination.

A new microprocessor-based control unit, designated IFF 2743, will be similar in function to the full-facility IFF 2723 but will be packaged in a smaller case and with increased code legibility as a result of a light emitting diode display.

Frequency: (transmitter) 1090 MHz
(receiver) 1030 MHz
Power output: 27 dBW (500 W) minimum under all service conditions
Receiver sensitivity: −76 dBm
Dynamic range: ≥50 dB
Sidelobe suppression: 3 pulse
Dimensions: (IFF 2720 transponder) ³/₈ ATR Short, case 3·52 × 7·625 × 12·562 inches (90 × 194 × 314 mm)
(IFF 2723 full-facility control unit) 5·75 × 3·75 × 3 inches (146 × 95 × 76 mm)
Weights: (IFF 2720 transponder) 10 lb (4·6 kg)
(IFF 2723 full-facility control unit) 2 lb (0·9 kg)
Qualification: STANAG 5017 Edition 2, ICAO Annex 10

STATUS: the IFF 2720 system in current production is Cossor's standard export IFF, and equips many types of strike, transport, interceptor, and trainer aircraft and helicopters. In particular, it is fitted to all BAe export Hawk strike/trainers.

IFF 3100 transponder

The IFF 3100 is a single-package transponder tailored to the RAF Tornado multi-role bomber. Claimed advantages over previous systems are small size, lower weight, and simpler installation. Although the component density is high,

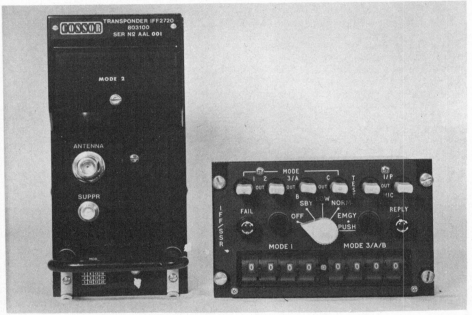

Transmitter/receiver and control unit of Cossor's IFF 2720 system

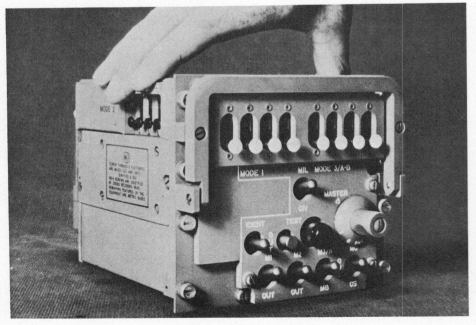

Cossor IFF 3100 transponder

reliability is ensured by the use of high-grade, close-tolerance circuits and a four-port circulator protects the output stages from the effects of any antenna mismatch. Open or short circuit conditions at the antenna do not damage the transponder.

Extensive integrity monitoring is incorporated, both during operation and when the test button is pressed. Checks cover receiver sensitivity, receiver centre frequency, mode decoding, reply coding and transmitter power level.

Interrogation modes are 1, 2, 3/A, B and C and the reply capability 4096 codes for modes 1, 2, 3/A and B. The x-pulse is included and there are 2048 codes for mode C.

Dimensions: 5·75 × 5·2 × 6·5 inches (146 × 132 × 165 mm)
Qualification: ICAO Annex 10, STANAG 5017 Edition 2
Weight: 11·7 lb (5·3 kg)

IFF 3500 interrogator

The IFF 3500 airborne interrogator contains advanced video-processing circuits for degarbling, defruiting, decoding and for echo- and multi-path suppression, all within a single unit. It employs monopulse techniques to achieve high accuracy in the measurement of target bearing and incorporates an automatic code-changing system (IFF 3502) to enhance security and eliminate the possibility of incorrect code setting.

The transmitter employs P2 emphasis to provide antenna beam sharpening. P1 and P3 are transmitted on the antenna sum channel and P2 on the difference channel. Selectable three or six decibels of emphasis is available. Both passive and active decoding are provided and two channels of passive decoding enable comparison during the overlap period between code changes. Active decoding provides serial readout of the 4096 reply codes. Coding information in either serial or parallel format ensures commonality with all in-service transponders.

Manual and continuous automatic built-in test circuitry checks transmitter power, interrogation coding, receiver sensitivity, defruiting/decoding, bearing accuracy and integrity of the transmission feeders.

Frequency: 1030 MHz
Power output: (P1, P3) 30·5 dBW (P2) 0, +3 or +6 dBW above P1 power
Spurious outputs ⩾76 dB below 1 W
Receiver frequency: 1090 MHz
Sensitivity (decoding): −80·5 dBm
Dynamic range: 60 dB

Principal unit of Cossor IFF 3500 interrogator

Indonesia's Hawk, an export version of the British Aerospace strike/trainer, has the Cossor IFF 2720 transponder

Spurious responses: 60 dB down outside the pass-band
Bearing resolution: dependent on antenna configuration, but around 5% of angle between intersection points of control and interrogate patterns
Dimensions: 1ATR Short case to ARINC 404A

Weight: 45·5 lb (20·7 kg)
Reliability: 1000 h mtbf
Qualification: compatible with NATO STANAG 50.7 Edition 3

STATUS: in production for RAF Tornado ADV, AEW Nimrod and F-4 Phantoms.

Plessey

Plessey Avionics & Communications, Vicarage Lane, Ilford, Essex IG1 4AQ

PTR446A transponder

The PTR446A miniature transponder identifies aircraft in response to ground radar interrogation and covers both civil and military modes. Great emphasis was placed on reliability combined with small size and low weight and these qualities have been achieved by the use of specially designed microelectronic circuits. A digital shift register replaces conventional delay lines in the encoder and decoder circuits, so providing time delays independent of temperature. Integrated circuits are used for the logic and video processing circuits and the logarithmic response intermediate frequency amplifier. Taken together, these techniques have reduced weight and size by a factor of six in comparison with other equipment, according to Plessey. Decoder, encoder and associated switches are located in the control unit, thus reducing the number of interconnecting wires to five and substantially reducing installation weight.

Two units of Plessey's PTR446A transponder

There are six versions of the PV447 control unit associated with this transponder, providing the following capabilities:

PV447 — Mode 1 or 3A/B and Mode C or off
PV447A — Mode 1 or 2 and Mode C or off
PV447B — Mode 2 or 3A/B and Mode C or off
PV447C — Mode 1 or 3A and Mode 2 or off
PV447D — Mode A or B and Mode C or off
PV447E — Mode A/B or off and Mode C or off

An alternative to these is the PV442 control unit which meets the full requirements of NATO STANAG 5017, ICAO Annex 10, ARINC 532D and FAA TSO C74 Class 1. Operational modes are 1, 2, 3A and C with Mode B if required. Three pulse sidelobe suppression is employed.

Power output: 24·7 dBW
Pulse rate: 1200 replies/s, each containing up to 15 reply pulses
Triggering sensitivity: –72 to –80 dBm
Dimensions: (transponder) 2¼ × 5 × 10 inches (57 × 127 × 254 mm)
(PV447 control unit) 5¾ × 2¼ × 4 inches (146 × 57 × 102 mm)
(PV442 control unit) 5¾ × 3¾ × 4 inches (146 × 95 × 102 mm)
Weights: (transponder) 3·7 lb (1·7 kg)
(PV447) 1 lb (0·48 kg)
(PV442) 2·25 lb (1·02 kg)

STATUS: in production.

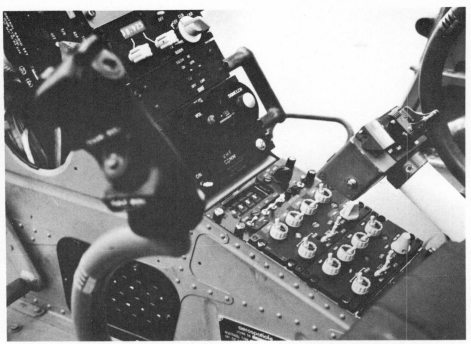

Control unit of Plessey PTR446A in communications pedestal of Gazelle helicopter

UNITED STATES OF AMERICA

AIL

Eaton Corporation, AIL Division, Deer Park, New York 11729

AN/APX-103 IFF transponder

This identification, friend or foe interrogation system was developed for the US Air Force/ Boeing E-3A Sentry AWACS airborne warning and command system now operational with the US Air Force. It is being prepared for service with NATO and has been ordered by Saudi Arabia. The AN/APX-103 tags echoes received by the surveillance radar aboard the E-3A as representing either friendly or hostile aircraft. The interrogator section of the equipment can query other aircraft selectively in either conventional air traffic or military modes. The receiver/ processor section automatically decodes the identities and locations of co-operative aircraft and feeds digital data to the E-3A's central command and control computer, which also stores primary radar data. Both types of data may be called up subsequently by operators aboard the early-warning aircraft.

STATUS: in production and service.

Boeing E-3A AWACS early warning aircraft equipped with AIL's AN/APX-103 IFF interrogator for the US Air Force, NATO and Saudi Arabian Air Force

Bendix

Bendix Avionics Division, 2100 NW 62nd Street, Fort Lauderdale, Florida 33310

TPR 2060 air traffic control transponder

The Bendix TPR 2060 is a lightweight, compact air traffic control transponder designed for light aircraft and general aviation. It responds automatically to mode A and mode C interrogations and, with a suitable encoding altimeter input, will transmit aircraft altitude information with the normal reply pulses. A mode B capability is also optionally available for use in areas employing B mode interrogation.

The TPR 2060 features special DME suppression circuitry to prevent interference between the transponder and DME installations when the antennas for the two systems are sited in close proximity. The system also permits transmission of a special identification pulse for a 20 second period on use of an ident button on the front panel. A reply lamp remains lit during this time to reassure the user that the transponder is in fact identing.

Self-test facilities are incorporated. During self-test operation, the unit's coding and decoding circuits are exercised in the same manner as they would be during actual radar interrogation. The unit, which may be panel, console or roof mounted, is in a single case and is of large scale integrated circuit-type construction.

Dimensions: 1·75 × 6·31 × 8·46 in (445 × 160 × 215 mm)
Weight: 2·6 lb (1·18 kg)

STATUS: in production and service.

Collins

Collins Avionics Divisions. Rockwell International, 400 Collins Road NE, Cedar Rapids, Iowa 52406

TDR-950 air traffic control transponder

Collins' TDR-950 is a panel-mounted air traffic control transponder for light aircraft. It responds to interrogation on 4096 codes, covering modes A and C, but can be converted to mode B or to B/C coverage. When used with a suitable encoding altimeter, it has an altitude reporting capability of up to 62 000 feet (18 900 metres). The system is interrogated on a frequency of 1030 MHz, responding with a nominal output of 250 watts on the 1090 MHz frequency.

The TDR-950 employs LSI-MOS integrated circuitry, with all encoding and decoding functions being carried out by a chip. Self-test facilities are incorporated.

Dimensions: 6·25 × 1·625 × 8·15 in (159 × 41 × 207 mm)
Weight: 2 lb (0·91 kg)

STATUS: in production and service.

TDR-90 air traffic control transponder

The TDR-90 is an air traffic control mode A/C transponder with 4096 codes and an altitude-reporting capability of up to 126 000 feet (38 400 metres) when used with an encoding altimeter. It is a remotely controlled system designed primarily for general-aviation.

The system has a transmitter output power of 325 watts nominal (250 watts minimum) on a response frequency of 1090 MHz. Positive sidelobe suppression facilities are incorporated in order to provide a cleaner 'paint' on the interrogator's trace. Two-way mutual suppression provision is also made to avoid interference with distance-measuring equipment. Another feature is the employment of a strip-line diplexer to control receiver front-end noise while retaining high sensitivity and frequency stability, regardless of antenna matching.

A built-in test facility for both the transmitter and receiver functions is included. Test signals are injected at just above the minimum sensitivity level to ensure that receiver, decoder, encoder and transmitter are all functioning correctly.

The system's CTL-90 control unit has two-knob code selection, ident, self-test, standby and altitude-reporting on/off controls. An optional system selection switch can also be incorporated for use in dual installations. The display is of the gas-discharge type. The TDR-90 electronic unit can however interface with most conventional transponder controllers as well as the CTL-90 unit.

Dimensions: 1/4 ATR Short
Weight: 3·5 lb (1·59 kg)

STATUS: in production and service.

621A-6 air traffic control transponder
The Collins 621A-6 transponder is an air-transport system operating in modes A, B and C on 4096 codes. Space is also available for mode D operation and it has a provision for X-pulse. Mode C altitude-reporting extends to 126 700 feet (38 600 metres) when the transponder is used with an altitude digitiser or an air-data computer. The unit operates on the normal receiver and transmitter frequencies of 1030 and 1090 MHz respectively and the transmitter output power is nominally 700 watts. The system is remotely controlled.

Solid-state digital microcircuitry construction is used throughout the 621A-6. Delay-lines are replaced by digital delay-circuits and all timing functions are performed by digital shift-register techniques. Front-end noise has been eliminated by using a strip-line duplexer which also permits both receiver and transmitter to operate through a single antenna. A lightweight quarter-wave cavity is employed in the transmitter section. A ferrite isolater between this and the antenna ensures constant loading regardless of antenna matching and contributes to frequency stability.

Self-testing is carried out by an internally generated signal which is injected into the receiver at just over the minimum sensitivity level. This checks receiver sensitivity and frequency, decoder tolerance, monitor circuits and the selected mode together with mode C. The test can be performed from either the controller or at the transponder itself.

During normal operation, the monitor circuits maintain a continuous check on transmitter frequency, transmitted power, transmitted code, reply pulse spacing and internal timing circuits. If a fault is detected, a visual warning is given. A facility for continuous automatic self-test when the aircraft is beyond the range of normal ground interrogation is available optionally.

Maintenance monitors isolate any malfunctions to the antenna and its cables or to the transponder unit itself. Latching maintenance monitors on the front panel change from black to yellow if a malfunction occurs and remain yellow until reset by maintenance personnel. A separate test connector on the back panel facilitates ground maintenance and is used to interface the unit with automatic test equipment.

Dimensions: 3/8 ATR Short
Weight: 13·6 lb (6·16 kg)

STATUS: in service.

621A-6A air traffic control transponder
Based largely on the design of the earlier 621A-6 model transponder (for which the 621A-6A is intended as a replacement) the later model exploits advances made in microcircuitry techniques to improve performance and reliability.

Compared with its predecessor, the 621A-6A has 190 fewer components and the functions of five cards in the earlier unit are now performed by a single card. Furthermore, technological improvements have resulted in reduction of 15 per cent in the initial purchase cost, and an increase of 100 per cent in mean-time between failures is also expected from the newer model.

The 621A-6A does however share a large number of components with the 621A-6 and the two systems are directly interchangeable so that retrofits are simple. There is, apart from the reduced number of parts, one significant design change, namely the addition of pulse-width adjustment facilities. This permits adjustment to offset the effects of ageing in the main transmitter valve, extending this item's working life with, consequently, a lower replacement burden.

Dimensions: 3/8 ATR Short
Weight: 13·6 lb (6·16 kg)

STATUS: in production and service.

Hazeltine
Hazeltine Corporation, Greenlawn, New York 11740

AN/APX-72 IFF transponder
The RT-859A/APX-72 system, which Hazeltine claims to be the most widely used identification, friend or foe transponder in the world, provides military and civil air traffic controllers with aircraft identification data, traffic information, and automatic altitude reporting with a suitable altitude digitiser. The system receives, decodes and replies to the characteristic interrogations of operational modes 1, 2, 3/A, C, and 4, the last of which uses a KIT-1A/TSEC computer. Pressurisation may be required.

Transmitter
Power output: 27 dBw nominal
Duty cycle: 1% max
Frequency: 1090 MHz

Receiver
Frequency: 1030 MHz
Dynamic range: >50 dB
Sensitivity: –90 dBv nominal

Decoder/coder
Mode C (altitude): decodes interrogations
Reply modes: (Mode 1) 32 codes
(Mode 2) 4096 codes
(Mode 3/A) 4096 codes
(Mode C) according to ICAO standards
(Mode 4) compatibility and full capability via a plug-in card

Mtbf: 300 h (a fully solid-state transmitter, now available, improves this figure)
Dimensions: 5·76 × 6·39 × 13·37 inches (146 × 162 × 340 mm)
Weight: 15 lb (6·8 kg)
Qualification: Mk 12, Stanag 5017, ICAO Annex 10, DOD AIMS 65-1000

STATUS: in production and service, more than 40 000 having been built. Users include all US Services and many foreign customers.

Hazeltine's AN/APX-72 IFF transponder

AN/APX-76A IFF interrogator
The Hazeltine AN/APX-76A IFF interrogator is an L-band interrogator for all-weather interceptors and other tactical aircraft, with full AIMS capability in Modes 1, 2, 3/A, and 4. Narrow antenna beam-width and reduction of 'fruit' are achieved via interrogation and receiver sidelobe suppression circuits in conjunction with special antennas having sum (mainlobe) and difference (sidelobe) suppression patterns.

Bracket-decoded video and discrete-coded video are displayed on the radar scope to provide unambiguous correlation between IFF and radar targets. The system comprises four units: transmitter/receiver, switch amplifier, interrogator control and electrical synchroniser. In addition dipoles are installed on the main radar antenna.

Transmitter
Frequency: 1030 MHz
Duty cycle: 1% max
High-power output: 2 kW

Receiver
Frequency: 1030 MHz
Sensitivity: –83 dBm

System weight: 37 lb (16·8 kg)
Mtbf: 225 h
Qualification: DOD AIMS 66-252 Modes 1, 2, 3/A, and 4

STATUS: in production and service with aircraft of the US Air Force and the US Navy. Types so fitted range from F-4, F-14 and F-15 to airborne early warning and anti-submarine warfare aircraft, and more than 4000 sets have been built.

AN/APX-104(V) IFF interrogator

Hazeltine claims that its AN/APX-104(V) air-to-air and air-to-ship interrogator set is the most technologically advanced, lightweight IFF interrogator yet produced. It provides positive identification and range and bearing measurements of other friendly aircraft and ships in an all-weather environment, and is a useful aid to setting up a rendezvous with a refuelling tanker, and for maintaining contact with other aircraft on a common mission.

The AN/APX-104 is intended as a replacement for earlier IFF offering substantially better performance in terms of reliability, space and weight. It provides full AIMS operation in Mode 1, 2, and 3/A, and also in Mode 4 when a KIR-1A/TSEC computer is fitted. The transmitter is a fully solid-state device employing high peak power transistor amplifier submodules paralleled to give the required output power. This provides better reliability and a significantly longer time between maintenance than electron tube/cavity amplifier transmitters. The elimination of high-voltage power supplies and the need for pressurisation permit significant volume and weight savings and also contribute towards the reliability. Incidentally, the local oscillator employs surface acoustic wave technology for simplicity, low production costs, good stability, and lack of field alignment requirements.

Frequency: (receive) 1090 MHz
(transmit) 1030 MHz
Decoding sensitivity: –83 dBm
Power output: 1·2 kW, switchable to half-power
Mtbf: 1000 h

STATUS: a private venture. Qualification tests have been completed.

Components of Hazeltine AN/APX-76A IFF system, including one short dipole antenna normally mounted on scanning antenna

Solid-state IFF transceiver for Hazeltine's AN/APX-104(V) IFF system

Displays

CANADA

Computing Devices

Computing Devices Company, a division of Control Data Canada Ltd, PO Box 8508, Ottawa, Ontario K1G 3M9

AN/ASN-99 projected-map display

This projected-map system, with company designation PMS 4-1, has an interface for use with a navigation computer, and comprises two units, a projected-map display PMD 2-2 and an electronics assembly unit EAU 1-2. The display unit contains 27 feet (8·28 metres) of maps in 35 mm colour-film format in a replaceable cassette, film drive system, optical elements, display screen, and controls. The assembly unit contains the electronics for processing the digital bit-serial and word-serial input commands from the aircraft navigation computer.

There are eight operating modes: Normal, North-up, Manual, Data, Test, Decentre, Hold, and Landing. In Normal mode, the projected-map image is orientated with aircraft track vertically up. Aircraft position is shown at the centre of the circular display, or at the origin of the radar graticle depending on the Decentre button illumination. In North-up the map is orientated with true north vertically up. On selection of Manual, the maps can be slewed in the event of computer failure navigation. The

Presentation of Computing Devices' projected-map display and overlaid annotations

Electronics assembly for Computing Devices' projected-map display

Data mode causes key chart information to appear on the screen. When the Test position is selected a test pattern appears on the screen. The Decentre facility causes the aircraft position to move from the centre of the circle to the origin of the radar graticle, or vice-versa. The Hold position stops map movement, and the film can then be slewed to a desired position. Selection of the Landing mode results in the automatic display of the nearest airfield approach chart.

Map scales and coverage: 1/500 000 and ½ million, 1000 × 800 nm and 2500 × 1500 nm
Capacity: 200 frames of 35 mm film
Accuracy: displayed position error 0·0276 inch cep (0·142 mile at 1/500 000 scale)
Brightness: visible in 10 000 ft-candles ambient illumination at 31° to the normal at the screen
Film slew rates: frame advance (row change) 3 sec max, data chart access 3 sec max.
Reliability: mtbf 100 h
Service life: 10 000 h or 10 years
Dimensions: (display) 7 × 6 × 15³/₅ inches (177 × 152 × 397 mm)
(electronics unit) 9⁴/₅ × 5³/₄ × 15¹/₄ inches (249 × 146 × 387 mm)
Weight: (display) 21 lb (9·55 kg)
(electronics unit) 21⁴/₅ lb (9·91 kg)

Control unit for Computing Devices' projected-map display

Power: 294 VA at 115 V 400 Hz three-phase, 140 VA at 21 V 400 Hz, 2 VA at 5 V 400 Hz, 0·2 A at 28 V dc
BIT: go/no-go check provided by test pattern. Also continuous automatic checking of both units by monitoring circuits in the electronics unit, and appearance of a fail flag in the event of a fault. Fault isolation of failed unit by latching indicators on the electronics unit on selection of test mode

STATUS: in production. More than 1300 sets supplied to number of countries for tactical aircraft, helicopters, and military transports.

PMS 5-6 projected-map display

This system is similar to the PMS 4-1, but takes its signals from air-navigation sensors and ground-based aids. Scale combinations are 1/50 000 and 1/250 000, which are particularly useful in helicopter operations. The full map display is augmented by 16 data-charts. Simultaneous map and numerical display of present position permit the crew to pass back to base accurate co-ordinates, or to update computed present position.

STATUS: in production for Italian Air Force G222 transport.

FRANCE

SFENA

Société Francaise d'Equipments pour la Navigation Aérienne, BP 59, 78141 Vélizy Villacoublay Cédex

WGD-1 and WGD-2 windshield guidance displays

In 1976 SFENA, in conjunction with the Air France subsidiary Air Inter, introduced on the latter's fleet of Dassault Mercure short-haul transports the world's first head-up display for commercial aircraft. The system has been operational with the airline since that time, permitting take-off and landing in Category III weather minima, significantly below the levels that would otherwise be allowed.

From the experience with this system SFENA has developed two newer devices, the WGD-1 and WGD-2 windshield guidance displays. The WGD-1 provides guidance during the ground roll at take-off and landing in poor visibility, permitting a reduction in runway visual range minima at take-off and, in conjunction with a Category III automatic landing system, during landing. It is also useful in the event of sudden loss of visibility (eg when traversing a patch of fog on the runway), engine failure, or unbalanced reverse-thrust operation. The WGD-2 provides guidance for manual visual approaches on to runways without instrument landing systems, for missed approach procedures, and duplicates the WGD-1 functions during take-off and landing roll. In effect the WGD-2 acts as an airborne visual approach slope indicator providing the crew with a

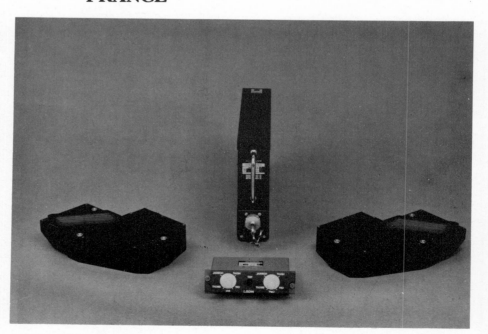

SFENA's WGD-1 windshield guidance display provides ground-roll guidance during take-off and landing

precisely defined touchdown point.

Both systems use solid-state symbol generation with incandescent lamps for illumination and fibre-optic image transmission. Each system comprises a computer unit and an optical head mounted in the glareshield. The guidance

symbols are collimated (ie focused at infinity so that they appear to be a long distance ahead of the aircraft) and reflected into the crew's line of vision from the interior surfaces of the windscreen. The advantages of the WGD system over a conventional head-up display with combiner glass are (says SFENA): unrestricted visibility with no limitation in pilot head movement; reduced training requirements; lower cost; improved reliability; and ease of installation.

WGD-1 symbology comprises a vertical command bar that moves to left or right of a fixed symbol representing the runway centre-line and displayed only below a fixed radio height. The crew follows the command bar with nose-wheel steering or on the rudder pedals. The symbol is governed by instrument landing system localiser deviation, runway heading error, and yaw rate. WGD-2 symbology employs a trajectory symbol in the form of a horizontal command bar that is manually aligned with the desired touchdown point on the runway, and a horizontal speed command symbol that has to be 'flown' on to the trajectory symbol. It switches to WGD-1 symbology during the landing and take-off roll.

WGD-1 computer
Type: 8-bit microprocessor with BIT
Dimensions: 1/4 ATR Short
Weight: 3 kg

WGD-1 optical head
Dimensions: 100 × 225 × 50 mm
Weight: 3 kg
Field of view: dependent on airframe geometry

WGD-2 computer
Dimensions: 3/8 ATR Short or 3 MCU
Weight: 6 kg

Optical head unit for more advanced WGD-2, standard option for Airbus A300 and A310 transports

WGD-2 optical head
Dimensions: 360 × 220 × 65 mm
Field of view: dependent on airframe geometry
Weight: 6 kg

STATUS: the WGD-1 has been certificated by the FAA for Lufthansa's Boeing Advanced 737-200s. The WGD-2 is the standard option on the Airbus Industrie A300 and A310 wide-body airliners.

Thomson-CSF

Thomson-CSF, Division Equipments Avioniques, 178 boulevard Gabriel Peri, 92240 Malakoff

TMV 980 A head-up/head-down display

The TMV 980 A is an integrated head-up/head-down display system designed for the new Dassault Mirage 2000 multi-role fighter. It comprises a digital computer and processor to generate the display symbology and help with flight and weapon-aiming computation, and three display units: a VE 130 cathode ray tube head-up display, a VMC 180 interactive multifunction head-down display, and a VCM 65 complementary monochrome cathode ray tube for electronic support measures information.

The head-up display is a 76 mm high resolution, high brightness cathode ray tube with collimating optical system based on a 130 mm lens providing a wide total field of view and a wide instantaneous field of view; the binocular instantaneous field of view is increased in elevation by the use of a twin-glass combiner that transmits some 80 per cent of the light incident upon it. Automatic brightness control, with manual adjustment, permits symbols to be read in an ambient illumination of 100 000 lux. The system provides continuous computation of both tracer line in the air/air mode and impact and release points in the air/ground mode.

The main head-down display presents, in red, green, and amber on a 127 × 127 mm cathode ray tube radar, information such as a map, 'synthetic' tactical situation, and range scales; raster images from television or forward-looking infra-red sensors; tactical data from the system itself or from an external source via a data-link; and some flight information such as aircraft attitude. The unit has two video inputs (radar and television) together with a link from the computer/processor carrying the colour calligraphy for superimposition on the radar or television pictures.

Future growth envisaged includes a second head-down display for the two-seat Mirage 2000 together with a TMV 544 sight repeater providing a collimated television image of the

Thomson-CSF TMV 980 A integrated head-up/head-down display (above) for new Dassault Mirage 2000 fighter (below)

view through the front cockpit head-up display. A helmet sight or display could also be integrated into the TMV 980 A to improve target discrimination and off-boresight target designation.

Weight: (electronic unit) 9 kg
(hud) 13 kg
(VMC 180 head-down display) 14 kg
(VCM 65) 4 kg

STATUS: in production.

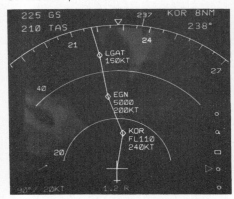

Presentation of the Thomson-CSF electronic horizontal situation indicator showing planned track between bad weather and waypoints

Four of the six Thomson-CSF electronic instrument displays are seen in this view of the Airbus Industrie A310 flight-deck

EFIS electronic flight instrument system for Airbus Industrie A310

In the autumn of 1979 Thomson-CSF, with Bendix and West Germany's VDO Luftfahrt-werke, won the competition against Collins and Sperry to supply the electronic flight instrument colour cathode ray tube display system for Europe's Airbus Industrie A310 wide-body airliner. This version of the top-selling A300 short-haul transport is one of the three all-digital airliners that will be coming into service in 1983, the others being the Boeing 767 and 757, all with multi-purpose cathode ray tube engine and flight instruments. The change to digital equipment, together with the cathode ray tube instruments replacing the traditional electromechanical types, has resulted in such an increase in flexibility of use that in each case the flight engineer can be dispensed with, and the aircraft certificated for two-pilot operation.

The electronic flight instrument display system designed for the A310 uses six shadow-mask colour cathode ray tubes. Each pilot has on the flight panel in front of him a primary flight display cathode ray tube which replaces the former attitude director indicator, and a navigation display cathode ray tube which now supplants the earlier horizontal situation indicator and weather-radar displays. Two more displays located on the centre panel are used to present flight information such as the take-off and landing drills and schematic diagrams of the hydraulic or electrical systems for example, to check their correct operation. One of them is normally reserved for warnings, the other for systems. All six cathode ray tubes are of identical size, 127 mm square, and are interchangeable with one another, so economising on the spares to be kept.

Since the contract was awarded, Thomson-CSF has debugged and refined the system. Colour stability and visibility of the displays in high ambient lighting conditions have been demonstrated, the latter at 100 000 lux instead of the 85 000 lux specified. No fewer than nine colours are used. On the primary flight display, seven of them are employed for symbology, to avoid confusion, and two (blue and brown) represent sky and Earth. To ensure a uniformly smooth sky, unmarred by the raster scan, slightly out of focus imaging was arranged and has proved particularly acceptable to the eye. The three-dimensional effect that pilots have long been used to with electromechanical primary flight displays has been achieved by masking certain symbols when they would

Airbus Industrie A310, extensive revision of A300, with virtually new avionics suite, including EFIS system

normally be occulted by others mounted further forward in the instruments. By means such as these, Thomson-CSF and Airbus Industrie have sought to offer not only a presentation that is instantly recognisable as that of a typical primary flight display, but one that can capitalise on the vast quantity of digital data now for the first time available and circulating within an aircraft on the data-bus. To this virtually classic presentation, therefore, Airbus Industrie and Thomson-CSF have added the following information to the periphery of the instrument: a moving speed scale along the left-hand edge with its various standard symbols generated by computer; a pressure altitude readout along the right-hand edge; an autopilot and autothrottle mode annunciator along the top edge; and radio altimeter heights along the lower edge of the instrument.

For the navigation display, classic compass-card symbology has been retained, again in conjunction with the flexibility that digital sensing and processing affords. This has been used in an innovative way by creating new symbology to suit each phase of flight, in the form of an electronic map display with five distance scales marking the course to be followed with radio waypoints and their identifying codes. On selection by the pilot, a three-colour weather map can be superimposed on

the navigation map, resulting in the saving of the panel space that would be required by a separate weather cathode ray tube, and the more comprehensible integration of weather and navigation information. The warning and system status cathode ray tubes on the centre panel show information not previously available to flight crews. They warn of failures not only by special alarms but also present the crew with synoptic displays of failed systems, with indications of vital actions to be followed to meet the difficulty and the effects on other systems.

Each of the two sets of flight instruments is driven by its own symbol generator, with a third standby unit that can switch to either set in the event of failure of its own generator. The two centre-panel instruments are driven by a pair of interchangeable symbol generators.

STATUS: in production for the Airbus Industrie A310-200 short-medium haul 200/250-seat transport. By the time of the 1981 Paris Air Show Thomson-CSF had delivered sets to the forward-facing crew cockpit simulator at Toulouse, to the flight-control test-rig, and to the first A310. Twelve sets were due for completion by the end of 1981, 16 in 1982, and 30 the following year. Production has been set provisionally at ten sets per month thereafter.

GERMANY, FEDERAL REPUBLIC

Teldix

Teldix GmbH, PO Box 105608, Grenzhofer Weg 36, D-6900 Heidelberg 1

Reflective-optics head-up display

Teldix, in a technology-demonstrator programme with Carl Zeiss, has sought to overcome the principal drawback of traditional head-up displays, restricted fields of view, by replacing the conventional lens and flat combiner glass with a suitable reflection-optics system. The improved field of view, says Teldix, would result in a wider weapons-delivery envelope so that, for example, weapons could be released in a much sharper turn than now possible and their release ranges extended.

In the wide-angle head-up display, or W-HUD as Teldix calls it, light from the head-up display's cathode ray tube is reflected by a concave mirror onto a large aspherical combiner that can still show only modest weight increases over current systems. In a typical installation the field of view could be increased from present-day values of 20 to 25 degrees in azimuth and elevation to 40 to 60 degrees in azimuth and 25 to 30 degrees in elevation. The optical geometry of the system requires a special cathode ray tube with a spherical screen, but the electronic picture could be produced by a standard symbol generator.

Format display: stroke symbols and alphanumerics for flight-control, navigation, weapons delivery. Blending with tv raster optional

Combiner dimensions: 400–500 mm wide × 200–250 mm high, depending on cockpit configuration

Line width: (stroke): 1·5 mrad

Aiming accuracy: 2 mrad near optical axis

Brightness: (stroke) 5000 cd/m^2
(tv raster) 1500 cd/m^2

Self-test: BITE

Weight: 115 kg

STATUS: technology development and demonstrator programme financed by the West German Ministry of Defence. The system was shown and demonstrated publicly for the first time at the 1981 Paris Air Show.

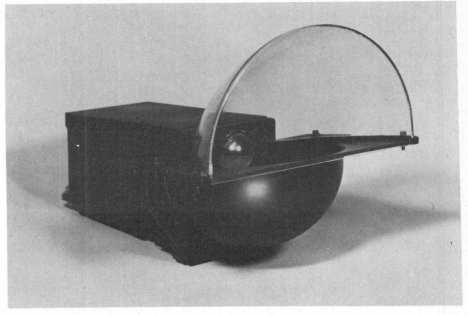

Curved combiner glass of Teldix reflective-optics head-up display

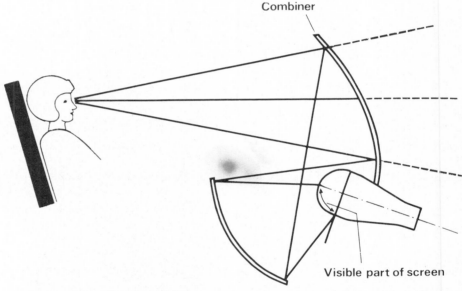

Combiner

Visible part of screen

Ray geometry for the Teldix reflective-optics head-up display

Teldix is a partner of Smiths Industries and OMI in the design and production of head-up display for Europe's trinational Tornado bomber

ISRAEL

Elbit

Elbit Computers Ltd, Advanced Technology Center, PO Box 5390, Haifa 31051

Multi-function display

This is essentially a monochrome armament control and display panel for combat aircraft with limited cockpit space, showing video information from a variety of sensors. The display cathode ray tube is surrounded on the face of the panel by 28 push-buttons for the selection of different functions.

Display size and type: 101 × 101 mm, P43 phosphor, raster scan selectable at 875 and 525 lines, 60 Hz
Brightness: 200 ft-lamberts (685·2 cd/m²) minimum
Contrast: seven shades of grey

Contrast ratio: 7:1 minimum
Video bandwidth: 30 Hz – 20 MHz to 3 dB levels
Dimensions: 175 × 147 × 279 mm
Weight: 8 kg
Power: 20 W at 28 V dc, 70 W at 115 V 400 Hz
Reliability: 1500 h mtbf per MIL-HDBK-271A

STATUS: in production.

Elbit's multi-function display

IAI

IAI Israel Aircraft Industries Ltd, Ben Gurion International Airport

Gyro-stabilised sight

This system, designed for light aircraft, helicopters and armoured vehicles, is a collaborative effort between IAI's Electronics Division and Tamam Precision Industries. The device is actually a variable magnification periscope, the optical head being outside the vehicle, free to scan in elevation and azimuth, and gyro-stabilised to eliminate image movement due to aircraft movement or vibration. It can be controlled by means of a joystick, slaved to an external designator or the designator can be slaved to the sight.

A television camera permits viewing by both eyes, and enhances penetration of haze, and a video recorder permits a flight to be run through for post-mission critique or training purposes. An infra-red goniometer can be attached to the sight for accurate launch and automatic guidance of missiles against armoured targets. Likewise a laser rangefinder permits target ranging, and the system can be integrated with a low-light television camera for passive, undetectable night vision.

Scan angles: 360° azimuth, –30° to +40° elevation
Stabilised tracking rate: up to 10° in both axes
Magnification: ×2·5, ×10
Field of view: 20°, 5° depending on magnification
Electrical pick-off accuracy: 0·1-0·2 mrad
Weight: 25 kg

STATUS: in production.

SWEDEN

Saab-Scania

Saab-Scania AB, Box 1017, S-551 11 Jonköping

RGS 2 and RGS 2A weapons sight

The RGS 2 and 2A lead-computing rate-gyro optical weapons sights are modular devices in which all but the most essential optical elements have been eliminated from the sight head and housed in a remote computer and gyro unit, thus enabling the smallest possible glareshield installation. The remote unit can drive a second sight head, either in side-by-side or fore-and-aft installations, and isolation of the computer unit facilitates their synchronisation.

Inputs to the computer are pitch, roll and yaw rates, with manually pre-set values to adjust sight sensitivity and the characteristics of the weapons being carried. In a more advanced configuration of the sight a roll angle input is added to stabilise the aiming mark to the true vertical reference, and a height input to compensate for altitude variations. Further refinements are possible when speed, pitch and drift angles can be fed into the system. As a result of the variations possible given the wide differences in sensor availability between combat aircraft, the basic RGS 2 system has developed into a family, the new RGS 1 fixed-sight device, the simplest member, being at one end of the scale and the RGS 3 and 4 at the other end.

The RGS 2A is a low-cost modification of the RGS 2 in which the lead-computing air-to-air mode has been replaced by a fixed aiming mark and the sight designed for air-to-ground modes. In both systems the aiming mark is controlled by a spherical mirror that is moved by a servo system driven by command signals from the computer.

RGS 2 facilities comprise stadiametric ranging with a throttle-mounted range-selector switch permitting weapons to be fired at up to four different ranges; a 'snap-shoot' switch for instant selection of a target on the boresight; a roll-stabilised aiming mark for tactical approaches; and automatic compensation for crosswind and target movement.

Saab-Scania RGS 2 three-unit optical sight

The RGS 2A includes stadiametric ranging for different firing ranges; roll-stabilised aiming mark for tactical approaches; manual compensation for crosswind; and automatic compensation for speed error, pitch angle and height variations.

The three-unit system comprises an optical sight head with electrical and mounting provisions for a recording camera, computer unit and controller. It complies with MIL-E-5400N Class 2 and weighs 8·8 kg.

STATUS: the system is certificated for BAe Hawk, Saab 105, Aeritalia MB339, MB Veltro 2, MB326 and CASA C-101 strike-trainers, and Fiat G-91R3 and F-5A light attack aircraft.

Installation of Saab-Scania RGS 2 sight in British Aerospace's private demonstrator Hawk strike-fighter

Helios helicopter weapons sight

The Helios is a roof-mounted weapons sight, for helicopters with air/ground weapons, with magnifications of ×3 and ×12 and provision for laser target marking and a thermal imaging system for night vision. It can accept commands from a helmet sight or from an avionics system and generates accurate position and rate outputs to weapon systems. It can also be slaved to directions given by a laser-warning receiver, hostile fire indicator, or radar-warning receiver. Future versions envisaged would have a laser illuminator as an easy add-on feature and a medium-size forward-looking infra-red camera could be incorporated if the stabilised optical sight head were replaced by a thermal-imaging head.

For mechanical and safety reasons the weight of roof sights has to be kept to a minimum and the Saab unit weighs only 19 kg

as a result of careful design, the use of rate-integrating gyros and remote location of the electronics. The optical design (by UK company Pilkington, which also builds the optical train) is characterised by low vignetting and low field curvature. The eyepiece arm protruding into the cabin area can be stowed close to the roof when not in use and the eyepiece height is adjustable to suit the observer/gunner.

The sight is composed of a number of modules with interfaces which ensure that minimal adjustments are necessary when a module is changed. The system itself comprises five units: the roof-mounted sight, the electronic unit, the control panel, the control unit and the line-of-sight indicator.

Magnification/field of view: ×3/18° and ×12/4·6°
Line-of-sight variation: (elevation) ±25° (azimuth) ±120°

Saab-Scania Helios lightweight modular helicopter sight

Slew rate: up to 100°/s
System weight: 27·6 kg

STATUS: in production.

UNITED KINGDOM

Ferranti

Ferranti Ltd, Electronic Systems Department, Ferry Road, Edinburgh EH5 2XS

Helmet-mounted display system

In October 1980 Ferranti delivered to the Royal Aircraft Establishment (RAE) at Farnborough an advanced, comprehensive helmet-mounted display system for development and evaluation. It was first tested in a flight simulator at RAE and later flown aboard a Westland Sea King helicopter there for airborne trials.

The system was developed as a research tool by the company's Pointing and Displays Group at Edinburgh under a UK Ministry of Defence contract.

The helmet-mounted display system is intended for both fixed-wing aircraft and helicopters in conjunction with sighting and surveillance equipment for weapons or sensor control, but also has sea and land applications. It is well adapted for operations in fast-patrol boats, large cruisers (as an alternative target acquisition system for sophisticated weapons), or as part of a low-cost protection system for armed merchantmen. In land applications it is suited to visual target acquisition and for the pointing of guns and missiles. It can be either vehicle-mounted or infantry operated and used for battlefield surveillance and air defence.

The system comprises an electronics unit driving a three-axis magnetic field generator under the direction of a pilot's control unit, a small magnetic-field sensor mounted on the helmet to measure the direction of the pilot's line of sight, and an optical sight with light source, aiming mark and simple optics to indicate the line of sight to the observer. It can also incorporate cueing arrows to draw the observer's attention to important data and alpha-numerics.

As an alternative to the optical sight, a miniature cathode ray tube can be mounted on the helmet. This has optics giving a wider field of view for presentation of more complex imagery.

The electronics unit contains the power supply and signal conditioning, digital processing, and interface electronics needed to measure continuously the pilot's line of sight in azimuth, elevation, and roll with respect to a datum direction, usually the fore-and-aft axis of the aircraft or missile.

The control unit provides on/off switching, a function switch for mode-changing, display-lighting intensity control for independent adjustment to the brightness of the sighting reticle and discrete cueing signs, and a light to indicate that helmet sensor alignment is required.

The cathode ray tube and associated wideband video amplifier are designed to give the best imagery for general surveillance and target

Ferranti helmet-mounted sight system showing sensors and display projector

discrimination. The television raster can be matched to existing or upcoming sensor line-standards. The picture is roll-stabilised to avoid pilot disorientation. The field of view is 40 degrees wide, and imagery within it is collimated. A high data-refresh rate is provided for compatibility with rapid head-slewing, so that the picture is space-stabilised. Imagery is presented in four grey scales.

Electronics unit
Dimensions: 1/2 ATR Short
Weight: 24 lb (10 kg)

Control unit
Weight: 1 lb 11 oz (0·77 kg)

Magnetic field generator
Weight: 2 oz (0·07 kg)

Helmet sensor
Weight: 0·04 lb (0·02 kg)

Optical sight
Weight: 6 oz (0·16 kg)

STATUS: in development and evaluation.

Isis weapon-aiming sights

For many years Ferranti has been producing a family of optical weapon-sights under the name Isis, for the aiming of guns, rockets and bombs in the close-support role, and for guns and missiles in air-combat.

The members of this family vary in complexity, from devices needing no inputs except electrical power, to integrated systems with inputs from a number of navigation and air-data sensors. Ferranti claims that adequate accuracy is coupled with exceptionally low cost of ownership.

The operation is basically the same for all members of the Isis family. All of them project a gyro-stabilised and collimated reticle pattern into the pilot's sight-line by means of a combining glass. The positioning of this reticle in relation to the weapon or gun datum is commanded by the response of a two-axis rate-gyroscope to aircraft rate of turn, and to which is attached a mirror. This reflects light from a fixed source shining through a reticle so projecting the image onto a combining glass ahead of the pilot.

In the ground-attack role the reticle is depressed below the armament datum through an angle that compensates for the combined effects of the weapon's ejection velocity, angle of attack of the airframe in the intended release conditions, the gravity 'droop' of the weapon during its flight, and the surface wind if known. These angles may be set up by the pilot to suit the intended attack speed, dive angle, and slant range at weapon release, or he may prefer to use a pre-determined set of parameters, one set for each type of weapon carried.

In air combat the unit measures the rate of turn of the sight-line and scales it to target range, thus producing a first-order lead-angle solution.

To assist steering and enable the pilot to make allowance for sudden manoeuvres on the part of the target, most sights include a second, fixed reticle representing the gun axis.

Typical sight, Isis D-126R

Number of units: sight head, control unit, electronic unit, and throttle unit
Weight: (sight head) 9 lb (4·1 kg)
(control unit) 11 oz (0·3 kg)
(electronics unit) 4 lb 6 oz (2 kg)
(throttle unit) ½ lb (0·2 kg)
Sight head dimensions: 9½ × 5¼ × 8 inches (242 × 132 × 203 mm)

STATUS: generally in production and service, with new versions being planned.

AF533 Rufus roof observation sight

Since 1967 Ferranti and the optical company Avimo of Taunton have collaborated in the design of a variety of sight-line stabilised weapon-aiming devices, the Ferranti AF533 Rufus roof-mounted helicopter sight being the latest joint project in this field.

This gyro-stabilised, monocular, periscopic telescope is designed for the gunner/observer in reconnaissance helicopters, particularly when scouting targets for anti-tank helicopters.

The device has a built-in interface for a laser designator and rangefinder, which, if of Ferranti design, can be installed at first-line level without the need for setting up or initial adjustment. It can also be adapted to operate with night-vision equipment, helmet sights, and weapons, and there are facilities for attaching a television recording camera for training or intelligence-gathering. The design of the optical system is such that the varying eye positions in different helicopter installations can be easily accommodated.

The gyro-stabilised head protrudes above the roof forward of the rotor mast, while the down-tube and eyepiece extend downwards from the roof so that the eyepiece falls into a comfortable viewing position. The down-tube is adjustable in height and retracts sideways, locking close to the roof when not in use. A control handle is extended by the operator and

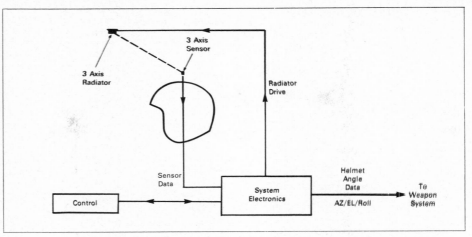

Computation of sight angles in the Ferranti helmet-sight

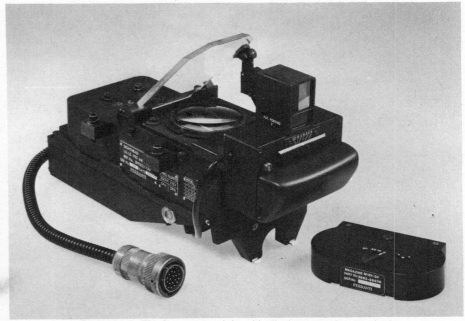

Ferranti Isis F95-3 weapons sight with display-recorder attached

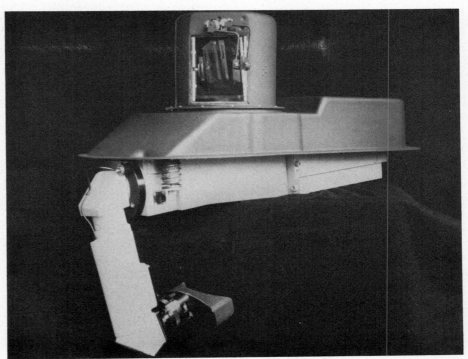

Ferranti/Avimo Rufus helicopter sight, showing rotating optical head and eyepiece

adjusted, in tilt, so that it can be used with his right forearm resting on his knee. A horizontal thumbstick is used to steer the sight-line and a direction indicator, to show its direction relative to aircraft heading, is mounted on the glare-shield in front of the pilot.

The sight provides a stabilised image of the chosen field of ×2·5 magnification for search and ×10 for identification and laser operation.

STATUS: in production.

COMED combined map and electronic display

The increasing effectiveness of surface/air guns and missiles obliges strike and interdiction aircraft to operate at the lowest possible heights above ground, making use of natural features such as valleys to shadow them from detection radars. But by so doing the pilot's 'look ahead' range is severely diminished and turning points, fix-points, and targets may be obscured by the same topography that affords protection. Head-up guidance provides a good method of giving aircraft-management information, but does nothing to help with navigation. To do this a topographical map is required, and in recent years this has taken the form of a projected moving map display, driven by the navigation system. The aircraft symbol is at the centre of the display, and the map is moved across the display area at a speed proportional to the map-scale and aircraft speed.

The maps used with current displays have a major disadvantage in that it is not possible to annotate them with planned route, target information and other tactical data; this has to be marked up on a hand-held map to be used as an adjunct to the moving map. This disadvantage has been overcome with the development by Ferranti of its Comed (combined map and electronic display) system, in which the map is electronically annotated with intelligence or navigation data appropriate to the particular sortie being flown. Using no more space than a conventional moving-map, Comed provides a colour topographical map display annotated with dynamic navigation information as required; this can include, typically, aircraft track and commanded track, present position, enroute fix-points, target position, locations of known hostile detection or anti-aircraft devices, and tactical information such as the delineation of forward edge of the battle area. In addition the system can print out alphanumeric information such as time-to-go to fix-point or target.

Comed's cathode ray tube and projection facilities also permit it to perform a number of other operational tasks. These can include the display of high-resolution, high-contrast symbology in raster form from radar, low-light television or forward-looking infra-red sensors. Electronics countermeasures threats can also be shown up. Tabular displays of weapons status, destination coordinates, or other tactical information can be shown on tabular 'forms' projected from images stored on the film. The system can also be programmed with a library of aircraft and engine check-lists. Another possibility is that Comed can act as a back-up primary flight information display, particularly for the horizontal situation and attitude director command indicators.

Comed interfaces with the main aircraft navigation computer via a MIL-STD-1553 serial digital data link. From a knowledge of film-strip layout, aircraft present position and demanded scale, the main computer calculates the appropriate map-drive words, and transmits them to Comed by way of the data-link. Within the display the information is converted into a form suitable for driving the map servoresolvers.

Comed uses standard 35 mm colour film up to 57 feet (17 metres) long. Typically this provides coverage of an area 100 nautical miles square at a scale of ¼ million, plus selected target areas at 1/50 000 and sufficient film frames for tabulated displays. Film replacement can be done through a side access-panel in under a minute.

Dimensions: (crt) 5½ inches square
(unit) 7¹/₁₀ × 8½ × 23¾ inches (180 × 218 × 604 mm)

STATUS: the system has been ordered by the Indian Air Force for its Sepecat Jaguar light bombers, made under licence by Hindustan Aeronautics. The first production Comed System is planned for delivery by the end of 1982.

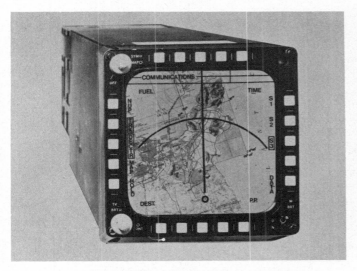

Ferranti's Comed combined map and electronic display

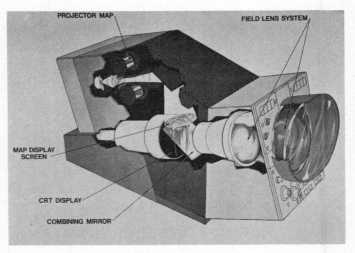

Optical arrangement of Ferranti Comed display

Marconi Avionics

Marconi Avionics Ltd, GEC-Marconi Electronics, Airport Works, Rochester, Kent ME1 2XX

Hudsight head-up display for F-16 fighter

The head-up display for the General Dynamics F-16 fighter, ordered in such large quantities by the US Air Force, four European nations (to replace their Lockheed F-104G Starfighters), and several other countries including Israel and Egypt, is the latest in a line of Marconi head-up displays that go back to the strike-sight of 1960 for the Royal Navy's Buccaneer carrier bombers. The F-16 takes a stage further the 'eyes outside' concept introduced by the US Air Force/McDonnell Douglas F-15 Eagle fighter, in which the pilot operates the engine throttle with his left hand and controls the aircraft with his right hand and feet without once taking his eyes from the combat situation going on outside, all his information coming through the head-up display.

The Hudsight system comprises an electronic unit (effectively the head-up display's 'brain'), a rate sensor unit and a pilot's display unit. The electronic unit is a small digital computer with a 16 K EPROM memory. The system in fact has spare capacity which has

The Marconi Avionics' head-up display dominates the F-16 instrument panel, with its prominent combiner-glass and selector panel

been used by General Dynamics to broaden the scope of the F-16's fire-control computer. The F-16 is the first aircraft to utilise a digital data highway as a convenient way of cutting down the number of dedicated information wires, an important consideration where continuing technical advances permit the same military effectiveness with progressively smaller airframes, so that finding space for the irreducibly large bundles of analogue cables becomes more and more difficult. The head-up display communicates with the attack radar, fire-control computer, air-data computer and inertial navigation unit through the MIL-STD-1553 data bus.

The pilot's display unit comprises the cathode ray tube assembly, the optical train and the control panel all mounted on a rigid chassis; the optical combiner glass is sufficiently strong to withstand the aerodynamic loads resulting from damage or loss of the canopy. Also incorporated in the optical module is a standby sight which is entirely independent of the principal electronic circuitry. The rate-gyro unit generates the pitch, roll and yaw rates and normal acceleration needed for accurate air-to-air weapon delivery.

The system operates in basically three modes: cruise flight management, ground attack, and air combat. The two combat modes are again sub-divided into further modes, each designed to provide the most effective information and guidance based on data from the F-16's fire-control computer. In the cruise mode the head-up display shows speed, height, heading, vertical velocity, velocity vector, climb/dive, Mach number, maximum available *g*, and range and time to destination. In the ground-attack mode there are five types of display: continuously computed impact point, strafe, dive-toss, electro-optical, and low-altitude drogue delivery for retarded weapons. Air-to-air symbology is divided into four configurations for the built-in gun and various types of air-combat missiles: snap-shoot (in which the system provides the guidance for transient gun-firing opportunities), smooth tracking (again for the gun), air-to-air missiles, and dogfight. In order to keep himself out of trouble from enemy fighters the pilot needs to keep a constant watch on his speed and height, and so in all these air-to-air modes energy-management scales may be selected in place of airspeed and altitude scales.

The system has built-in test facilities to isolate faults to a particular line-replaceable unit and predicted mean-time between failure for the head-up display is now 870 hours. Considerable growth capability is incorporated; for example, the system can be upgraded to include dual-mode day/night display in which the normal day-compatible stroke symbology is augmented by a raster scan that can provide an overlay of radar, low-light television or forward-looking infra-red imagery.

STATUS: like most other major airframe, engine and systems activities in this huge multinational programme (the F-16, in terms of financing, manpower, number of companies involved and political commitment, is the world's largest aviation effort) the head-up display is the basis of a complex international production agreement. Twin production lines to support the US Air Force and export aircraft have been established in the United Kingdom and the USA (at Marconi Avionics' Atlanta plant), while production for the NATO F-16s of Norway, the Netherlands, Denmark and Belgium is largely vested in those countries. The overall production and maintenance activity is under the control of the Marconi Avionics' Central Management Team at Rochester in the United Kingdom.

Lantirn holographic head-up display for F-16 and A-10

Despite the capability and growth potential of the standard F-16 head-up display, the US Air Force in the mid-1970s circulated requirements

General Dynamics F-16 fighter which carries Marconi Avionics head-up displays

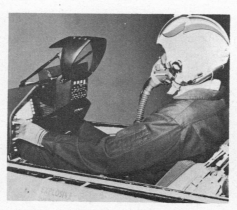

Installation of the Marconi Avionics holographic head-up display in the F-16

for an even more advanced system to industry. These, in conjunction with improvements in other areas would provide the basis for a substantial upgrading of the F-16's effectiveness at a relatively early stage in its career under the designation the Multinational Staged Improvement Program. The specific improvement sought was the adoption of a new imaging capability known as Lantirn (low-attitude navigation, targeting, infra-red for night) permitting the F-16 to operate in all weather and at night.

It was soon realised that the field of view needed for this new capability would be far greater than that available with existing head-up display technology developed to its limit. Present conventional head-up displays with lateral and vertical fields of 13·5 and 9 degrees respectively could be expanded to 20 and 15 degrees using standard optics, but this would still be inadequate.

Thus Marconi started developing a head-up display based on holographic techniques, using the principles of diffractive optics, and in July 1980 was awarded a General Dynamics development contract and initial production options totalling $103 million. The US Air Force by May 1982 had options on 603 holographic head-up displays for F-16s and 250 A-10 anti-tank aircraft as part of the Lantirn programme. The system combines the wide-angle display geometry made possible by using hologram technology with a Marconi-developed method of combining raster and cursive-scan symbol writing that greatly reduces the amount of equipment needed for day/night head-up display. The system will provide a field of view of 30 degrees laterally and 18 degrees vertically. On the electronics side, it will incorporate MIL-STD-1589B high-order language, MIL-

Close-up of the holographic head-up display. As in all head-up display optical systems, great rigidity is needed to prevent vibration and misalignment

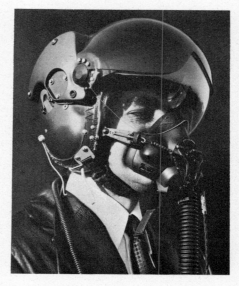

Optical projector for Marconi Avionics helmet-mounted display (facing page) visible inside clear visor panel

STD-1750A airborne instruction set architecture, and MIL-STD-1553B digital data bus standards.

STATUS: the programme is sponsored by the Aeronautical Systems Division of the US Air Force Systems Command for the General

Dynamics F-16 air-superiority fighter and the Fairchild A-10 battlefield bomber. The production phase will be shared with Marconi's two European partners in the current F-16 head-up display programme, Köngsberg Vapenfabrik of Norway and Odelft in the Netherlands. It is thought the Lantirn system will enter service in 1984. The first pre-production Lantirn head-up display was delivered to the US Air Force in March 1982 and flight evaluation with the F-16 was to begin a few months later.

Helmet-mounted display and optical position sensing system

To complement conventional head-up displays a number of companies are producing helmet-mounted display systems that, mounted on flying-helmets, provide a very much wider field of view to acquire and designate targets for sensors and weapons. Marconi Avionics is developing a special sight for combat aircraft pilots that is not only light in weight, to minimise discomfort, but is qualified for ejection at speeds up to 650 knots.

The company built its first helmet-mounted display system in 1969, choosing solid-state light-emitting diodes as the image source in conjunction with a simple optical projection system as being most suitable for high-*g* combat conditions. By 1974 the company was able to offer a helmet-mounted display system providing data on range, bearing, and the type of threat, as well as sighting and cueing information, and the system was flown in joint US Air Force/US Navy trials between 1976 and 1979. It used a 20 × 23 dot matrix as the display medium, but the most recent system employs a 32 × 32 dot format.

The helmet-mounted display system comprises two main parts: a helmet-mounted display (hmd) and a helmet optical position-sensing system (hops). It can be used simply as a flight information system, in which case the helmet optical position sensing facility is not needed. When used as a weapons sight, however, that facility is required to generate sight-line information for radar, missiles and other equipment. The helmet-mounted display is formed by the illumination or non-illumination of a 32 × 32 dot light-emitting diode matrix or a hybrid fixed-format/matrix array, the light being projected into the pilot's eyes by reflection from the inside surface of the visor.

Sight-line sensing implies the use of helmet optical position sensing, and calls for the fixing of at least four light-emitting diodes (and perhaps six or eight, depending on likely head movements) on each side of the helmet. The illumination from them is detected by three so-called V-slit cameras mounted one behind and two ahead of the pilot and containing linear charge-coupled photo-diodes. A remote electronics unit associated with the helmet optical position sensing system processes the signals from the cameras to produce pitch, roll, and azimuth sight-line angles with respect to the aircraft fore and aft axis and these are presented on the visor 'screen' either by matrix or fixed-format array.

Applications of the helmet-mounted display/helmet optical position sensing system include display of flight-information, energy management, threat warning, missile lock-on and guidance, navigation fixing, ground-target designation, TOW and HOT anti-tank missile designation for helicopters, and commands for helicopter gunfire using a steerable gun-turret.

Voltage levels on the helmet are low and reflection of the display from the spherical visor means that the position of the visor is not critical.

Hmd field of view: 10° typical
Hmd helmet installation weight: 0·15 kg
Hops weight, extra to hmd weight: 0·093 kg (8 leds) 0·06 kg
Hmd brightness: 20% contrast ratio against 27 408 cd/m² background, with 10% neutral density sun visor

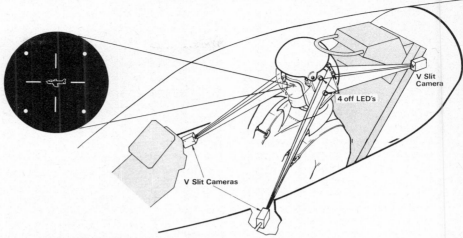

Sensing of head angles with respect to the aircraft fore-and-aft datum is accomplished with the Marconi Avionics system by means of three V-slit cameras in the cockpit

Installation of the Marconi Avionics Monohud

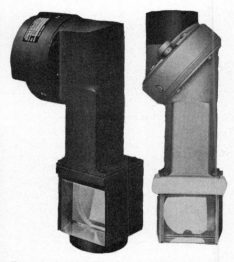

Two versions of the Monohud that have been built for trials; left, for Nasa, and right, for Britain's Royal Aircraft Establishment

V-slit camera
Dimensions: 50 × 50 × 35 mm
Weight: 0·38 kg
Hops accuracy: 0·5° cep

Electronic unit
Weight: 10·1 kg
Dimensions: ³/₄ ATR short for prototype

STATUS: development.

Monohud miniature head-up display

The number of commercial airline flights terminating in Category III visibility (ceiling below 30 metres and/or visibility less than 400 metres) is a very small proportion of all flights, and even in these circumstances some visual reference is usually possible. The classic difficulty facing the crew is the changeover from instruments (instrument landing system director and radio height) to visual reference as the runway comes into sight during the last few seconds of flight.

In order to ease the transition from instrument to visual flight, Marconi has started development of a head-up display for transport aircraft that integrates visual cues from the outside world with internally generated flight-path information from an instrument landing system coupled to autopilot. The company lists the following advantages of such an arrangement:

More precise approaches can be flown to runways without instrument landing systems, an important factor where terrain characteristics or the presence of buildings can create optical illusions;

The difficulty of crew take-over after disconnecting an autopilot with a fail-passive

autoland system can be considerably eased. The head-up display presentation permits the approach to be monitored and provides assistance either to continue the approach or to initiate go-around;

With a dedicated instrument landing system channel to provide the signals, the head-up display can completely and independently monitor a fail-operational autoland system;

A head-up display provides unambiguous warning of deviation from the approach path;

A head-up display gives instant warning of wind-shear;

A head-up display assists take-off and go-arounds by providing flight parameters in a form instantly available to the crew;

Landing precision is improved in manual landings, both in terms of runway touchdown dispersion and sink-rate.

More economic use of fuel is achieved by providing a guide to energy management at take-off and landing.

After long investigations of the difficulties of reconciling the largely conflicting installation and display requirements, Marconi settled on an optical system providing a display, for one eye only, contained in a tube mounted on the forward roof in the flight-deck and hinged so that it can be swung out of the way when not in use. By this means the weight and size of the system are kept to the minimum. When in use the end of the tube containing the combiner glass is about 102 millimetres from the eye and the instantaneous field of view (that is, without moving the eye) is a 20-degree circle. By moving the head very slightly this can be increased to 30 degrees. With development these figures could be increased to 20 × 30 degrees and 24 × 36 degrees, according to Marconi. The line width is 1·5 milliradians, writing speed 37 500 degrees per second or 24·38 metres per second. The collimation error is less than two milliradians and the display colour is green (P1 phosphor).

Electronics for the Monohud are incorporated in a 3·8 kg video drive unit, normally located in the flight-deck roof-space, and 6·3 kg symbol generator. The use of EPROM memory devices permits symbology changes without modification to the equipment.

STATUS: in development. Trials have been conducted by NASA and by the Royal Aircraft Establishment in a BAC One-Eleven.

Marconi Avionics hudsight head-up display for US Air Force/Nato F-16 fighter

Smiths Industries

Smiths Industries, Cricklewood, London NW2 6JN

EFIS electronic flight instrument system

The Smiths Industries' colour cathode ray tube EFIS system represents the United Kingdom's entry into the electronic flight-deck display market and will be competing with Thomson-CSF in France and Rockwell Collins, Bendix, and Sperry in the USA for selection on versions of the Airbus Industrie A310, Boeing 767 and 757, the 150-seat transports when they eventually appear and the more sophisticated business and corporate aircraft.

The system comprises two units: a symbol generator and a display unit. They would (except for customer-specified changes of form, ie box configuration) be common to all installations, whether aboard individual aircraft or on different types of aircraft, so that spares holdings and cost of ownership would be kept to a minimum.

The symbol generator is designed around ARINC 725 requirements and to ARINC 600 dimensional standards, housed in a 6 MCU box. This unit contains the generator for stroke and hybrid displays, processor, ARINC 429 digital interface, and display unit interfaces. To achieve the specified level of integrity each generator is designed to drive several display units (so that failures can be contained through redundancy) including head-up displays. For the same reason each generator accepts multiple inputs from each type of aircraft sensor. Reliability is enhanced by the use of large-scale integrated circuits and good thermal design.

The display unit has automatic brightness sensors, with their associated electronics, and incorporates a high-resolution, rugged shadow-mask multi-colour cathode ray tube displaying information in stroke or hybrid formats. Designed around ARINC 725 Form Factor D (unit face size 203 × 203 millimetres), the system can display all flight, navigation, and systems information including ARINC 708 weather radar maps.

STATUS: the system has been undergoing evaluation trials in a BAC One-Eleven at the Royal Aircraft Establishment, Bedford. These began early in 1981 with monochrome cathode ray tubes, but by the middle of the year colour cathode ray tubes had replaced them. Two major demonstration tours have been conducted, one to North America, the other to The Netherlands, West Germany and France. A special demonstration was also given to Airbus

The two Smiths Industries' EFIS displays are the principal flight instruments in the RAE's BAC One-Eleven civil avionics research aircraft at the Royal Aircraft Establishment, Bedford. The left-hand unit is the EFDS electronic flight-director system, the right-hand one the EHSI electronic horizontal situation indicator

The RAE's One-Eleven avionics research and trials aircraft

Industrie at Toulouse with the aim of providing an alternative European system to the Thomson-CSF system chosen for the A310 wide-body airliner.

5-50 Series head-up display/weapon-aiming computer

The 5-50 head-up display/weapon-aiming computer is based on a single advanced digital computer, similar to that supplied by Smiths Industries for the multi-national Tornado bomber, that can readily interface with any of the head-up displays manufactured by the company. It provides all necessary flight, navigation and weapon-aiming symbology, and performs accurate weapon-aiming computation. As with the earlier 4-40 system, the symbology presented to the pilot at any particular time is limited to information appropriate to the mode of flight at that time, so minimising display clutter.

Though designed to interface with other digital systems, the 5-50 can accommodate analogue inputs, eg from sensors fitted to non-digital aircraft, by means of an interface matching unit. This arrangement, says Smiths Industries, is particularly useful during preliminary flight trials, or as an economic way of providing a suitable interface when only a small production run is envisaged.

Using UK or MIL-standard symbology, the 5-50 system can operate in a number of modes associated with waypoint and target acquisition, and weapon-aiming. These modes are for air/ground, continuously computed release point and continuously computed impact point, and for air/air, continuously computed impact line and weapon release. Additional weapon-aiming modes are depressed sight-line, director-guided aerial gunnery and toss-bombing. A breakaway cue is provided in most bombing modes to indicate that immediate recovery from the delivery manoeuvre is necessary to avoid impact with the ground or damage caused by flying through the blast envelope.

Symbology: MIL-D-81641(AS), STANAG 3648, or as required
Inputs: 8 serial digital data channels and 32 discrete commands
Memory: up to 9 K words
Speed: (add/subtract) $3 \cdot 75 \, \mu s$
(multiply) $8 \cdot 63 \, \mu s$
(divide) $9 \cdot 13 \, \mu s$
Dimensions: $5 \cdot 05 \times 7 \cdot 74 \times 14 \cdot 97$ inches ($128 \times 197 \times 380$ mm)
Weight: 21 lb (9 kg)
Accuracy: (boresight) $0 \cdot 05$ mrad
(symbol positioning) $0 \cdot 22$ mrad up to 5° off axis, $0 \cdot 32$ mrad 5° – 10° off-axis

STATUS: the electronic unit for this system is in production for the multi-national Tornado bomber. By mid-1982, more than 300 units had been delivered; estimated total production is 1200 systems.

6-50 Series head-up display/weapon-aiming computer

The most recent Smiths Industries head-up display programme, the 6-50, builds on previous developments by the addition of facilities to include air data and navigation computation. The system incorporates many detailed improvements over earlier head-up display systems and is an economic, low-weight device suitable for a wide range of aircraft types. It is designed to interface with the data-bus systems now coming into general use in the newer aircraft, but can also be adapted as a retrofit for earlier, 'analogue' aircraft which are being upgraded for mission enhancement by the addition of modern avionics. There are many variations in packaging but in its simplest form the 6-50 comprises a display/multi-purpose processor driving a pilot's display unit through a control panel.

The processor generates the symbology for the display unit, performs the weapon-aiming

Smiths Industries is prime contractor for the Tornado hud, produced in collaboration with Teldix of West Germany and OMI of Italy

computations, and can be used for a variety of air-data and navigation tasks. The designation 6-50 embraces not only a range of packaging options, but also of performance and capability. The most comprehensive system offered at present can be contained in a 1 ATR (8 MCU) box weighing 12 kg and capable of generating additional symbology for a raster display and of interfacing with a wide range of sensors via a MIL-STD-1553 data-bus. Extensive weapon-aiming computation facilities and cursive display symbology are also available in a ¾ ATR box (6 MCU) weighing nine kg. At the other end of the scale, weapon-aiming computing and symbology for aircraft requiring only basic head-up display/weapon-aiming computer capabilities can be housed in a ½ ATR (4 MCU) unit of six kg.

Variations are also provided for the pilot's display unit. The simplest has a 102 mm collimating lens providing a 25 degree field of view for $11 \cdot 9$ kg. Another variant, this time in line-replaceable form and designed to MIL specifications for the Anglo-American AV-8B Harrier V/STOL fighter, has a 114 mm collimating lens, a 22 degree field of view, and weighs 17 kg.

The system has a precision dual combining-glass assembly and an electronically depress-

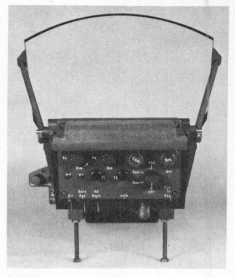

Smiths Industries' diffractive-optics hud, with wide-angle combiner-glass and control panel

ible standby sight. Special care in the mechanical design has resulted in a unit with the same rigidity as earlier units but weighing 20 per cent less. Compensation for windshield distortion is applied electronically and optically to suit the particular characteristics of the aircraft. The cathode ray tube is protected from the damaging effects of solar radiation by ultraviolet and infra-red filters in the optical train, and cathode ray tube brightness, display accuracy and deflection amplifier performance are monitored by a high-reliability built-in test system.

The standby sight is a precision light-emitting diode matrix on a glass substrate, the brightness of which can be adjusted by varying the intensity of the light-emitting diodes.

STATUS: in production, one version for the US Marine Corps. By mid-1981 Smiths Industries had built more than 1200 head-up displays of all types for versions of the Harrier, Jaguar, Saab JA37 Viggen, and Tornado.

Smiths Industries 6-50 hudwac as chosen for AV-8B V/STOL fighter

Diffractive-optics head-up display

In May 1982 Smiths Industries announced that an advanced technology diffractive-optics head-up display (dhud) designed and built in conjunction with Hughes Aircraft in the USA had successfully completed a series of flight

trials aboard a Jaguar light bomber at British Aerospace's Warton airfield. These trials, with the dhud working in association with a low-light television sensor, were conducted under a United Kingdom Ministry of Defence contract. According to the British company, they represent the world's first system demonstration to yield comprehensive analytical data on dhud performance in a combat aircraft, and the equipment is already attracting interest among air forces across the globe. The purpose of dhud technology is to increase the field of view over those of conventional head-up displays by perhaps 50 per cent, from 20 to 30 degrees. The Smiths Industries system incorporates a diffractive-optics combiner glass and optical module developed by Hughes, and is claimed to show substantial advantages over other dhud systems currently under development. A cross-transfer technology agreement is part of the working arrangement between the two companies.

Modular electronics and the wide-angle field of view made possible by diffractive-optics geometry permit greater compatibility with a broad range of electro-optical and low-light television sensors. The configuration of the dhud can be adapted to widely different but invariably cramped cockpit areas. Both companies anticipate mid-life upgrading and direct replacement of earlier systems as a major source of business in this new field.

The rationale for this collaborative effort is the complementary nature of the two companies' work in display technology. Hughes has extensive experience in cathode ray tube work by virtue of its large market for surface and airborne radars and has some of the most advanced and extensive optical-research facilities in the world to maintain its eminence in this field. Smiths Industries on the other hand is a major supplier of head-up displays (eg the Harrier/AV-8A/AV-8B family, Jaguar and Tornado) but does not have Hughes' optical research capabilities. Collaboration is seen as the way of exploiting each other's strengths and meeting the needs of US and European markets in politically and commercially acceptable ways. The Smiths/Hughes dhud is one of the very few truly collaborative avionics programmes, in the sense that it embraces not only production agreements, but research, design and development as well.

STATUS: in development. Tornado mid-term refits are a possibility for dhud application and the F-18 and F-14 are also possible candidates. Sweden has made a decision in principle to adopt a dhud system on its new JAS fighter project, based on experience with an experimental and bulky Hughes diffractive optics sight system flown aboard a Saab AJ37 Viggen fighter in the mid-1970s. Britain's P10 fighter project being pursued by British Aerospace at Warton would almost certainly have a dhud.

UNITED STATES OF AMERICA

Astronautics

Astronautics Corporation of America, 907 South First Street, Milwaukee, Wisconsin 53204

E-scope radar-repeater display for Tornado

This display portrays video signals from the Texas Instruments terrain-following and attack radar in the Mark 1, interdiction/strike version of the trinational Panavia Tornado variable-geometry bomber. In the terrain-following mode it shows a hyperbolic-shaped symbol representing the ground ahead, above which the aircraft symbol remains if the autopilot is functioning correctly. Topographical information can be shown in the ground-mapping mode. For check purposes, an operator-initiated test pattern can also be brought up on the screen.

STATUS: in production and service.

E-scope display for Tornado by Astronautics · Model 131A head-up display by Astronautics

Digital ECM display

This advanced display, originally designed for the US Navy/Grumman EA-6B electronic-warfare deck-landing aircraft, has such versatility that it can be used for a wide variety of airborne functions. It can, for example, accomplish computations and display for navigation and weapon-delivery. The unit contains a signal-data converter with a 16 K word digital computer.

STATUS: in production and service. In addition to the EA-6B, the system was also selected for its US Air Force land-based equivalent, the General Dynamics EF-111.

Electronic moving-map display

By early 1982 this system was under development for the General Dynamics F-16 fighter and the Fairchild A-10 anti-tank aircraft. It generates on a cathode ray tube a moving topographical map from stored film, on which can be superimposed appropriate navigation, target, or other data. This information can be presented in stroke or alphanumeric form. The system can also display information from the joint tactical information distribution system, when that facility becomes operational.

The optical system contains an electronically controlled zoom facility so that the image scale can be changed, the film-movement speed being adjusted to compensate. The display can be shown in film-negative form, making it more suitable for night operations. The system uses standard 150 mm film, or 35 mm format for scales of 1/50 000 or larger. A MIL-STD-1553 digital data-bus interface is provided for communication with other equipment, eg navigation computer.

STATUS: in development.

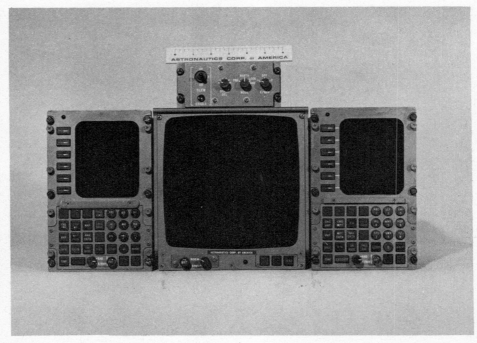

Astronautics' area-navigation electronic chart for L-1011, flanked by two control/display units

Area-navigation control/display for L-1011 TriStar

This system, installed on forward sections of Lockheed L-1011 throttle boxes so that it can be viewed and operated by both pilots, comprises a centrally-mounted electronic automatic chart, with identical control/display units situated alongside, and a control unit and an electronic unit for the chart display. The control/display units present pages of alphanumeric information as required, while the electronic chart shows pictorially the route being flown, with positions marked of waypoints, destination airfields and navigation beacons.

The control unit has a scale selector that can shrink or expand the navigation situation shown on the chart, a slew control that can move the display bodily about the screen or rotate it (the alphanumeric symbols, however, remaining upright), mode selector permitting north-up, track-up, or look-ahead display, and an rnav switch that can select data from one or other of the navigation computers on the aircraft. The chart can store 1000 symbols and display 500 of them at any one time, and can also act as a back-up to the control/display units, presenting commanded alphanumerics as required.

The control display units can display 12 lines of information with 17 characters per line, and can access and display data stored in an aircraft computer, such as route details, flight plans, en-route nav-aid frequencies, and terminal area procedures, eg standard departures and approaches.

The system incorporates continuous failure monitoring and self-test, and the computing section furnishes a self-test pattern for complete 'end-to-end' checks.

Dimensions: (control unit) $5^4/_5 \times 2^7/_{10} \times 6^3/_{10}$ inches (147 × 69 × 160 mm)
(chart) $8^1/_2 \times 9^1/_{10} \times 15^1/_{10}$ inches (216 × 231 × 384 mm)
(electronic unit) $4^9/_{10} \times 7^7/_{10} \times 14^1/_2$ inches (124 × 196 × 368 mm)
(cdu) $5^4/_5 \times 9 \times 12$ inches (147 × 229 × 305 mm)
Weight: (control unit) $1^1/_2$ lb (0·68 kg)
(chart) 27 lb (12·25 kg)
(electronic unit) 14 lb (6.35 kg)
(cdu) 18 lb (8·17 kg)
Power: (chart) 200 VA
(electronic unit) 90 VA
(cdu) 90 VA

STATUS: in production and service.

Model 131A head-up display
This device was produced for the US Army/Bell AH-1G light attack helicopter. It is located in the forward (gunner's) compartment.

Exit aperture: 3 inches
Total field of view: 20°
Instantaneous field of view: 8·5° monocular (15·5° binocular) at $18^1/_3$ inches from combiner glass
Standby reticle: fixed, red colour
Computer: microprocessor, 500 nsec/instruction, full arithmetic capability
Memory: 4096 – 16 bit words semiconductor ROM, 256 – 12 bit words semiconductor RAM

STATUS: in service.

Austin
Polhemus Navigation Sciences, The Austin Company, 3650 Mayfield Road, Cleveland, Ohio 44121

Advanced helmet-mounted sight
In the spring of 1979 a contract for the development of an advanced helmet-mounted sight system was awarded jointly to Polhemus Navigation Sciences (an Austin subsidiary) and Honeywell by the US Air Force. When used in tactical aircraft, the device measures the pilot's line-of-sight to the target in relation to the fore-and-aft axis of the airframe. These sight angles are then processed into a form suitable for the direct control of weapon delivery equipment and other sensors on the aircraft. The system thus couples the wide field of view of the human eye with the high resolution but generally small fields of view of aircraft sensors. It thus closes the man-in-the-machine loop by integrating the pilot's natural versatility in visual and motor skills with the specialised accuracies of weapon-aiming and delivery equipment.

At the heart of the system is a small magnetic-field sensor mounted inside the standard-issue helmet. This senses in three axes the controlled magnetic field produced by a transmitter mounted inside the cockpit area. The voltages induced in the receiver are amplified and processed by a sight electronics assembly to determine the helmet sight angles. To ensure coincidence between the helmet and the pilot's line-of-sight, a small optical generator projects a virtual-image aiming reticle on the helmet's visor, and the pilot adjusts the position of his head so that the reticle aligns with the target. The reticle also contains cueing information that allows the pilot to be directed from external data.

Austin's advanced helmet-mounted sight system for two-seat aircraft, with magnetic-field transmitter, sensors attached to helmets and control units

Dimensions: (control panel) $5^3/_4 \times 3 \times 3^3/_4$ inches (146 × 76 × 95 mm)
(sensor) 1 × 1 inches (25 × 25 mm)
(transmitter) $^1/_2 \times ^1/_2 \times ^5/_8$ inches (12·7 × 12·7 × 15·9 mm)
(sight electronics unit) $6^3/_{10} \times 8^2/_5 \times 10^2/_5$ inches (160 × 213 × 264 mm)

Weight: (control panel) 1 lb (0·45 kg)
(sensor) 1 oz (0·03 kg)
(sight electronics unit) 24 lb (10·9 kg)
Accuracy: forward cone 0·1° or 2 mrad, max error 0·5° or 10 mrad

STATUS: in development for the US Air Force.

Collins
Collins Avionics Divisions, Rockwell International, 400 Collins Road NE, Cedar Rapids, Iowa 52406

EFIS-700 electronic flight instrument system for Boeing 767 and 757
Aviation history was made when, in December 1978, Boeing placed with Collins the world's first production orders for primary flight instruments based on cathode ray tubes rather than electromechanical indicators. They were designed initially for the new-generation Boeing 767 and 757 transports, which made their first flights respectively in October 1981 and February 1982.

The EFIS-700 electronic flight instrument system comprises an electronic attitude director indicator and electronic horizontal situation indicator for each of the two pilots in these transports, and each pair of instruments has an associated mode-control panel. They provide all the functions associated with earlier electro-mechanical devices, and in addition show map and flight-plan data, weather patterns, radio height, automatic flight control modes and flight-path information.

These instruments utilise bright, three-gun, rugged shadow-mask cathode ray tube technology permitting no fewer than nine colours: the traditional red, amber, blue and white associated with cathode ray tubes is augmented by magenta, yellow, green, cyan and chartreuse. When used in conjunction with contrast enhancement filters, the high resolution cathode-ray tubes provide bright displays

The Collins EFDS-85 combines the electronic flight director at top and the electronic horizontal situation indicator (below) into a single instrument

that are readable under all flight-deck lighting environments.

The three-gun system permits redundancy so that, in the event of one or even two guns failing, the system reverts to monochrome with no loss of information.

Each of the two pairs of EFIS instruments in an aircraft is driven by its own symbol generator, but a third, standby, generator is retained as a 'hot-spare' that can be switched in as necessary on failure of a dedicated unit. These symbol generators utilise the Collins CAPS-6 sixth-generation data-processor, implemented with the latest medium-scale integration and bipolar bit-slice LSI 2900 devices.

The electronic attitude director indicator presents primary attitude information, together with pitch and roll steering commands. Secondary data is also shown, eg ground-speed, autopilot and autothrottle mode and many others. In order to keep the display as uncluttered as possible information is switched out as soon as it is not needed; eg instrument landing system and radio height symbols are absent during cruise, appearing only during the final approach.

The electronic horizontal situation indicator depicts the horizontal position of the aircraft in relation to selected flight data and a map of the navigation features in the vicinity of the aircraft at any given time. Aircraft track, trend vector information and desired flight plan indicate any position error. This allows rapid and accurate manual or automatic flight-path correction. Other information can also be displayed, such as wind-speed and direction, lateral and vertical deviation from a selected vertical profile, and time to the next navigation waypoint. Weather patterns can also be superimposed on the navigation picture.

STATUS: following competitive selection of the Collins EFIS in late 1978 for the Boeing 767 and 757, the company was contracted to build 600 sets of equipment for deliveries beginning in mid-1982.

EICAS engine indication and crew alerting system for Boeing 767 and 757

Collins' EICAS engine indication and crew alerting system was chosen by Boeing in March 1980 as standard equipment on the new-generation 767 and 757 airliners. The EICAS system, developed by the company's Air Transport Division, comprises two multi-colour cathode ray tube displays, two computers and a single selector panel. The display unit is identical to the electronic horizontal situation indicators on the pilot's display panels, though rotated through 90 degrees in its function as an engine indicating system.

Each aircraft has two EICAS cathode ray tubes mounted one above the other on the centre instrument panel where they can be monitored by the two pilots. For those aircraft operated by three-man crews, a third cathode-ray tube is installed on the flight-engineer's station. On the centre panel the top EICAS cathode ray tube is programmed to display primary engine information (engine pressure ratio, fan speed and exhaust gas temperature) as electronic symbols representing traditional circular scale and pointer instruments, together with cautionary information (eg wheel-well overheat showing dangerously high tyre temperatures, perhaps as a result of brake binding or failure of a yaw damper in the flight-control system). The lower EICAS display shows lower priority information such as compressor speed, fuel flow and oil temperatures, pressures and tank contents.

In the case of failure of one EICAS display, priority information can be switched to the other cathode-ray tube, and the dual-redundant computer installation permits both cathode-ray tubes to be driven from one unit.

The multi-colour cathode-ray tube displays in the EICAS configuration are seven inches wide and six inches high, and both are driven by

The world's first new-generation digital airliner is the Boeing 767. It will have the Collins EFIS-700

Rockwell Collins EFIS-700 electronic display as an attitude director indicator

Rockwell Collins EFIS-700 electronic display as a horizontal situation indicator, with expanded forward sector

one of the two computers, the other acting as a 'hot' spare.

STATUS: the first engineering model for the 757 was delivered to Boeing in mid-1980 and the company has contracts for 300 sets of 757 equipment for deliveries beginning in early 1983.

EFDS-85 electronic flight director system for general aviation

As a result of its work on television-type flight-instrument displays for new-generation commercial transport aircraft such as the Boeing 767 and 757, Collins has developed equivalent systems designed for business, corporate and commuter aircraft. A prototype system was shown for the first time at the National Business Aircraft Association's annual meeting in September 1980, and two months later at the Commuter Airline Association of America simulated instruments were exhibited on the Embraer 120, Saab-Fairchild 340 and Shot 360 stands.

A full EFDS-85 system comprises dual electronic attitude director indicators and electronic horizontal situation indicators (one set of equipment for each pilot), a multi-function display on the centre panel to be shared by the pilots, and mode controls. The electronic attitude director indicator and electronic horizontal situation indicator cathode ray tubes can display all the information traditionally associated with electromechanical flight-director instruments, and in addition weather-radar patterns, navigation maps, crew checklists, performance data and navigation waypoints. Presentation is very flexible; for example the EHSI-85 can show conventional horizontal situation indicator information on a circular

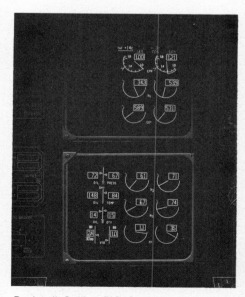

Rockwell Collins EICAS engine instruments display for the new twin-engined Boeing 767 and 757 airliners

scale or it can expand just the forward sector of the display and show weather maps.

The EFDS-85 cathode-ray tube incorporates a three-gun assembly, a shadow-mask, a faceplate with phosphor coating and a glass envelope to enclose the elements. The in-line electron gun assembly provides improved convergence and mechanical rigidity and the high-resolution shadow mask gives four to six times better resolution than that of a domestic television set because the phosphor dots are so much closer together. The displays use both stroke and raster writing; the high-intensity stroke writing of symbols, in conjunction with contrast-enhancement filters, enables displays to be read even in full sunlight. Primary colours are red, blue and green with easy synthesis of several derivative colours, including white.

The company is also developing a version of the EFDS-85 for medium to heavy commercial helicopters and, again, a prototype was shown at the 1980 NBAA Convention. By late 1981 the presentation and equipment configuration had still to be finalised, but the most likely scheme for the attitude director indicator involved symbology appropriate to collective pitch command, flight-guidance mode annunciation, an instrument landing system gate, a perspective runway and digital airspeed. The horizontal situation indicator would have superimposed navigation and weather-radar situations and an interface with the Collins FMS-90 flight-management system or the LRN-85 long-range navigation system, enabling navigation waypoints, distance, time-to-go, course lines between waypoints and wind vectors to be shown.

A Collins multi-function cathode ray tube display was being considered for the display of tabular navigation data from the FMS-90 or the LRN-85 in addition to the use of the EFIS system to show some of this information.

STATUS: in production with deliveries due to begin about the same time as those of the EFIS-700 electronic flight instrument systems for the Boeing 767 and 757, ie in mid-1982. One of the first applications is the Saab-Fairchild 340 34-seat commuter airliner which will have a pair of EFIS instruments for each pilot driven by two display generators. By early 1982 it was being claimed that the life-cycle costs of these EFIS instruments were already better than those of mechanical flight directors. A Collins electronic horizontal situation indicator is also standard on the US Coast Guard SRR short-range recovery helicopters.

CAI-701 electronic caution annunciator indicator

The CAI-701 is yet another Collins cathode ray tube instrument chosen by Boeing as standard equipment for customers preferring electronic to mechanical displays on their 767 and 757 new-generation medium and short-haul airliners. The units use bright-display technology to provide three primary colours (red, green and blue), permitting the traditional combination of red and amber for caution annunciation, and a full colour spectrum, including white, for message display.

Central to the operation of the unit is a high-speed controller that executes a set of instructions to update the cathode ray tube display. The unit accepts data in alpha-numeric form from a high-speed ARINC 429 digital data bus and combines both symbol generator and display into one box. The cathode ray tube is fed from a triple-gun assembly so that in the event of one gun failing the system reverts from colour to monochromatic, though with full retention of information.

STATUS: in production.

General Electric

General Electric Company, Aircraft Equipment Division, Binghampton, New York 13902

AN/ASG-26A lead-computing optical sight

General Electrics has developed from its background of earlier equipment an improved lead-computing optical sight that enables air/air combat to be conducted without the need for continuous target tracking. This capability is incorporated in the system for the US Air Force/McDonnell Douglas F-4E fighter. The equipment comprises a head-up display, two-axis lead-computing gyroscope, gyro mount and lead-computing amplifier. For airborne targets the system displays gun and missile fire-control information by means of a servoed aiming mark. Against ground targets the pilot adjusts the aiming mark manually to control gunnery, rocket and bombing displays.

The aircraft's own manoeuvres generate rate and acceleration signals in the gyro lead-computer, range-to-target is measured by radar, and angle of attack, air density and airspeed (needed for trajectory correction) are supplied by the air-data computer. With these parameters fed into the system, the aiming reference is displaced so as to produce the appropriate lead angle and gravity corrections. Analogues of roll angle and range are also projected on to the combining glass. In ground attack modes other sensors generate corrections for drift and offset bombing is also possible.

STATUS: in development. General Electric has been, or is, engaged on lead-computing optical sight programmes for the F-101, F-104, F-105, F-111, and F-5 combat aircraft, and for the F-5E (export version of Northrop's top-selling F-5 lightweight fighter) the company provides the AN/ASG-29 lead-computing sight (see below).

AN/ASG-29 lead-computing optical sight

This sight has been specially produced for the US Air Force/Northrop F-5E fighter chosen in 1970 as a winner of the IFA (International Fighter Aircraft) competition to equip the air forces of several friendly south-east Asian countries such as South Korea and Taiwan. The AN/ASG-29 has a family relationship with the AN/AGS-26A but comprises only two units, a pilot's display and lead computer. The system

McDonnell Douglas F-4E version of the Phantom is one application of General Electric's AN/ASG-26A optical sight

provides guidance for air/air and air/ground weapons delivery.

Total field of view: (azimuth) 12°
(elevation) 14°
Instantaneous field of view: 7·5°
Collimating lens aperture: 4 inches (102 mm)
Weight: (sight head) 6·9 kg
(lead computer) 7·9 kg
(mounting base) 0·8 kg
Reliability: mtbf claimed to be more than 300 h

STATUS: in production and service.

Head-up display for Northrop F-5G Tigershark

In November 1981 General Electric announced that it had been contracted by Northrop to build a new head-up display for the private venture F-5G Tigershark tactical air-defence fighter being developed by the West Coast company.

The system builds on the 1500+ lead-computing optical sight units that General Electric has supplied for the many earlier F-5Es and F-5Fs that serve around the world, and particularly on a demonstrator head-up display that the company has been showing to potential customers.

The new head-up display is the Tigershark's primary flight instrument, permitting the pilot to keep before him all important flight and mission information without having to glance down into the cockpit. It forms part of the integrated digital avionics package that has been specially developed for the F-5G, and which will help make it an attractive alternative to the General Dynamics F-16 fighter for countries that cannot afford the latter or are barred politically from acquiring it.

STATUS: in development. First flight of the Tigershark was scheduled for September 1982 and fully qualified, combat-capable aircraft will be available by July 1983.

Kaiser Electronics

Kaiser Electronics, Kaiser Aerospace & Electronics, 2701 Orchard Parkway, San Jose, California 95134

AN/OD-150(V) integrated cockpit display for F/A-18 Hornet

In late 1976 Kaiser was awarded contracts to develop and produce both the head-up display and the multi-purpose display group for the US Navy/McDonnell Douglas F/A-18 Hornet single-seat carrier-borne air-superiority and ground-attack fighter. The three head-down displays operate in an all-stroke, in-retrace, linear raster, weapon, air/air and ground-map radar (with synthetic aperture and Doppler beam-sharpening capabilities), horizontal situation and tabular alpha-numeric modes. Contact push-buttons are situated around the edges of the displays for mode selection and data call-up.

STATUS: in production for the US Navy/US Marines F/A-18 Hornet and due to enter service during 1982. This programme for 1366 aircraft is, after the US Air Force/General Dynamics F-16, the Western world's numerically largest aircraft production effort, worth about $40 000 million.

Multi-purpose display for F-16 AFTI advanced fighter technology integrator

In conjunction with prime contractor General Dynamics, Kaiser has developed a multi-purpose cathode ray tube display group for the advanced fighter technology integrator (afti) research aircraft based on an extensively modified F-16 fighter. The purpose of the afti programme is to bring together in a suitable flying testbed as many as possible of the technologies that have been developed since the designs of the current high-performance fighters (F-15 Eagle, F-16 Fighting Falcon and F/A-18 Hornet) were frozen in the late 1960s and early 1970s. Just as the General Dynamics YF-16 and Northrop YF-17 lightweight fighter technology demonstrators of around 1972 brought together what was then possible in airframe, propulsion and avionics (and went on to become, with consequently low technical risk, the F-16 and F-18 fighters), so afti will demonstrate what is currently achievable and the risk levels involved.

The display group consists of two programmable display generators and two multi-purpose interactive control/display units. The design is based on the radar/electro-optical display in production for the F-16 fighter and the electronic attitude director indicator for the Hughes AH-64 armed attack helicopter. It incorporates, according to Kaiser, new but unspecified technology to provide a flexible, high performance raster display specially for the afti programme.

STATUS: research. Due to fly in the afti demonstrator in mid-1982.

AN/AVA-12 attitude display for F-14 Tomcat

The AN/AVA-12 is the primary attitude and director display for the US Navy/Grumman F-14 Tomcat swing-wing carrier-borne air-superiority fighter. An unusual feature is that the vertical situation display cathode ray tube, though electrically independent of the head-up display, is mechanically integrated with the latter in a common box. The vertical display, or attitude director, uses a high-brightness cathode ray tube to produce 525-line raster television and stroke writing. Another unusual feature is that there is no separate combiner glass for the head-up display, the inside face of the windscreen acting as the combiner element.

STATUS: in production and service.

Kaiser Electronics integrated hud/hdd display for the F/A-18 Hornet fighter. The hud has a dual combiner-glass

Kaiser Electronics multipurpose cockpit display for the F-16 afti demonstrator

Kaiser Electronics head-down radar display for the F-16 fighter

Head-up display/weapon-aiming system for Alpha Jet

Kaiser supplies the head-up display/weapon-aiming system for the West German version of the Franco-German Alpha Jet light strike/trainer, built in collaboration by Dassault and Dornier. The three-unit system comprises a computer/symbol generator, sweep driver unit and pilot's display unit. The core element is the symbol generator, which processes information from sensors such as the air-data computer and vertical gyro to produce symbols for the head-up display. Spare storage capacity in the computer permits a range of optional functions including comprehensive navigational computations.

STATUS: in production and service.

Electronic attitude director for YAH-64

The electronic attitude director for the US Army/Hughes Helicopter YAH-64 attack helicopter is a high-performance 875-line television raster system for use in high ambient light conditions and provides the pilot with attitude commands and other information. The two-unit system, comprising symbol generator and display unit, can overlay symbology on to an externally generated 875-line video. The symbol generator contains digital microprocessor circuitry for accuracy and flexibility and for compatibility with the MIL-STD-1553 data-bus in this aircraft.

STATUS: development completed and 30 sets delivered for the YAH-64 trials programme.

Night/adverse weather head-up display for A-10

The head-up display system for the two-seat version of the US Air Force/Fairchild Republic A-10 battlefield bomber and tank-buster incorporates folded-optics geometry in the pilot's display unit to bring the image closer to the pilot and so provide a wider field of view. The system provides low-altitude terrain-avoidance commands, target acquisition signals and weapon-delivery alerts as stroke symbology superimposed on outside-world forward-looking infra-red raster imagery for low-altitude attack in poor weather or at night. The Kaiser head-up display comprises a computer/symbol generator, raster adapter/terrain processor, sweep drive unit, an option selector, and pilot's display.

Recognition of the need for an all-weather version of the A-10 led to the construction of a specially equipped two-seater and its first flight in May 1979. This single evaluation A-10 successfully completed Fairchild and US Air Force trials.

STATUS: evaluation.

Inertial navigation head-up display for A-10

This is an updated version of the Kaiser head-up display for the US Air Force/Fairchild Republic A-10 tank destroyer, incorporating inertial navigation data from the A-10's INS system. The head-up display contains a MIL-STD-1553 multiplex data-bus system with the ability to compute total velocity vectors and other algorithms.

STATUS: in service.

AN/AVA-1 electronic attitude display for A-6

The AN/AVA-1 attitude indicator is of historical significance in being the very first aircrew display to use a cathode ray tube for flight information (cathode ray tubes had of course been in use since 1940/41 to show airborne radar data). As the pilot's primary flight instrument on the US Navy/Grumman A-6 carrier-borne low-level all-weather conventional/nuclear bomber, it provides pitch and roll attitude, navigation and weapon delivery information and terrain-clearance commands through electronically generated symbology on a high-brightness cathode ray tube.

The significance of this equipment lay in the implications for airworthiness, failure of weapon-delivery cathode ray tube and consequent loss of target acquisition symbols may be embarrassing during a bombing mission, but loss of attitude information could be potentially disastrous and so a much higher order of integrity is required.

STATUS: in service.

AN/AVA-12 vertical situation display by Kaiser Electronics for the F-14 Tomcat fighter

Kaiser Electronics' attitude director for the Hughes YAH-64 attack helicopter

Head-up display for AH-1S Cobra

The head-up display for the US Army/Bell Helicopter AH-1S Cobra attack helicopter is a lightweight, low-profile conventional cathode ray tube that superimposes aiming information for the multi-barrelled gun and Hughes TOW anti-tank missile onto the pilot's forward field of view. The programming flexibility and spare capacity of the microprocessor-controlled symbol-generator permits other functions to be incorporated, such as the derivation of flight commands for nap-of-earth flying and laser tracking and pointing information.

STATUS: in service and production.

Pilot's display unit for the Kaiser Electronics hud in the Cobra AH-1S light attack helicopter

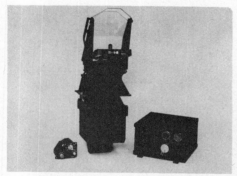

The Kaiser Electronics hud for the Fairchild A-10 tank-destroyer

Symbol generator, combined hud/control unit, and sweep driver for the Alpha Jet strike-trainer, by Kaiser Electronics

Kaiser Electronics attitude display for the Grumman A-6 sub-hunter

Loral

Loral Electronic Systems, 999 Central Park, Yonkers, New York 10704

AN/ASA-82 anti-submarine warfare display

This tactical information system is the primary data display for the US Navy Lockheed S-3A Viking carrier-borne anti-submarine aircraft, being the link between the four-man crew and the array of electronic sensors. It presents high-speed, high-density data in the form of alphanumerics, symbols, vectors, conics and other appropriate formats from acoustic and non-acoustic submarine detection devices such as sonobuoys.

The system comprises a general-purpose digital computer and display generator driving five display units. The tactical co-ordinator (who directs the anti-submarine warfare mission in the S-3A) and the sensor operator are provided with identical displays showing forward-looking infra-red raw radar, scan-converted radar and tactical data, magnetic anomaly detector and acoustic data. Taken together, this information enables the tactical co-ordinator to anticipate the impending situation and accordingly make decisions. The sensor operator has in addition an auxiliary readout unit that presents a continuous display of the acoustic environment, the monitoring of which is his principal function. The pilot's display cathode ray tube shows overall command and control information that permits him to monitor the entire tactical situation by way of, for example, sonobuoy position, time-to-waypoints, aircraft position and track, navigation and predicted target positions. The lower part of the display area presents various cues and alerts and information describing the appropriate sequences of action. The co-pilot's display repeats the tactical situation presented on the pilot's cathode ray tube, but in addition can show outputs from non-acoustic sensors. The pilot's and co-pilot's cathode-ray tubes are readable in an ambient brightness of 8600 foot-candles, and the symbology is shown in eight shades of grey on all cathode ray tubes except that of the pilot which has two shades. The total weight of the AN/ASA-82 system is 268 lb (121·8 kg).

STATUS: in service with Lockheed S-3A Vikings of the US Navy and continuing in production for Lockheed CP-140 Aurora version of the P-3 Orion for the Canadian Armed Forces.

Integrated tactical display

A multi-purpose television and random-write alpha-numeric, graphic and video display for military applications, this system is essentially a colour version of the AN/ASA-82 described above. It is suitable for 'stand alone' operation by virtue of a built-in TI-9900 microprocessor, display generator and 8 K × 16 RAM/ROM refresh memory. It presents on a 16-inch (406 mm) screen, information from electro-optical, radar and acoustic sensors in different formats, each of which can be overlaid with tabular, graphic situation, TV raster and raw radar symbology.

Loral says that the system shows significant improvements in performance, reliability, maintainability, and potential for growth over current displays. Local data storage, an internal refresh function and the ability to program through the microprocessor all help to off-load the host computer, increase the quantity of data

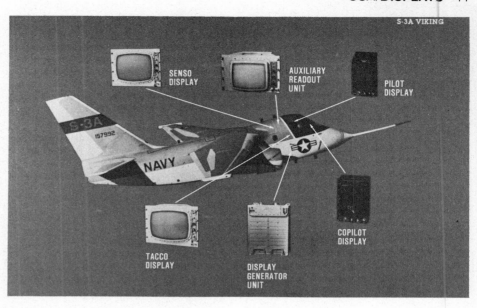

Seven principal units of AN/ASA-82 tactical display in US Navy /Lockheed S-3A Viking sub-hunter

that can be displayed and allow a more flexible operation of the equipment. The use of colour reduces clutter in high-density displays and increases the operator comprehension.

Active display size: 9 × 12·7 inches (229 × 323 mm)
Spot size: 15 mils
Addressability: 1024 × 1024
Random write speed: 100 000 inches/s
Linearity: 1%
Dimensions: (display) 18 × 14·5 × 22 inches (457 × 368 × 559 mm)
Weight: (display) 73 lb

STATUS: in limited production for some aircraft types with classified electronic warfare functions.

AN/APR-38 control indicator set for Wild Weasel F-4G

As increasingly sophisticated methods were used to bring down American aircraft operating over Viet-Nam in the late 1960s, the US Air Force devised counter-methods; one in particular was to send aircraft specially equipped to seek out and destroy North Vietnamese ground radars used for anti-aircraft gun and missile acquisition and guidance. This mission was code-named Wild Weasel and was flown principally by US Air Force/McDonnell Douglas F-4G Phantom two-seat fighter-bombers.

Loral produces the control indicator set for the AN/APR-38 reconnaissance tactical system for the F-4G, consisting of four displays, a digital display processor, and associated controls. The system analyses and presents data from radar emitters in four types of display: plan position, panoramic frequency analyser, real-time pulse video and homing (elevation and azimuth).

The plan position control indicator selects emitters by type, or combination of types, and displays their ranges and bearings from the aircraft. It then indicates the range and bearing of the radar posing the greatest threat and superimposes on this the destruction footprint of the Standard anti-radiation missile so that the crew can judge whether or not they are within launch range. The panoramic analysis

Loral's tactical integrated display

and homing unit contains two displays, one of amplitude versus frequency, the other showing emitter and missile situation relative to the aircraft or, alternatively, real-time signal amplitude versus time. It also controls access to the missile for reprogramming.

Computer: serial, 400 kHz bit rate, 400 word/frame, 16 bit word-length, 50 Hz refresh rate
Ppi display: range versus bearing marks at 10, 25, 50, 100, and 200 miles (16, 40, 80, 161, 322 km)
System weight: 100 lb (45·45 kg)
Reliability: 177 h mtbf

STATUS: in production and service.

Martin Marietta

Martin Marietta, Orlando Aerospace, PO Box 5837, Orlando, Florida 32855

Mast-mounted sight

This is basically a magnifying periscope to assist helicopter crews in target detection and acquisition. The title is derived from the configuration of the equipment, in which the optical head, with its objective lens and mirror system, is mounted above the rotor-drive mast and is concentric with it. The periscope tube and optical train pass directly through the hollow rotor mast, emerging at the gunner's position in an eyepiece or interface for other sensors such as forward looking infra-red and low-light television.

By contrast with the other type of helicopter sight, in which the optical head is mounted in front of the rotor mast, the mast sight has two advantages: it can scan an uninterrupted 360 degree field of view (the other type has a blind region to the rear caused by the obstruction of the mast), and with the sensing head much higher in relation to the helicopter, can see targets with less exposure of the airframe, thereby increasing survivability.

A unique 'soft mount' gimbal system, based on a well-proven centre-post design, provides good isolation from vibration in an area notorious for high-level out-of-balance forces, and is the outcome of work on a previous, daylight-only sight that underwent a 1000 hour test and evaluation programme.

The system integrates weapon guidance equipment for the Hot and Hellfire anti-tank missiles.

STATUS: in production for Italy's Agusta A129 multi-mission combat helicopter.

Sperry

Sperry Flight Systems, Defense Systems Division, PO Box 29222, Phoenix, Arizona 85038

Multi-function display for F-16

In June 1981 Sperry announced that it had won a $4 million contract to develop an advanced multi-function display system for the US Air Force/General Dynamics F-16 Fighting Falcon air-superiority fighter. The system will form a part of the F-16 multi-staged improvement programme that is already under way to introduce as soon as possible new-technology equipment perfected since the F-16 design was frozen in the mid-1970s.

The system comprises two four-inch (102 mm) square cathode ray tube displays together with a digital, programmable display generator for each single-seat aircraft (the two-seat F-16Bs that make up about 15 per cent of the total Fighting Falcon fleet will have four displays, but still feed from the one signal generator). The displays present the pilot with navigation, radar and weapon-aiming information. They use raster-scan techniques to provide clear and sharp presentations of sensor information, symbol overlays and alpha-numerics in all ambient light conditions. Each display has 24 push-button keys arranged around the bezel for interactive control and data entry.

The programmable display generator uses advanced task-sharing techniques for high-speed operation and symbol-generation. It is a software-based device with two processors and 32 K words of core memory, permitting rapid changes of mode and the use of several sets of symbology. Data from the F-16's fire-control computer is fed to the system via the MIL-STD-1553 digital data bus while video information comes from its electro-optical sensors.

The system's digital design permits rapid reprogramming to meet changing needs and this flexibility includes the ability to integrate future avionics, including the Lantirn (low-altitude navigation and targeting infra-red for night) and joint tactical information and distribution systems. The display system incorporates comprehensive test and diagnostic routines and there are growth provisions for the generation of colour symbols and for the MIL-STD-1750A instruction set architecture.

Display colour: phosphor green P43
Symbol contrast ratio: 7:1 at 10 000 ft-lamberts (34 260 cd/m²)
Brightness: (display screen) 3000 ft-lamberts (10 278 cd/m²)
(filtered) 300 ft-lamberts (1028 cd/m²)
Line width: 0·0075 inch (0·19 mm)
Linearity: 2%
Usable screen area: 4 × 4 inches (102 × 102 mm)
Jitter: (symbology) < 2·5 mils
(raster video) 3 mils
Video inputs: 9 total
(electro-optical weapons) 525 line and 875 line
Video oututs: 6 total, 2 independent channels, 3 outputs/channel
Dimensions: (display) 5·62 × 5·62 × 12·11 inches (143 × 143 × 308 mm)
(display generator) 6·28 × 7·63 × 17·64 inches (159 × 194 × 448 mm)
Weight: (display) 13 lb (5·9 kg)
(generator) 28 lb (12·7 kg)

Sperry's multi-function display for the F-16 fighter

STATUS: initial production. The $4 million contract awarded by General Dynamics called for the delivery of eight pre-production sets by early 1982 and agreed options for the manufacture of up to 2250 sets beginning in autumn 1983 and continuing till 1989.

Multi-purpose colour display

The Sperry full-colour, high-resolution airborne display system has been designed for a wide variety of applications. Proposed for the joint tactical information and distribution system on the US Air Force/McDonnell Douglas F-15 air-superiority fighter, this colour display incorporates advanced display techniques to produce high-contrast presentations of flight, engine, navigation and weapon-aiming information.

A hybrid unit that can produce both raster and stroke-driven displays, the system accepts inputs from two separate symbol generators so that it can show map, vector, or other navigation information, together with an alpha-numeric overlay. Up to seven different colours in two intensities can be displayed at one time, with an automatic contrast control that adjusts brightness so that the display is readable even in direct sunlight, equivalent to 10 000 foot-candles.

The display control unit uses advanced microprocessor architecture to perform sensitive computations. Near-perfect colour tracking throughout the brightness range is obtained by processing brightness and projection data, and other process-functions include focus setting, deflection gain setting, automatic calibration, display orbiting and built-in test. The system can be used as an electronic flight instrument display showing basic attitude information and flight-director commands in traditional format, with mode annunciation, airspeed and altitude; navigation display, with weather maps overlaid on moving-map and navigation information; engine display, with engine parameters shown on the cathode ray tube as traditional circular-scale instruments; warning/status displays to show systems' or other faults or problems needing crew attention; and as a systems display showing schem-

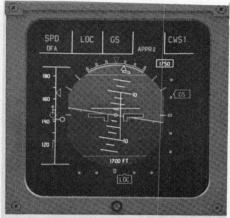

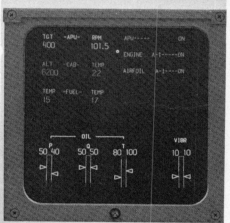

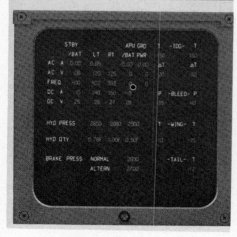

Three possible presentations on the Sperry multi-purpose colour display: top, as a flight director, and middle and lower, as systems status indicators

atic line diagrams and checklist information on demand.

Writing speed: (stroke) 25 000 inches/s (635 m/s)
(raster) 125 000 inches/s (3175 m/s)
Accuracy: 1% diagonal, or 2 mm
Jitter: 0·007 inch (0·18 mm)
Phosphor: P-22
Refresh rate: (raster) 30 and 60 Hz
(stroke) 60 Hz

Display size: 5 × 5 inches (127 × 127 mm)
Display unit dimensions: 6·7 × 6·25 × 14 inches
(170 × 159 × 355 mm)
Weight: 25 lb (11·36 kg)

STATUS: development completed.

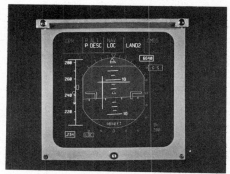

Sperry EFIS instrument for general aviation. Basic presentation is the flight director, with flight modes arrayed alongside

EDZ-800 electronic flight instrument system for general aviation

Television-screen flight instrument and systems displays, initially developed for the new-generation, all-digital airliners such as the Boeing 767 and 757 and the Airbus Industrie A310 have begun to make their appearance at the upper end of the general aviation industry rather sooner than many observers had expected. Sperry's effort in the general-aviation electronic flight instrument system field is represented by the EDZ-800 family of attitude director indicators and horizontal situation indicators replacing the conventional electromechanical attitude and horizontal situation indicators, and available with five and six inch (127 and 152 mm) displays. Initially, Sperry has duplicated the symbols and formats of the earlier electromechanical instruments, but, in addition, routes weather radar information to the horizontal situation indicator, and some symbols appropriate to other sensors have also been added. Thus a dedicated weather radar display is no longer needed, producing an· immediate and substantial space-saving in the traditionally overcrowded flight-decks of general-aviation aircraft. But the system also has great capacity for growth and flexibility for changes in presentation to meet the needs of integrated performance or flight-management systems.

Although the two electronic flight instrument system (EFIS) displays on each pilot's flight panel are normally dedicated to specific functions (either horizontal situation or attitude director indication), it will be possible to combine these displays electronically so that a single cathode ray tube will be able to show basic information from both in the event of the failure of one display. The EFIS attitude director indicator display also permits unwanted information to be automatically dropped from view. For example, the glideslope and localiser symbols disappear when the navigation receivers are tuned to stations en route and the radio altimeter readouts vanish above the system's maximum operating altitude. The horizontal situation indicator is also very flexible in use and can operate in three navigation modes. In the first of these, the display can show weather radar returns on an expanded, partial compass card presenting just the forward 90 degree sector. In the second mode, the weather is removed, leaving just a partial compass card with navigation mapping or standard courses and heading information. The third presentation gives a standard, full compass card horizontal situation indicator with all appropriate alpha-numeric navigation data. Separate electronic flight instrument system control units for the two pilots permit the different modes to be selected.

STATUS: production was beginning early in 1982 for the first two applications, the Grumman Gulfstream 3 and Canadair Challenger corporate jets.

ASZ-800 electronic air-data for general aviation display

Sperry has integrated into two electronic displays all the flight-deck air-data readouts appropriate to heavy turboprops and corporate business aircraft. They simplify the crew monitoring tasks, reduce panel space needs and add a number of capabilities not available in conventional electromechanical instruments. These include a programmable reference speed window on the airspeed display and an integrated altitude alert/preselect on the altimeter. The standard vertical-case cathode ray tubes can be programmed for either conventional circular-scale or vertical-tape format presentations.

The circular-scale airspeed indicator incorporates a pointer and counter-drum airspeed display and dedicated Mach number, true airspeed and total or static air temperature. Programming the V-speed window permits the pilot to enter the appropriate reference speeds for each flight based on aircraft performance curves and weather. V_l, V_r, and V_{ref} speeds can be entered via a set/select button and value knob on the instrument bezel. A red V_{mo} arc appears on the periphery of the airspeed scale.

Like the cathode ray tube airspeed indicator, the altimeter incorporates several related functions to simplify readability. The electronic symbology duplicates the conventional scale and drum height readout with dedicated windows for barometric pressure in millibars or inches of mercury. It also includes a vertical speed readout with a maximum digital reading of 500 feet per minute (152 metres per minute), though the pointer window can read up to 9999 feet per minute (3047 metres per minute).

Apart from these scale and drum formats, two types of vertical scale presentations are also available for each of these two-air-data instruments.

STATUS: in production.

ARINC 700 symbol generator

Sperry's symbol generator, designed to interface with other ARINC 700 equipment in commercial transport aircraft, takes in information from a variety of sensors and computers and provides outputs to colour displays. Converting basic aircraft parameters to a pictorial display requires a significant data-processing capability, and Sperry's answer is a 16-bit processor designed for display applications with a special memory hierarchy scheme. The system incorporates the company's raster/stroke display technology stemming from a Sperry patent more than a decade ago on hybrid symbol generation. Bold symbology, such as sky/ground shading and weather radar returns, is painted using raster-scan techniques. Smooth moving high-resolution graphics such as a navigation display, on the other hand, calls for stroke writing. A special timing scheme permits the raster and stroke symbol generators to drive two displays simultaneously, each with its own writing characteristic. When one display is being stroke-written, the other is in the raster-scan mode. The methods are then switched to generate each composite display at a frequency of 80 times per second so as to eliminate flicker.

Another refinement is symbol priority. It is accepted that certain information needs to remain always in view without interference by other symbology. To ensure this, the company has developed an innovative technique providing several levels of priority. Symbols move smoothly into and out of these priority areas as if from behind mechanical windows, resulting in a smooth, uncluttered presentation.

STATUS: in production.

Sundstrand

Sundstrand Data Control Inc, Overlake Industrial Park, Redmond, Washington 98052

Head-up displays

Sundstrand is now active in providing head-up displays for commercial and some military transport aircraft. The company makes both electromechanical and electronic head-up displays, its philosophy being that where fewer than six discrete symbols are required the former are more appropriate, the latter being more suitable for applications needing greater numbers of independently moving symbols and a larger collimating lens.

Electromechanical head-up display for Boeing 737

A typical electromechanical head-up display is the example supplied to British Airways as an approach and landing aid for its Boeing 737s. This system provides each pilot with flight-path bar and pitch scales, cues for centreline guidance, and speed error, the latter using a 'fast', 'on-speed' or 'slow' symbol with a different colour for each. All inputs are in the form of analogue signals, and analogue com-

Sundstrand electromechanical head-up display for Boeing 737 airliner

puting techniques produce the signals that move the mechanical symbols behind the collimating lens.

STATUS: installation and evaluation programmes have been carried out on 747, 737, 727 and DC-8 aircraft, the HS Trident for the People's Republic of China, and C-130 Hercules and CH-3 helicopters of the US Air Force.

Computer for Sundstrand electronic head-up display

Electronic head-up displays

The electronic head-up display in the form chosen by McDonnell Douglas as a standard option for the DC-9-80 commercial transport, provides a much larger number of modes than its electromechanical counterparts. Lateral guidance through localiser deviation and course error, together with airspeed, are provided during the take-off roll. At rotation it continues to give lateral guidance, now via magnetic heading, and vertical guidance by means of pitch attitude, plus airspeed. During the subsequent climb-out it shows attitude information in pitch, roll and heading together with height, speed and take-off/go-around command. In a missed approach situation it gives pitch and roll attitude, and heading, as well as height, speed and 'go-around' command.

Guidance during approach and landing falls into four modes. Above the decision height the system gives lateral guidance via localiser deviation and course error, flight-path guidance through glideslope deviation, pitch, roll and yaw attitudes, and speed and height. Between decision and flare heights the same information is presented, together with decision height and flare height messages. Between flare height and touch-down the information remains the same, except for the addition of safe roll-out limits. During the roll-out phase lateral guidance is given through localiser deviation and course error.

The display unit fits into an overhead cavity in the roof panel where it is normally stowed and swung out for use. It is fed with signals from a digital processor and controlled by a panel-mounted control unit.

Pilot's display unit
Calibration: calibrated during manufacture and contains automatic boresight feature

Dimensions: 14 × 15 × 9 inches (356 × 381 × 229 mm)
Weight: 20 lb (9·1 kg)
Reliability: estimated 4000 h
Image generation: crt 2 × 2 inches (51 × 51 mm) PI phosphor
Image brightness: 15 000 ft-lambert (51 390 cd/m²) max
Field of view: (horizontal) 30° (vertical) 26°
Accuracy: 3·5 mrad over entire field
Display integrity: fail-safe through auto alignment and self-test. Display illumination interrupted upon fault detection

Processor: stored program digital processor with 20 Hz sampling and computation iteration rate, 50 Hz crt image refresh rate
Inputs: attitude, angle of attack, speed
Self-test: equipment and software completes self-checking sequence with each computation iteration. Display flashes to warn of faults
Weight: 15 lb (6·8 kg)

Control unit
Functions: flight-path angle select, press-to-test, barometric altitude select
Reliability: estimated mtbf 36 700 h
Face size: 5·75 × 1·88 inches (146 × 48 mm)

STATUS: in production.

Warning Systems

FRANCE

TRT
Télécommunications Radioélectriques & Téléphoniques, 5 avenue Réaumur 92350 le Plessis Robinson

APS-500 ground-proximity warning system
The TRT APS-500 ground-proximity warning system complies with ARINC 594 recommendations and is compatible with all ARINC 552/552A radio altimeters. TRT is a well known manufacturer of radio altimeters, and the APS-500 is complementary to the company's own altimeters, so that customers can have an integrated system provided by one manufacturer. The APS-500 system is of fully digital, solid-state design based on a micro-processor and, says TRT, may be considered as a second-generation ground-proximity warning system.

Operational modes are mode 1: excessive rates of descent; mode 2 and 2B: excessive closure rate to terrain; mode 3A and 3B: negative climb rate following take-off and altitude loss under 700 feet (210 metres); mode 4: insufficient terrain clearance on approach with incorrect configuration; mode 5: excessive glideslope deviation.

In all modes the warning takes the form of an imperative aural alarm (the characteristic 'whoop, whoop') and the words 'pull up' repeated, with the exception of mode 5 in which the word 'glideslope' is annunciated.

At the heart of the APS-500 is a single-chip, 16-bit micro-processor with a solid-state memory using random-access memory devices for data storage and reproms for programme store. Complementary metal-oxide silicon integrated circuitry is widely employed throughout. A time-shared operating sequence allows self-monitoring at a rate of 30 times a second, irrespective of mode. The system employs 32-bit words for critical algorithms which require high accuracy; for example, the filtering and

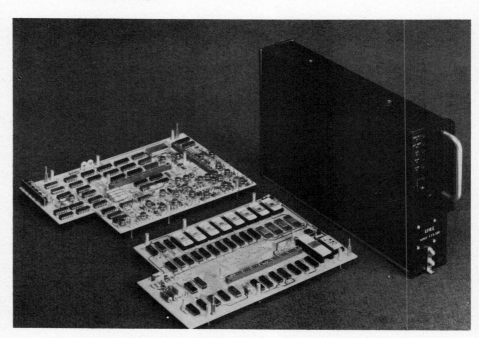

TRT's APS-500 digital ground-proximity warning system with circuit boards

integration routines. This factor, together with the high sampling rate, reduces the probability of spurious alerts. Decision algorithms are adapted to each warning mode requirement.

The APS-500 comprises a wired chassis with plug-in interconnecting printed circuit board, an input-output peripheral plug-in printed circuit board, a digital computer plug-in printed circuit board, and a front panel with solid-state displays, and the power-supply section.

Built in self-test facilities are provided for in-aircraft test and servicing, and automatic test

equipment is available for bench repair and maintenance.

A notable aspect of the APS-500 is that only 50 per cent of the 16-bit memory is at present employed, so there is considerable growth potential to meet any future mode-change requirements.

Dimensions: ¼ ATR Short
Weight: 5·7 lb (2·6 kg)

STATUS: in production and service.

GERMANY, FEDERAL REPUBLIC

Diehl
Diehl GmbH and Co, Heinrich-Diehl-Strasse 2, 8505 Rothenbach

Warning panels for Alpha Jet
A series of warning panels with the type numbers 3189-12, 3189-21, 3189-31, 3190-21, 3191-12 and 3191-31 has been developed by Diehl and is designated for French, Belgian and West German versions of the Franco-German Alpha Jet strike-trainer. The panels provide visual and aural warning of system failures to both forward and rear cockpits and also contain switches for control of battery, generators and inverter.

In the event of the failure of a system which is not critical to aircraft safety, the panel illuminates an amber master warning light which is external to the main unit while simultaneously illuminating a captioned amber light (indicating which system is failed) on the warning unit's front panel. If the failed system is flight-critical, a red master-warning and a red panel caption lamp illuminate and an aural alarm is also sounded. An external switch enables the master warning lamps and the aural warning to be cancelled and the system to be reverted to standby. The captioned lamp however, remains lit until the system fault has been rectified.

The brightness level of all warning lamps is

Diehl warning system panels for Alpha Jet

adjustable to meet both day and night ambient lighting. A test facility permits functional checks of all electronic circuitry and warning lights. The 3190-21 model provides critical failure (red) warning on six channels but has no non-critical (amber) warnings. All other models provide critical failure (red) warning facilities on five channels and non-critical (amber) warnings on 15 channels.

Dimensions: (3189-12, 3189-21, 3189-31) 115 × 118 × 147 mm
(3190-21) 91 × 88 × 124 mm
(3191-12, 3191-31) 115 × 118 × 125 mm

Weight: (3189-12, 3189-21, 3189-31) 1·15 kg
(3190-21) 0·55 kg
(3191-12, 3191-31) 0·85 kg

STATUS: in production and service.

UNITED KINGDOM

Marconi

Marconi Avionics Ltd, Airport Works, Rochester, Kent ME1 2XX

AD2610 ground-proximity warning system

The AD2610 ground-proximity warning system (gpws), which is manufactured under licence from Litton Aero Products of the USA, provides warning of potentially hazardous ground-proximity for modes 1, 2A, 2B, 3, 4 and 5. It provides swept audio tone warning plus the voice command 'pull up' when activated, with the exception of mode 5 when the command 'glideslope' is substituted. Outputs for visual warning lights are also incorporated.

The system is digitally based and micro-processor controlled. Any changes required by regulatory authorities may be accommodated with minimal disruption to aircraft operations.

Two levels of self-test are incorporated; a crew-operated ground check test is used for pre-flight serviceability testing, while during flight, automatic self-monitoring is carried out for confirmation of performance. Analysis of in-flight warnings and isolation of faulty systems, whether in the ground-proximity warning processor or in the associated sensor systems, are facilitated by latching magnetic front-panel indicators.

In flight, the ground-proximity warning system indicator will log a system failure while the remaining indicators log warnings for the five modes. On the ground, the ground-proximity warning system indicator will log a computer failure and the remaining indicators will log which sensor (flaps, gear, radio height, baro rate or glideslope) has failed. The system has a guaranteed mean time between failures of 10 000 hours.

An optional item called the barometric altitude rate computer is available for use with the AD2610 ground-proximity warning system. This system, the BARC (AA26101-1), is manufactured by Harowe Systems of the USA.

Dimensions: (gpws) ¼ ATR Short
(BARC) ¼ ATR Short Dwarf
Weight: (gpws) 6·5 lb (2·95 kg)
(BARC) 3 lb (1·35 kg)

STATUS: in production and service.

Rosemount

Rosemount Engineering Company Ltd, Durban Road, Bognor Regis, West Sussex PO22 9QX

871 series ice-detector

The Rosemount 871 series ice-detector is a Type 1 system, as defined in MIL-D-8181, which provides a signal when ice forms on the sensor element. All forms of ice can be detected over a 360-degree area around the element which is immune to airborne contaminants such as oil, grease or dust and insects. Owing to the element's low temperature recovery factor and high collection efficiency, it normally accretes ice more rapidly than its surrounding surface, and so gives warning in advance of icing build-up.

STATUS: in production and service.

Icing-rate indication system

Rosemount's model 871FFI ice-detector employs an aspirated head to induce an airflow over the sensing probe at a rate which remains constant over a helicopter's operating air-speed range. The indicator instrument used in conjunction with the detector is the 512AE1 icing-rate meter panel. This incorporates a liquid water content indicator with press-to-test warning facility and a failure warning flag.

STATUS: in production and service.

Smiths Industries

Smiths Industries, Aerospace and Defence Systems Company, Cricklewood, London NW2 6JN

Wind-shear indicator

The Smiths Industries combined vertical-speed and wind-shear indicator has been developed with the backing of the United Kingdom Department of Industry and with the co-operation of British Airways. The system is a self-contained electronic instrument which requires no ground-based data inputs and which provides a continuous indication of prevailing wind shear conditions.

Essentially, the instrument is a conventional presentation vertical-speed indicator with an additional pointer driven by a signal derived from either the true air-speed and vertical speed outputs of the aircraft's air-data computer, or from simple barometric rate and air-speed transducers. The additional pointer displays the rate-of-change of the total energy of the aircraft, hence giving an immediate indication of possible wind-shear. Under normal, stable approach conditions, both pointers track together, the vertical speed indicator pointer covering the energy-rate pointer, which can therefore be ignored. If however wind-shear causes a change in indicated air-speed, then the pointers diverge due to kinetic energy changes. This occurs several seconds before the aircraft's performance is seriously affected and provides the pilot with sufficient warning to correct the flightpath by adjustment of thrust and attitude until the pointers again coincide.

If the aircraft is fitted with an autothrottle system, the instrument allows independent monitoring of autothrottle performance. Although recovery from significant wind-shear may require unfamiliar power and attitude changes, Smiths claims that the instrument shows if the power applied is correct to re-establish the appropriate flightpath and speed.

STATUS: in development, at operational proving stage.

Smiths Industries vertical speed/energy-rate indicator presentation

Sperry

Sperry Gyroscope, Downshire Way, Bracknell, Berkshire RG12 1QL

Cable-warning system

The Sperry cable-warning system is designed to alert helicopter pilots to the proximity of power, telephone, or other cables which can be almost invisible from the air. It is suitable for military aircraft flying low-level missions such as nap-of-the-earth operations to avoid detection, or for civil aircraft in crop-spraying and pipe-line inspection.

The system is said to give adequate warning and indication of a potential cable hazard in time for a pilot to take avoiding action. Typically, says Sperry, an aircraft flying at 100 knots towards a cable 500 metres away needs ten seconds warning and the system will provide at least this.

Operation is based on the detection of radiated energy emitted from live power-transmission cables. The system is designed primarily to reduce collisions with high-voltage transmission lines, the cause of most major accidents. The system comprises two units and their associated aircraft wiring. A sensor/processor containing the signal-processing and display circuitry is installed at the most suitable location in the aircraft for detection of cables, commensurate with the lowest degree of engine-generated local signal interference.

The panel-mounted display is augmented by an aural alarm to ensure that attention is drawn to it.

The display itself consists of a circular presentation divided into 12 segments of 30 degrees each. In the centre is an amber warning light which illuminates to indicate that a cable is within detection range. Simultaneously, the aural alarm is sounded. As the aircraft approaches the cable, so segments illuminate, in red, to indicate the direction of the cable. The display shows whether the cable is ahead, obliquely placed or parallel to the aircraft, so indicating the direction in which the crew should look, and what evasive action is required. The system is of particular value at night-time or in poor visibility. Built-in test circuits are provided to permit rapid functional checking. Calculated mean times between failures are 6000 hours for the display and 8500 hours for the sensor.

Dimensions: (display) 1·96 × 1·96 × 2·36 inches (50 × 50 × 60 mm)
(sensor) 5¹⁄₁₀ × 5 × 4¹⁄₂ inches (130 × 150 × 115 mm)
Weight: (display) 1·32 lb (0·6 kg)
(sensor) 5·5 lb (2·5 kg)

STATUS: in production.

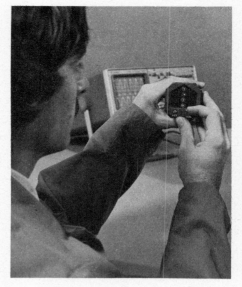

Sperry Gyroscope cable-warning indicator

UNITED STATES OF AMERICA

Bonzer

Bonzer Inc, 90th and Cody, Overland Park, Kansas 66214

Voice-adviser flight-profile information system

The Bonzer voice-adviser flight-profile information system is designed for flight-crew alert purposes to provide voice announcement of radar altitude below 1000 feet (300 metres), passing through levels at 100 foot (30 metre) intervals during descent; descent below decision height; deviation from instrument landing system glideslope and localiser flightpaths; and landing-gear status. The system depends on inputs from radar altimeter, instrument landing system receiver and landing-gear position switches. The system can be installed in any aircraft equipped with instrument landing system receivers, and Bonzer Mark 10, Mark 10X or impatt radar altimeters.

In operation, the system first voices the words 'radar altitude' when the aircraft has penetrated the radar-altimeter range. The next announcement indicates that the aircraft is passing through the 1000 foot (300 metre) level with the words 'radar altitude, 1000'. Thereafter the system advises the flight-crew of their radar altitude in numerical terms alone, e.g. '900 ... 800 ... 700). The voice advisor responds to pre-set decision-height 'bugs' on the altimeter with the verbal warning 'minimum, minimum'. It will continue to respond with the same words if the bug is subsequently reset. On deviation from instrument landing system flightpaths the voice advisor informs the crew with the appropriate wording – either 'glideslope, glideslope' or 'localiser, localiser' and automatically re-sets itself to provide any further necessary warnings when the deviation is corrected. Below 500 feet (150 metres), the system will state the actual radar altitude with the appropriate warning if the landing gear is not down and locked, e.g. '500, check gear'. Warnings are assigned priority in the following order: decision height, radar altitude, glideslope deviation, localiser deviation, landing-gear not extended.

The system has a controllable volume setting and built-in self-test facilities which may be used on the ground or during flight at any altitude. The unit is normally remotely located with the exception of the combined volume control/test button which is panel-mounted. With an impatt altimeter system, the unit can provide outputs for a flight director, autopilot, ground-proximity warning systems, dual indicators, combination barometric/radio altimeter indicator trips, and chart recorders. An electronically-reproduced feminine voice is used to provide verbal warnings. The system is of solid-state construction.

Dimensions: 2 × 4 × 6 inches (51 × 117 × 168 mm)
Weight: 2 lb (0·9 kg)

STATUS: in production and service.

Intercontinental Dynamics

Intercontinental Dynamics Corporation, 170 Coolidge Avenue, P O Box 81, Englewood, New Jersey 07631

Radio-barometric altimeter with voice terrain advisory

Termed a 'talking altimeter', a radio/barometric altimeter with voice terrain advisory (rad/bar) keeps the flight-crew informed of vertical flight progress with particular reference to altitudes within 2500 feet (760 metres) of the ground. The unit combines the functions of standard barometric instruments with those of radio altimeters, but its notable quality is that it provides voice output warning of potential hazards in certain discrete flight regimes. In some cases, the unit may replace conventional ground-proximity warning systems whose functions they tend to cover. Rad/bar instruments typically provide warnings in low-altitude flight whenever the aircraft-ground separation altitude decreases, during en-route flight, and advise of increasing ground-proximity while flying within 2500 feet of rapidly rising terrain.

Intercontinental manufactures a range of rad/bar models with differing options, though the principles of operation are broadly similar. The visual instrument presentation is closely allied to that of a conventional barometric altimeter with drum and pointer scales. A window showing radio altitude is set into the face of the instrument. Models giving either three or four digit radio altitude are available.

Intercontinental Dynamics rad/bar 'talking altimeter' and machmeter

The radio-altitude window digits read altitudes to above 2500 feet, after which the window goes blank.

Following take-off the voice-warning facility remains silent unless there is a negative rate of climb or unless the aircraft traverses rapidly rising ground, in which case, a voice warning 'terrain' would be annunciated. Similarly, during en-route flight, the same warning would be announced if the aircraft descended below 2500 feet or if rapidly rising ground were below.

Before entering the final approach pattern at the destination airfield, the crew can programme the rad/bar for the correct decision height. Using a decision height control on the cockpit instrument, the pilot illuminates the radio altimeter readout digits and by rotating the control knob can set the desired decision height into a memory, as indicated by the digits. During approach, the voice output will announce 'terrain' as the aircraft descends through 1000 feet (300 metres) and continues to announce radio altitude at 100 foot (30 metre) decrements. At decision height, the word 'minimum' is announced.

The system also provides a glideslope deviation warning mode in which any significant departure below the correct path causes the call 'glideslope' to be announced at two-second intervals. If the divergence continues, the call is repeated, at twice the previous volume level and one-second intervals. These 'glideslope' annunciations are interrupted by altitude calls whenever the two coincide. Glideslope calls cease once the aircraft descends through decision height whereupon the 'minimum' call is heard. The minimum (i.e. decision height) call takes priority over all other announcements, but the altitude calls continue as the aircraft continues to descend below decision height.

During all procedures the barometric altitude drum continues to function normally and the radio-altitude digital readout shows increasing or diminishing altitude below 2500 feet (760 metres), in ten-foot (three metre) increments.

Various models of the Intercontinental rad/bar system, with differing options, are available. Some units provide digitised data output for automatic in-flight altitude reporting, altitude-change rate signals for vertical-speed indicators, and barometrically corrected altitude synchros for altitude alerters. Decision height warning lights, which illuminate coincidentally with the call 'minimum', are incorporated in certain units. Barometric settings may be presented in windows showing inches of mercury (Hg), millibars or both. Self-test facilities, for in-flight or ground use, are also incorporated. Intercontinental rad/bar models

can interface with all existing ARINC and many non-ARINC radio altimeter transmitter-receivers. Instruments cover altitude ranges from minus 1000 to plus 50 000 feet and can be lit in either white or blue/white according to configuration selected. The systems comprise panel-mounted instruments in standard ARINC 408 cases and remotely located radio altimeter-converter units which supply radio altitude information to the indicator units. Aural output may be channelled through standard intercommunication systems using cabin speakers or pilot headsets.

Dimensions: (panel unit) 3 ATR case (converter) 3/8 ATR Short
Weight: (total system) 8·3 lb (3·8 kg)

STATUS: in production and service.

Safe Flight

Safe Flight Instrument Corporation, PO Box 550, White Plains, New York 10602

6500/6501 wind-shear warning system

Safe Flight's 6500/6501 wind-shear warning system is a computer-based device which, using conventional sensing elements, resolves the two orthogonal components of a wind gradient with altitude and provides a threshold alert that an aircraft is encountering a potentially hazardous situation. The vectors concerned are termed downdraught drift angle and horizontal wind-shear.

Horizontal wind-shear is derived by subtracting ground-speed acceleration from air-speed rate. The latter term is obtained by passing air-speed analogue data from the air-speed indicator or the air data computer through a high pass filter. Longitudinal acceleration is sensed by a computer integral accelerometer, the output of which has been summed with a pitch attitude reference gyro to correct for the acceleration component due to pitch. A correction circuit is employed to cancel any errors due to prolonged acceleration. This circuit has a 'dead-band', equivalent to 0·2 degree of pitch, which prevents correction for air-speed rates of less than 0·1 knot a second. Summed acceleration and pitch signals are fed through a low-pass filter, the output from which is summed with the air-speed rate signal to give horizontal wind-shear.

The vertical computation (downdraught drift angle) is developed through the comparison of measured normal acceleration with calculated glidepath manoeuvring load. Flightpath angle is derived by subtracting the pitch-attitude signal from an angle of attack signal sensed by the stall-warning flow sensor. This is fed to a high-pass filter, thence to a multiplier to which the air-speed signal has been applied. Thus the flightpath angle rate, corrected for air-speed, provides the computed manoeuvring load term. This is compared in a summing junction with the output of a normal computer-integral

accelerometer and the failure of the two values to match is the indication of acceleration due to downdraught. The acceleration, when integrated, is the vertical wind velocity and is further divided by the air-speed signal to compute the downdraught angle.

The outputs of both horizontal and vertical channels are determined solely by the atmospheric conditions, and ignore manoeuvres that do not increase the total energy of the aircraft. In the horizontal channel, wind-shear correctly compensated by increased engine thrust will show no change in air-speed in the presence of an inertial acceleration as thrust is applied. If, however, the shear goes uncorrected, an acceleration (or deceleration) will become apparent. Similarly, in the case of the vertical component, a vertical displacement compensated by the crew will show a positive flightpath angle rate (in a downdraught) with less than the computed incremental normal acceleration. Correspondingly, in a downdraught for which the downdraught drift angle is allowed to develop, a negative angle rate at a near-constant 1 g will result in the same computation and output. This is of vital importance to the crew as it eliminates the possibility that their actions in anticipating or countering wind-shear might well mask the condition as far as the warning system is concerned.

Both downdraught drift angle and horizontal wind-shear signals are combined and the resulting output fed through a lowpass filter to the system computer. This provides two output signals, a discrete alert, and, through a voice-generator, an audio alert. Warning output is set at a threshold of minus 3 knots a second for hsa and minus 0·15 rad downdraught drift angle or for any combination of the two components which acting together would provide an equivalent signal level. According to Safe Flight, any wind condition requiring additional thrust equivalent to 0·15 g to maintain glidepath and air-speed will result in a non-stabilised approach.

The company has paid particular attention to the elimination of spurious alerts. A cross-over

Safe Flight wind-shear computer

network is employed to sense zero cross-overs of the combined wind-shear warning signal and this is sampled every 25 seconds. If the warning signal does not pass through a band close to zero, the network automatically provides failure indication, alerting the crew to the fact that the unit is inoperative.

Dimensions: 1/4 ATR
Weight: 6 lb (2·72 kg)

STATUS: in production and service.

Sundstrand

Sundstrand Data Control Inc, Overlake Industrial Park, Redmond, Washington 98052

Digital ground-proximity warning system

The Sundstrand digital ground-proximity warning system is designed for service aboard advanced aircraft equipped with ARINC 700 avionic systems. It provides the flight-crew with back-up warning for six potentially dangerous situations. Alerts and warnings are provided by steady or flashing visual indications and by audible warnings. Each audio warning is also annunciated to identify the particular situation. The Sundstrand ground-proximity warning system is unusual in that the annunciated

warnings provide a relatively precise description of the hazard, rather than the 'pull up' and 'glideslope' warning words provided by most systems, and for this reason the full range is quoted: –

Mode 1 –Excessive descent rate
    alert: 'Sinkrate, sinkrate'
    warning: 'Pull up, pull up'
Mode 2 –Excessive closure rate
    alert: 'Terrain, terrain'
    warning: 'Pull up, pull up'
Mode 3 –Alert to descent after take off
    alert: 'Don't sink, don't sink'
Mode 4 –Alert to insufficient terrain clearance
    alert: 'Too low terrain, too low terrain'
    alert: 'Too low gear, too low gear'
    alert: 'Too low flap, too low flap'

Mode 5 –Alert to inadvertent descent below glideslope
    alert: 'Glideslope, glideslope'
Mode 6 –Alert to descent below minimums
    alert: 'Minimums, minimums'

Alerts and warnings repeat continuously until the aircraft is manoeuvred out of the situation. A particular advantage of the variety of voice alerts is that its operationally orientated warnings permit confirmation by cross-checking of the panel instruments. Diagnosis of flight 'squawks' can thus be quickly carried out and corrected. The speed of the ground-proximity warning system envelopes has been increased, providing longer warning times.

The Sundstrand ground-proximity warning system contains a number of features to assist in test maintenance and repair procedures. These include a non-volatile memory which stores both steady-state or intermittent faults occurring over the last ten flight sectors and which can be erased only when the unit is removed from the aircraft for bench work. The accepted test procedure is programmed within the computer and a simple test fixture is all that is required to re-address computer output data back into the computer itself. An alpha-numeric display on the front of the unit can be used to isolate faults and indicate specific line replacement units which require replacement. On the bench, faults can be isolated to board level.

The ground-proximity warning system complies with ARINC 600 standards, and its subcomponents are grouped by circuit function on plug-in/fold out removeable printed circuit boards with easily removed captive hardware.

Dimensions: 2 MCU form-factor

STATUS: in production and service.

Mark II ground-proximity warning system

Sundstrand's Mark II ground-proximity warning system computer is designed for aircraft wired to ARINC 594 standard and is thus suitable for service in a wide cross-section of commercial, military or business aeroplanes. The Mark II model is claimed to be the first ground-proximity warning system to use a Mach/air-speed input and therefore to have a much faster response time than previous ground-proximity warning system computers. It was the first such system to offer voice-alerts which specifically identified each warning mode, and the first to offer a warning mode for minimum approach conditions.

Generally, warning modes for the Mark II unit are the same as for the Sundstrand digital ground-proximity warning system (see previous entry) although there are differences between the warning times and the warning envelopes themselves.

Sundstrand Mark II ground-proximity warning system computer

Dimensions: ¼ ATR Short
Weight: 8 lb (3·63 kg) max.

STATUS: in production and service.

Sundstrand digital ground-proximity warning system computer

Navigation Systems

Area Navigation

UNITED KINGDOM

Decca

Racal-Decca Navigator Ltd, Burlington House, Burlington Road, New Malden, Surrey KT3 4NW

RNAV

This new system was being certificated at the end of 1981. Comprising a navigation computer and cdu, it accepts information from any or all of the following sensors: VOR/DME, Decca Navigator, omega/vlf, and Doppler. Although applicable to both military and commercial fixed-wing aircraft and helicopters, the system is designed principally for civil airways operation.

Decca's RNAV can accommodate up to 100 stored waypoints and up to 30 stored routes, with a data-retention facility. It provides serial digital output of aircraft and waypoint position and other information to displays such as the Sperry Data Nav III, roll and steering commands to an autopilot, and cross-track deviation to an hsi. Data is entered on the ground by a tape-loader or by the cdu keyboard.

The display can indicate present position in latitude and longitude or Decca co-ordinates, heading to steer to next waypoint and eta there, desired track, actual track and ground-speed, windspeed and direction, bearing and distance to any waypoint from aircraft or other waypoint, magnetic variation, position of any waypoint or beacon in latitude and longitude, and aircraft altitude.

Cdu presentation for Decca area navigation system

Type 80791A control/display unit
Dimensions: 146 × 114 × 220 mm
Weight: 2·5 kg

Type 80792A Navigation/control unit
Dimensions: 4MCU
Weight: 4·5 kg

Type 80790A Mk 32 receiver
Dimensions: 4MCU
Weight: 4·5 kg

Type 80750A antenna amplifier
Dimensions: 125 × 80 × 100 mm
Weight: 0·5 kg

STATUS: certification was being sought at the end of 1981, and pre-production units were being delivered in January 1982. The first customers are British Airways and Bristow Helicopters, for installation on their respective helicopter fleets.

UNITED STATES OF AMERICA

Foster

Foster Airdata Systems Inc, 7020 Huntley Road, Columbus, Ohio 43229

RNAV 511

A low-cost full-time navaid for general aviation purposes, the RNAV 511 is claimed to have the lowest work-load factor of any system available, and interfaces with most commercially available VOR/DMEs, cdis, hsis, and autopilots. Designed in accordance with the recommendations of an FAA study on RNAV utilisation, the system meets all criteria for light aircraft avionics of its type for single-pilot IFR and VFR operation.

The system is based on two-waypoint storage as being optimum for light aircraft operation especially by one pilot. Microprocessor computation provides continuous updates of range and bearing to the active waypoint, but an alternative display of ground-speed and time-to-waypoint may be selected providing a useful mode for flight-planning and fuel-management.

For the private pilot, RNAV applications vary from simplifying cross-country flights and finding airports and airfields without co-located navigation beacons, to skirting with minimum fuel and time penalties terminals and restricted zones and penetrating complex terminal areas,

RNAV instrument approaches, or simply as range advisories during ILS, ADF, VOR, and MLS procedures.

Dimensions: (panel-mounted cdu) 6·25 × 2·375 × 8·75 inches (158·75 × 60·33 × 222·25 mm)
(steering adaptor) 6 × 4 × 1·375 inches (152·5 × 101·6 × 34·9 mm)
Weight: (cdu) 2·5 lb (1·1 kg)
(steering adaptor) 1 lb (0·4 kg)
Power: 0·7 A at 28 V dc
Max waypoint offset: 199 nm
Max distance to next waypoint: 199 nm

STATUS: in production.

AD611 RNAV

This designation describes a family of modular RNAV systems that can be specified or built up according to need, and aimed at the professional market. They are assembled from various combinations of four types of identically sized unit: range selector, waypoint setter, and the horizontal display and data entry modules that together comprise the memory waypoint system.

The simplest system is the one-waypoint RNAV, comprising a range/mode selector and a manual waypoint setter. As with all Foster

RNAV equipment, range and bearing to the waypoint are entered by means of thumb-wheels, easing operation in rough air. Despite its simplicity, the one-waypoint system has found wide acceptance, particularly among operators with limited panel space.

Since these manual waypoint setters have no memory, the two next most advanced variations (with two and three waypoints) are obtained by adding, respectively, one and two more such units. The two-waypoint system is acknowledged to be the minimum needed for most RNAV operations, while the combination handling three meets the challenge of IFR flying.

A ten-waypoint system, claimed to be the most straightforward RNAV available, uses a range display mode selector, and the two units of the non-volatile memory waypoint option.

Finally, an 11-waypoint system can be assembled by adding a manual waypoint setter to the ten-waypoint system just described.

STATUS: in production. During 1980 the US Navy installed 177 AD611/A-T sets on its Beech T-34C trainers. Although RNAV has enjoyed growing popularity for civil applications since the mid-1970s, this turboprop T-34C is believed to be the first military primary trainer to have this capability. The AD611/A-T is a specialised version of the AD611.

VNAV 541 and 541/A

In order to initiate climb or descent to a new level at the correct time and range, the company provides the VNAV 541 and 541/A vertical navigation guidance computer. These devices compute the triangulation problem involving height and distance to the destination airfield, a new flight level, or waypoint-crossing altitude, to provide a commanded flight profile. All the pilot has to do is to enter the required altitude at the airfield or flight level, select the flight-path angle (1·5, 3, or 4·2 degrees), and then fly the aircraft so that the altimeter readings match the demanded readouts on the VNAV; the system provides the basic vertical guidance.

Descent angle: (VNAV 541) 1·5° cruise descent, 3° approach descent, designed for unpressurised aircraft
(VNAV 541/A) customer-specified cruise descent, 3° approach descent for pressurised aircraft and jets
Dimensions: 3·255 × 1·3 × 5·5 inches (82·7 × 33 × 139·7 mm)
Power: 2 A at 28V dc

STATUS: in production.

AD611 area navigation system by Foster

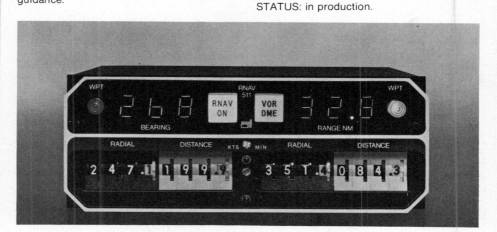

Foster's RNAV 511 low-cost area navigation system

JET

JET Electronics and Technology Inc, 5353 52nd Street, Grand Rapids, Michigan 49508

DAC-2000

The DAC-2000 is a three-dimensional navigation system for en route and terminal position-fixing, giving information and commands for both horizontal and vertical guidance. The heart of the system is a digital computer based on National Semiconductor's M-16 microprocessor, which can instantly calculate or pre-program climb or descent angles from zero to 10 degrees. A solid-state, non-volatile memory permits the pilot to enter, recall, or cancel up to 180 waypoints. Once programmed, the system provides commands for straight-line routing RNAV, together with profile climbs and descents by way of the VNAV facility, the total capability constituting so-called 3-D navigation.

The DAC-2000 system comprises three units: a cdu, digital navigation computer, and control annunciator. Optional equipment includes a frequency controller, offset control unit and instrument switching relay. The frequency controller ensures that the NAV receiver is automatically tuned whenever a waypoint is selected, the offset control unit permits the waypoint to be shifted or offset 99 nautical miles in any direction, while the relay provides the switching functions between the computer and the navigation sensors.

Certification: in accordance with FAA Advisory Circular 90-45A in horizontal and vertical modes, with complete autopilot/flight director coupling.
Dimensions: (cdu) 5·75 × 4·5 × 7·82 inches (146 × 114·3 × 198·6 mm)
(computer) 4.9 × 7·66 × 15·78 inches (124·4 × 194·5 × 400·8 mm)

(switching relay) 5·5 × 4·05 × 4·72 inches (139·7 × 102·9 × 119·9 mm)
Weights: (cdu) 3·7 lb (1·68 kg)
(computer) 10·5 lb (4·77 kg)
(switching relay) 3·8 lb (1·73 kg)

STATUS: in production.

DAC-7000

This is an advanced version of the DAC-2000 that has all the features of the -2000 and in addition is compatible with Bendix or Rockwell-Collins weather radars, has remote programming by using a Texas Instruments TI-59 hand-held calculator, automatic waypoint advance and automatic course change. The first-named is a potentially valuable benefit; waypoints connected by a course line can be brought up on the weather radar and can be changed to give new courses in order to avoid the most severe weather. These new and automatically programmed waypoints are instantly tabulated for quick reference.

The substantial improvements in capability over the simpler system are reflected in the respective prices: around $33 000 against about $20 000 at January 1981 prices.

STATUS: in production.

Cdu presentation for JET DAC-7000 area navigation system

Doppler Navigation

GERMANY (FEDERAL REPUBLIC)

Litef

Litef (Litton Technische Werke), Lorracher Strasse 18, Postfach 774, 7800 Freiburg

Navigation system for West German Air Force Alpha Jet

This West German-based Litton subsidiary has been building 190 sets of its own design of Doppler radar for the German version of the Franco-German strike/trainer. The system uses the Teledyne-Ryan Electronics AN/ANP-220G Doppler velocity sensor, built under licence in Germany, in conjunction with a Litef navigation computer and control/display unit.

With the addition of aircraft attitude and heading as correction factors, the system computes a number of quantities for navigation. A unique filter permits a very accurate instantaneous velocity to be derived for weapons delivery. The computer provides signals to the cdu for the following outputs: present position, up to ten waypoints or target co-ordinates, four Tacan co-ordinates (including channel number and station height), five targets of opportunity, automatic correction for magnetic variation, fix and update facilities (using Tacan or known points), ground-speed and drift, wind speed and direction, range and bearing to waypoints or targets, time-to-go to waypoints or targets, automatic back-up calculation for failures, and diagnostic test results.

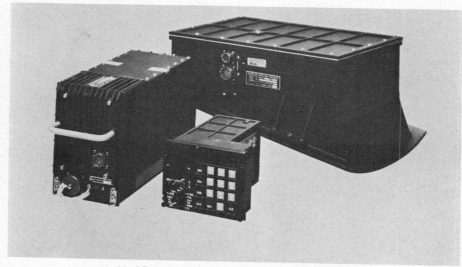

Litef Doppler system for West German Alpha Jet

Accuracy: 0·4% cep
Range: 40 – 600 knots, 0 – 50 000 ft (16 400 m) over land with pitch and roll up to 45°. Accurate navigation between ±80° in lat
Weight: sensor 9 kg
processor 6·7 kg
cdu 2·8 kg

Power (total): 186 W at 28 V dc
Reliability (predicted): sensor 8000 h
computer 6000 h
cdu 12 000 h

STATUS: in production for West German version of Alpha Jet.

UNITED KINGDOM

Decca

Racal-Decca Navigator Ltd, Burlington House, Burlington Road, New Malden, Surrey KT3 4NW

Type 71

Developed and marketed in the late 1960s, the Type 71 is designed for helicopters and smaller fixed-wing aircraft with speeds up to 300 knots.

Type 9305 velocity sensor for Decca 71 Doppler

Position and bearing indicator for Doppler 71

The system is based on the Decca Type 9305 antenna/electronics unit that provides Doppler velocities to a navigation computer and display system such as the company's TANS (tactical air navigation system), and outputs are provided for flight instruments, eg a hover meter, a ground-speed/drift meter, and an automatic chart display.

Antenna/electronics unit Type 9305

The Type 9305 velocity sensor employed in the Type 71 Doppler is a fixed (ie non-stabilised) unit mounted on the underside of the aircraft and comprises two antennas (one transmitting, the other receiving) and associated tracking and amplifier circuits. Extensive use is made of analogue and digital integrated circuits. The use of a fixed antenna system reduces size and weight and, for helicopters, is more suitable since it is undesirable to attempt stabilisation of such a relatively large mass at the hover.

Dimensions: fits into recess 16 × 16 × 5 inches (406 × 406 × 127 mm)
Weight: 36·5 lb (16·5 kg)

TANS computer Type 9447D

The cdu for the 71 series Doppler is a small general-purpose computer with integral keyboard and display that can accomplish a variety of functions with a fixed program. It is optimised for helicopter applications, accepting inputs from the Type 71 Doppler and air-data and attitude/heading systems. The TANS cdu controls and displays present position, bearing and distance/heading to steer and time-to-go to any one of ten waypoints or targets, the vectoring of moving waypoints (eg ships), pitch and roll corrections, automatic wind calculation, automatic reversion to air data signals at Doppler failure, and sea surface motion compensation. TANS can also display other information such as track and ground-speed, true airspeed, drift angle, and Doppler velocities in three axes. X and Y output signals are available to drive an automatic chart display, and an analogue output of track error can drive a steering indicator. The fixed program can accommodate up to 16 000 words of eight bits.

Dimensions: 5·75 × 6·39 × 9·72 inches (146 × 162 × 247 mm)
Weight: 13·2 lb (6 kg)

Position bearing and distance indicator Type 80475

The position bearing and distance indicator is a panel-mounted instrument that can be chosen in place of the TANS cdu to display aircraft position and bearing from any one of ten waypoints, using outputs from Series 70 or 80 Dopplers and from a heading sensor. Built into a 4ATI instrument case, it displays aircraft position, or bearing and distance to one of up to ten waypoints, or ground-speed and drift angle. It also indicates track error to the chosen waypoint. The general-purpose processor in the unit can accommodate up to 1500 words of eight bits.

Dimensions: ARINC 4ATI case 4·18 × 4·18 × 7·68 inches (106 × 106 × 195 mm)
Weight: 4·5 lb (2 kg)

Automatic chart display Type 1655

This is a simple and economical means of fixing aircraft position on a standard chart. Position is given by the intersection of cross-wires that move in accordance with position computed by either a TANS cdu or a position bearing and distance indicator.

Type: knee- or table-mounted, stowed when not in use
Scales: choice of 5: 1:50 000, 1:100 000, 1:250 000, 1:500 000, 1:1 000 000
Chart area: 10 × 10 inches (254 × 254 mm)
Dimensions: 12 × 12 × 2⁷/₁₀ inches (306 × 306 × 67 mm)
Weight: 7 lb (3·2 kg)

Ground-speed/drift meter Type 9308

This standard aircraft indicator-size unit shows ground-speed within 3·5 knots at 100 knots or 5

Type 1655 automatic chart display

knots at 300 knots. Drift is registered within 0·5 degrees of true value. If the signal is lost, the last-measured ground-speed and drift are displayed.

Hover meter Type 9306

This standard aircraft indicator-size unit displays along-heading velocity over the range –10 to 20 knots with an error not greater than one knot, and vertical velocity range plus or minus 500 ft/minute with a maximum error of 40 ft/minute.

STATUS: in production. The Type 71 Doppler is installed on RAF Puma and SH-3D and Royal Navy and British Army Lynx helicopters.

Type 72

This system, also designed in the late 1960s, is virtually identical to the 71 series but is scaled for fixed-wing aircraft with much higher performance.

The basic velocity sensor is the Type 9307 antenna/electronics unit. A complete, self-contained Doppler navigation system can be built up in one of two ways: first, in conjunction with a TANS Type 9447D cdu and an optional Type 80568 distance/time-to-go to waypoint indicator, and second, with a TANS 9447G cdu and optional TANS Type 80767 remote indicator. In both cases the system can be used with a type 9308 ground-speed/drift meter, and needs heading, attitude, and air data information. Weights, dimensions, and accuracies are generally similar to those quoted for the Type 71 Doppler.

STATUS: in production for RAF Tornados.

Type 80

The Type 80 Doppler is a new navigation system that can be used alone or as an integrated Doppler/inertial navigator. Unlike the Type 71 and 72, it accommodates the speed requirements of both helicopters and high-performance aircraft into one unit, though scaling to the particular speed range needed is carried out during manufacture to optimise it for helicopter or fixed-wing applications.

The system is based on the Type 80561 non-stabilised antenna unit which provides Doppler velocities to a range of possible navigation equipment. Thus the sensor could be used in conjunction with a Type 1655C automatic display (the simplest system), a Type 80475 position bearing and distance indicator, or a combination of position bearing and distance indicator and chart. In each case a compass heading input is required and a Type 80564B hover meter could be added for helicopter applications.

Velocity sensor Type 80561

Printed antennas transmit and receive the three beams, the solid-state cw microwave source generating 20 mW of energy at 13·325 GHz. Velocity outputs to associated computers and displays can be provided in pulse form or in digital form to ARINC 429. The digital-output versions are equipped with microprocessor control and can be integrated with the Litton LR 80 inertial system. BITE enables the system to be checked for serviceability with a high level of confidence.

Range: (helicopter version) –50 to 350 knots (along heading), ±100 knots (across heading), 0 – 25 000 ft (7620 m)
(fixed-wing version), 50 – 1000 knots (along heading), ±100 knots (across heading), 0 – 60 000 ft (18 290 m) (max beam length)
Accuracy: error to 95% probability, ignoring compass inaccuracies, within 0·5% of distance flown along track, and within 0·75% across track for drift angles less than 20°

Type 9308 ground-speed/drift meter presentation

Type 9306 hover meter

Reliability: mtbf is 3000 h
Dimensions: 15 × 14 × 3·15 inches (381 × 356 × 80 mm)
Weight: 19 lb (8·6 kg)
Power: 60 W at 28 V

Position bearing and distance indicator Type 80475
See reference under Type 71 Doppler above

Automatic chart display Type 1655
See reference under Type 71 Doppler above

Hover meter Type 9306
See reference under Type 71 Doppler above

STATUS: in production. Applications include British Army SA.341 Gazelle helicopters, and RAF Hawk trainers. The system has also been chosen by the UK company Airship Industries for its SKS-500 ten-seat airship.

TANS tactical air navigation system
In the 1970s Decca developed a versatile, low-cost control and display unit as an adjunct to its Navigator and 70 and 80 series Doppler radars for helicopters and fixed-wing aircraft. Intended primarily for military applications, TANS is a small, general-purpose digital computer with integral keyboard and display that can be set up to fulfil a variety of functions. The designation TANS refers to a family of equipment, each member of which may be tailored to customers' needs.

System characteristics for all TANS units
Dimensions to ARINC format: 5·75 × 6·39 × 9·72 inches (146 × 162 × 247 mm)
Weight: 13·2 lb (6 kg)
Power: <50 W at 28 V dc
Computer: fixed program with capacity for 16K words of 8 bits, and capable of storing details of up to 10 way-points.
Specific models are:

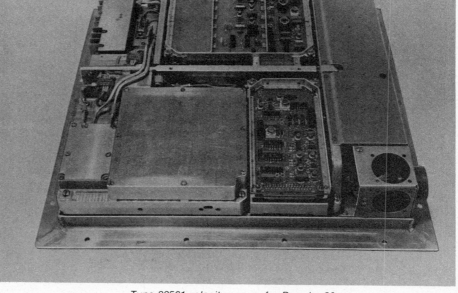

Type 80561 velocity sensor for Doppler 80

Type 9447D
Designed for use with Doppler series 70 radars in helicopters or fixed-wing aircraft, inputs comprise Doppler velocities and compass heading. Air data and pitch/roll attitude are optional inputs. The system can also feed a TANS remote indicator and a track error display.

Type 9447F
This TANS unit goes with the hyperbolic Decca Navigator Mk 15 and 19 receivers, as well as with the series 70 Dopplers, when outside the coverage of a Decca Navigator ground chain. A variant is the Type 9447F-14, having the additional capability of working with series 80 Doppler. Both can also feed a TANS remote indicator and track error display.

Type 9447G
This cdu is programmed to accept inputs from both series 70 and 80 Dopplers, an air-data system, compass heading, and pitch and roll. It can also feed a TANS remote indicator and a track error display.

STATUS: all in production.

Position bearing and distance indicator for Doppler 80

Decca TANS Type 9447F indicator

UNITED STATES OF AMERICA

Kearfott

Kearfott Division, The Singer Company, 1150 McBride Avenue, Little Falls, New Jersey 07424

AN/ASN-128 Doppler radar

Kearfott's AN/ASN-128 lightweight is the US Army's standard airborne Doppler navigator and comprises three units: receiver-transmitter-antenna, signal data converter and computer display unit. A steering hover indicator can also be included as an optional extra. With inputs from heading and vertical references, the system provides aircraft velocity, present position, and steering information from ground level to above 10 000 feet (3000 metres).

Propagation: 4-beam configuration operating fm-cw transmissions in J-band. Beam-shaping eliminates need for land/sea switch. Single transmit-receive antenna uses full aperture in both modes to minimise beamwidth and reduce fluctuation noise
Number of waypoints: 10
Volume: (velocity sensor including signal data converter) 1037 in³ (16 993 cm³)
(cdu) 224 in³ (3670 cm³)
Weight: (velocity sensor including signal data converter) 23 lb (10·43 kg)

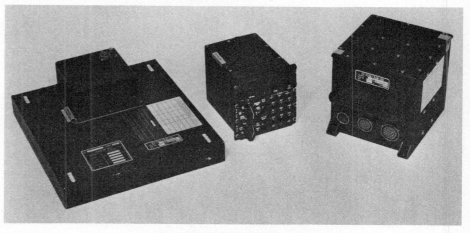

Kearfott's AN/ASN-128 Doppler

(cdu) 7 lb (3·8 kg)
(hover indicator) 2 lb (0·9 kg)
Self-test: localisation of faults at line-replaceable level by BITE
Reliability: (complete system) >2000 h
(velocity sensor) >4500 h

STATUS: in production. The system, complete with hover indicator, is also being built for Federal German Army BO 105 and PAH-1 helicopters, and has been chosen for its CH-53 helicopters and for the Federal German Navy's BHS helicopter.

Teledyne

Teledyne Ryan Electronics, 8650 Balboa Avenue, San Diego, California 92123

Teledyne Ryan Electronics was an early pioneer in the design and manufacture of Doppler radars and navigation systems, its work in this field going back to 1946 with a target-seeking guidance system for one of the first air-air missiles. TRE was chosen to build the first Doppler spacecraft landing radar and radar altimeter, flown successfuly during 1966 on the Surveyor lunar soft-landers. It has since built every Western spacecraft landing radar and radar altimeter, including those of the Apollo manned lunar-landers and the later Viking Mars-landers, representing 24 separate and successful flights. The company says that, during the last decade, it has been awarded more official Doppler designations than any other firm in this field.

AN/APN-217 Doppler velocity sensor

This transmitter/receiver will be the primary navigation sensor for the US Navy's Sikorsky SH-60B Seahawk LAMPS Mk III twin-engined utility and asw helicopter, and deliveries of a slightly modified version, the AN/APN-217(J) have been made to Japan for its Sikorsky HSS-2 Sea Kings. The US Navy is also planning to install this radar on its Sikorsky RH-53D minesweeper helicopters, and a retrofit is being considered for its SH-2 and SH-3 helicopters. The US Marines Corps plans to have the APN-217 on its Sikorsky CH-53 night vision equipped helicopters.

The AN/APN-217 is a microprocessor-controlled, solid-state, single-unit radar with a unique cw space-duplex design that avoids modulation losses, eliminates altitude 'holes', and bypasses the limitations associated with modulated systems. At low altitudes the

APN-217 is claimed to be 100 times more sensitive than modulated radars. A combination of six special features prevents acquisition or tracking of antenna vertical sidelobe returns, making the system suitable for coupled transitions from forward flight to hover and viceversa.

Outputs: dc signals to afcs and flight instruments such as hover indicator, digital outputs to MIL-STD-1553 provide velocities to navigation computer, and one version of APN-217 has digital outputs to ARINC 575
Accuracy: (digital) 0·3% of ground-speed, 0·3 knot in heading and drift, 35 ft/minute vertically. (analogue) 2% of ground-speed, 0·5 knot in heading and drift (afcs)
5% ground-speed, 1 knot in heading and drift 70 ft/minute vertically (flight instruments)
Display: outputs are compatible with a number of displays, including CP-1251 cdu, and with ground-speed/drift and hover instruments.

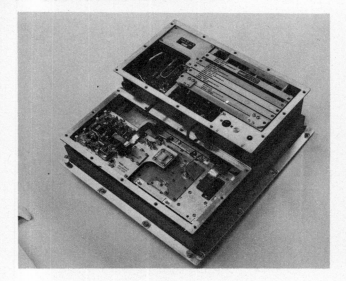

Teledyne's AN/APN-217 Doppler velocity sensor

Teledyne AN/APN-218 Doppler velocity sensor

Dimensions: 16⁹/₁₀ × 16³/₁₀ × 6⁹/₁₀ inches (429·3 × 414 × 175·3 mm)
Weight: 34 lb (15·5 kg)
Controls: land/sea select. System in SH-60B is controlled via MIL-STD-1553 bus
Power: 55 W at 28 V dc
Speed range: –40 to 250 knots along track, ±100 knots drift, ±4500 ft/minute vertically
Reliability: demonstrated mtbf 3750 h
Self-test: built-in test facility detects 90% of predictable failure modes

STATUS: final development in late 1981. Deliveries of Lot 1 production scheduled to begin October 1982.

AN/APN-218 Doppler velocity sensor

This Doppler velocity transmitter/receiver is a high-performance nuclear-hardened radar chosen by the USAF as its common strategic Doppler for fixed-wing aircraft. Competitive development of the APN-218 began in 1976. Production awards followed in 1978 and deliveries began the following year.

The performance and reliability of the system stems from two characteristics that it shares with the parent APN-213 Doppler: use of a continuous-wave space-duplex transmit-receive system that is typically 7–10 dB more efficient in its use of transmitter power than a modulated system, and a high-gain, narrow-beamwidth planar array antenna for high accuracy and sensitivity.

The four features that commended it to the USAF as its CSD are nuclear-hardness, high performance, choice of ARINC 575 or dual-redundant MIL-STD-1553 data-bus interface, and integral radome.

The system can be used as a source of velocity information to other equipment, or in conjunction with a ground-speed and drift indicator or cdu as a self-contained navigation system. The optional cdu combines the functions of navigation computer and control/display unit, and contains an incandescent alpha-numeric display panel and a keyboard for entering data and selecting operational modes. Up to ten waypoints can be accommo-

dated, and a non-volatile scratch-pad memory holds critical information during power transients or interruptions.

Speed: 96–1800 knots
Altitude: 0–70 000 ft (21 330 m)
Beam geometry: 4 beams shared
Terrain: land or smooth sea
Power: 170 VA at 115 V 400 Hz
Transmitter power: 1·5 W
Frequency: 13·325 GHz
Mtbf: 3634 h calculated
Weight: sensor 70·2 lb (31·9 kg), ground-speed and drift indicator 3·4 lb (1·5 kg), cdu 8·5 lb (3·9 kg)
Dimensions: sensor 28¹/₅ × 25²/₅ × 6⁷/₁₀ inches (716·3 × 645·2 × 170·2 mm),
gdsi 5³/₄ × 3 × 6¹/₁₀ inches (146·1 × 76·2 × 154·9 mm),
cdu 5³/₄ × 6 × 6¹/₂ inches (146·1 × 152 × 165 mm)
Self-test: built-in test both continuous and commanded

STATUS: in production as retrofit equipment for B-52 bombers and KC-135 tankers. Also on USAF/Lockheed C-130s and C-141s, Boeing E-3A AWACS, Rockwell B-1s, General Dynamics FB-111s, and Lockheed S-3A Vikings.

AN/APN-220 Doppler velocity sensor

This system can be used either as a single-unit velocity sensor providing outputs to other aircraft systems, or with a cdu and hsi as a self-contained navigation facility. As a velocity sensor to provide data to the navigation and weapons-delivery systems, it was chosen by the West German Air Force for its Franco-German Alpha Jet strike/trainers. Versions of the equipment have also flown on remotely-piloted vehicles such as Teledyne's BGM-34C, and on helicopters, and was designed for applications in which size, weight, performance, and reliability are critical. The APN-220 family evolved from a small, lightweight Doppler sensor originally designed for the US Army and was subsequently qualified by that service and later by the USAF and the West German Air Force. TRE claims that the APN-220 is the smallest Doppler velocity sensor available anywhere.

An optimum velocity range and near-zone rejection are offered for each application.

Ground-speed/drift indicator for AN/APN-218

Cdu presentation of AN/APN-218 Doppler

System characteristics of typical fixed-wing application

Output: heading, vertical velocity, and ground-speed/drift, to aircraft systems eg afcs or to CP-1251 and hsi

Number of waypoints: up to 10, entered via front-panel keyboard

Range: (typical system) –40 to 600 knots forward, 150 knots in drift and 5000 ft/minute (1533 m/minute) vertically. Height, up to 50 000 ft (15 330 m)

Accuracy: (over land) 0·25% + 0·2 knot (0·25% + 37 km/h)

(over sea) 0·3% + 0·2 knot

Dimensions: (velocity sensor) 16·78 × 11·46 × 4·45 inches (426 × 291 × 113 mm)
cdu 6 × 5·75 × 6·5 inches (152 × 146 × 165 mm)

Weight: sensor 21·3 lb (9·66 kg)
cdu 8·5 lb (3·86 kg)

Power: sensor 28 W at 28 V dc; cdu 30 W at 28 V dc

Reliability: 2150 h mtbf demonstrated in Lockheed S-3A Viking

Self-test: BITE diagnostic program locates faults at first-line level to 95% confidence

STATUS: in production.

Doppler velocity sensor for AN/APN-220

Hyperbolic Navigation

Loran (long-range navigation) is the ultimate development in a line of ground-based radio-navigation aids going back to equipment devised in the Second World War to support precision bombing operations over Europe. Loran, a so-called hyperbolic navigation system, was developed in the USA and for very many years was the standard navaid over the North Atlantic and North Pacific.

There are basically two Loran systems, A and C. Loran A has been discontinued, but Loran C and its variation Loran D remain in wide-spread use and will probably be maintained until the end of the century. Apart from the traditional areas of coverage, (North Atlantic and North Pacific) Loran chains have been established to serve areas of special interest, notably the Gulf of Mexico, where nearly 1000 helicopters serve the world's most extensive off-shore oil industry.

UNITED KINGDOM

Decca

Racal-Decca Navigator Ltd, Burlington House, Burlington Road, New Malden, Surrey KT3 4NW

ADL-81 Loran C/D

The ADL-81 was the first in a new family of Decca series 80 equipment when it was introduced in the early 1970s. Since Loran is a US system, and by far the greatest market is in the USA, Decca produces and markets its equipment through a US subsidiary, NSI (Navigation Systems Inc) of Silver Spring, Maryland.

The design combines C-MOS logic and temperature-compensated frequency references into a high-performance system that provides time difference read-outs to 100 ns on a cdu, or 50 ns via a computer interface. The system, comprising receiver, cdu, and antenna coupler, automatically acquires, synchronises, and tracks signals from all Loran C and D chains. Identification of signals during acquisition is of high integrity and does not result in false identification in the absence of signals from the selected chain.

Decca's ADL-81 Loran system

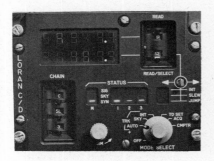

Close-up of ADL-81 control/display panel

Accuracy: 100 ns
Sensitivity: minimum signal level 10 μV/m
Selectivity: (search bandwidth) 5 kHz
(track bandwidth) 20 kHz
Power: 50 VA at 115 V 400 Hz
Dimensions: (receiver) ½ ATR short
(cdu) 5¾ × 4½ × 4 inches (146·1 × 114·3 × 101·2 mm)
(antenna coupler) 4 × 4 × 2 inches (101·6 × 101·6 × 50·8 mm)

Weight: (receiver) 15 lb (6·8 kg)
(cdu) 2 lb (0·9 kg)
(antenna coupler) 1 lb (0·45 kg)
Reliability: mtbf 600 h

STATUS: in production. An earlier model, the Type ADL-21, equips all transport aircraft of RAF Strike Command.

UNITED STATES OF AMERICA

ONI

ONI Offshore Navigation Inc, 5728 Jefferson Highway, PO Box 23504, Harahan, Louisiana 70183

ONI-7000 Loran C navigator

ONI claims that its new Type 7000 Loran navigator represents a breakthrough in signal-processing technology, enabling it to operate in regions where other systems are unusable. It also says that the system is the only one capable of providing coast-to-coast coverage over the USA.

The ONI-7000 automatically tracks up to eight stations in as many as four chains simultaneously, thereby providing a degree of redundancy not previously available. It also incorporates a number of features designed to reduce pilot work-load, including a flight-planning computer mode, automatic waypoint sequencing, pre-stored routes and automatic correction of magnetic variation.

Waypoints can be defined by latitude and longitude, Loran time differences, or as an offset radial/distance from a previously defined waypoint. Any number of routes can be defined and stored in non-volatile memory, utilising any number up to 200 stored waypoints. The cdu employs fibre optics for better oblique and daylight visibility. Outputs are provided to CDI and remote annunciators, and optional outputs comprise signals to HSI, RMI, radar, and TAS. Optional inputs include compass, IAS, and encoding altimeter.

The system consists of three units: antenna, receiver/computer, and control/display unit.

ONI-7000 Loran C navigator

Dimensions: (receiver/processor) 7½ × 7⅝ × 12⁹/₁₆ inches (190·5 × 193·5 × 319·5 mm) (cdu) 5¾ × 4½ × 6½ inches (146 × 114·3 × 165·1 mm) (antenna) 6½ × 3½ inches (base) × 14½ inches long (146 × 88·9 × 368·3 mm)
Weight: (receiver/processor) 13·1 lb (5·95 kg) (cdu) 5 lb (2·27 kg) (antenna) 1·5 lb (0·68 kg)

STATUS: production deliveries began in September 1981 to helicopter and fixed-wing aircraft operators in the USA, Canada, and Mexico. IFR approval for en route use in US national airspace was anticipated shortly afterwards, with approvals for North Atlantic and North Pacific following in early 1982.

Flite-Trak 100 telemetry system

In early 1979 ONI proposed a joint programme to the FAA whereby they could evaluate the use of Loran C to pinpoint the position of helicopters operating beyond the reach of air traffic control surveillance radars in the crowded airspace of the Gulf of Mexico. The idea was to telemeter back to the air traffic control centre at Houston position information derived from Loran C equipment aboard helicopters servicing the many offshore oil installations in the area. This information was then to be displayed on computer-graphics terminals adapted to resemble as far as possible conventional air traffic control radar screens.

The US Government financed development of the FAA displays, while ONI launched and funded its own airborne telemetry equipment, at the same time developing a commercial version of the FAA system to enable helicopter operators to track their own fleets. The FAA system became operational during 1981, and ONI began delivering equipment to the commercial companies which are participating in the FAA's evaluation programme. In August 1981 ONI delivered the first Flite-Trak 600 dispatcher terminal, with colour graphics display, to Pemex, the Mexican national oil company. That operator is using the system to control its fleet of helicopters in the Gulf of Campeche. These aircraft are being equipped with ONI-7000 Loran C receiving equipment and Flite-Trak 100 telemetry sets. Another company, Transco, is using the same equipment in conjunction with its new, automated Gulf of Mexico helicopter communications network. Flite-Trak systems were being delivered during late 1981 to other major oil and gas companies operating helicopter fleets.

Flite-Trak 100 telemetry system

ONI's Flite-Trak 600 terminal for air traffic control centres and private operators

Inertial Navigation

FRANCE

Sagem

Société d' Applications Générales d'Electricité et de Mécanique, 6 avenue d'Iéna, 75783 Paris Cedex 16

Sagem is one of the major French producers of navigation systems, its inertial and nav-attack equipment being represented by the Uliss family. Specific systems and applications are:
Uliss 45: KC-135 tankers of French Air Force
Uliss 47: Mirage F.1 CR (reconnaissance) and Mirage F.1 EJ for Jordan
Uliss 52: Mirage 2000 (DA and N fighters), French Air Force
Uliss 53: Atlantic New Generation, Aéronavale
Uliss 54; Mirage IV bombers of Armée de l'Air
Uliss 80: Super Etendard carrier fighter for Argentina
Uliss 81: Alpha Jet and Mirage V for Egypt, and Alpha Jet for Cameroon
Uliss 82: Jaguar light bombers for India

More than 80 per cent of the components and sub-assemblies are common to all members of the Uliss family, the principal differences residing in specific interfaces and computation functions.

Uliss 52

This navigation system is designed for high-performance combat aircraft. Avionics information and commands are distributed by digital multiplexed data-bus. It comprises three units: a UNI 52 inertial navigator, a PCN 52

control/display box and an SMC 52 mode selector.

The platform has four gimbals for full manoeuvrability, and carries three tuned-rotor gyros and three dry accelerometers. The computing section employs LSI integrated circuits and hybrid micro-electronics, performing at 300 000 operations a second. The EPROM memory is insensitive to transients or interruptions in the power supply.

Accuracy: 1 nm/h cep
Power: 250 VA at 115 V 400 Hz
Self-test: comprehensive system that isolates and signals failure on platform-mounted annunciator
Dimensions: (navigator) 386 × 194 × 191 mm
(cdu) 165 × 162 × 146 mm
(mode selector) 165 × 38 × 146 mm
Weight: (navigator) 15 kg
(cdu) 3·7 kg
(mode selector) 0·8 kg

STATUS: in production for Dassault Mirage 2000 air-superiority fighter.

Uliss 81

The Uliss 81/82 combines into a single box all the functions of an inertial navigation and a weapon delivery system. It comprises three units: a UNA 81 inertial/attack box, a PCN 81 cdu, and a PSM 81 mode selector. The platform

has a four-axis gimbal system with tuned-rotor gyros and the computer section operates at 300 000 operations a second, in conjunction with an EPROM memory.

The system provides aircraft position, velocity, and altitude information; computation of navigation and steering information to waypoints; position updating by navigation fixes; weapon delivery computation (ballistic, determination of release point, ripple spacing of weapons, safety pull-up information, and head-up display information for target acquisition and commands for 'blind' release of stores); attack modes; and air-data computations.

Accuracy: position 1 nm/h cep, weapon-delivery >5 mrad
Power: 220 VA at 200 V 400 Hz
Dimensions: (navigator) ¾ short ATR, 386 × 194 × 191 mm
(cdu) 209 × 116 × 153 mm
(mode selector) 148 × 41 × 123 mm
Weight: (navigator) 15 kg
(cdu) 3·5 kg
(mode selector) 1 kg
Self-test: comprehensive built-in test system indicating about 93% of faults, fault isolation to module level, annunciation on front panel of platform unit

STATUS: in production for Jaguar, Alpha Jet, and Mirage V.

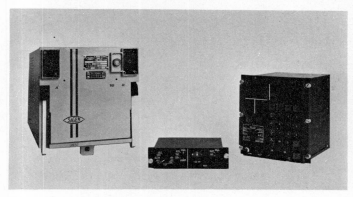

Sagem's ULISS 52 IN system for Mirage 2000 fighter

Sagem ULISS 81 IN equipment

Sfim

Société de Fabrication d' Instruments de Mesure, 13 avenue Marcel Ramolfo Garnier, 91301 Massy

26SH strapdown inertial navigation system

The 26SH system is designed to meet the requirements of asw, anti-tank, combat, scout and observation helicopters, in conjunction with a Doppler radar to minimise long-term errors. Other inputs comprise magnetic heading and barometric and radio height information. The strapdown system employs high-precision gyros and accelerometers rigidly mounted to the structure of the containing box.

The system is contained in three boxes: an inertial reference unit, a navigation and control/

display unit, and a flux-valve heading sensor. The inertial reference unit carries a pair of two-axis tuned rate-gyros, and three slaved, pendulous accelerometers. The computing section employs C-MOS PROM for protection against power-supply transients, and EAROM for storage. Sfim claims that the 'GAM' gyros employing Hookes' joint principle are the cheapest and simplest such devices to produce and eliminate all the mechanical components associated with conventional gimbals, and there are no bearings, motors or slip-rings.

Readiness: performance 90 s after switch-on
Interfaces: most information available in ARINC 429, MIL 1553, or GINA data-bus format, though certain data can be supplied in analogue form to feed flight or other instruments

Accuracy: (all accuracies to 95% cep)
(heading) 0·3°
(pitch and roll) 0·2°
(velocities) 1·5 m/s in horizontal axes, 1 m/s vertically
(present position) 0·3'
Alignment: automatic, and needing no external reference
Computer: (instructions) 84
(memories) PROM (16K words of 16 bits), RAM (4K words of 16 bits), EAROM (1K words of 16 bits)
Dimensions: (iru) ¾ ATR short 190 × 194 × 214 mm
(cdu) 130 × 145 × 205 mm
(magnetic sensor) 125 × 89 × 81 mm
Weight: (iru) 9 kg
(cdu) 2·7 kg
(magnetic sensor) 0·7 kg

Power: (total) 105 W at 28 V dc
Certification: meets French standards 7304, 2021D, 2025, and 205 for combat helicopter category

STATUS: in production for, and in service with, Puma and other French combat helicopters.

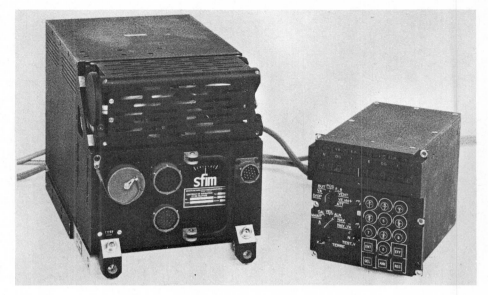

Two of three units comprising Sfim 26SH IN system

UNITED KINGDOM

Ferranti

Inertial Systems Department, Ferranti Ltd, Silverknowes, Ferry Road. Edinburgh EH4 4AD

FIN 1060 digital inertial nav-attack system

The designation FIN 1060 refers to a family of digital inertial nav-attack systems based on the company's FIN 1000 range of equipment. The most recent member of the family is the FIN 1064, chosen by the Ministry of Defence (PE) for the RAF Jaguar mid-term avionics update. During 1982/83 this equipment will replace the late 1960s NAVWASS (navigation and weapon-aiming sub-system) in these light bombers.

The first system was delivered in December 1980 to British Aerospace at Warton for rig-testing, and several sets of equipment were installed in Jaguars for flight trials beginning in the summer of 1981. Meanwhile in March 1981 the Edinburgh company was contracted to build 100 sets, equal to about half the RAF's fleet of Jaguars.

The principal advantages over the earlier system are a reduction in size and weight and a significant increase in computer storage capacity. Five boxes are replaced by one, with a reduction in volume of two-thirds and a weight-saving of 50 kg. The four-fold increase in computer capacity, from 16K to 64K, means that the aircraft can accommodate training, strike and reconnaissance roles without needing to re-program for each role. This limitation had increased as roles had become more

demanding over the years, and new and more complex equipment was fitted to enhance performance eg the Ferranti laser ranger and marked-target seeker. Pre-flight alignment times have also been reduced; the crew can select the so-called full-performance alignment in 12 to 15 minutes depending on the temperature of the platform, or some accuracy can be sacrificed to achieve readiness in 2½ minutes in a quick-reaction situation.

The two operating modes are navigation and weapon-delivery. In the second mode the system can be used in visual conditions to provide, on a head-up display or weapon sight, information and commands for direct attack, offset attack employing ips (initial points, or navigation references some distance from the target), or guns and unguided rockets. It can also be used in poor or zero visibility for blind bombing, and for air-to-air combat, using either stadiametric or radar ranging.

Several reversionary modes are available. Poor performance, eg through curtailed alignment, can be partly compensated by switching in air-data information. Should inertial capability fail completely, the computer section of the system can rely solely on air-data information to compute navigation functions.

The system comprises two units: an inertial nav-attack box (the inu) containing platform, computer, power supplies, and interface electronics, and a cockpit display unit for control/display functions.

Inertial nav-attack unit

The platform has four gimbals for full manoeuvrability, mounting three Ferranti single-axis floated rate-integrating gyros, and three Ferranti single-axis viscous-damped force-feedback accelerometers. The computer processor contains an arithmetic logic unit, with 12 byte registers, under the control of a ROM microprogram defining the instruction set. The machine can address up to 64K 8-bit program words. The data word length is normally 16 bits, with 32-bit working available.

Cockpit display unit

This unit comprises a ten-digit keyboard assembly for inserting data into the system, two main alpha-numeric displays, and control switches.

Number of waypoints: 10
Navigation accuracy:
(normal alignment) 0·86 nm/h cep, 2·5 knots 1 sigma
(rapid alignment) 1·92 nm/h cep 3·5 knots 1 sigma
Self-test: comprehensive software permits automatic self-testing of system; also checks serviceability of other systems such as Tacan, hud, and HSI, so enabling crew to judge practicability of continuing a mission in the event of failure

Ferranti FIN 1064 IN system configured for Jaguar

FIN 1064 cdu installed on port cockpit coaming of Jaguar

Dimensions:
Inu: 428 × 335 × 247 mm
Cdu (standard unit): 146·5 × 130 × 153·4 mm

STATUS: in production for RAF Jaguar, British, German and Italian Tornados, Harrier, Nimrod MR2, Mitsubishi F1, and Ariane European launch vehicle.

FIN 2000

A prototype of this fully aerobatic IN system, which incorporates a new miniature gimballed platform, was shown at the 1980 SBAC display at Farnborough. Smaller and lighter than previous Ferranti equipment, the system is designed for retrofitting on existing aircraft or for new designs in which space is at a premium.

One of the design goals was a very rapid response time: the FIN 2000 can be aligned and ready for navigation within two minutes of switching on. It is suitable for operation worldwide, either as a self-contained navaid or in conjunction with other navigation equipment. All navigation and attitude information is available in digital form from the INU, and can be supplied to the pilot on an optional control/display unit.

The INU contains the inertial platform, a powerful dedicated computer, and the associated interface electronics. The miniature inertial platform is stabilised by a pair of two-axis dry-bearing oscillogyros, and inertial velocities are derived from three accelerometers mounted on it. Stabilisation of the platform and the derivation of velocities are controlled by the computer, which also processes the inertial information to provide present position, velocities, and attitude outputs. The extensive use of digital techniques permits considerable flexibility in meeting individual customer needs.

Number of waypoints: 10
Alignment: 1 mode for all conditions. 2-minute readiness time from switch-on, with indication of alignment status
Outputs: present position in lat/long, attitude, ground-speed and track; range, bearing and time-to-go to destination; drift angle and heading
Test feature: system checking via BIT modes
Accuracy (design goals): position 1 nm/h (1850 m/h) RMS, velocity 2·5 ft/s (0·76 m/s) RMS, heading 0·25° RMS, attitude 0·15° RMS, all in first 2 hours of operation

INU
Dimensions: 195 × 200 × 385 mm
Weight: 18 kg
Power: 160 W at 28 V dc

Cdu (optional)
Dimensions: 147 × 153 × 140 mm
Weight: 2·5 kg
Power: 40 W at 28 V dc

STATUS: in development. Pre-production models have been on flight trials at the Royal Aircraft Establishment, Farnborough.

NAVHARS navigation, heading, and attitude-reference system

This system combines a high-quality inertial platform with a powerful, dedicated mini-computer to produce information for navigation, weapon-aiming and aircraft management. A key feature is the choice of a pair of two-axis dry-bearing oscillogyros, developed by Ferranti, in place of the three single-axis devices traditionally employed for platform stabilisation. They use rotating beams as the gyroscopic mass. Changes in angle between the planes of rotation of the beams and the platform reference surface are used to measure platform displacement in two axes. Other aircraft sensors provide Doppler radar velocities, TAS, and flux-valve magnetic heading.

The 8/16P minicomputer was developed specifically by Ferranti to process the data generated by the platform and other aircraft equipment. A two-minute self-alignment can be accomplished on land or at sea. During sea

Ferranti FIN 2000 platform/electronics unit and cdu

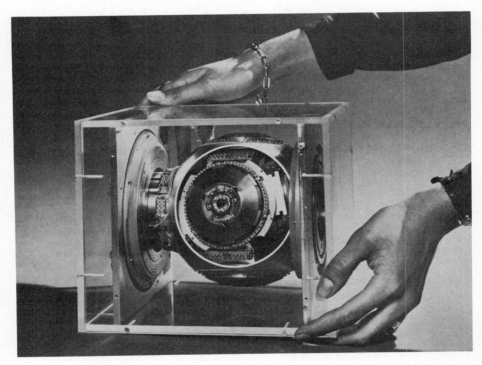

FIN 2000 gimbal system

Ferranti NAVHARS three-unit IN system for RN Sea Harrier

alignment certain compensations are made for ship's motion. Once airborne, alignment is further refined over a period of five minutes by damped, second-order erection techniques. Doppler velocities are corrected for sea-state.

The system can display on command up to ten waypoints, present position in latitude and longitude or grid co-ordinates, range and bearing to waypoint or destination, destination, speed, and track, magnetic or true heading, on-top and Tacan position-fixing, wind speed and direction, drift-angle and ground-speed, and fuel remaining.

Inertial platform unit
Dimensions: 8·345 × 8·464 × 13·06 in (212 × 215 × 331·7 mm)
Weight: 26·25 lb (11·9 kg)

Processor unit
Dimensions: 10·28 × 7·84 × 15 inches (261·1 × 199·1 × 381 mm)
Weight: 30 lb (13·64 kg)

Cdu
Dimensions: 5·77 × 6 × 5·47 inches (146·6 × 152·4 × 138·9 mm)
Weight: 5·5 lb (2·5 kg)
Power: 200 V 3-phase 400 Hz + 28 V dc for switching and lighting

STATUS: in production for Royal Navy/British Aerospace Sea Harrier V/STOL close support aircraft.

Sperry Gyroscope
Sperry Gyroscope, Downshire Way, Bracknell, Berkshire RG12 1QL

Laser-gyro IN
In October 1981 a strapdown laser-gyroscope inertial navigation system, designed and built by Sperry Gyroscope, began airborne trials in a Comet IV at Britain's Royal Aircraft Establishment, becoming the first such equipment of British origin to fly. It marked a milestone in the RAE's laser-inertial development programme, and represents more than three years' work by the Bracknell company.

Work formally began in October 1978 when the Ministry of Defence (PE) awarded Sperry a contract to develop a laser strapdown IN system for military aircraft. The demonstration equipment flown in 1981/82 comprised a three-axis laser system, three conventional accelerometers, a digital processor, and a display unit. Follow-on development contracts were expected in early 1982.

There is no attitude-stabilised platform as in a conventional electro-mechanical IN or attitude-reference system. Instead, attitudes in the three axes are measured by integrating rates of rotation provided by three ring-laser units mounted at right-angles to each other. Each unit contains two laser beams that travel around a closed path in opposite directions. Rotation of the unit in the plane of travel of the beam creates a path difference between the two which is sensed as a phase difference. The information is computer-processed to derive change of attitude from the horizontal.

The accelerometers are likewise mounted rigidly to the inside of the case, and their signals have to be corrected by computer to compensate for changing aircraft attitude to derive inertial velocities.

Advantages claimed for laser-gyro IN equipment compared with current electro-mechanical systems are lower first cost resulting from the smaller number of components and therefore lower labour costs, lower maintenance costs as a result of greater reliability and smaller costs when attention is needed, improved fleet reliability stemming from the same advantages, and substantially reduced warm-up times.

STATUS: development.

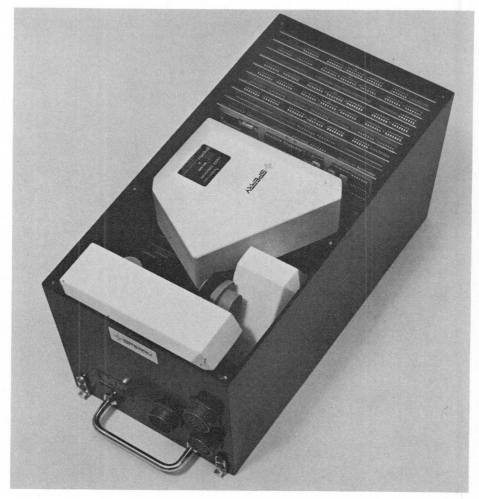

Mock-up of Sperry Gyroscope's laser-IN platform

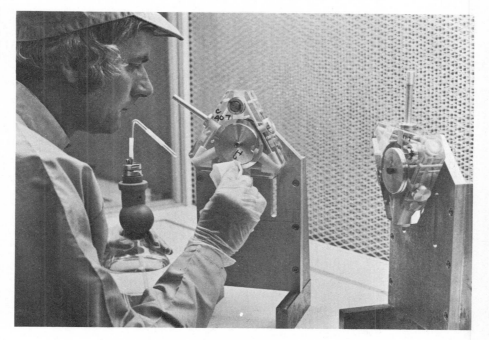

Laser-gyro rate sensors being tested at Sperry Gyroscope

UNITED STATES OF AMERICA

Honeywell

Aerospace & Defence Group, Honeywell Inc, Honeywell Plaza, Minneapolis, Minnesota 55408

In January 1981 Honeywell delivered to Boeing for the new-generation 757 and 767 commercial transport aircraft the world's first production laser-gyro inertial reference system. These two airliners represent the first production applications of the ring-laser gyro, heart of the laser inertial reference system. The Honeywell Avionics Division at Minneapolis was claiming then to be the only plant in the world in quantity production of ring-laser gyros, a position that stems from a Boeing contract calling for the delivery of 400 inertial reference systems up to the end of 1984, with an option on a further 200. Each system consists of three ARINC 700 series digital inertial reference units.

A Honeywell system installed aboard an Air France Boeing 747 for flight trials demonstrated in early 1981 an accuracy of 0·26 knots on its initial 7 hour 51 minute flight from Chicago to Paris. By comparison, the system specification called for 2 knots. Residual velocity on landing at Paris was only three knots.

In August 1981, the company delivered to McDonnell Douglas for test and evaluation on the AV-8B Harrier V/STOL fighter the first military laser-gyro inertial navigation system. Previously, towards the end of 1980, the USAF had completed testing a Honeywell ring-laser gyro navigation system on a Lockheed C-141 transport at the Central Inertial Guidance Test Facility at Holloman AFB, New Mexico. Results from these tests indicated a circular error probability of 0·88 nm/h and a velocity error of 2·9 ft/s.

Apart from commercial and military aircraft applications, Honeywell also produces laser gyros for missiles and spacecraft.

Honeywell has been developing ring-laser gyros for more than 18 years, and the first-generation strapdown systems employing them began flying in 1974. Standards of performance equivalent to INS equipment's was first demonstrated in flight trials at Holloman AFB in the following year, and have since been confirmed in subsequent flight tests by the US Navy, NASA, McDonnell Douglas, Boeing, and Honeywell. Environmental immunity during simulated combat manoeuvres and carrier operations has been demonstrated. More than 600 hours in nine different aircraft types had been flown up to the end of 1980 with no flight failures or resulting maintenance.

Ring-laser gyro strapdown inertial reference system for Boeing 757/767

The world's first production ring-laser gyro inertial reference system was chosen by Boeing as part of the avionics package common to both the 200-seat 767 and 170-seat 757 airliners. The 767 is some five months ahead of the 757 in production and was due to enter service with its first customer, United Airlines, in August 1982.

The strapdown configuration is so called because the gyro-stabilised platform of current conventional IN and attitude-reference systems is replaced by three ring-laser gyro units mounted rigidly and at right-angles to one another. The laser gyro detects and measures angular rates of motion by measuring the

Honeywell's laser-IN facility

frequency difference between two contra-rotating laser beams made to circulate (hence the term 'ring') in a triangular cavity by mirrors. When the units are at rest the distances travelled by each beam are the same, and the frequencies are the same. When the unit rotates, one path lengthens while the other shortens and so a frequency difference is established, proportional to the rate of rotation of the unit. The difference is measured and processed digitally in ARINC 704 format as aircraft attitude in pitch, roll, and yaw.

The accelerometers are likewise mounted rigidly in the box, at right-angles to one another. Their signals are therefore related to aircraft axes, and have to be processed to convert them to an external inertial reference frame.

Honeywell claims that the laser gyro irs will cost significantly less than an ARINC 56 IN system, and less than any other combination of devices capable of offering the same functions and performance. A strapdown system has no

moving parts to wear, fail, or become misaligned; no gimbals, torque motors, spinmotors, slip-rings, or resolvers, and no scheduled maintenance, realignment, or recalibration requirements are anticipated.

Outputs: primary attitude information to displays and afcs, linear accelerations and angular rates to afcs, velocity vector to afcs, wind-shear detection and energy management, magnetic heading for displays and afcs and long-range navigation data
Specification: ARINC 704
Size: 10 MCU
Weight: 44 lb (20 kg)
Power: 96 W
Accuracy (10 h flight): position 2 nm/h (95% probability), velocity 12 knots (95% probability)
Reliability: mature mtbf predicted to average > 4500 h
Self-test: BIT (initiated and continuous) will detect 95% of failures with 95% confidence level

STATUS: in production.

Kearfott

Kearfott Division, The Singer Company, 1150 McBride Avenue, Little Falls, New Jersey 07424

SKN-2416 system for USAF/General Dynamics F-16 multi-national fighter

In April 1975, after months of intensive competitive evaluation, the USAF chose Kearfott's SKN-2416 INS for its new F-16 lightweight fighter programme, and awarded the company funds for full-scale development. This award was the third major contract for the SKN-2400/2600 navigation and nav/attack system. In 1973 France had chosen it for the Dassault Super Etendard carrier-based fighter, and in the following year Sweden followed suit for the JA-37 interceptor version of the Saab Viggen multi-role fighter.

The system assumed international importance when, in July 1975, Europe's 'second tier' countries Belgium, Denmark, The Netherlands, and Norway, chose the F-16 as their F-104G Starfighter replacement. It has become part of the largest-ever international collaborative defence programme.

The SKN-2416 is a prime sensor for aircraft velocity, attitude and heading, and also the principal source of navigation information. It comprises three units: an INU (inertial navigation unit), fire-control/navigation panel, and an emergency battery and mounting.

The INU is a self-contained navigator consisting of an inertial platform and ten electronics cards packaged into a single line replaceable unit. The heart of the unit is an inertially stabilised platform with four gimbals for all-attitude operation, vertical and azimuth gyros, and three sub-miniature, pendulous linear accelerometers. The INU operates in conjunction with a cockpit-mounted control unit for entering, changing, or recalling digital computer data; displaying navigation, weapons delivery, and operational status data; controlling electrical power to the INU, fire-control computer, and target identification laser; and providing control facilities to the INU.

The system can be updated or driven with present position information (latitude and longitude), velocities in INS co-ordinates, gimbal angles about INS axes, and gyro torquing rates (all in MIL-STD-1553 format). Outputs, also on the data-bus format, are: present position (latitude, longitude, altitude), aircraft attitude (pitch, roll, heading true or magnetic), aircraft velocity (horizontal and vertical), and steering information (track angle error).

There are five operating modes: stored heading alignment, normal alignment, navigation, attitude reference, and calibration. The first of these is a rapid-align mode which permits entering navigation to an accuracy of better than three nm/h (4560 m/h) circular-error probability within 90 seconds provided the aircraft has not been moved during shutdown (the computer stores the last heading recorded before shutdown). In the normal mode the system aligns itself within eight .minutes, with a corresponding accuracy of one nm/h (1850 m/h).

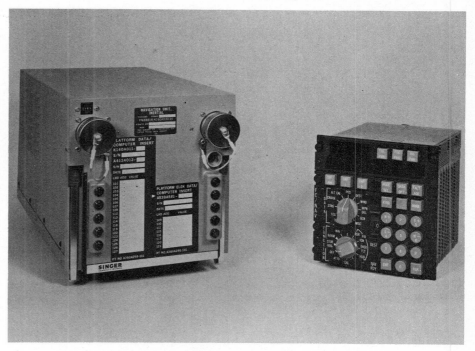
Singer-Kearfott SKN-2416 INS for GD F-16 fighter

INU
Dimensions: 7½ × 7½ × 15⅕ inches (190·5 × 190·5 × 386·1 mm)
Weight: 33 lb (15 kg)
Interface: MIL-STD-1553 data bus
Performance: accuracy better than 1 nm/h (1850 m/h)
Rapid-align time: 9 minutes at 0°F (–17·8°C)

Fire-control/navigation panel
Dimensions: 5¾ × 6 × 7³/₁₀ inches (146·1 × 152·4 × 185·4 mm)
Weight: 8·25 lb (3·75 kg)

INU battery and mounting
Type: Nickel-cadmium
Endurance: up to 10 s during interruption of primary power supply. Battery charged through INU during normal operation

STATUS: in production.

Litton

Litton Aero Products, 26540 Rondell Street, Calabasas, California 91302

Litton claims to be the world's largest supplier of aircraft IN equipment. Some 18 000 such systems have been built, representing (says the company) about 80 per cent of the total free world deliveries. Aero Products was set up as a separate division in 1969 to concentrate on aircraft navigation equipment.

The feasibility of inertial navigation for aircraft was demonstrated shortly after the Second World War with a system weighing more than 2000 lb (900 kg). Soon after Litton Industries was formed, inertial technology received initial R & D support and within two years a 200 lb (90 kg) system was demonstrated. Initial production contracts were awarded in 1957, and first-generation INS found its way onto the US Navy's WF-2, A-6A, E-2A, P-3, the USAF's F-4 and F-111, and NATO'S F-104G Starfighters. More than 9000 systems were delivered over the 15 years since

Litton system AN/ASN-130 for F-18 and AV-8B fighters of USN

Litton's AN/ASN-63 first-generation INS for F-4 fighter-bombers

1958, when the first production equipment was delivered.

Development work on second-generation INS, characterised by the introduction of digital technology, began in 1960, and led to the first production orders in 1967 for CAINS (carrier aircraft inertial navigation system), and some 1300 sets now equip S-3A, F-14A, A-6E, E-2C, and the USMC RF-4B. The US Army ordered Litton equipment in 1967 for its OV-1D and

RU-21 forward air control and reconnaissance aircraft, and in the following year the USAF chose a system for its B-52s.

McDonnell Douglas in 1970 placed an initial order for the third-generation equipment that had meanwhile been under development, and the F-15 Eagle became the first aircraft to benefit from 40 per cent improvements in size and weight. Since that time more than 1100 sets have been built for F-15, F-5E/F, Mirage,

LTN-72 INU opened to show platform and electronics cards

Canberra, Atlantic, Bandeirante, F-4, F-104, and in 1977 the Navy chose Litton for its new F-18 carrier fighter.

Similar acceptance of these successive families of IN equipment has been recorded by the airline industry.

LTN-72

The LTN-72 is a self-contained, all-weather, easy-to-maintain, world-wide navigation system for commercial aircraft that is independent of ground-based aids of any kind. Designed with the capability to incorporate RNAV, the INS provides continuous position, navigation, and guidance data.

The system comprises three units: mode selector, control/display, and inertial navigation box. The first is used to energise and align the system prior to flight and to select navigation or attitude reference modes of operation. The control/display unit permits the crew to enter present position and waypoint co-ordinates, select track steering, and display information generated by the system. The inertial navigation unit houses the gimbal structure with its gyros and accelerometers, associated electronics, power supply, and data converter. Ease of maintenance has been emphasised: for example, the principal mechanical elements – gyros and accelerometers – can be removed and replaced in 20 minutes using only screwdrivers. The gimbals are cantilevered, permitting the servo electronics to be mounted directly on the platform. This permits the use of flexible leads instead of slip-rings in some cases, improving reliability. The platform has only two slip-rings compared with four on a conventional platform. The LTN-72 has very extensive self-test and failure-detection facilities; the system complies with ARINC 561 in that the probability of an undetected failure in attitude during the last 30 seconds before touchdown is less than 1 in 10^6. Again, an analogue output test feature permits tests not only of the IN system but also of the flight instruments by driving them to various test readings.

INU
Dimensions: $10^1/_5 \times 8^5/_8 \times 19$ inches ($267 \times 219 \times 507$ mm)
Weight: 59 lb (26·8 kg)

Cdu
Dimensions: $5^3/_4 \times 4^1/_3 \times 6^1/_5$ inches ($146 \times 114 \times 157$ mm)
Weight: 5 lb (2·3 kg)

Msu
Dimensions: $5^3/_4 \times 1^1/_2 \times 2$ inches ($146 \times 38 \times 51$ mm)
Weight: 1 lb (0·5 kg)

Number of waypoints: 9 plus remote entry capability
Waypoint offset capability: up to 399 nm worldwide
Display: 7-segment incandescent numerals for all ARINC terms, together with display of INS parameters on flight director, HSI, or remote indicators

Litton's second-generation INS, AN/ASN-92, for F-14 swing-wing fighters

Third-generation Litton IN system, AN/ASN 109, for F-15 fighter

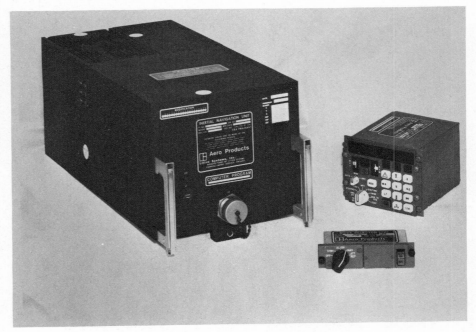

Three-unit LTN-72 INS for airlines

Inputs: self-contained inertial guidance. Avionics interface is designed to ARINC 561 and 575, and compatible with all flight directors and autopilots
Outputs: actual track, track angle error, cross-track and desired track, plus access to computer during flight for great circle distance computations.
Power: 115 V, 400 Hz

STATUS: in production.

LTN-72R IN/RNAV with VOR/DME update

The LTN-72R area-navigation system is a development of the LTN-72 with automatic radio position update, automatic omega position update, and triple-system mixing capability. It may be operated in the RNAV mode, using range and bearing information from selected Vortac (VOR/DME) stations or from a combination of omega transmitters, providing very high accuracy independent of time.

The system uses newly developed gyros, platform, and accelerometers, and a new, expansible C-4000 digital computer.

Number of waypoints: 9, plus remote entry capability
Waypoint offset capability: up to 400 nm worldwide
Display: as for LTN-72
Inputs: as for LTN-72 plus data on up to 9 VOR and DME stations
Outputs: the system is certificated for en route, terminal, and approach categories. It provides crosstrack, ground-speed, position (both updated and 'raw'), waypoints, Vortac position, frequency, and elevation, update code, magnetic variation, distance and time-to-go, wind speed and direction, desired track, status, self-test, and malfunction codes
Memory: UV-ROM (ultra-violet erasable only) for optimum protection with flexibility for change
Power: 115 V, 400 Hz
Weight: 61 lb (27·7 kg)

STATUS: in production.

LTN-72RL

The LTN-72RL is an advanced, world-wide area-navigation system that can automatically update itself by radio navigation fixes. It has a control/display unit that functions as an 'intelli-

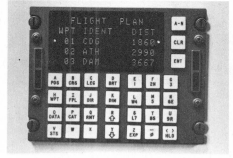

LTN-72RL cdu with multi-line led symbols

gent' data terminal and incorporates a 5-line × 16-character light-emitting diode display for presentation of operator-entered or computer-processed data. Waypoint data and VOR/DME locations can be pre-stored in the computer, and the system contains an algorithm (mathematical model) of magnetic variation that can be used to compute magnetic heading, track, and desired track, independently of the aircraft compass system; this algorithm is limited to latitudes between 60°N and 60°S. The pre-stored data base contains information specified by the operator on selected vhf navaids, airports, and some high-altitude waypoints. This bulk data is programmed in read-only memory.

A section of electrically alterable memory is allocated for particular waypoints or fixes not contained in the standard data bases; up to 160 routes, with an average of 20 waypoints per route, can be stored in this way and recalled for use at any time. Total number of waypoints in all routes is limited to 3200.

STATUS: in production. Certificated in July 1981 for Saudia Airlines' Boeing 747.

LN-39 INU

Litton's LN-39 inertial navigation unit is the sensing and data-processing device that has been chosen by the USAF as the basis of the standard inertial navigation system on the Fairchild A-10 anti-tank and battlefield bomber. The unit contains a P-1000C platform, stabilised by two G-1200 gyros and mounting three A-1000 accelerometers, and an LC-4516C general-purpose computer. An optional control/display unit controls the system and selects and displays data and status/failure information. Particular emphasis has been placed on reliability, and this is accomplished partly by the use of LSI/MSI and hybrid components, allowing a significant parts-count reduction.

INU

Dimensions: $7^1/_2 \times 7^3/_5 \times 18^1/_{10}$ inches (190·5 × 193·04 × 459·7 mm)
Weight: 38·2 lb (17·36 kg)

Cdu

Dimensions: $5^3/_4 \times 6 \times 7^3/_{10}$ inches (146 × 152·4 × 185·4 mm)
Weight: 8·2 lb (3·73 kg)

Power: 180 W (running), 550 W (starting)
Accuracy: 0·8 nm/h (1482 m/h) cep, 2·5 ft/s (0·86 m/s) (gyro compass); 3 nm/h (4560 m/s) CEP, 3ft/s (0.914 m/s) (stored heading)
Align time: 8 minutes (gyro compass), $1^1/_2$ minutes (stored heading), at 70° F (21°C)
LC-4516C computer: 16-bit single or 32-bit double precision, 65 536 words, semi-conductor RAM/ROM or EPROM 24K words total
Mtbf: Specified goal, 740 h over full military environment

STATUS: in production for USAF/Fairchild A-10 bomber and certain other test programmes. Firm contracts and options for the LN-39 represented 2500 sets by mid-1981.

Northrop

Electronics Division, Northrop Corporation, 2301 West 120th Street, Hawthorne, California 90250

NAS-26 stellar-inertial system

Northrop's involvement with stellar-inertial navigation equipment goes back to the late 1940s, when the new, first-generation jet bombers being planned or in production were in need of highly accurate, long-range navaids independent of ground beacons or give-away airborne transmissions. The long-term accuracy of the early inertial systems was far from adequate to meet navigation and weapon delivery needs, and so manufacturers developed ways to update them in flight. Northrop's answer was a star-tracker that could pin-point the position of the aircraft by reference to selected bright stars, and use this information to correct the gradual drift of the basic IN system.

A particular advantage of the star-tracking method was that the much higher cruising altitudes of the jet bombers (for better fuel efficiency and greater protection against interception) put them above almost all cloud and haze, where the stars were always visible.

Whereas the emphasis has now changed to low-level delivery, almost always below cloud levels (at least in Europe), stellar-monitored inertial systems continue to have wide-scale applications, eg in long-range military transport aircraft and strategic missiles.

The NAS-26 provides continuous position, velocity, and attitude information, and can interface with Doppler, Tacan, GPS (America's satellite-based global positioning system),

ground-mapping radar, altimeters, and missile systems. It comprises four units: an astro-inertial instrument, digital computer, control/display box, and power supply. The astro-inertial unit is the movement and position sensor, containing a three-gimbal, stable reference platform, with two-degrees-of-freedom star tracker, two gyros, and three accelerometers mounted on it. The tracker conducts a square search over large angular regions of the sky, picking up and identifying an average of three stars a minute by day or night. The computer is a fully qualified, general-purpose military instrument designed for airborne applications needing high data rates. The ephem-

erides (positions and visibilities) of 61 stars are stored in the computer's memory and permit celestial navigation anywhere in the world.

Accuracy: air- or ground-aligned: astro-inertial (occasional star tracking), better than 1000 ft (304.8m) cep after 10 h; pure inertial (no star tracking), less than 0·5 nm/h (926 m/h)
Attitude readout: better than 25″
Velocity: 0·5 ft/s (0·152 m/s) per axis typical error
Weight: 184 lb (83·64 kg)
Volume: 3·75 ft³ (0·11 m³)
Reliability: 800 h Mtbf

STATUS: in production.

Northrop's NAS-26 stellar/IN system being tested

Omega/VLF Navigation

CANADA

Canadian Marconi Company

Canadian Marconi Company, 2442 Trenton Avenue, Quebec, H3P 1Y9

More than 1200 CMC omega installations have been completed on over 90 aircraft types. CMC ON systems have been chosen by some 40 airline and charter operators, more than 300 general aviation operators, and over 18 military and government users.

CMA-719

This system was developed from the outset for military applications, though several commercial installations were completed before the CMA-734 and CMA-740 series were introduced to replace it. It was one of the first fully automatic airborne receiver and navigation systems to go on the market. Flight testing took place in 1970, and by late 1971 the system had been qualified and was in initial production.

STATUS: still in production for Canadian Armed Forces.

CMA-734

This fully automatic navigation system is based on the very-low-frequency signals transmitted from the eight omega stations around the globe, but an optional vlf receiver permits access to the US Navy's vlf radio network used to provide communications with its fleet of nuclear submarines.

The system was designed for aircraft in which small size and low weight are particularly important, such as business aircraft and light/medium helicopters. Standard features are: provision for heading and speeding inputs, course deviation indicator outputs (cross-track deviation and discrete signals, eg on/off), autopilot-steering roll command, automatic modal interference detector, choice of either hyperbolic or relative navigation mode, and updating from VOR range and bearing inputs. An optional digital-to-synchro converter unit extends the CMA-734's output capability to provide a synchro output with a choice of heading, track, drift angle, or track-angle error, to other instruments and equipment, eg HSI.

Number of waypoints: 9
Outputs: cross-track deviation, to/from indication, waypoint alert, DR warning signal, failure warning signal to autopilot, steering commands
Inputs: heading, speed
Power: 1·8 A at 28 V dc
Number of units: 3: antenna coupler, receiver/processor, control/display (digital/synchro unit makes a fourth)
Weight: 24 lb (10·9 kg)
Dimensions: receiver/processor ½ATR short; control/display 5¾ × 4½ × 6⅛ inches (146 × 114·3 × 155·6 mm); antenna coupler (suppressed plate) 2⅜ inches high × 7 inches flange diameter × 5 inches body diameter (60·2 × 177·8 × 127 mm)

STATUS: in production.

CMA-771

This system is a development of the CMA-740 conforming to ARINC 599 for omega. The basic version of the CMA-771 replaces the earlier CMA-740, providing an additional interface capability by means of plug-in modules. Incorporation of ARINC 561 and 575 digital data-bus circuitry permits integration with a wide range of other navigation systems, such as IN, digital air data, radio, Doppler, and flight-management. High-level discrete outputs and synchro analogue signals are also available to drive HSIs. The US Navy's very-low-frequency network is accessible through use of a plug-in module.

Two navigation modes are available: absolute hyperbolic, using the omega stations, or relative, employing a combination of omega and vlf. Either can be selected as the primary mode by the customer during installation. Thereafter the equipment will automatically employ the chosen mode for all navigation functions, though the alternative mode can be selected at any time through the control/display unit. The hyperbolic mode permits position ambiguities of up to 36 miles to be resolved under most conditions, and makes it particularly suitable for long over-water flights or where reference navaids are not available. Accuracy is dependent upon the number of signals usable at any time, and their geometry, but is independent of time.

The CMA-771 interfaces with a variety of other equipment, and so is appropriate to a broad range of applications. Autopilot and HSI interfaces that include a steering command signal, cross-track deviation, two synchro drive outputs, a variety of discretes (eg on/off, to/from, etc) and alerts qualify the CMA-771 as a primary navigation system. The equipment is also suited to applications such as maritime patrol or reconnaissance where integration with radar or a camera system may be specified.

Number of waypoints: 9
Inputs: heading, speed
Power: 125 VA at 115 V 400 Hz (including cdu and vlf receiver)
Number of units: 3: loop antenna coupler, receiver/processor, control/display
Weight: 35-40 lb (16-18 kg) depending on antenna configuration
Dimensions: receiver/processor ¾ATR long; control/display 5¾ × 4½ × 6⅛ inches (146 × 114·3 × 155·6 mm)

STATUS: in production

In development for the CMA-771 is an alpha-numeric control/display unit that will permit up to 30 waypoints to be stored and displayed together with their international identifying codes. The new system will simplify operating procedures since the 'alpha' section of the display can be used to give 'next step' instructions to the crew when it is not diplaying waypoints.

CMA-740

Utilising the CMA-734 receiver, interface, and computer, the CMA-740 is adapted for airline use and is packaged into a ¾ATR unit. It has now been superseded by the basic version of the CMA-771.

Cdu for Canadian Marconi's CMA 771 omega/vlf system

FRANCE

Crouzet

Crouzet SA, Division Aerospatial, 25 rue Jules Vedrines, 26027 Valence Cedex

Equinox Type ONS 200 omega

Designed for dual installation in commercial aircraft, the Type 200 omega comprises three units: an antenna coupler, receiver/processor and control/display unit. The system operates in conjunction with the world-wide vlf coverage provided by the eight omega stations.

Number of waypoints: 10
Inputs: ARINC 407 synchro heading, ground-speed and Doppler, ARINC 419 TAS from other omega system, ARINC 571 digital ground-speed and track, and ARINC 575 digital TAS
Outputs: 28 V dc discretes, ARINC 419 digital information to cdu and aircraft other equipment, ARINC 407 heading, steering signals to auto-pilot, track deviation

Receiver/processor

16-bit microprocessor, 16-bit address bus and data bus, cycle time 0·5 μs
Memory: (program) EPROM 12K words 16 bits (computation) RAM 4K words 16 bits, including 256 words in protected memory
Dimensions: ½ short ATR, 384 × 194 × 126 mm
Weight: 7·8 kg

CDU Type 10

Display: 13 numerals and 16 status lights
Dimensions: ATI 146, 146 × 114 × 164 mm
Weight: 2 kg

Antenna coupler

Type: ferrite receiver frames with built-in preamplifier
Frequency range: 10 – 14 kHz
Dimensions: (ARINC 580 type) 311 × 235 × 48 mm
(small 'brick' type) 176 × 160 × 44 mm
Weight: (ARINC 508 type) 3·3 kg
(small 'brick' type) 1·6 kg

STATUS: in production.

Equinox Type ONS 300 omega

This is a more advanced version of the ONS 200, using the same antenna coupler, but with a receiver/processor program memory that can store 18K words of 16 bits in place of the other unit's 12K words. This capability necessitates a cockpit display unit upgrade, from the Type 10 to the Type 20, with capacity for 30 waypoints instead of ten. Elsewhere the ONS 300 is identical to the 200.

STATUS: in production for Airbus Industrie A300, and chosen by Garuda and Tunis Air.

ONS Type 500 and 600

These are versions of the ONS Type 200 and 300 modified for military use. The antenna-coupler units are the same, but the Type 500 receiver/processor has 5K words of 16 bits for the memory program, while the Type 600 operates with 4K 16-bit words. They operate in conjunction with the Type 10 and Type 20 cdus.

STATUS: in production.

UNITED KINGDOM

Beattie-Edwards

Beattie-Edwards Aviation Ltd, 20 Norman-hurst Close, Three Bridges, Crawley, Sussex RH10 1YL

The prototype of this hand-held omega navigation aid was displayed at the Eighth Business and Light Aviation Show at Cranfield, England, in September 1981.

The control/display unit measures approximately 7 × 3½ × 2 inches (177 × 88 × 51 mm), and is powered by nickel-cadmium batteries; alternatively, it can be plugged into a panel-mounted cigar lighter. Latitude, longitude and other information is presented on a liquid-crystal display, and the co-ordinates of up to nine waypoints can be entered by a 16-button array. The system updates itself every ten seconds (the cycle time of the eight Omega stations). As a flight progresses the display shows cross-track distance and ground-speed as well as distance to the next waypoint and time of arrival there. When close to the next waypoint the display flickers to draw attention to its passage. Minimum and maximum speeds are 50 and 300 knots.

The antenna is designed to fit on top of the dash panel or glare shield of light aircraft without causing any interference to compass or radio systems.

STATUS: in development. Initial market cost was estimated at the end of 1981 to be around £1000, or about a fiftieth of the cost of a transport-aircraft omega system.

Beattie-Edwards' omega system

UNITED STATES OF AMERICA

Global Navigation

Global Navigation Inc, 2144 Michelson Drive, Irvine, California 92715

Global Navigation, a part of Sundstrand Corporation, was one of the companies that pioneered the use of the US Navy's world-wide network of eight vlf radio stations for commercial aviation. These transmitters radiate 50 to 1000 kW in the 16 to 24 kHz band for communicating with nuclear submarines.

GNS-500

In 1974 the company introduced the GNS-500, a system that combined vlf reception with microprocessor data-handling to provide ten position updates per second. This facility was particularly suited to the new generation of fan-engined, long-range business and executive aircraft, since pilots could now programme their flights by entering the co-ordinates of departure points, waypoints, and destinations.

GNS-500A

In 1976 the omega network of very-low-frequency transmitters, also eight in number, was introduced, and in May of that year was approved by the FAA for IFR en route navigation. In anticipation of this development, Global had been perfecting the GNS-500A, which could receive the new stations as well as the vlf transmitters, and also in May 1976 this equipment was FAA certificated for direct, en route operation in domestic US airspace.

GNS-500A Series 2

In response to customer requests, improved equipment was introduced to offer non-volatile retention of waypoints and present position, automatic computation of magnetic variation, and a continuous clock to maintain Greenwich mean time and date.

Number of waypoints: 10
Outputs: position, track, distance and time to destination, ground-speed, HSI, wind speed and direction
Inputs: heading, true airspeed, ARINC 571 bus interface with IN, radar, RNAV, and navigation-management systems
Power: 5 A at 28 V dc
Number of units: 3: control/display, receiver/computer, optional equipment
Weight: 38 lb (17·27 kg)
Approval: North Atlantic, domestic en route IFR, as sole over-water navaid, and in remote areas world-wide.

GNS-500A Series 3

The Series 3 was introduced in 1980 to answer the need for more waypoints and a faster and clearer presentation of data from any combination of up to five letters and numbers, on nine flight plans of up to 20 waypoints each. Information is now presented on a sunlight-readable CRT providing a 14-character by 8-line matrix. In the Series 3 system all data is organised into one or other of three sections:

navigation, data, and flight plan. Each section in turn contains information laid out on a number of 'pages'.

Number of waypoints: 127
Outputs: position, track, distance and time to destination, ground-speed, HSI, wind speed and direction
Inputs: heading, true airspeed, ARINC 571 bus interface with IN, radar, RNAV, and navigation-management systems
Power: 5 A at 28 V dc
Number of units: 3: control/display, receiver/computer, optional equipment
Weight: 39 lb (17·72 kg)

GNS-500A Series 3B

This refinement was being introduced in September 1981, and features trip planning, fuel planning, and offset waypoint facilities. In the trip-planning mode, the system computes distance and time for each leg as well as total distance and time for the flight. When the fuel-planning mode is selected, the system computes time and distance to the reserve point based on ground-speed and manually inserted values for fuel remaining, fuel flow, and desired reserve. It also displays fuel efficiency in terms of nautical miles flown per 100 lb (45·36 kg) of fuel consumed. The offset-waypoint facility makes it possible for the crew to set up an artificial waypoint based on distance and bearing from a real waypoint, so avoiding the necessity of having to overfly the actual waypoint.

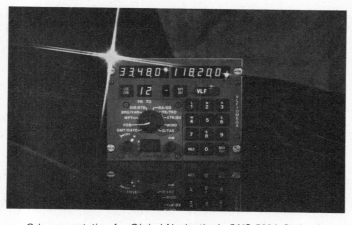

Cdu presentation for Global Navigation's GNS-500A Series 2

Global's GNS-500A Series 3 cdu

Litton

Aero Products, Litton Industries, 6700 Eton Avenue, Canoga Park, California 91303

LTN-211

In 1979 the FAA certificated the Litton omega system, in dual installation form, as meeting all requirements for sole navaids over the North Atlantic on Olympic Airways' B-707s, Icelandair's B-727s, and the DC-8s of Airlift International and Intercontinental Airways. Previously omega had been authorised only in conjunction with other navigation systems, eg Doppler.

The LTN-211 airborne equipment in conjunction with the omega world-wide network of very-low-frequency stations, or the US Navy's chain of vlf transmitters, provides a limited-error display of guidance information for long-range, great-circle navigation. It is programmed to adopt under certain circumstances a back-up, dead-reckoning mode of operation based on available aircraft velocity and heading. Switchover to the dead-reckoning mode is accomplished automatically when the number and quality of signals received falls below the requirement needed for acceptable position tracking and navigation.

Complete Litton LTN-211 omega/vlf system

When the initial position, time and date have been entered, the system automatically chooses the stations to be used on the basis of measured signal/noise ratios of the transmissions being received at the time. This method yields the best position accuracy since it considers both the quality of the stations selected as well as their propagation stability, and uses those stations least likely to be affected by diurnal signal changes. The basic LTN-211 uses signals of the three standard omega frequencies, ie 10·2 kHz, 11·33 kHz, and 13·6 kHz. Capability to process the fourth omega frequency, 11·05 kHz, is available as an option. In the omega/vlf system, this fourth frequency is used in conjunction with a vlf converter to process signals from the US Navy transmitters.

The system comprises three units: receiver/processor, control/display, and antenna coupler.

Receiver/processor
Dimensions: 7½ × 7⅗ × 19⅗ inches (190·5 × 193·5 × 49·8 mm)

Weight: 26 lb (11·8 kg)
Power: 62 W

Control/display
Dimensions: 5¾ × 4½ × 6⅕ inches (146 × 114·3 × 158 mm)
Weight: 3·8 lb (1·72 kg)
Power: 24 W

Antenna
Dimensions: 18 × 10½ × 1¾ inches (457·2 × 105 × 44·5 mm)
Weight: 8 lb (3·63 kg)
Power: 1 W

Number of waypoints: 9
Inputs: signals from all 8 omega stations and the US Navy vlf chain, in ARINC 599 format including TAS input for computation of wind speed and direction
Computer: TMS 9900 second-generation microprocessor, with 16 bits/word, and directly addressing 24 576 words of UV-ROM reprogrammable memory

Outputs: ground-speed, track, heading, drift angle, cross-track, track-angle error, present position in lat/long, waypoints, distance and time-to-go, wind, desired track, station status and signal quality, GMT, date, self-test, and malfunction codes. All outputs are to ARINC standard. Digital outputs, together with auto-pilot command steering signals and synchro outputs to the HSI can be provided
Failure detection: C9000 computer/processor has extensive BIT facilities within itself, as well as controlling the BITE section of the LTN-211 system. Probability of BIT system detecting a fault anywhere in the navigation system is above 99%.

STATUS: in production. LTN-211 equipment has been installed aboard Boeing 707, 727, 737, 747, McDonnell Douglas DC-8, DC-10, Grumman GII, GIII, Lockheed C-130, P-3, L-1011, Jetstar, Dassault Falcon, British Aerospace HS-125, Cessna Citation and Fokker F-27.

Rockwell
Collins Avionics Divisions, Rockwell International, 400 Collins Road NE, Cedar Rapids, Iowa 52406

LRN-85 omega/vlf
Designed for the world-wide navigation of commercial, corporate and military aircraft, the LRN-85 is one of the most comprehensive omega/vlf systems currently available. By comparing all available signals from omega and vlf ground stations against a precise rubidium frequency standard, the system can navigate a direct route between departure and arrival points, eliminating dependence on vortacs, and providing continuous position information to the crew.

As well as maintaining 100 per cent duty cycle on the omega transmitters, the system receives both data sidebands from the seven world-wide vlf stations.

A notable feature of the system is the mass memory, which can store the co-ordinates of all the world's vortacs and major airports, and the waypoint memory, with capacity to hold the co-ordinates of up to 100 pilot-entered waypoints. For the majority of airports, therefore, the crew need only enter the identifying codes of the start and destination airports, and the LRN-85 computes a direct course between them. Should this still not be adequate to define a route or series of regularly used routes, the crew can insert the appropriate waypoints and destinations and for convenience designate them by alpha-numeric codes.

Two dedicated micro-computers provide the LRN-85 with power and flexibility to drive or command many functions, eg provide desired track, cross track, and distance-to-go to a horizontal situation indicator; automatic and manual waypoint advance; roll command with turn anticipation; data-bus feed for automatic transmission of waypoint information to a second LRN-85 or other navigation system;

Control/display unit for Collins LRN-85 omega/vlf navigator

internally computed magnetic variation; true and magnetic display of bearing, desired track, and wind speed and direction; computed offset waypoint; automatic computation of diurnal shift; and test mode for station de-select and for computer diagnostics.

The system comprises five units: E-field vlf blade antenna or H-field loop, antenna coupler, optional equipment unit (containing the atomic frequency standard, a battery, and additional interfaces), receiver/processor, and control/display unit. In its ARINC 599 version, the

optional equipment unit is incorporated into the receiver/processor.

Dimensions: cdu 5¾ × 4½ × 6¼ inches (146 × 114 × 159 m)
receiver/processor 7½ × 7⅔ × 19⅔ inches (190 × 193 × 498 mm)
Accuracy: 2 nm cep with minimum of 2 usable stations
Qualification: ARINC 599

STATUS: in production and service.

Sperry
Commercial Division, Sperry Flight Systems, PO Box 2111, Phoenix, Arizona 85036

IONS-1020 and IONS-1030 inertial omega navigation system
Sperry Flight Systems has teamed with Canadian Marconi Company to develop a combined attitude/heading reference and omega system designated IONS-1020 (inertial omega navigation system). The equipment is based on the Sperry SRS-1020 strapdown AHRS and includes a new omega/vlf receiver being developed by Canadian Marconi. It will be compatible with any inertial reference system

designed to ARINC 561. A follow-on system designated IONS-1030 is also planned, to be compatible with IN equipment to ARINC 704.

The two companies foresee a large market for IONS equipment since, they claim, it will combine in one system the best features of inertial navigation and omega at less than half the cost of a pure IN system. Perhaps even more important is the saving in cost of ownership, projected at 35 to 50 cents per hour for maintenance, equivalent to about one-tenth that of IN.

The IONS-1020 contains both the omega/vlf navigation system and the strapdown AHRS elements in a 1 ATR long computer/processor/receiver line-replaceable unit, physically and

electrically interchangeable with an ARINC 561 INS sensor. The two other units which make up the system are the loop-antenna coupler and the control/display unit. As an ARINC 561 replacement, the AHRS/omega equipment provides position and waypoint navigation while at the same time supplying pitch, roll, heading, and acceleration data to flight-control and flight-instrument systems.

Heart of the computer/processor/receiver is the Sperry SDP-175 digital processor and interface electronics, which handle data from the gyros and accelerometers and from the omega/vlf circuits.

IONS-1020
ARINC specification: 569/599
Dimensions: processor/receiver, 1 ATR long; cdu, 5¾ × 4½ × 6⅛ inches (145 × 114·3 × 155·6 mm)
Weight: 43 lb (19·6 kg)
Mtbf: receiver/processor, 5000 h; cdu, 9000 h

Performance: flight-test demonstrates accuracies better than 0·1° in attitude, 10 knots ground-speed, and 0·5° in heading
STATUS: in development. By March 1981 the system had flown some 500 h aboard McDonnell Douglas DC-9-80, Boeing 727, and Lockheed L-1011 TriStar.

IONS-1030
ARINC specification: 704/599
Dimensions: receiver/processor, 10 MCU; cdu, 5¾ × 4½ × 6⅛ inches (146 × 114·3 × 155·6 mm)
Weight: approx 38 lb (17·27 kg)
Mtbf: to be determined

STATUS: projected.

Tracor
Tracor Aerospace, 6500 Tracor Lane, Austin, Texas 776414

Model 7800 omega/vlf
The Model 7800 is the latest and most advanced system of its kind to be built by this manufacturer. Developed from earlier equipment, it has the additional capability of receiving vlf signals without sacrificing performance when working in conjunction with omega stations. The system listens to omega and vlf simultaneously, identifies the best combination of signals, and automatically selects the most reliable navigation mode.

The Model 7800 can operate in its primary hyperbolic or circular mode, a secondary Rho-Rho-Rho mode, or in the unlikely event that the signals degrade to the stage where only two omega and/or vlf stations are usable, a tertiary Rho-Rho mode.

The antenna and receiver electronics are sensitive enough to respond to signal strengths below normal thresholds. A unique adaptive noise-blanker and frequency cross-coupler combine to improve the signal-to-noise ratio up to 20 dB. Another feature of the Model 7800 permits it to 'work through' modal interference events, a propagation phenomenon of omega that can cause serious navigation errors. Appropriate circuitry detects the presence of such interference and if necessary cuts out signals from the offending station.

The heart of the 7800 system is the Tracor Super SIMP MIL-specification microprocessor, first used in an airborne application in teleprinters aboard the USAF/Boeing E-4 airborne command post. To complement this unit is an ultraviolet-erasable, electrically re-programmable read-only memory, permitting fast and economical program changes. With a memory combination of 32K 12-bit ROM and 48K 8-bit RAM, the 7800 system is claimed to contain one of the most powerful omega/vlf computers available.

The system comprises three units: receiver/processor, control and display, and antenna coupler.

Tracor's Model 7800 omega/vlf system

Receiver/processor unit
Dimensions: ¾ ATR long
Weight: 29 lb (13·15 kg)
Power: 55 W typical, choice of 28 V dc or 115 V 400 Hz
Cooling: convection
Mtbf: >4000 h
Altitude: tested to 55 000 ft (16 760 m)
Certification: TSO C94

Control/display unit
Dimensions: 4½ × 5¾ × 6¼ inches (114·3 × 146 × 158·7 mm)
Weight: 4 lb (1·82 kg)
Power: 9 W typical
Environment: sealed unit
Display: sunlight readable
Mtbf: > 10 000 h
Certification: TSO C94

Antenna coupler unit
Configurations: 3 'E' field: ARINC 599, internal brick, or external teardrop, 3 'H' field omega/ADF, omega/ADF Boeing configuration, or omega blade
Certification: TSO C94

Cdu presentation for Model 7800 equipment

STATUS: in production for some 30 airlines and numerous corporate aircraft users. A recent customer was the Indian Air Force, for whose AN-12, HS-748, and Canberra fixed-wing and Mi-8 helicopter fleets the equipment is to provide the primary means of navigation.

Satellite Navigation
UNITED STATES OF AMERICA

Collins

Collins Avionics Divisions, Rockwell International, 400 Collins Road NE, Cedar Rapids, Iowa 52406

Navstar Global Positioning System

The Navstar GPS is a space-based system under development by the US Air Force that will substantially improve the effectiveness of under-sea, surface, and airborne vehicles. When the GPS system becomes fully operational in 1987 it will consist of a network of Navstar satellites circling the Earth, transmitting information to thousands of receiving sets so that users will be able to locate their positions anywhere on the Earth to an accuracy of ten to twenty metres. The information will include altitude, latitude and longitude, velocity to an accuracy of 0·1 metre per second, and time to a few nanoseconds.

The Navstar satellites will orbit the Earth at an altitude of 10 898 kilometres, each one broadcasting continuous time and position information. A limited constellation of five satellites is being used for development testing, to be augmented in 1984 by the first of the GPS operational satellites.

Collins' Government Avionics Division established itself as one of the leaders in this project when it successfully participated in the concept validation phase by building what was called the generalised development model under contract to the US Air Force Avionics Laboratory.

To cater for all classes of users with the greatest commonality of equipment, the Earth-based system comprises no fewer than 19 modules or units. For any given class of vehicle however, a smaller number will generally suffice to provide a service. For example, the Grumman A-6E, General Dynamics F-16, and Boeing B-52 would share three units in common – a controller to direct the antenna towards the satellite, a controlled-reception antenna, and a five-channel receiver. The A-6E and F-16 would in addition have a fixed-reception antenna, and the A-6E would further have an interface and a control/display unit. For all vehicles, the Collins system has in common 82 per cent of all software, 75 per cent of line-replaceable units, and 94 per cent of all shop-replaceable units.

For tactical aircraft GPS will facilitate rendez-vous, target positioning, weapons delivery, and recovery of aircraft. In addition it will have an advanced anti-jamming capability that will permit reliable and high-quality performance under the most severe conditions. The GPS tactical air equipment will permit continuous signal tracking during all manoeuvres, and will be integrated into other systems so that GPS information can improve mission effectiveness and expand the weather capability.

The air-breathing segment of the US strategic deterrent will likewise benefit from an improvement of its penetration and survival capabilities. The world-wide common grid position system available from GPS will improve strategic targeting and accuracy of weapons delivery. Stand-off weapons such as the US Air Force/Boeing AGM-86B air-launched cruise missile will also be more effective as a result of accurate and continuous position-updating.

STATUS: in full-scale development of user equipment for the GPS Joint Program Office at the US Air Force Space Division of US Air Force Systems Command.

Full set of customer equipment for Rockwell Collins Navstar satellite navigation system

Beacon Navigation
FRANCE

TRT
Télécommunications Radioélectriques et Télé-
phoniques, Defence and Avionics Commercial
Division, 88 rue Brillat-Savarin, 75640 Paris
Cedex 13

TAC-200 Tacan
This low-cost digital Tacan builds on the
company's experience in digital avionics including radio altimeters, distance-measuring
equipment, air traffic control transponders,
and ground-proximity warning systems. Software analysis of the timing and amplitude of all
received pulses permits good co-channel and
adjacent channel performance with no echo
susceptibility.

The transmitter is fully solid-state, and
signal-processing by micro-processor software
instead of discrete circuits reduces the number
of components. Extensive use of integrated
circuits keeps the power consumption low. The
two-unit system comprises the IRT-200 trans-
mitter/receiver and the INT-200 distance-
measuring equipment indicator.

Acquisition time: 1 s
Interrogation rate: 20 (track), 40 (search),
automatically varied by micro-processor
Sensitivity: better than –90 dBm
Accuracy: 0·02 n miles, 1°
Memory time: 10 s

STATUS: in production.

Principal units of TRT's TAC-200 Tacan navigation system

UNITED KINGDOM

Marconi Avionics
Marconi Avionics, Christopher Martin Road,
Basildon, Essex SS14 3EL

Type AD2780 Tacan
A follow-on from the Type AD2770 supplied for
the European multi-national Tornado bomber,
the AD2780 is also proposed for military
applications and will, says Marconi Avionics, be
lighter, smaller, and less expensive than its
predecessors with extensive use of large-scale
integration and micro-processing technology.
The system provides slant range and relative
bearing to a standard Tacan station, range rate
(which approximates to ground-speed when
not used in conjunction with the Marconi RNAV
system), time-to-go to waypoint or station,

ARINC 429 serial data output, and 252 channels
in x and y modes.

Frequency range: (transmitter) 1025 – 1150 MHz
(receiver) 962 – 1213 MHz
Range rate output: 0 – 999 knots with accura-
cies ±15 knots for 0 – 300 knots, and ±5% for
300 – 999 knots
Time-to-station output: 0 – 99 minutes
Dimensions: 5 × 6 × 12½ inches (127 × 153 ×
318 mm)
Weight: 7·7 lb (3·5 kg)
Tracking speed: 0 – 1900 knots, 0 – 20°/s
Memory: (range) 10 s
(bearing) 4 s

STATUS: in development.

*Transmitter/receiver for Marconi Avionics'
Type AD2780 Tacan*

Marconi Avionics
Marconi Avionics Ltd, Airport Works, Rochester, Kent ME1 2XX

AD 380 and AD 380S automatic direction finders
Latest in a number of automatic direction-
finding systems developed and produced by
Marconi, the AD 380 is designed to ARINC 570
and covers the frequency range 190 to
1799 kHz in 0·5 kHz steps while its variant, the
AD 380S, embraces the spread 190 to 1599·5
kHz. The latter, however, additionally covers the
international maritime distress frequency of
2182 kHz with the ability to tune to plus or
minus 0·5 kHz on either side of the nominal
frequency, rendering it particularly suitable for
search and rescue operations. This difference
apart, both systems are designed to the same
standard.

AD 380 systems have automatic, crystal-
controlled frequency selection and are of all-
solid-state construction with instantaneous

electronic tuning. Built-in test facilities are
incorporated.

A range of controller options are available,
permitting single, dual or programmable operation. Standard controllers come in three
versions, all of which provide selection of
frequency and mode adf, ant, test or bfo,
together with volume control. The decade
frequency selectors display the operational
frequency in inch (6·3 mm) high numerals.
Frequency controls comprise three concentric
knobs which operate the logic frequency
circuitry.

The first type of controller is the G4032E, a
single-frequency version, and the second type,
the G4033E, is used for tuning two receivers.
Mode facilities in each case are selected by
toggle switches, and each controller has a test-
button. The third controller type is the G4034E,
designed for rapid retuning of a single receiver
to either of two frequencies which can be then
selected by operation of a transfer switch when
a white bar is displayed across the figures of
that frequency not in use. Mode facilities are

selected by a rotary switch.

A programmable controller, the AA-3809,
conforms to the dimensions, form-factor and
electrical requirements of ARINC 570. It provides a four-channel preselect facility and can
control both versions of the AD380. Six push-
buttons permit instantaneous selection of frequencies previously entered into the memory
store, which retains information when the
equipment is unpowered.

The displayed frequency is always that to
which the receiver is tuned and the operator
can change stored information while in flight
or, alternatively select the N button which
allows normal operation of the controller. A
brightness control, which is independent of the
main aircraft panel lighting system, is incorporated. Normal mode facilities are also
available for pilot operation.

Dimensions: (receiver) ¼ ATR Short
Weight: (receiver) 10 lb (4·5 kg)

STATUS: in production and service.

AD2620 integrated navigation and control system

The AD2620 is a micro-processor-governed integrated navigation and control system based on the earlier AD260 navigation computer. It provides not only navigation facilities but also control of sensors such as Tacan and vor/dme and, in addition, can compute true airspeed from basic air-data information. Tas provides smoothing for Tacan and vor/dme signals and dead-reckoning navigation in the absence of Tacan and dme/vor. It can operate in either polar or cartesian co-ordinates and will automatically tune a Tacan or dme/vor when waypoints are selected. Other facilities include the provision of full slant-range correction and outputs to an afcs or weapon-aiming system. 'Fix' and 'Mark' facilities are also available. Data, entered by a control and display unit keyboard, is retained when power is cut off.

The system comprises three units, a navigation computer unit, a control and display unit, and a remote-readout unit. Each controls a micro-processor to allow simple interfacing with its companion units to form an ARINC 429 digital-data highway. The inclusion of a microprocessor in the control and display unit also permits a rapid display response to keyboard input while at the same time allowing display formatting to take place within the control and display unit itself. The three units have considerable spare computer capacity for future development and growth.

Features of the navigation computer unit include an expanded memory capacity for both programme and in-flight data acceptance, improved interface circuit design for better performance and higher integrity, and a new form of mechanical construction which makes each module a self contained entity and allows testing down to component level with simple equipment. A built-in test programme monitors all input and output signals as well as the computer software.

Dimensions: (NCU) 7³/₅ × 5 × 12¹/₂ inches (194 × 127 × 318 mm)
(CDU) 3³/₄ × 5³/₄ × 6¹/₂ inches (95 × 146 × 165 mm)
(RRU) 1¹/₅ × 3¹/₅ × 5⁷/₁₀ inches (31 × 80 × 144 mm)
Weight: (NCU) 7·7 lb (3·5 kg)
(CDU) 4 lb (1·8 kg)
(RRU) 0·75 lb (0·34 kg)

STATUS: in production and service.

AD2770 Tacan

The AD2770 Tacan navigation and homing aid is suitable for all types of aircraft. It provides range and bearing information from any selected ground Tacan station or from any suitably equipped aircraft, and is available in a number of forms offering outputs in digital or analogue form or a combination of the two. Output signals may be provided to drive range/bearing or deviation indicators on a pilot's panel or to interface directly with a computer.

The system has full 252 channel x and y mode capability and operates up to 300 nautical miles range with a range accuracy of better than 0·1 nautical miles. Bearing accuracy is said to be better than 0·7 degrees on normally strong input signals. It comprises basically two units, a transmitter-receiver, hard-mounted in the avionics bay, and a panel-mounted remote control unit. A switching unit, also installed in the avionics bay, is required when two antennas are fitted. A mounting tray with a cooling air blower is necessary if a cooling air supply is not available from the aircraft's own air-conditioning system.

The transmitter-receiver section with a digital interface only is contained in a ³/₄ ATR Short case with a front doghouse. Versions with analogue outputs are accommodated in a case of longer dimensions to house the additional circuitry. Signal-processing circuitry is largely digital in the interests of system reliability, and continuous integrity-monitoring techniques are used to eliminate the risk of erroneous outputs.

Range and bearing analysis and output formats are prepared in a general-purpose computer module which is referred to as the analyser. This allows both range and bearing signal processing to use the same circuitry, with a consequent reduction in the number of components. The range system uses a parallel search method, said to be unique, which by making use of all signal returns achieves a very rapid lock-on.

The AD2770 system operates in three modes, receive (giving bearing information only), transmit-to-receive, and air-to-air (providing range information only). Transmitter frequency range is from 1025 to 1150 MHz with an output of 2·5 kW peak pulse power. Receiver frequency coverage is from 962 to 1213 MHz. The tracking speed range is from 0 to 2500 knots.

The basic design of the AD2770 system is flexible both electrically and mechanically, and alternative configurations with appropriate form factors, output characteristics and mechanical and electrical interfaces can be provided for new aircraft or for retrofit.

Dimensions: (transmitter-receiver standard unit) 7³/₅ × 7¹/₂ × 15 inches (194 × 191 × 380 mm)
(control unit) 2¹/₄ × 5³/₄ × 3¹/₄ inches (57 × 146 × 83 mm)
(antenna switch) 2⁷/₁₀ × 5¹/₁₀ × 2¹/₅ inches (69 × 130 × 56 mm)
Weight: (transmitter-receiver standard unit) 31 lb (14 kg)
(control unit) 1 lb (0·45 kg)
(antenna switch) 0·55 lb (0·25 kg)

STATUS: in production and service.

UNITED STATES OF AMERICA

Bendix

The Bendix Corporation, General Aviation Avionics Division, 2100 NW 62nd Street, Fort Lauderdale, Florida 33310

ADF-2070 adf system

The Bendix ADF-2070 is a panel-mounted unit designed for the general aviation sector. It is claimed to be able to receive signals from exceptionally long ranges and has two sensitivity settings, extended range reception and conventional adf which is used primarily on approach. This extended range facility is provided by a 'coherent detection' feature which results in good reception characteristics with high immunity from thunderstorm and other static interference. Continuous digital tuning is employed to ensure lock-on to the desired frequency.

The system also features a blade antenna which serves both the communications radio and the adf systems. The adf sensor is installed in the base of the blade and feeds signals to the receiver via a small amplifier which is mounted adjacent to the antenna blade but within the aircraft skin. It is claimed that this configuration provides nearly twice the gain of other combin-ation-antenna units, results in a reduction in cable length, and no impedance-matching requirement.

Provision is made for correction of quadrantal error either on the ground or while airborne. The receiver output can drive either a standard adf indicator with rotatable azimuth card or an HSD-800 horizontal situation indicator.

Dimensions: 1³/₄ × 6¹/₄ × 9¹/₅ inches (45 × 159 × 234 mm)
Weight: 5·79 lb (2·63 kg)

STATUS: in production and service.

Collins

Collins Avionics Divisions, Rockwell International, 400 Collins Road NE, Cedar Rapids, Iowa 52498

51Z-4 marker-beacon receiver

The Collins 51Z-4 marker-beacon receiver automatically provides aural and visual indication of passage over airways and instrument landing system marker-beacons. The system is approved for Category II approaches in a number of Collins all-weather avionic systems certifications. Operating at a frequency of 75 MHz, the receiver sensitivity can be varied between two pre-adjusted levels through a cockpit HI-LO switch. The HI position is used to gain early indication of a marker beacon. The LO position is then subsequently used closer to the beacon for a sharper position fix. Alternatively, the receiver sensitivity is continuously variable from the cabin or flight-deck by means of a potentiometer.

The system is of all-solid-state construction and contains triple-tuned circuitry for the rejection of spurious signals generated by television and fm broadcast transmitters. It is designed for three-lamp indication but may be easily modified for single-lamp operation by removal of a resistor and wiring the three-lamp outputs together. In either type of operation, outputs can operate two sets of indicator lamps in parallel.

An optional self-test facility causes the internal generation of 3000, 1300 and 400 Hz marker signals which are detected in sequence by the receiver. The indicators light in order and the corresponding aural tones are also generated.

The unit is suited to retrofit installation since it is mechanically and electrically interchangeable with a number of other marker-beacon receivers, including the Collins 51Z-2 and 51Z-3 units. An associated marker-beacon antenna, the Collins 37X-2 system, is also available. Designed for operation with the 51Z-4 and other compatible receivers, the antenna is plastic-filled and sealed to reduce the effects of precipitation static. This unit, which weighs less than 1 lb (0·45 kg), can be mounted without cutting into the airframe, and has negligible drag.

Dimensions: ³/₄ ATR Short Low
Weight: (without self-test option) 3 lb (1·36 kg) (with self-test option) 3·23 lb (1·47 kg)

STATUS: in production and service.

TCN-40 lightweight Tacan

The TCN-40 brings digital Tacan technology to training, communications, utility, and other secondary-role aircraft and helicopters. While fulfilling the need for lower weight, size, and cost, the basic navigation functions are comparable with those of the company's full-capability systems. According to Collins, digital technology brings with it the following advantages: instantaneous channelling with no mechanical misalignment problems; nominal search and lock-on times of 1 and 3 seconds respectively for distance functions and bearing; accuracies of 0·1 nautical miles for distance measurement and 1 and 2·5 degrees for obi and rmi bearings; elimination of the 40 degree lock-on error by the use of special monitoring circuits; and the avoidance of interference with iff and atc transponders by the use of mutual suppression.

In addition to a multi-function digital distance display and ARINC-standard 2-out-of-5 cockpit control, the TCN-40 system comprises two remote units: the DME-40 distance measuring and BRG-40 bearing computers. Considerable output versatility is incorporated.

Transmitter frequency band: 1025–1150 MHz
Receiver frequency band: 962–1213 MHz
Number of channels: 252
Transmitter power: 300 W peak
Receiver sensitivity: –82 dBm
Memory after signal loss: (distance) 12 s (bearing) 3 s
Range: 250 nm
Track rate: 999 knots 20°/s
Dimensions: (DME-40) ARINC ½ATR Short Low; 4⁹/₁₀ × 3½ × 14 inches (124 × 89 × 356 mm) (BRG-40) identical with DME-40
Weight: (DME-40) 7·17 lb (3·26 kg) (BRG-40) 6 lb (2·72 kg)
Qualifications: FAA TSO C66A, RTCA DO-138

STATUS: in production and service.

TCN-150 air-to-air Tacan (AN/ARN-118(V) and AN/ARN-139(V))

The AN/ARN-139(V) transforms an aircraft into a flying Tacan station by providing air-to-air bearing and distance information, inverse Tacan operation, and selectable range ratios. These capabilities are added to all the standard features of the AN/ARN-118(V) Tacan. Rendezvous with an ARN-139(V) equipped aircraft or ship simplifies critical missions by providing a reliable read-out of bearing and distance during the approach.

The ARN-139(V) is derived from the ARN-118(V), of which more than 15 000 units have been built. The TCN-150 has two sides: side one is an SRN-118(V) with modified automatic gain control, and side two is an ARN-139(V) with air-to-air distance and bearing transmission, and inverse Tacan operation. The latter function allows a tanker aircraft, for example, to read the bearing and distance to a Tacan-equipped aircraft in need of air-refuelling. Inverse Tacan also enables a pilot to determine the bearing to a DME-only ground station. The selectable range-ratio capability permits the pilot, by means of a switch, to limit all replies to within four times the range of the nearest aircraft, or to concentrate on aircraft more than 30 times the distance of the nearest aircraft.

Transmitter frequency band: 1025–1150 MHz
Receiver frequency band: 962–1213 MHz
Transmitter power: 500 W minimum, 750 W typical
Receiver sensitivity: –92 dBm
Modes: (AN/ARN-118(V)) x/y channels, receive, transmit/receive, air-to-air receive, air-to-air transmit/receive
(AN/ARN-139(V)) x/y channels as side one, together with x/y channels, inverse beacon, inverse air-air, inverse transmit-receive, inverse receive
Range: 390 nm
Track rate: 3600 knot, 20°/s
Accuracies: (distance) 0·1 nm (digital) 0·2 nm (analogue)
(bearing) 1° (digital) 1·5° (analogue)
Reliability: mtbf designed for 1000 h
Dimensions: (side one) 11⁷/₁₀ × 8⁹/₁₀ × 20½ inches (297 × 225 × 521 mm)
(side two) 10 × 7⁷/₁₀ × 19²/₅ inches (254 × 196 × 494 mm)

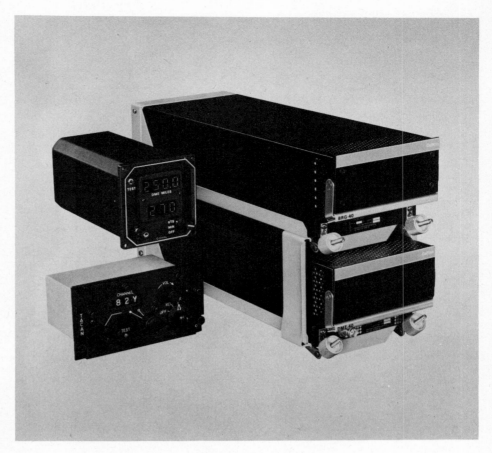

Rockwell Collins TCN-40 Tacan system, with channel selector, range and bearing readout, and receiver

Rockwell Collins TCN-150 flying Tacan station, showing side one and side two equipment

(C-10059/A and C-10994 control units) each 5³/₄ × 3 × 3¹/₅ inches (127 × 76 × 81 mm)
Weight: (side one) 41½ lb (18·8 kg)
(side two) 69¹/₈ lb (31·3 kg)
(control units) each 2 lb (0·9 kg)

STATUS: in production. One of most recent applications is US Air Force/McDonnell Douglas KC-10 Extender tanker derivative of DC-10 commercial wide-body airliner. The AN/ARN-118(V) was awarded the original US Air Force production contract in 1975, and this equipment is standard with that service and with the US Coast Guard. It is also becoming the standard for the US Army and Navy as well, and has been chosen by more than 35 countries.

Gould

NavCom Systems Division, Gould Government Systems, 4323 Arden Drive, El Monte, California 91731

AN/ARN-84 MicroTacan

Operational features of the ARN-84 Tacan include x-y position fixing, air-to-ground link with any standard Tacan station, air-to-air bilateral ranging, receive-only (bearing and beacon identification), radio frequency data link, air-to-air bearing and inverse mode. The system provides automatic antenna switching together with digital and analogue range and bearing outputs to an aircraft computer. The solid-state design gives a level of reliability previously unmatched according to Gould, and there are no 40 degree false lock-ons even in

the absence of North burst data. The range is 300 nautical miles.

STATUS: in production. The system is fitted to the US Navy/McDonnell Douglas F-18 Hornet deck-landing fighter.

AN/ARN-130 MicroTacan

This system incorporates features that significantly improve the utility of the MicroTacan series, including two-way air-to-air bearing and range measurement, full y-mode operation, and solid-state coupler. Precision digital and analogue outputs are simultaneously available from the transmitter-receiver, and a separate converter is not needed for retrofits. A single-tube power has increased the reliability and reduced power source dissipation so that the

Transmitter-receiver and control unit for Gould's AN/ARN-84 MicroTacan

mean time between failures is now 1500 hours, and there is continuous in-flight performance monitoring. The range is 399 nautical miles.

STATUS: in production and service.

Map Displays/Readers

GERMANY, FEDERAL REPUBLIC

Teldix

Teldix GmbH, PO Box 105608, Grenzhofer Weg 36, D-6900 Heidelberg 1

KG 10 map display for helicopters

This stowable system has been developed to ease crew work-load in all-weather helicopter operations, particularly at low level. It is also suitable for low-speed fixed-wing aircraft and land vehicles fitted with a source of navigation data. A notable feature of the KG 10 is that it can use unprepared standard maps, folded as convenient, and clamped between the edge frames of the display area. Navigation information to the system can be taken from a variety of sources in the aircraft, eg Doppler, and is converted into signals to energise x and y cursor drive-motors by means of a microprocessor chip in the display unit. All electronics, including the M6800 microprocessor, are carried on a single circuit board in the display unit, adjacent to a control/display panel. Power supplies, navigation information, and BITE signals are fed to the unit by a free multicore cable to whatever navigation computer is fitted to the aircraft.

There are four operational modes: nav, fix, map, and grid. In the first, the cursors move over the map in accordance with information furnished by the navigation computer. In the second, the crew can correct the position defined by the cursor intersections by external position fixes. The third mode is chosen when a map is inserted and the system has to be aligned or 'zeroed'. In the fourth mode, the grid co-ordinates of any point on the map can be displayed, eg for reporting.

Dimensions: 354 × 270 × 32 mm
Weight: 2·4 kg
Map display area: 248 × 248 mm
Position display: intersection of x and y cursors
Power: 7 W at 28 V dc
Appropriate UTM charts: 1:50 000, 1:250 000, 1:500 000
Radio charts: 1:10 nm, 1:18 nm
Navigation computer interface: ARINC 575

STATUS: in production.

Teldix KG 10 portable map display

SWEDEN

Bofors

Bofors Aerotronics, S-181 81 Lidingo

ANF data processor/moving map

This microprocessor-based navigation computer and moving map is a new-generation, flexible system intended for ground-attack aircraft, strike-trainers and helicopters. It consists basically of a computer with interfaces for a variety of navigation inputs, eg from air-data and magnetic heading up to full IN equipment, driving a full-colour moving map. Bofors claim accuracies of just over one per cent of distance travelled using air-data and heading information, relying on continuous updating and the system's 'learning' capability.

The display uses two colour slides of mission maps taken just before flight (thus including appropriate annotations), with a 6 × 6 camera and developed in 15 minutes by a special procedure devised for the system. The first

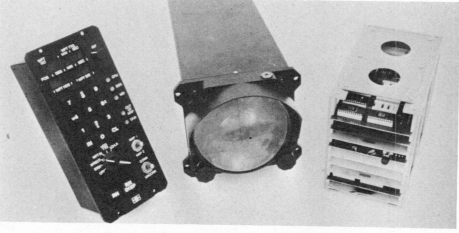

Bofor's ANF moving map for ground-attack aircraft

slide is of a 1:500 000 scale en route map, and the second is a target map with a scale of 1:50 000 (or whatever scale is desired), and both are contained in a magazine which, after briefing the pilot, is inserted into the display system. The first map is used in transit to the target, the pilot switching to the second as he approaches the target.

STATUS: in development. The system was flight-tested aboard a Viggen in 1980 and refinements to incorporate more facilities are planned.

ANF moving map display on centre console of JA 37 light bomber

UNITED KINGDOM

Marconi Avionics
Marconi Avionics Ltd, Airport Works, Rochester, Kent ME1 2XX

Automatic map reader
This is a knee-mounted, portable, lightweight, microprocessor-controlled device for use with standard aeronautical charts or maps and particularly useful in tactical situations. It is at present designed to interface with a Doppler or IN computer, and uses the navigation information from one of these systems in the form of UTM co-ordinates.

Aircraft position is indicated by the intersection of radial and spiral lines etched on two ten-inch (250 mm) perspex discs driven by a micro-computer and stepper motors contained in the lower half of the clamshell casing. In addition to multiple scales the device can store and display up to 32 programmed waypoints or targets.

A separate interface unit contains the navigation matching electronics and power supply. The design permits the operator to fold a map and to clamp it in the display head in any orientation.

Dimensions: (display head) 290 × 270 × 50 mm
(interface unit) 190 × 165 × 100 mm
Weight: (display head) 2·4 kg
(interface unit) 1·9 kg
Power: 40 W at 28 V dc

STATUS: in development. Equipment has been flight-tested in the UK, USA, and West Germany.

Marconi's knee-mounted automatic map-reader

Propulsion Management Systems

CANADA

Canadian Marconi

Canadian Marconi Company, Avionics Division, 2442 West Trenton Avenue, Montreal H3P 1Y9

CMA-730 engine instruments

Canadian Marconi's vertical-scale solid-state engine instruments have been installed in a number of corporate and military aircraft and helicopters. These include the Dassault Falcon 10 and 20, Canadair Challenger, and Gulfstream 2 in the executive category, and the Canadair CF-5 Freedom Fighter and Grumman OV-1D Mohawk military aircraft. Helicopters to be equipped with the instruments include the Bell 214ST, Hughes YAH-64 Apache, Sikorsky UH-60A Black Hawk, Aérospatiale Puma, and Westland Sea King (a particular specimen at Britain's Royal Aircraft Establishment).

The Falcon 10 instrumentation comprises five vertical-scale instruments which give readouts of fan speed, exhaust-gas temperature, compressor speed, fuel flow, and fuel quantity. Each value is indicated by a vertical column composed of small lights which may be coloured appropriately; blue for low, amber for medium, and red for danger. A digital reading below each column can be incorporated if desired. Other functions such as engine oil temperature and pressure can be given, and helicopter applications use vertical scales for rpm and torque readings.

An additional instrument is the selectable parameter digital display, which gives digital readings of a chosen function. This may be fan speed, compressor speed, exhaust gas temperature, fuel flow, fuel used, or fuel quantity. This digital readout complements the vertical analogue display, and is claimed to increase monitoring precision. It provides redundancy for the vertical scales. The function to be displayed is selected by a rotary switch, and a light shows to indicate which function is active. Used together with the sensor interface circuits which provide dc input signals, the display contains a digital voltmeter, binary-coded decimal encoder, and seven-segment decoder drives. The level of brightness is chosen with a manual dim control, and this selected level sets the mean operational threshold for the automatic dimming circuit. Optional over-range warning lights can be provided to record critical values that have been exceeded.

The company has built redundancy into the instrument by using parallel solid-state circuits and dual power supplies. Reliability is claimed to be better than that of mechanical instruments, since there are no moving parts. Increased resolution can result from the use of the complementary digital readouts, and scale expansion and contraction are available. Parallax compensation and viewing angle reading errors are avoided, and the displays are visible

Canadian Marconi's CMA-776 engine status display system

in bright sunlight, with a high contrast ratio between display and panel. The electronic elements consist of plug-in modules with printed circuit-boards, and no hermetic sealing is needed. Individual light-sources can be replaced within ten minutes. Even if a segment fails, the readability remains unimpaired.

Measurement accuracy: ±0·5% full-scale deflection
Response: 1 s average
Design standard: (civil) DO-138
(military) MIL-E-5400 Class 1A
Approval: TSO

STATUS: in production.

CMA-776(M) engine health monitor

The CMA-776 status display system is a caution/warning system with optional checklist readout. The two most important current faults are displayed to the crew on a two-line gas-discharge display. Reduced pilot workload, increased safety, and reduced panel space are cited as advantages for the CMA-776. Engine performance monitoring can also be incorporated. By removing the control and display unit, the CMA-776 can become a pure engine health monitor, when it receives the designation CMA-776(M). All engine parameters are sampled at 30 second intervals, and all out-of-limit conditions are recorded whenever they occur. The result is extracted from the

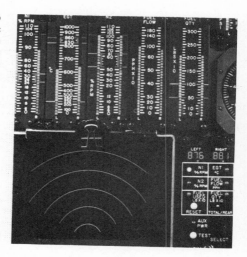

Canadian Marconi CMA-730 vertical-scale engine instruments as fitted in a Dassault Falcon 10 executive jet

computer's solid-state memory by connection of a printer or digital cassette when the aircraft lands. Environmental and flight data is recorded alongside the engine parameter sampling.

STATUS: in production.

FRANCE

Snecma-Elecma

Elecma Division Electronique de la Snecma, 22 quai Gallieni, 92150 Suresnes

RN-1435 full-authority digital engine control

Snecma-Elecma had delivered three prototype RN-1435 full-authority digital engine controllers to Turboméca by Spring 1982. The system has been tested on the Turmo engines of a Super Puma, and tests on other types of helicopter engines are anticipated. The RN-1435 is being developed under a French Government contract in close co-operation with engine-manufacturer Turboméca, which designed the fuel-flow regulator. The computer provides the various speed regulation phases of the free turbine and gives overspeed protection. In the case of a twin-turbine helicopter, the two computers can exchange data to balance the loads on the engines and to provide redundancy and failure protection.

STATUS: in development.

GERMANY, FEDERAL REPUBLIC

Bodenseewerk

Bodenseewerk Gerätetechnik GmbH, Postfach 1120, D-7770 Überlingen

Digital thrust-control computer

Bodenseewerk is responsible for the thrust-control computer section of the digital flight-control system on the Airbus A310 twin-engined wide-body airliner. The French company Sfena is in overall charge of the programme, and UK firm Smiths Industries is also involved. The thrust control computer is an extension of Bodenseewerk's activities in that area, which started with the autothrottle on Lufthansa's Boeing 707 fleet, and continued with the Airbus A300 autothrottle. The company delivered the first A310 unit in June 1981.

The thrust-control computer controls the fuel flow, and hence engine thrust, according to the aircraft speed selected by the pilot. It calculates the maximum admissible engine rpm or pressure ratio, and controls it during take-off according to circumstances (runway length, gradient, atmospheric conditions). Furthermore, it provides protective functions assuring the maintenance of four safety limits: maximum flight speed, maximum admissible engine speed, minimum flight speed, and maximum angle of attack.

The digital electronics are housed in an Arinc case. The computer is a redundant system using up to date technology micro-processors, with dissimilar software to guard against the danger of a software failure not being detected.

Bodenseewerk's digital thrust-control computer is on the new Airbus Industrie A310 wide-body airliner

The mechanical clutch units and the actuator are identical to those of the A300, and of proven reliability.

STATUS: in production for A310.

SWEDEN

SATT

SATT Electronics AB, Box 32006, S-12611 Stockholm

Data acquisition unit

The Swedish Air Force has been using the SATT engine data acquisition unit on its Saab JA37 Viggen fighters since 1978. The unit measures engine parameters such as pressures, temperatures and rotational speeds, and passes the data to a recording device for processing on the ground so that trends can be established. The use of this system has allowed the Swedish Air Force to extend the time between overhauls of the Volvo Flygmotor RM8B engine, saving money and improving the fleet's operational readiness.

The data acquisition unit is flexible enough to accept information from sensors with four alternative types of output: analogue dc, digital, variable resistance and tacho-generator. The unit can also accept a signal to identify the individual engine. Eleven analogue and five digital channels can be accepted. The analogue data is amplified, filtered and fed to an analogue multiplexer for channel selection. It is then converted to digital format and passed on to the internal data-bus. Digital signals need amplifying and formatting only for the data-bus. The data-bus leads to the data recorder.

Each channel is scanned at 2000 Hz and two 15-bit words are placed on the tape. One word contains channel number and status and the other gives the channel data. The recorder can be single- or multi-channel and SATT provides interface units to suit different types of recorder. The resolution and accuracy of the data must be high for the system to be effective and this is achieved by built-in calibration facilities and the provision of five auxiliary reference channels.

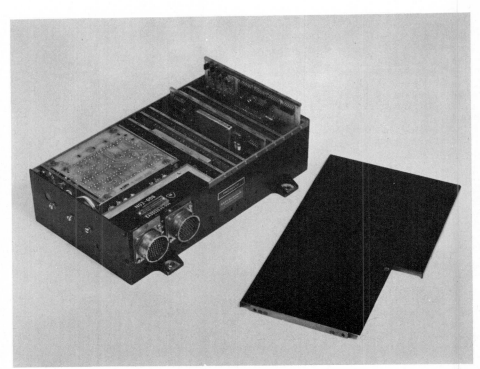

Satt's data-acquisition unit flies on the Viggen fighter

Dimensions: 300 × 170 × 81 mm
Weight: 4·2 ± 0·2 kg
Power supply: 50 W, 28 V dc
Accuracy (full-scale): 0·3% between –20° and +60°C; 0·6% between –40° and +80°C

Resolution (full-scale): 4096 bits
Specification: Conforms to appropriate parts of MIL-STD-461/462

STATUS: in production in seven-channel analogue version for JA37.

UNITED KINGDOM

DSIC

Dowty/Smiths Industries Controls (DSIC), Arle Court, Cheltenham, Gloucester GL51 0TP

Digital powerplant control

The DSIC Fadec (full-authority digital engine control) flew for the first time on a Harrier V/STOL fighter in March 1982 and has been demonstrated on a PT6 testbed at Pratt & Whitney, Canada. DSIC, a marriage between fuel-system specialist Dowty and electronics company Smiths Industries, uses improved acceleration, weight and maintainability as its main platforms for marketing Fadec.

DSIC minimises the number of components, especially hydro-mechanical ones, and uses large-scale integrated circuits. Both philosophies result in improved reliability over conventional hydro-mechanical systems. Maintainability is also better because setting-up adjustments are fewer, in service 'tweaking' is removed and fault diagnosis is facilitated by micro-processors.

The system uses primary and reversionary electronic channels which are completely independent but operate a common metering valve via a stepper motor. The primary channel controls fan speed, and limits engine speed and temperature but the reversionary channel may be simpler if required. The system is software-programmable, in common with all digital devices, so consistent engine handling is the result of complex schedules rather than hardware 'tweaking'. Changes in engine conditions (especially deterioration with age) and different operating environments and fuels may be catered for by modifying the software.

DSIC claims high integrity for its Fadec since the two channels monitor each other, and reversion to the second channel in the event of primary channel failure is automatic. Although it describes common-mode failure as a 'very rare possibility', DSIC has included a manual hydro-mechanical back-up using the same pump and distribution system, at least in the single-engined Harrier installation. For the helicopter application DSIC provides a read-out of engine state on its Fadec control box, to give accurate engine diagnostics.

STATUS: in development.

DSIC's Fadec digital powerplant control on Rolls-Royce Pegasus test-bed

Lucas Aerospace

Lucas Aerospace, Engine Management Division, Shaftmoor Lane, Hall Green, Birmingham B28 8SW

ECC-200 engine control computer

In 1981 Lucas completed a four-year development programme for its ECC-200 full-authority digital engine control (Fadec) system, culminating in bench-testing on Avco Lycoming's 800 shaft horsepower Advanced Technology Demonstrator Engine. The ECC-200 is a single-channel device which controls fuel flow and intake geometry for helicopters, and Lucas is continuing to develop it with a view to full production.

Dimensions: 7 × 3½ × 6½ inches (178 × 89 × 165 mm)
Weight: 5·5 lb (2·5 kg)
Mounting: on engine

STATUS: in development.

ECC-300 engine control computer

Avco Lycoming has chosen the ECC-300 for the latest versions of its LTS101 family of turboshaft engines, starting with the LTS101-750 series. The system modulates fuel flow via a hydro-mechanical metering system governing compressor and turbine speeds, limiting exhaust-gas temperature, and featuring automatic starting and surge recovery. In multi-engined installations the EEC-300 performs torque limiting and load sharing.

Dimensions: 7 × 3½ × 6 inches (178 × 89 × 152 mm)
Weight: 5 lb (2·3 kg)
Mounting: airframe or avionics bay
Qualification: FAR Part 33

STATUS: in pre-production for the Avco-Lycoming LTS101-750 shaft-turbine engine for light helicopters.

ECC-400 engine control computer

This is similar to the ECC-300 in concept and has been developed for the auxiliary power unit of the McDonnell Douglas AV-8B V/STOL

Main engine control unit made by Lucas for the Turbo-Union RB.199 powerplants in multi-mission Tornado bomber

fighter, also supplied by Lucas. As well as replacing the hydro-mechanical fuel control, the ECC-400 sequences and monitors the auxiliary power unit, permitting in-flight power generation for the first time.

Dimensions: 6½ × 5 × 4½ inches (165 × 127 × 114 mm)
Weight: 4·5 lb (2 kg)
Mounting: airframe

STATUS: in pilot production for the US Navy.

ESC-102 engine supervisory control

Lucas co-produces the ESC-102 in conjunction with Marconi Avionics for the Rolls-Royce RB.211-535 turbofan engines which power the new-generation Boeing 757 airliner. The controller is built in two halves, each of which has its own computer, inputs and outputs. The two units are housed in the same case and the computers are interconnected.

The first control channel trims the hydro-mechanical fuel system to achieve the desired engine pressure ratio, and hence thrust. It interfaces with engine transducers, the air-data computer, flight-deck instruments, and on-board maintenance system. The second part of the ESC-102 prevents the engine from over-speeding or over-heating, using its own engine transducer inputs. It also stores and displays system faults. If one channel loses its data it has access to the other channel's information via the interconnect, providing redundancy.

Lucas says that the ESC-102 reduces pilot workload and increases engine life because there are fewer thrust excursions above the set value.

Dimensions: 12 × 8 × 5 inches (305 × 203 × 127 mm)
Weight: 17 lb (7.72 kg)
Mounting: engine or avionics bay

STATUS: in production.

Full-authority digital engine control (Fadec) for civil powerplants

The civil full-authority digital engine control (Fadec) under development by Lucas is aimed initially at the Anglo-Japanese-US RJ500 20 000 lb (89 kilonewtons) thrust turbofan engine for which work is also being done by DSIC. The system is duplex and provides engine pressure ratio, setting and protection against over-speeding and over-heating.

Lucas claims that its Fadec saves fuel by being more flexible in coping with different ambient conditions and by controlling thrust more accurately. To do this the system compares actual engine parameters with a computer model of the engine's characteristics. Deviations and trends are displayed in the cockpit and on the engine maintenance panel.

Dimensions: (controller) 10½ × 8 × 4 inches (267 × 203 × 102 mm)
(monitor) 10½ × 8 × 5 inches (267 × 203 × 127 mm)
Weight: (controller) 12·2 lb (5·5 kg)
(monitor) 15·5 lb (7 kg)
Mounting: fan casing (system is fuel cooled)

STATUS: in development.

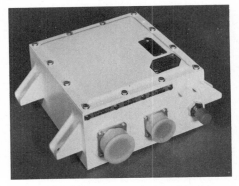

Lucas ECC-300 computer designed for Lycoming's LTS101-750 turboshaft engine

Lucas CUE-300 main engine controller for RB.199 engines in the Tornado

Full-authority digital engine control for military powerplants

The RB.199 augmented turbofan engines for the tri-national Tornado bomber are the first target for the Lucas military Fadec, known as the digital engine control unit. This system weighs the same and occupies the same space as the analogue main engine control unit currently fitted in Tornado and described below, but Lucas claims it has better control accuracy and system reliability. The ability to make changes at software level is also of great importance.

The digital engine control unit has the built-in test equipment and self-monitoring common to digital systems, and transfer to the safety channel is done at module level. Thus working modules from both channels can be combined rather than using one channel or the other. The engine is monitored via an interface to the air-data system.

Dimensions: 12 × 7½ × 7½ inches (305 × 191 × 191 mm)
Weight: 33 lb (15 kg)
Mounting: avionics bay

STATUS: in development. To be introduced on the RB.199 Mk 103 from May 1983.

CUE-300 main engine control unit

The Turbo Union RB.199 augmented turbofan engines which power the European Tornado bomber are controlled by this analogue electronic unit. Dry-thrust operation uses one of two channels with automatic channel switching if the built-in test equipment detects a failure. The second channel has a separate simplex reheat control fitted with built-in safety networks to prevent a dangerous situation arising from a failure.

The main engine control unit uses twin electro-mechanical metering devices to modulate the fuel flow according to its calculations, and maintains the engine within its operating envelope. The system operates three actuators for reheat control, two to control the fuel and one to operate the variable nozzle.

Dimensions: 12 × 7½ × 7½ inches (305 × 191 × 191 mm)
Weight: 33 lb (15 kg)

STATUS: in production.

Self-adaptive full-authority digital engine control

Lucas is conducting a feasibility study for the US Army to find out the potential improvement for helicopter engine controls arising from the self-adaptive capability of digital systems. The company is looking at improving the turboshaft's power response and stability and improving the helicopter's performance and safety when a defect arises in the control systems, engine or airframe.

STATUS: feasibility study.

Active remote module (ARM)

Lucas is developing a system to display information obtained from a helicopter's full-authority digital engine control system. The active remote module has an IM 6100 microprocessor to decode the data which includes engine speeds, torque and temperatures. The data is displayed digitally but fault and diagnostic lights are included. The system could be used for engine health monitoring and as a maintenance aid. The active remote module is so designed that any failure within it does not affect other systems through its interfaces with them.

STATUS: development.

Lucas civil-engine Fadec is aimed initially at the Anglo-Japanese-US RJ.500 powerplant

ESC-102 supervisory control system for Rolls-Royce RB.211-535, principal engine for new Boeing 757 airliner, co-produced by Lucas and Marconi Avionics

Marconi Avionics
Marconi Avionics Ltd, Flight Controls Division, Airport Works, Rochester, Kent ME1 2XX

Engine supervisory control ESC-102
Marconi Avionics is responsible for half of the system for the Rolls-Royce RB.211-535 engine, developed for the new-generation Boeing 757 airliner. Lucas Aerospace is the prime contractor. See under Lucas Aerospace for further details.

Full flight regime autothrottle
The Marconi autothrottle is an optional installation on Boeing 747 wide-body airliners, and by early 1982 was in service with more than 40 airlines. As well as the production line in Rochester, Marconi's Atlanta facility in the USA also supplies autothrottles. The full flight-regime autothrottle system computer is programmed with the aircraft's engine pressure ratio, speed and Mach number control laws which permit it to control the throttles automatically in one of those three modes, selectable on a control/display unit on the flight-deck.

For take-off, climb and go-around, the system controls engine pressure ratio. Speed or Mach control may be selected in cruise, and speed control is used in the descent, hold and approach phases. During autopilot-coupled landing the throttles are retarded automatically. All four throttles are controlled to the same position.

The system accepts angle of attack, flap position, total air temperature, airspeed, Mach number and aircraft attitude as inputs. As well as the three primary modes, sub-modes are provided for aircraft and engine protection. These are control to a safe minimum speed, flap placard-speed protection, engine over-boost protection, and constant throttle hold during the critical take-off regime. In the engine pressure ratio mode the crew can select the engine pressure ratio for take-off, climb, cruise, or go-around, and the appropriate engine pressure ratio is acquired and maintained, though the crew can modify its limit if required. In the speed mode a commanded speed set on the aircraft's autopilot control panel is acquired and maintained while

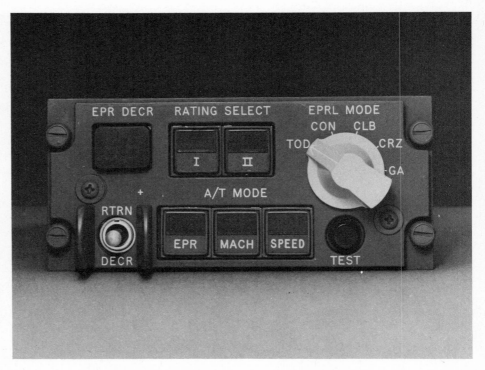

Marconi Avionics full flight regime autothrottle for Boeing 747

in the Mach mode the existing Mach number is kept constant. In the latter two modes, engine pressure ratio protection is provided as a sub-mode.

As well as governing the 747's throttles the full flight regime autothrottle system sends the selected airspeed for display on the airspeed indicators, and generates a fast/slow indication on the attitude-director indicators. Total air temperature and engine pressure ratio limits are displayed to the crew and the target engine pressure ratio is shown on each engine pressure ratio indicator. Built-in test equipment is provided and the system status and warning indications are provided on the flight mode annunciator panels.

The autothrottle system increases engine life and saves fuel, responding rapidly to changes in the aircraft configuration or flight conditions and avoiding unnecessary throttle excursions. Pilot work-load is also reduced.

STATUS: in production for the Boeing 747 wide-body airliner.

Autothrottle system for Boeing 747
The full flight regime autothrottle system is the standard fit for all current production airframe and engine versions of the Boeing 747 wide-body airliner. It provides complete autothrottle control to engine limits for all phases of flight, and contributes to extended engine life and minimised maintenance. Airspeed-select and Mach-hold functions provide accurate approach and fuel-saving cruise performance. Protection features include engine limit, safe flap-speed and minimum speed facilities in approach and cruise. The system comprises an autothrottle computer and an integrated mode selection panel.

STATUS: in service.

Smiths Industries
Smiths Industries Aerospace & Defence Systems Co, Winchester Road, Basingstoke, Hampshire RG22 6HP

STS 10 autothrottle-speed control
The STS 10 is a full flight regime autothrottle with speed control computation and was certificated for the Boeing 727 and 737 in 1980-81. Recent 737 customers such as British Airways and Lufthansa have chosen the STS 10, which is also available as a retrofit item. The system operates the throttles of each engine individually to take account of difference in engine age and control-run length. The required throttle settings are obtained from one of the advisory performance data computers now available.

During climb and go-around, the STS 10 controls the pitch bar of the flight director, and there is a speed hold mode for cruise, during which the STS 10 drives a fast/slow indicator. The system interfaces with the autopilot to optimise the flight path.

As well as reducing flight-deck workload, the STS 10 is claimed to save fuel and increase engine life by reducing the number and magnitude of throttle excursions and by keeping the powerplant continuously at the optimum engine pressure ratio.

Principal components of the Smiths Industries' digital fuel-management system

The STS 10 has built-in test equipment and also 'freezes' during take-off. It protects the engine from stall and flameout and still functions correctly if an engine fails.

STATUS: in production as an option on Boeing 727 and 737 airliners.

Series 200 engine limiter

This system monitors jet-pipe temperature and controls it by limiting fuel flow. Its applications include the Rolls-Royce Viper turbojets powering the Aeritalia MB.326 strike-trainer, HS125 executive jet, and Soko Jastreb strike-trainer. Other engines to have had Series 200 limiters include Avon, Orpheus and Pegasus.

Dimensions: 6¾ × 6 × 2 inches (170 × 150 × 50 mm)
Weight: 4·12 lb (1·87 kg)
Temperature range: 3 datums between 350° and 1000° C can be externally selected
Accuracy ±2½° C

STATUS: in production.

Series 500 engine limiter

The Series 500 is used for the Rolls-Royce Pegasus engine powering British Aerospace Harrier V/STOL fighters. It limits jet pipe temperature and compressor shaft speed.

Dimensions: 6¾ × 6 × 2 inches (170 × 150 × 50 mm)
Weight: 4·5 lb (2 kg)
Temperature: 4 datums between 350° and 800° C can be externally selected, each capable of fine adjustment by ±7·5° C
Speed range: Pre-set between 10 400 and 12 700 rpm
Accuracy: ±2° C and ±1% of rpm

STATUS: in production.

Series 600 engine limiter

This jet-pipe temperature limiter is used for civil and military Rolls-Royce Spey engines on aircraft including the maritime patrol Nimrod, BAC One-Eleven, Gulfstream 2, Fokker F.28, Buccaneer and Trident. The Series 600 has the facility to compensate for changes in intake temperature.

Dimensions: 8½ × 6 × 2½ inches (215 × 150 × 65 mm)
Weight: 4·1 lb (1·88 kg)
Temperature range: Pre-set between 450° and 650° C. Setting may be fine-tuned by ±25° C
Accuracy: ±1½° C

STATUS: in production.

Series 800 engine limiter

The Series 800 limits jet-pipe temperature, compressor delivery temperature and mass flow, and is engine-mounted on the Rolls-Royce/Allison TF41 Spey engine which powers some versions of the LTV A-7 Corsair. A time-profiled datum shift is provided in which the upper temperature limits are increased temporarily during sudden power demands such as take-off or go-around.

Dimensions: 10 × 7½ × 3¼ inches (250 × 190 × 80 mm)
Weight: 8 lb (3·63 kg)
Temperature range: Pre-set between 0 and 1000° C
Accuracy: ±2½° C and ±0·25% of rpm

STATUS: in production.

Series 900 engine limiter

Rolls-Royce and other engine manufacturers are evaluating the Smiths Industries Series 900 which may be used as a jet-pipe temperature and/or mass-flow limiter. The principal benefit from this latest limiter is its low-cost, low-density design which uses single-sided circuit boards to allow room for future growth. Double-sided boards may however be used if space is at a premium.

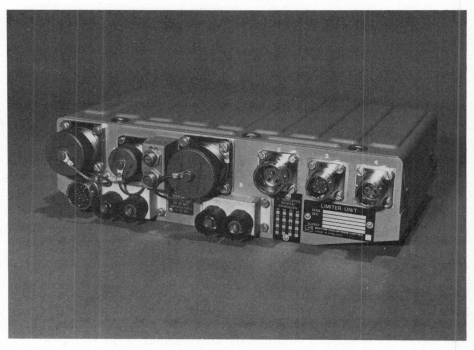

Smiths Industries' Series 800 temperature limiter for the Rolls-Royce/Allison TF41 engine in some versions of the US Air Force/LTV A-7 Corsair bomber

Dimensions: 6½ × 4¾ × 2¾ inches (150 × 115 × 65 mm)
Weight: 2·2 lb (1 kg)
Temperature range: Pre-set between 0 and 1000° C and fine-tuned to ±25° C
Accuracy: ±2° C over ambient range –26° to +70° C
±4° C over range –55° to +125° C

STATUS: development completed.

Fuel gauging and management systems

Smiths Industries has supplied analogue fuel gauging systems of varying complexity for more than 75 different types of fixed-wing aircraft and helicopter. Of these aircraft the following are still in production in 1982: Airbus A300, HS125, Bandeirante, Nomad, Casa C212, One-Eleven, HS748, Skyvan, Jetstream, Harrier/Sea Harrier/AV-8A, Jaguar, Alpha Jet, Hawk, Strikemaster, Viggen, Agusta A109, Gazelle and Puma. The A310's digital system, one of the most recent, is treated separately below.

The basic Smiths fuel gauge is an ac or dc driven moving-coil or servo indicator which may have twin pointers to show the contents of pairs of tanks. A fuel-level warning device set by the pilot is optional. A more advanced installation may have digital readings on light-emitting diode displays and may incorporate a summation facility to give total fuel level.

The company has found that metallic tubular tank probes are the best contents sensors, combining the most reliable results with low weight. They are protected against corrosion and are anodised to reduce the risk of microbiological growths.

Any fuel system is tailored to the individual aircraft type and in the design stages Smiths Industries undertakes a computer study of tank geometry and the effect of changes in wing loading. The company now offers a self-compensating probe which incorporates capacitors to allow temperature variation in the fuel. Infra-red compensation may be provided. The fuel level is detected by a float, capacitance measurement or thermistors (which provide the most compact solution).

As well as the measurement of fuel contents, Smiths offers several optional fuel management functions. Automatic refuel/defuel is provided by pre-selectors at the refuelling point. Instead of relying on the flight-deck indications, which are too far away to be of use on large aircraft, the refueller sets the amount

of fuel required and the pumps are stopped automatically at the correct level.

Another fuel management function is centre of gravity indication. The pilot sets into the system, before take-off, the actual centre of gravity and it monitors fuel consumption from the tanks throughout flight, giving a continuous centre of gravity readout. Allied to this are systems which control fuel distribution between tanks to keep the centre of gravity within limits as fuel is burned off. Smiths also has a system that pumps fuel between tanks in flight to counteract trim changes, either for vertical take-off or supersonic aircraft. Such a system has been applied to the Harrier family of V/STOL fighters.

Lastly, the company provides fuel flow-rate indication equipments which may be used to show the mass of fuel consumed by one or more engines and fuel remaining. Current aircraft weight may also be computed and shown.

Typical configurations
High-performance strike aircraft: Servo indicator, centre of gravity position indicator, separate bridge amplifier (weighing 0·7 lb (0·3 kg)), and self-compensating probes. The system is designed for severe conditions including large temperature variations in the tanks.
Executive jet: Moving-coil indicator, separate bridge amplifier, and uncompensated probes. No fuel management is needed.
Transport aircraft: Moving-coil indicator with combined amplifier, refuelling pre-selector, infra-red compensated probes and separate immersed reference unit. Full temperature and density compensation is needed, and fuel management speeds up the refuelling operation.
Helicopter: Servo indicator with combined amplifier, infra-red compensated probes, and separate immersed reference unit.

STATUS: in production.

Digital fuel management

Smiths Industries has developed and flight-tested a digital fuel management system which can be applied to any civil or military aircraft. The only specific application by early 1982 is the Airbus Industrie A310 200-seat wide-body airliner, for which a digital fuel quantity indication system is produced by Smiths in collaboration with Intertechnique of France.

The digital system reduces attitude errors and performs temperature compensation within the processor so that linear tank probes can be used. Smiths says that fewer probes are needed, thereby increasing accuracy. Fuel is sampled by a densitometer as it is on-loaded. The computer is connected to the aircraft's attitude sensor so that changes from straight and level flight may be taken into account when measuring fuel levels.

Built-in test equipment is included and the system is duplex. Input and output data is sampled and abnormal measurements are rejected, so indicator fluctuations and errors are reduced. The processor can control the refuelling operation, including fuel distribution between tanks and cut-off.

The system specific to the A310 and the forward-facing crew cockpit A300 uses digital fuel readout and shows aircraft weight and total fuel weight. Intertechnique designed the probes, which are characterised to suit the individual aircraft, and to compensate for tank shape variations. Smiths Industries make the contents and totaliser indicators, whereas production of the processors is shared between the two companies.

Flexibility is always one of the advantages of digital systems and this one may be adapted to changes in operating procedures and tank configurations simply by modifying the software. The system is also suitable as a retrofit in existing A300s without changing the tank probes since the software can correct the differences.

A Smiths Industries fuel-gauging and management system is installed in Italy's Agusta A109 helicopter

STATUS: in production.

UNITED STATES OF AMERICA

Astronautics

Astronautics Corporation of America, PO Box 523, 907 South First Street, Milwaukee, Wisconsin 53201

Analogue engine instruments

Astronautics produces circular-scale engine instruments for various US military aircraft. Temperature, engine pressure ratio, jet nozzle position and fuel-flow are provided and the company makes a mechanical event timer which counts the number of times three specified engine temperature limits are exceeded.

Thermocouple temperature indicator:
(temperature range) 0° – 1200° C
(diameter) 2 inches (51 mm)
(length) 6 inches (152 mm)
(weight) 1·5 lb (0·68 kg)
(power) 10 W 115 V, plus 2·5 W at 5 V for integral lighting
(specification) MIL-STD-I-27552B (USAF)

Synchro temperature indicator:
(temperature range) 0° – 1200° C
(diameter) 2 inches (51 mm)
(length) 4¼ inches (108 mm)
(weight) 1 lb (0·45 kg)
(specification) MIL-STD-I-25685

Synchro thrust indicator:
(engine pressure ratio range) 1·2 – 3·4
(other indications) cruise and take-off epr in windows
(diameter) 2¼ inches (57 mm)
(case) 2⅜ inches (60 mm) square
(length) 4¾ inches (121 mm)
(weight) 1·5 lb (0·68 kg)
(specification) MIL-STD-I-25859A (USAF)

Nozzle position indicator:
(nozzle range) 5 positions between open and closed
(diameter) 2 inches (51 mm)
(length) 1¾ inches (44 mm)
(weight) 0·6 lb (0·27 kg)

Fuel-flow gauge:
(fuel-flow range) 0 – 5000 lb/h (0 – 2268 kg/h)
(diameter) 2 inches (51 mm)
(length) 6⅛ inches (155·6 mm)
(weight) 1·5 lb (0·68 kg)

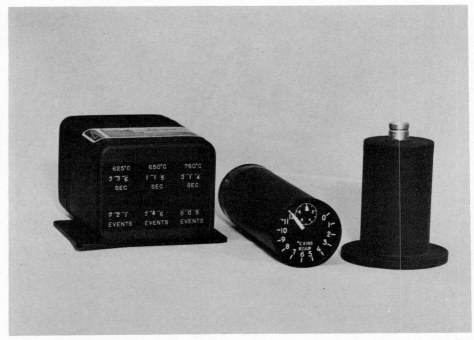

Astronautics analogue engine instruments (left to right): over-temperature timer, exhaust-gas temperature indicator, and converter

(power) 8 W 115 VA plus 2 W at 5 V for integral lighting
(specification) MIL-STD-I-27182A (USAF) or MIL-STD-26299D

Engine over-temperature timer
(display) indicates duration of limit exceeded and number of occasions for each temperature specified
(dimensions) 3½ × 3 × 5½ inches (89 × 76 × 131 mm)
(weight) 2·25 lb (1·1 kg)

STATUS: in production and service.

Astronautics vertical-strip engine indicator for single powerplant, showing (from left) pressure ratio, turbine inlet temperature, fuel flow and rpm

Linear engine instrument

Astronautics make this vertical-scale engine performance indicator for US military aircraft. Five indications are combined on to one square display. These are fuel-flow rate, fan or compressor rotational speed, turbine inlet temperature, engine pressure ratio and oil pressure. Each parameter is indicated by a moving tape read against a fixed scale. Each channel is monitored automatically and an error warning flag is shown if a parameter is outside limits. A counter on the back of the instrument measures time and is calibrated from zero to 9999.

Fuel-flow range: 0 – 12 000 lb/h (0 – 5443 kg/h). Readable to within 50 lb/h up to 5000 lb/h, and within 500 lb/h up to 12 000 lb/h (at a distance of 28 inches (71 cm))
Rotational speed range: 0 – 110% rpm. Readable to within 5% from 0 – 10% rpm and 30 – 70% rpm; within 1% from 10 – 30% rpm; and within 0·5% between 70 – 110% rpm
Temperature range: 0 – 1200° C. Readable to within 50° up to 500° C, and within 10° between 500 – 1200° C

Epr range: 1 – 2·5. Readable to within 0·05 epr
Oil-pressure range: 0 – 100 lb/inch² (0 – 7·03 kg/cm²). Readable to within 5 lb/inch² (0·35 kg/cm²)
Dimensions: 4¹/₁₀ × 4½ × 7²/₅ inches (104·1 × 114·3 × 188 mm)
Weight: 8·5 lb (3·9 kg)
Power: 115 V 45 W, plus 6 W at 5 V for integral lighting
Specification: Naval Air Systems XWS-3164

STATUS: in production and service.

Bendix

The Bendix Corporation, Energy Controls Division, 717 North Bendix Drive, South Bend, Indiana 46620

EH-L2 digital electronic fuel control

Bendix has developed this full-authority digital engine control for future airliners, and the system has been test flown on a Boeing 747 during an evaluation known as the Electronic Propulsion Control System programme. The EH-L2 is designed to give improved performance over current systems, while retaining good fuel consumption. A version of the control is being adapted for testing in the Advanced Technology Engine Gas Generator programme at Pratt & Whitney's Government Products Division.

The EH-L2 comprises a full-authority digital electronic controller and complementary fuel-flow handling unit. It incorporates micro-electronic-based computation including selective or complete redundancy as appropriate for high reliability and fault tolerance. The EH-L2 has built-in high accuracy quartz pressure sensors, and Bendix has proved its tolerance to shock, vibration, electro-magnetic interference, electro-magnetic pulse interference, and temperature extremes.

The company has also developed an advanced system for digital full-authority control of complex variable-cycle engines. The design is intended for future advanced military aircraft, which are likely to have variable-cycle engines to suit supersonic flight, while improving subsonic performance. This engine control has also been used by Detroit Diesel Allison for its Joint Technology Demonstrator Engine and ATEGG programmes.

STATUS: in development.

Propulsion data multiplexer (PMUX)

The Bendix propulsion multiplexer (PMUX) monitors and collects a variety of engine data including temperatures, pressures, speeds, fuel flow, variable geometry parameters within the powerplant, cockpit discrete (ie on/off signals),

Bendix EH-L2 digital electronic fuel control system

and engine identification. This compact engine-monitoring system provides data appropriate to examining engine condition trends, tracking limited-life equipment, and improving engine life forecasts. The data generated by PMUX can improve engine performance evaluation, leading to reductions in fuel consumption and further savings from more extensive use of on-condition rather than scheduled maintenance. Additional savings can result from the early detection of potential failure of engine components and the elimination or reduction of secondary engine damage.

STATUS: in development.

Bendix propulsion data multiplexer

Endevco

Endevco Corporation, Dynamic Instrument Division, Rancho Viejo Road, San Juan Capistrano, California 92675

Microtrac engine vibration monitor

Microtrac is a micro-processor-based engine vibration monitor which is basic equipment on the new-generation Boeing 757 airliner and optional on the 747 and 767 transports. The use of micro-processors permits the device to be reprogrammed to suit most aircraft/engine combinations. Vibration is detected by a piezoelectric accelerometer mounted at a sensitive region in the engine. The piezoelectric crystal generates a voltage proportional to the vibration level and the signal is passed to a processor in the avionics bay.

The Microtrac processor uses a narrow-band digital filter, controlled by the output from a

tachometer on the engine, to isolate the vibration frequency. The narrow band-width means that the signal-to-noise ratio of the final vibration indication is high, and the filter's transient response is fast so that it can detect resonances as the engine accelerates or decelerates.

The tuning accuracy does not need to be high, as it must be with analogue filters, so no regular adjustment is needed. The filter's accuracy is determined by the number of bits and the signal-to-noise ratio. Built-in test equipment is provided, and only one programmable read-only memory needs to be changed to adapt the system to a different engine type. Microtrac has an ARINC 429 digital data-bus output so the engine vibration can be displayed and recorded by the air-data system. An analogue output to conventional instruments can be provided. The built-in test

equipment results can also be put on to the ARINC 429 bus, and up to 17 fault indications can be stored for display to maintenance crews.

In-flight engine balancing is an optional possibility with Microtrac. The amplitude and phase of the once-per-revolution vibration of the Microtrac accelerometer is presented to the micro-processor which selects the weight and mounting position required to balance the rotor more accurately. This data is stored in an expanded memory, for later display to ground-crew, and it may be passed into the ARINC 429 bus for a direct cockpit readout. The in-flight balancing can be accomplished more accurately than possible on the ground and repeated ground run-ups are avoided. Time and fuel are therefore saved. Microtrac is housed in a ³/₈ATR box.

STATUS: in production.

Views vibration imbalance early-warning system

Endevco produces analogue engine vibration monitors for the Boeing 747 wide-body transport and Canadair Challenger and Mitsubishi Diamond corporate jets; all aircraft certificated to FAR Part 25 (which embraces the latter pair) now need engine vibration monitoring. Views is an adaptation of the Challenger system to suit the Garrett AiResearch TFE731 engine, but Endevco has not pursued the anticipated TFE731 retrofit market because of opposition to vibration monitoring by the engine manufacturer.

Views uses a piezoelectric accelerometer mounted on the engine carcass. The solid-state computer amplifies and filters signals from the accelerometer and converts them into a cockpit readout of engine imbalance. There is an optional warning light which is activated when the engine vibration level is more than 1·1

inches per second for three seconds. Endevco stresses that the main function of Views is to establish vibration trends so that maintenance can be scheduled in time to prevent permanent damage. Views will also display the sudden increase in vibration that accompanies the loss of a fan, compressor or turbine blade.

Signal-conditioner dimensions: $2^{3}/_{10} \times 7^{7}/_{10} \times 12^{1}/_{2}$ inches ($60 \times 195 \times 319$ mm)
Display unit dimensions: $5^{3}/_{4} \times 2^{3}/_{5} \times 2^{9}/_{10}$ inches ($146 \times 67 \times 73$ mm)
Display range: 0–2 inches/s per channel
Accelerometer: 1 Endevco Model 6222M8 per engine
Frequency response: 5–5500 Hz
Specification: FAR Part 25
Total weight (2 engines): 12·7 lb (5·8 kg)

STATUS: similar systems in production for Boeing 747, Challenger and Diamond.

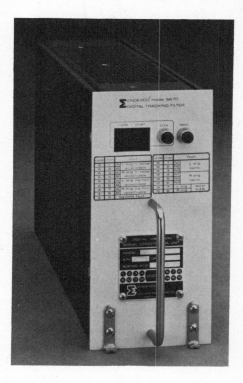

Endevco's Microtrac narrow-band tracking filter

Garrett

AiResearch Manufacturing Company of California, Division of the Garrett Corporation, 2525 West 190th Street, Torrance, California 90509

Engine bleed-air control system

Garrett manufactures an electronic bleed airflow sensing and control system. High-temperature cooling-effect airflow-sensors with directly heated platinum elements provide accurate, responsive flow signals to solid-state electronic control circuitry. The electronics may be packaged in a separate box or as an integral part of the sensor. The circuitry can be designed to operate valves, indicators, or annunciator lights, as required.

The flow sensor is designed specifically for engine applications, and is impervious to contaminants typically found in such an environment. The system is used in performance monitoring and in engine flow-sharing applications. It enables more efficient use of the large quantities of bleed-air required of modern engines for various purposes, improving fuel efficiency and lengthening engine life.

Temperature range: 167 to 260° C
Flow range: 10 to 250 lb/minute (4·5 to 113 kg/minute) for a 3 inch (76 mm) duct
Response time: Typically 3 s
Controlling accuracy: ±5%

STATUS: in production.

Engine performance indicators

Garrett makes a series of engine performance indicators for business and commuter aircraft, typical examples being the British Aerospace HS 125-700, Gates Learjet 35/36, and Lockheed JetStar 2 business jets. The indicators provide a continuous analogue display of critical engine variables, such as fan and core rotation speeds, turbine temperature and fuel flow. Configurations with circular or square cross-section are available.

Solid-state circuitry is employed throughout. The indicators are unaffected by the large fluctuations in electrical supply during engine starts.

Power requirement: 120 mA nominal, 450 mA max at 12 to 34 V dc, plus 250 mA at 5 V dc for lighting
Input signals: (rotation) monopole pulse (turbine temperature) chromel-alumel thermocouple per NBS Monograph 125

IAI's Westwind 2 executive jet has Garrett TFE731 fan engines fitted with Garrett's engine performance reserve controller system

(fuel flow) second harmonic selsyn mass flow transmitter, 115 V ac 400 Hz reference
Rpm accuracy: (0–55° C ambient) ±0·25% at 100% rpm
(–30–+70° C ambient) ±0·5% at 100% rpm
Temperature accuracy: (0–55° C ambient) ±5° C at 900° C
(–30–+70° C ambient) ±10° C at 900° C
Fuel flow accuracy: (0–55° C ambient) ±1% of full-scale
(–30–+70° C ambient) ±2% of full-scale
Rpm range: 0–110% rpm
Turbine temperature range: 100–1000° C
Fuel flow range: 0–2300 lb/h
Response: 3 s full-scale slew
Temperature range: (Operating) –30–+55° C
(short-term) –30–+70° C
(ambient extreme) –65–+70° C
Shock: 6 g for 0·011 s
Vibration: 5–3000 Hz, 0·02 in amplitude, limited to ±1·5 g

STATUS: in production.

Engine performance reserve controller

The Garrett automatic engine performance reserve (apr) controller is used currently on aircraft powered by the Garrett TFE731 turbofan engine. The controller detects a loss of thrust on take-off from either engine through the comparison of fan speeds, a five per cent difference being considered the nominal threshold. Following detection of a power decay, the system automatically boosts the remaining engine's maximum thrust by 5·48 per cent (at 21° C ambient temperature). This is achieved by raising the high-pressure spool maximum speed by 1 per cent and the maximum operating temperature by 25° C. Manual apr control is available to give thrust boost should the system fail.

Weight: 4·75 lb (2·2 kg)
Dimensions: $6^{3}/_{10} \times 2^{9}/_{10} \times 9^{3}/_{10}$ inches ($160 \times 74 \times 236$ mm)
Power: 14 W at 28 V dc

STATUS: in production for TFE731-powered aircraft.

Propeller synchrophaser

The Garrett synchrophaser controls the phase relationship between the blades of two aircraft propellers. By this means noise and vibration can be minimised and the objectional beat frequencies associated with propellers operating at different rotational speeds be eliminated. The synchrophaser is designed to operate with twin turboprops and electronic engine fuel controls. It works with existing speed detection systems, and requires no rigging adjustments, calibration, or indexing. It is compatible with engines operating closed-loop on torque and temperature. The system uses a master-and-slave engine concept with a full 120 degree phase angle authority for three-bladed propellers.

Dimensions: $2^1/_3 \times 5 \cdot 05 \times 8 \cdot 05$ inches (57 × 128 × 204 mm)
Weight: $1 \cdot 75$ lb ($0 \cdot 79$ kg)
Power: 28 V dc with MIL-STD-704 Category B transient immunity
Authority limit: $2 \cdot 4\%$ maximum speed separation
Steady-state phase-angle accuracy: $<5°$ (typical)
Operating range: cruise and max continuous speeds
Vibration: RTCA DO-160-PAR. 8 curves
Electro-magnetic interference: MIL-STD-6181D

STATUS: in production.

Engine synchroniser

When two common-structure mounted engines operate at slightly different rotational speeds, a third beat frequency is set up in the structure. This can be a source of discomfort to passengers. Garrett is producing an electronic engine synchroniser to reduce this problem. It interfaces with the engine through the existing electronic fuel control, and requires no modifications to the aircraft wiring. The synchroniser may be applied to two, three, or four engines without modification. It works by nominating one engine as master and adjusting either fan (N_1) or compressor (N_2) speeds of the other powerplants so that all speeds are closer to each other. The trim authority varies with power lever setting. The flight deck switches allow on/off and N_1/N_2 selections, and a synchronisation indicator is optional.

Max acquisition time: 30 s
Synchronisation accuracy: (N_1 control) 11 rpm (N_2 control) 18 rpm
Weight: $4 \cdot 6$ lb ($2 \cdot 1$ kg)
Dimensions: $9^2/_5 \times 7^3/_{10} \times 3^3/_5$ inches (239 × 186 × 91 mm)

STATUS: in production.

Full-authority engine control for Garrett ATF3 powerplant

The controller for the Garrett ATF3 turbofan engine, which powers the Dassault Falcon 200 executive jet and HU-25A Guardian, executes closed-loop control of the powerplant. Its solid-state electronic design includes a continuous integral monitoring system. Failure detection is included, and manual backup control is selected automatically if there is a problem. Dual redundancy is provided for closed-loop protection against hard-over failures. External test points are provided for rapid fault isolation. Inputs include low-pressure and high-pressure compressor speeds, inlet pressure and temperature, turbine discharge temperature, and power-lever position. The unit controls the engine fuel valve and bleed valve positions.

Dimensions: $8 \times 17 \times 7$ inches (203 × 431 × 178 mm)
Weight: 19 lb ($8 \cdot 6$ kg)
Environmental: MIL-STD-5007
Cooling: natural convection
Inputs: 10
Outputs: 9

STATUS: in production.

Full-authority engine control for Garrett TFE731 powerplant

Garrett is developing a full-authority engine fuel control system for the TFE731 turbofan, operating according to an engine acceleration/temperature schedule. It provides closed-loop exhaust-gas temperature control during acceleration, along with on-speed governing, and integral bleed-air control based on exhaust-gas temperature. Over-temperature protection is given at all times. The equipment has the same features as those for the ATF3: solid-state design, failure detection with automatic switch to manual back-up, dual redundancy, continuous integral monitoring, and external test points.

Typical Garrett engine performance indicators, as fitted in the Gates Learjet 35 executive jet

Dimensions: $12^3/_5 \times 3^3/_5 \times 7^3/_5$ inches (320 × 91 × 193 mm)
Weight: $8 \cdot 6$ lb ($3 \cdot 9$ kg)
Cooling: natural convection
Inputs: 7
Outputs: 4

STATUS: in development for TFE731-5 powerplant.

Full-authority engine control for Garrett TPE331 powerplant

The Garrett full-authority engine controller under development for the TPE331 turboprop has the following features: temperature-limited automatic starting; isochronous propeller governor speed control; closed-loop torque and temperature limiting; single red-line exhaust-gas temperature correction; continuous speed and fuel governing; power-level matching; and simplified engine rigging.

Dimensions: $7^3/_5 \times 5^2/_5 \times 11^9/_7$ inches (194 × 137 × 301 mm)
Weight: $10 \cdot 2$ lb ($4 \cdot 6$ kg)
Operating ambient temperature range: –54 to +71° C
Cooling: natural convection and radiation
Power: 28 V dc
Reliability: 6000 h

STATUS: in development.

General Electric

General Electric Company, Aircraft Equipment Division, Aerospace Instruments and Electrical Systems Department, 50 Fordham Road, Wilmington, Massachusetts 01887

DJ 288 engine performance indicator

General Electric produces the DJ 288 standby engine performance indicator for the US Navy/McDonnell Douglas F-18 Hornet fighter, and has designed a similar system for the AV-8B Harrier II V/STOL light attack aircraft. Five engine parameters are displayed electromechanically for the F-18's two engines. Digital values of rpm, exhaust-gas temperature and fuel flow are given, below which are pointer displays of nozzle position and oil pressure. The DJ 288 can use ac, dc, or pulse-pair types of sensors, and General Electric is seeking to expand its applications. The system replaces ten conventional 2 inch-dial instruments with a $3^7/_{10}$ by $5^3/_4$ inch display.

The digital displays use magnetic wheels which are driven by latch-decoder drivers. These provide a pre-set, variable update rate. The analogue presentations are controlled by dc torque motors using the servo-driven nulling position to ensure smoothness of display. The self-contained power-supply operates on a high-efficiency fly-back converter principle

The US Navy's F/A-18 Hornet fighter uses General Electric DJ 288 standby engine indicators

The world's fastest operational aircraft, the US Air Force/Lockheed SR-71, has a General Electric fuel-quantity measurement system

incorporating overload protection and operating over a wide input voltage range. The internal built-in test circuitry is powered by an external command signal. The 'canned' and 'display' built-in test sequence provides a visual demonstration check of the correct operation of more than 90 per cent of the component failures. The 'canned' built-in test provides an accuracy check on all channels, which is held for a nominal 8 seconds, following which the magnetic wheels sequence through all indexable digits at the rate of one numeral per second. The unit continues to cycle through these canned and display built-in test sequences until the command signal is removed, at which time normal operation is resumed.

Input power: 15 W at 28 V dc, plus 1·8 W for lighting
Operating temperature range: –54 to +71° C
Vibration: 10 *g* sinusoidal
Electro-magnetic interference: MIL-STD-461 and 200 V/metre (RS03)
Dimensions: $3^{7}/_{10} \times 5^{3}/_{4} \times 7^{4}/_{5}$ inches (94 × 146 × 198 mm)
Weight: 5·4 lb (2·4 kg)
Mtbf: 2700 h
BIT failure detection rate: > 99%
Rpm range: 2–110% ±1%
Exhaust gas temperature range: 0–999° C ±5° C
Fuel-flow range: 300–15 000 lb/h ±100 lb/h
Nozzle-position range: 0–100% ±5%
Oil-pressure range: 0–200 psi ±10 psi
Full-scale response: (rpm, egt, ff) 2 s

STATUS: in production.

Integrated engine instrumentation
General Electric claims to have pioneered the concept of integrated engine display systems, now standard options on the Boeing 747,

McDonnell Douglas DC-10, Airbus A300, and Lockheed L1011 TriStar wide-bodied airliners. The displays are offered in round or vertical configurations. The two 2-inch round and 2 ATI configurations are common for the airliner applications, while early examples of the vertical tape displays were supplied to the US Navy for the Grumman A-6E Intruder and EA-6B Prowler attack and electronic-warfare aircraft.

The round instruments typically have both pointer and electro-mechanical counter. They are servo driven, either by standard ac servoes

or by General Electric's Accutorque drive. This is a gearless, direct-drive torque motor used for pointer displays, either alone or in combination with digitally encoded individual magnetic counter wheels when high accuracy is needed. Accutorque may also be used with a minimum of gearing to provide a numerical readout on a mechanical counter, with no complex analogue to digital conversion or digital circuits. General Electric has supplied this type of instrumentation to measure almost all primary and secondary engine characteristics: engine pressure ratio, fan and compressor rotation speeds,

General Electric integrated engine instrumentation, with lcd displays

exhaust-gas temperature, turbine inlet temperature, torque, propeller rpm, fuel flow, pressure (fuel, oil, air, and hydraulic), temperature (oil and fuel), and oil and fuel quantity. The vertical-scale instruments also use Accutorque, and may incorporate two, three, or four tapes per parameter.

The latest General Electric development in this area is the incorporation of liquid-crystal displays (LCDs) for the digital information, replacing electro-mechanical counters. The projected three-crew Boeing 767 new-generation wide-body airliner was to have used these counter/liquid-crystal display instruments, but the two-crew version now specified by nearly all 767 customers uses a cathode ray tube for primary engine information. However, both 767 and 757 use General Electric liquid-crystal displays for the standby engine instrumentation. They need very low power (measured in microwatts) and are lighter and more reliable. The data is displayed in white on a black background. General Electric has achieved completely modular construction, with separate mechanical and electrical assemblies for motor/potentiometer, gear-box, set counter, numeric display/printed circuit board, and plug-in circuit modules.

Pointer/counter electro-mechanical displays
Dimensions: 2·27 × 2·27 × 6 to 13 inches (58 × 58 × 152 to 330 mm) (depth depends on complexity)
Weight: 1·2 – 1·8 lb (0·5 to 0·8 kg)

Accuracy: Typically ±0·5% full-scale; ±0·1% full-scale for some digital instruments
Response time: Typically 3 or 4 s full-scale

Vertical scale instruments
Dimensions: (dual-tape) 1¾ × 5¾ × 6 inches (44 × 146 × 152 mm) or 1¼ × 5½ × 8 inches (32 × 140 × 203 mm) typical
3- and 4-tape instruments higher and deeper
Weight: (dual-tape) 2 lb (0·9 kg)
(three-tape) 3 lb (1·4 kg)
(four-tape) 4 lb (1·8 kg)
Accuracy: Typically 0·5% full-scale; 0·3% for EPR
Response: (cold start) < 3 s
(after warm-up) < 1 s
Mtbf: (dual-tape) 3500 h demonstrated

STATUS: in production.

Fuel-management systems
General Electric's mass-flow measurement system, introduced in 1951, senses the mass-flow rate of fuel to the propulsion system or in refuelling operations to another aircraft. The output can be displayed in terms of flow rate and/or fuel consumed, or used as an input into other system elements. Recent technology improvements include the design of a flow-transmitter in which the impeller is driven by the flow of fuel, eliminating the impeller motor. An

extension of this approach led to the production of a flowmeter in which fuel provided fluid torque to drive the rotor. The output is then a pulse which is proportional to the mass flow rate. No input power is required.

STATUS: in production.

Liquid-level measurement systems
General Electric has developed a liquid-level transducer which provides an output proportional to the level of fluid in a tank. The primary application is the measurement of oil quantity, but hydraulic fluid and fuel quantities may also be measured. Civil applications include Airbus A310; Boeing 707, 727, 737, 747, 757, and 767; McDonnell Douglas DC-8, DC-9, and DC-10; and Lockheed TriStar. In the military field, aircraft using these systems comprise the Northrop F-5E/F Tiger, Lockheed P-3C Orion, Grumman A-6/EA-6B, McDonnell Douglas F-15 Eagle and F-18 Hornet, and General Dynamics F-16 Fighting Falcon. The Boeing KC-135 Stratotanker and McDonnell Douglas KC-10 Extender tankers also use them for air/air refuelling, and the Mach 3 Lockheed SR-71 reconnaissance aircraft uses the system to measure fuel quantity. Development activities include the completion of a fluidic transmitter and work on an advanced non-intrusive flow transmitter.

STATUS: in production.

Hamilton Standard
Hamilton Standard division of United Technologies, Bradley Field Road, Windsor Locks, Connecticut 06096

AIC-12 air intake control
By 1981 Hamilton Standard had delivered more than 3000 air-intake controls for the US Air Force/McDonnell Douglas F-15 Eagle air superiority fighter. The system controls the F-15's intake ramps to achieve sub-sonic flow to the engine via a normal shock wave, minimising the energy loss due to the shock. The system also controls the bypass doors which allow the correct flow of air into the compressor at the right pressure.

A single control unit measures free-stream and intake static and total pressures, total air temperature and the aircraft's angle of attack. This data is used by the micro-processor element to calculate the optimum inlet throat area, and control signals are generated to drive the ramps and bypass doors via servo-actuators. A byproduct of the calculation is the aircraft Mach number, which is fed to the EEC-90 electronic supervisory control detailed below.

Dimensions: 11 × 9 × 7½ inches (279 × 229 × 190 mm)
Weight: 16·2 lb (7·3 kg)
Power: 65 W

STATUS: in production for F-15.

Hamilton Standard provides the EEC-81 temperature limiters for the twin P&W TF30 engines in F-14 fighter

EDM-110, EDM-111, and EDM-112 PMUX propulsion multiplexers
Hamilton Standard supplies propulsion multiplexers for the Pratt & Whitney JT9D-7R4 (EDM-110) and General Electric CF6-80 (EDM-111) big turbofan engines, and the PW 2037 (EDM-112) 37 000 lb (165 kilonewton) thrust medium-size turbofan. The applications for these are Boeing's 747, 767 and 757, and Airbus Industrie's A310. The propulsion multiplexer collects analogue data from the engine and air-data system, converts it to digital format, samples it, and changes it into a multiplexed data stream. The data is then transmitted to a ground-based diagnostic unit.

The propulsion multiplexer is programmed with the engine's mechanical and thermo-

dynamic limits, and parameters going beyond these limits are stored in a semiconductor memory for later diagnostic investigation. This is intended to help solve any problems as they arise. Data accepted by the propulsion multiplexer includes engine pressures, temperatures, fuel-flows, vibration levels, oil condition and positions of variable-geometry features such as intake guide-vanes and active-clearance controls.

Dimensions: 16 × 14 × 4 inches (406 × 356 × 102 mm)
Weight: 18 lb (8·1 kg)
Power: 24 W

STATUS: in production.

Hamilton Standard EPR-101 pressure-ratio transmitters fitted in Lockheed L-1011 TriStar wide-body airliners

EEC-81 temperature limiter

The Hamilton Standard EEC-81 temperature limiter is an analogue electronic unit which has been in production for the Pratt & Whitney TF30 engines powering the US Navy/Grumman F-14 Tomcat carrier-borne air-superiority fighter since 1969. More than 1800 units had been delivered by 1981. The unit receives turbine inlet temperature signals from engine thermocouples and compares them with pre-set datum temperatures. The resulting temperature-error signal drives a fuel-control trim motor.

Dimensions: 10 × 4 × 6 inches (254 × 102 × 152 mm)
Weight: 5·9 lb (2·7 kg)
Power: 30 W

STATUS: in production.

EEC-90 engine supervisory control

Hamilton Standard claims that the US Air Force/McDonnell Douglas F-15 air-superiority fighter was the first high-performance military aircraft to have micro-processor-based engine controls. The EEC-90 was designed for the Pratt & Whitney F100 augmented turbofan engine that powers the F-15 Eagle and the F-16 Fighting Falcon air-superiority fighters, and more than 3000 were delivered in the first ten years of production which began in 1972. Flexibility, accuracy and growth potential are cited by Hamilton Standard as advantages over earlier analogue systems.

The EEC-90 uses a general-purpose digital processor to compare the aircraft and engine performances with stored data which gives the desired performance under the relevant operating conditions. The system then drives the engine's hydromechanical fuel control and changes engine geometry schedules, such as exhaust nozzle position, to achieve the desired performance. The EEC-90 is mounted on the engine and is protected against adverse temperature, pressure and vibrations. Its function is essentially one of fine-tuning engine performance.

Dimensions: 17¹/₅ × 11³/₅ × 6¹/₅ inches (437 × 295 × 157 mm)
Weight: 24·5 lb (11·1 kg)
Power: 160 W

STATUS: in production.

EEC-103 engine supervisory control

Pratt & Whitney's JT9D-7R4 commercial turbofan engine comes with a Hamilton Standard EEC-103 supervisory control, so the digital unit is to be found on the Boeing 767 and Airbus Industrie A310 and A300–600 medium/short-haul airliners. It is similar in function to the EEC-90 described above but its smaller volume indicates a more recent introduction and its function is mainly to control fuel; there are few if any variable-geometry control features. Built-in test equipment is incorporated and fault information is passed into the ARINC 429 databus to assist fault diagnosis.

Dimensions: 14³/₄ × 13⁹/₁₀ × 3³/₄ inches (375 × 353 × 95 mm)
Weight: 21 lb (9·5 kg)
Power: 50 W

STATUS: in production for the A310 and Boeing 767 airliners.

EEC-104 full-authority engine control (FAEC)

Hamilton Standard claims that 80 per cent of the world's jet airliners fly with their engine controls and they expect their analogue full-authority engine control system to be the first such system to enter revenue service. The EEC-104 has been chosen for Pratt & Whitney's PW2037 turbofan, which is optional on Boeing 757s delivered from late 1984.

According to Hamilton Standard, the full-

Hamilton Standard's EEC-103 supervisory control has been chosen for P&W JT9D engines in Boeing 767 new-generation airliner

Hamilton Standard's EEC-104 full-authority analogue engine control system will be optional on later Boeing 757 airliners. Note shockmount on near corner

Hamilton Standard's EEC-118 multi-application engine control system is proposed for the Rolls-Royce Gem turboshaft for the Westland Lynx helicopter

authority engine control allows 'set and forget' operation, since the throttle setting is a reflection of a thrust demand. The EEC-104 trims fuel-flows to the engine automatically so no throttle 'tweaking' is needed. As well as controlling fuel flow the dual-channel unit takes care of engine starting, variable stator positioning and active clearance control. A constant-speed idle function is provided and the system controls engine acceleration and deceleration. Protection against over-heating and over-speeding is provided. The result is simpler operation, more consistent and accurate thrust settings and improved engine life.

Dimensions: 5 × 13½ × 19⅕ inches (127 × 343 × 488 mm)
Weight: 37·1 lb (16·8 kg)
Power: 50 W

STATUS: in development. Flight testing to start in 1983 on Boeing 747 testbed.

EEC-118 MACS multiple-application control system

Hamilton Standard has drawn on its long experience in engine control to produce a standard electronic engine control which is suitable for turboprop and turboshaft engines producing up to 5000 shaft horsepower (3700 kilowatts), and turbofans rated at up to 5000 lb (22·2 kilonewtons). MACS is an integrated electronic and hydro-mechanical unit designed to reduce operating cost and increase engine performance and life.

The system has already been chosen for the Pratt & Whitney Canada JT15D (Mitsubishi Diamond, Peregrine and SIAI-Marchetti S.211), the PW100 (de Havilland Canada Dash 8, Embraer EMB-120 Brasilia and Aérospatiale/ Aeritalia ATR-42) and the Rolls-Royce Gem (Westland 30, Lynx and Agusta A129 helicopters). For the turbofan, MACS provides automatic thrust ratings, climb without throttle movement, engine synchronisation and over-speeding protection. Turboprops also have torque limit display, propeller speed control and automatic trim. The EEC-118 offers fast-response power-turbine governing, torque sharing and limitation, and temperature limitation.

Dimensions: 7½ × 7½ × 2¹/₁₀ inches (191 × 191 × 53 mm)
Weight: 4·5 lb (2 kg)
Power: 11 W

STATUS: in production.

Engine pressure ratio transmitter (EPRC)

Hamilton Standard introduced this high-temperature four-bellow engine pressure transmitter in 1965, following production of earlier systems for the USA's Convair B-58 Hustler supersonic bomber and France's Super Caravelle airliner and Mirage fighter family. Over 2000 EPRCs have been delivered for the Pratt & Whitney powered General Dynamics F-111 fighter-bomber.

Four bellows are positioned at 90 degree intervals in a cruciform layout in the engine

Hamilton Standard PSC-101 synchrophaser used to synchronise propeller phasing on Lockheed C-130 Hercules transport

Hamilton Standard AIC-12 air-intake control for McDonnell Douglas F-15 fighter

intake and exhaust. A change in pressure tends to move the bellows, which are then maintained in their original position by a feedback servo system. The signal required to return the bellows is also used to drive a cockpit dial which reads the ratio of exhaust pressure to inlet pressure. The configuration minimises the effect of temperature changes.

Diameter: 5²/₅ inches (137 mm)
Length: 7⁷/₁₀ inches (196 mm)
Power: 35 W at 115 V, 400 Hz

STATUS: in service.

EPR-100 engine pressure-ratio transmitter

Essentially the civil version of the EPRC, this system entered production for the Pratt & Whitney JT9D-powered Boeing 747 in 1968 and more than 2000 had been delivered up to 1981. The EPR-100 works in the same way as the EPRC, measuring the ratio of turbine exhaust pressure to compressor inlet pressure.

Dimensions: 9¾ × 7²/₅ × 5¾ inches (248 × 188 × 146 mm)
Weight: 5·5 lb (2·5 kg)
Power: 115 V 400 Hz

STATUS: in production.

EPR-101 engine pressure-ratio transmitter

This transmitter works in a similar manner to the EPRC and EPR-100, but calculates the pressure ratio electronically rather than mechanically. It comes in various configurations to suit the aircraft upon which it is mounted: DC-10, Airbus A300 and TriStar wide-body airliners. The CF6-powered DC-10 has three engine pressure-transducers and an ambient ram pressure-inducer. The DC-10 with JT9Ds has two extra ambient transducers. The A300 systems have no ambient pressure transducers.

The unit for the Lockheed L-1011 TriStar is called the integrated exhaust pressure ratio transmitter (IEPRT) because the exhaust pressure ratio is determined by integrating two separately measured pressures.

Dimensions: (DC-10 and A300) 4⁹/₁₀ × 3¼ × 8¹/₁₂ inches (125 × 83 × 205 mm)
(TriStar) 6½ × 8¾ × 10¹/₅ inches (164 × 222 × 260 mm)
Weight: (DC-10 and A300) 3·95 lb (1·78 kg)
(TriStar) 7·25 lb (3·3 kg)
Power: 115 V, 400 Hz

STATUS: in production.

EPR-102 engine pressure-ratio transmitter

This is a solid-state integrated exhaust pressure ratio transmitter and has been chosen for the RB.211-535 turbofan engine that powers the new generation Boeing 757 airliner. The EPR-102 calculates engine intake pressure for the engine control system and engine pressure ratio for the cockpit indicator. Vibrating cylinder pressure transducers convert the frequency of vibration to a digital pressure reading according to calibration equations stored in programmable read-only memories. Effects on the calibration are taken into account using built-in temperature measurements.

An Intel 8085 micro-processor calculates the engine pressure ratio from these values, and converts it, together with intake pressure into ARINC 429 data-bus format for transmission around the aircraft. The software, contained in programmable read-only memories, is trimmed to individual engine characteristics by a five-bit code.

Dimensions: 9¹/₁₀ × 5⁷/₁₀ × 3³/₅ inches (231 × 145 × 91 mm)
Weight: 5·6 lb (2·5 kg)
Power: 13 W

STATUS: being flight-tested and in early production.

ETT-100 tachometer transmitter

Hamilton Standard supplies this tachometer transmitter which measures fan rotational speed on the Pratt & Whitney JT9D big-fan engines powering the Boeing 747, DC-10 and Airbus A300. It is mounted on the fan casing and comprises an eddy-current detector which generates a small voltage as each fan blade passes. The signals are amplified and the time interval between pulses gives the revolutions per minute.

Dimensions: 3¾ × 4⁹/₁₀ × 2¾ inches (95 × 125 × 70 mm)
Weight: 1·12 lb (0·51 kg)
Power: 2·8 W

STATUS: in production.

PSC-100, PSC-101, PSC-102 solid-state propeller synchrophasers

The first type of propeller synchrophaser built by Hamilton Standard employed thermionic tube technology and is still used aboard the Lockheed C-130 Hercules military transport and P-3 Orion and CP-140 Aurora submarine-hunters. The unit synchronises all four engines at 100 per cent rpm and controls the propeller phase to minimise sound level.

The solid-state synchrophasers which have superceded them have the usual benefits of later technology: less weight, volume, power consumption and crew workload, and greater reliability. Aircraft noise levels are also lower because the phasing is controlled more accurately. As well as performing the functions of the earlier synchrophasers, the solid-state device also provides dynamic speed control for power changes and aircraft manoeuvres, again reducing noise. The solid-state PSC-100 is in production for the Grumman E-2 Hawkeye

carrier-borne early-warning aircraft and C-2 Greyhound deck-landing transport, while the PSC-101 is built for the civil and military Hercules. The PSC-102 was proposed for the now-shelved L-400 Twin Hercules project.

Weight: 10 lb (4·5 kg)
Power: 42 W

STATUS: in production.

PSC-103 digital propeller synchrophaser

Hamilton Standard is developing this digital synchrophaser for the next generation of

commuter airliners and it has been selected for the de Havilland Canada Dash 8 and Embraer EMB-120 Brasilia. Its functions are similar to those of the solid-state analogue synchrophasers described above, but control accuracy and reliability are better. The digital unit is also considerably smaller and lighter and consumes less than half the power of the earlier units.

STATUS: in development.

Kollsman

Kollsman Instrument Company, Daniel Webster Highway South, Merrimack, New Hampshire 03054

Electro-luminescent displays

For nearly 40 years scientists have attempted to imitate fireflies and glow-worms by producing electro-luminescent devices. In the 1960s evaporated thin-film phosphors were developed that, while holding much promise, were characterised by low brightness and short life. During the past ten years significant progress has been made by a number of companies, including Kollsman, through materials and process improvements. As a result of continuing development, ac thin-film electro-luminescent displays are seen to be viable alternatives to servo-mechanical displays, light-emitting diodes, liquid crystal displays and cathode ray tubes. Matrix displays are now possible with resolutions greater than those of conventional television tubes and brightness levels greater than 1000 foot-lamberts (3426 cd/m²) are now practicable, compared with 100 foot-lamberts (342·6 cd/m²) of a typical television tube. Life-times greater than 40 000 hours have been achieved and efficiencies are said to be greater than those of more traditional displays, with outputs of two to three lumens per watt. Kollsman says that the manufacturing method that is has developed gives a cost-effective display, competitive with other equivalent systems.

In the manufacturing process various layers of thin films are vacuum-deposited on a glass substrate. The first layer, deposited directly on the rear surface of the glass (the front surface forms the face of the instrument) is a transparent conductor which provides one of the electrical connections. The second and fourth layers are insulators needed to prevent electrical breakdown. The third layer is the phosphor, the source of light. The fifth layer is a light-absorbent or 'black' material to improve contrast. The final layer is the back conductor

The Kollsman vertical-strip engine instrument display, configured for a twin-engined aircraft, shows percentage rpm, exhaust-gas temperature, and fuel flow rate

and the entire assembly is sealed to prevent contamination from the external environment. Light is produced in the phosphor layer by applying an alternating voltage to the conducting layers, emission occurring as a result of the excitation of activator ions by injected electrons.

One of the major considerations in using any type of display in an aircraft instrument panel is the ability of the crew to be able to read it in the presence of direct sunlight. In the Kollsman system this is achieved by preventing the sun's radiation from being reflected from the display. Reflection can be reduced to one per cent by using a 'black' layer which not only absorbs

most of the sunlight but permits lower brightness levels to be used, reducing the power needed to drive the system and increasing its reliability.

Based on this technology the company is developing vertical scale and round dial instruments and flat screen, alpha-numeric displays. A vertical scale instrument can offer 35 elements per inch, integral lighting, sunlight readability, manual dimming, silk-screen scales for simple customising, modular design, and dc, analogue, or parallel binary interfaces with other equipment and sensors.

STATUS: in development.

Northrop

Northrop Corporation, Electronics Division, 2301 West 120th Street, Hawthorne, California 90250

TEMS turbine engine monitoring system

The US Air Force is evaluating 25 Northrop turbine engine monitoring systems in A-10 Thunderbolt II tank-destroyers serving with 917 Tactical Fighter Group at Barksdale AFB, Louisiana. The system has been flight tested in six A-10s so far and initial trials were conducted in T-38 Talons of the US Air Force Air Training Command. TEMS is part of an attempt by the US Air Force to improve its aircraft availability. It monitors and records engine parameters for later analysis to determine whether the aircraft is serviceable and to schedule maintenance.

The system records engine data such as temperatures, pressures and vibration, in a processor which weighs less than 10 lb (4·5 kg). After landing, the data is transferred to a diagnostic display unit small enough to be transported around the flight-line on a bicycle. The transfer is completed in a few seconds, after which maintenance personnel can immediately tell whether the engine is serviceable.

The data is then transferred to a data-collection unit which can record flight information for up to ten aircraft. The data can be studied in greater depth in the workshop and may be stored on disc and transferred via telephone to other locations such as the engine factory or a command unit. TEMS monitors the engine parameters continuously but the processor is programmed according to the particular mode or regime required. The regime

can be when parameters over-step normal limits, upon pilot command, or during critical flight phases such as start-up, take-off and climb-out. Data from several flights, normally six, can be stored.

TEMS is part of the US Air Force on-condition maintenance philosophy to improve aircraft availability and reduce life-cycle costs. Engine defects or potential defects are detected more rapidly, unwanted engine removals are reduced and some maintenance actions can be omitted if the engine is still performing well. In addition, fewer ground run-ups are needed and the TEMS data is also used to trim the engine's fuel-control system, again reducing run-up time and saving fuel. TEMS can be modified to suit other aircraft and engines, civil as well as military.

STATUS: in operational flight-test.

Safe Flight

Safe Flight Instrument Corporation, White Plains, New York 10602

Autopower autothrottle

Safe Flight has developed an advisory speed control system called Speed Control of Attitude and Thrust (SCAT) which gives the pilot attitude guidance for rotation and climb-out and thrust guidance for the approach. The company also makes a fuel performance computer to give the pilot his optimum speed for various specific ranges. Autopower is an autothrottle system that moves the throttle levers according to the outputs from SCAT or the fuel performance computer, according to the phase of flight.

Autopower takes care of the throttles during busy times of the flight, especially on the approach where changes in aircraft configuration (flaps, slats, and undercarriage operation) mean that thrust has to be adjusted frequently. Built-in test equipment switches Autopower off if it malfunctions and the pilot can monitor the situation by observing a fast/slow indicator on the flight director driven by the SCAT or fuel performance computer. SCAT is programmed not to allow the aircraft speed command to fall below the minimum safe speed for the configuration in use. The Autopower throttles allow individual throttle 'tweaking' without disengaging the system and is 'transparent' – the pilot over-ride needs no more force than usually employed.

STATUS: in production and service.

Silver Instruments

Silver Instruments, 2346 Stanwell Drive, Concord, California 94520

Fueltron fuel-management system

Fueltron is the name of a complete fuel-management computer system aimed at adding economy, convenience and safety to twin and single-engined general-aviation aircraft and helicopters powered by piston and turbine engines. Large digital displays enable rapid, precise setting of engine power for any flight condition, together with more accurate mixture-leaning than is possible with an exhaust-gas temperature readout. The system is available in a large number of configurations to suit various applications. It can pay for itself in 500 to 1000 hours of operation in a typical twin, through savings in fuel, time, and engine wear. Complete Fueltron packages are available for Bell JetRanger and LongRanger, Cessna 337, and Piper Aerostar. Installation kits can be supplied for aircraft produced by Beech, Cessna, Mooney, Piper, and Gulfstream, as well as for the Douglas DC-3/C-47.

All electronics are in a single, panel-mounted unit. Incandescent displays are used to give digital indication of fuel remaining (which may be translated into time remaining based on flow rate), and flow rate for the engine or engines. The computer calculates how much fuel has been burned by integrating fuel flow throughout the flight, and fuel used can be displayed instead of fuel remaining. A warning light indicates when time remaining falls below 48 minutes.

Fueltron receives its data from the digital output of a flow transducer located in the fuel line to each engine. This output, which is directly proportional to fuel flow, is combined in the Fueltron computer with precise time data from a quartz crystal clock. The system's accuracy is claimed to be 1 per cent or better, and Silver says that fuel economy can be improved by up to 10 per cent when setting engine power and mixture by Fueltron rather than exhaust-gas temperature. Fuel flow for a given climb rate can be reduced, and up to several times more spark-plug life can be achieved.

Indicator and computer

Weight: 1·7 lb (0·77 kg)
Power required: compatible with any voltage between 10–30 V dc. Current drain (max) 1·4 A at 14 V, 0·7 A at 28 V
Accuracy: ±1%
Environmental: TSO C44A

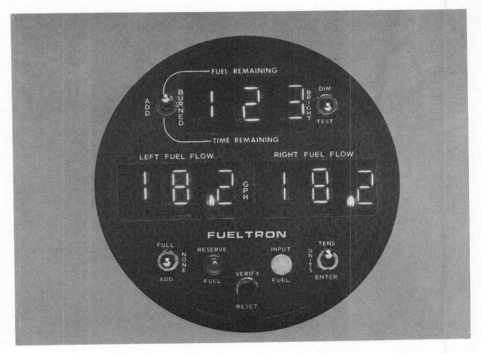

Pilot's display for the Silver Instruments Fueltron fuel-management computer

Mtbf: 3000 h plus
Dimensions: 3¼ × 3¼ × 6¹/₇ inches (83 × 83 × 156 mm)

Standard flow transducer
Flow range: 1½–60 US gal/h per engine
Temperature range: –65 to +125° C
Pressure drop: 0·3 psi (0·1 kg/cm²) at 15 US gal/h
Max working pressure: 200 psi (68·2 kg/cm²)
Max (burst) pressure: 2000 psi (682 kg/cm²)
Life expectancy: >10 000 h
Weight: 0·31 lb (0·14 kg) per engine

Turbine aircraft PT Series flow transducer
Flow range: 8 ranges from 1000–20 000 lb/h max
Temperature range: –90 to +300° C
Pressure drop: 1 psi (0·34 kg/cm²) at 500 lb/h (typical)
Max pressure: 5000 psi (1700 kg/cm²)
Life expectancy: 10 000 h
Weight: 9 oz (0·25 kg) per engine

STATUS: in production for wide variety of general-aviation aircraft.

Fuelgard fuel-management system

Fuelgard is a simplified version of Fueltron, suitable for light single-engined aircraft. An incandescent display gives a digital readout of fuel flow with 2 per cent accuracy. A switch allows the display of fuel used as an alternative. Like Fueltron, Fuelgard allows accurate mixture leaning for piston engines, and fuel savings are expected to be about 10 per cent. The total fuel used remains in the equipment's memory, a useful facility where more than one flight is made before refuelling. Fuelgard can be installed in fixed-wing aircraft or helicopters with a single reciprocating or turbine engine.

Dimensions: 2⁹/₁₀ × 1 × 4²/₅ inches (74 × 25 × 112 mm)
Weight: 14 oz (0·4 kg)
Voltage: 14 V or 28 V dc
Flow rate range: 1½–60 US gal/h
Accuracy: ±2%
Approval: TSO c44A

STATUS: in production.

Simmonds Precision

Simmonds Precision, Instrument Systems Division, Panton Road, Vergennes, Vermont 05491

Fuel-management systems

Simmonds Precision has generated over 100 patents since 1950 for its aircraft fuel-management systems, based on technologies including capacitance measurement, micro-processors, and solid-state electronics. Simmonds Precision makes conventional fuel-quantity gauging systems for the Boeing 727 and 737 and Lockheed L1011 TriStar airliners, as well as for military aircraft such as the McDonnell Douglas F-15 Eagle and F/A-18 Hornet, and the General Dynamics F-16 Fighting Falcon. Helicopter applications include the Sikorsky CH-53 and UH-60A Black Hawk.

For many years, fuel gauging was exclusively electro-mechanical in nature. Probes installed in fuel tanks sent impulses to a signal conditioner which amplified the signal and transmitted it to a flight-deck indicator. This type of reliable system represents the mainstay of Simmonds' fuel gauging business and has done so for decades.

In recent years the company has focused its research and development attention on the intrinsic speed and power offered by advances in micro-computer technology. It has thus developed a family of small, light, reliable, efficient micro-computers which can be installed directly in the aircraft. By substituting an intelligent computer for the electro-mechanical system, Simmonds has produced advanced fuel-gauging systems which are more accurate and more reliable. The introduction of the micro-processor permits the use of more simplified tank probes with non-linear characteristics, along with solid-state circuitry and digital instrumentation. Accuracy is enhanced because the computer compensates for variables such as aircraft attitude and differences in fuel densities and temperatures. In addition, the computer can detect, remember, and advise the groundcrew of any malfunction.

Fuel gauging innovations such as these have given rise to complete fuel-management and centre of gravity control systems – the first Simmonds example was built for the Rockwell B-1 bomber. The centre of gravity monitoring and automatic shifting of fuel from tank to tank optimises the aircraft's weight distribution for improved fuel efficiency and safety, and may be essential for variable-geometry aircraft. It also reduces trim drag. Simmonds has produced such a system for the Lockheed L1011 TriStar. Computerised fuel-management systems can monitor all operating parameters of the fuel system, and compute throttle settings for maximum efficiency. Some also incorporate autothrottles.

Conventional fuel-gauging systems are based on a capacitance measurement principle which uses an assumed fuel density in converting fuel level measurements into gallons or pounds. But fuel stocks throughout the world may vary widely in density, depending on the characteristics of the original crude oil. Thus Simmonds has developed a fuel-type detector for the Northrop F-5G Tigercat export fighter.

Fuel quantity in the General Dynamics F-16 fighter is measured by Simmonds fuel-gauging equipment

This system senses the type of fuel in the tanks, and adjusts the cockpit displays for the appropriate density.

STATUS: in production.

Anti-Submarine Warfare Equipment

CANADA

CAE Electronics

CAE Electronics Limited, PO Box 1800, Saint-Laurent, Montreal, Quebec H4L 4X4

AN/ASA-64 submarine anomaly detector

CAE is working on a product improvement programme in support of AN/ASA-64, the submarine anomaly detector originally built for the US Navy/Lockheed P-3C Orion. The improved version is to have more processing power and a variant for helicopters is proposed. CAE says that the latest type has been evaluated by the Royal Navy, RAF and the US Navy. The AN/ASA-64 is a magnetic anomaly detector (mad) processor which identifies and marks local distortions in the earth's magnetic field induced by the presence of submarines. The operator is alerted by both visual and aural alarms, reducing the experience required by the operator and needing less monitoring of the mad system, so that he has more time to devote to other sensors.

Dimensions: (control unit) 4 × 5¾ × 3½ inches (102 × 146 × 90 mm)
(ID-1559 processor) 9 × 5⁹/₁₀ × 6 inches (229 × 150 × 153 mm)
Weight: (control unit) 1·5 lb (0·68 kg)
(ID-1559 processor) 6·5 lb (3 kg)
Operating temperature range: –54 to +55° C
Power requirement: 115 V 20 W
Specification: MIL-E-5400

STATUS: in service.

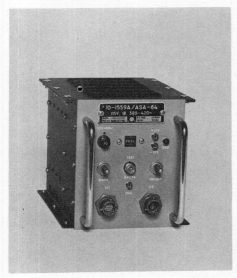

CAE AN/ASA-64 submarine anomaly detector

AN/ASA-65(V) mad nine-term compensator

CAE developed this semi-automatic magnetic anomaly detector (mad) compensator to improve the effectiveness of mad on aircraft having only manual compensation for aircraft interference on the earth's magnetic field. Previously, fixed permalloy strips and copper coils were mounted in the mad boom to create induced and eddy-current fields equal and opposite to those caused by the aircraft. These compensators had to be custom designed for each individual aircraft, took a long time to adjust on flight-test and did not cater for the changes which take place during the aircraft's life. CAE also says that new, more sensitive mad equipment needs greater precision than fixed compensators can provide.

The AN/ASA-65(V) compensates for permanent interference after only five minutes flying, compared with about an hour needed for manual compensators. The system allows for manoeuvre interferences after 30 to 45 minutes, improving mad detection range, especially when frequent manoeuvres are performed and when conditions are turbulent.

The all-solid-state AN/ASA-65(V) is compatible with all current mad systems, and internal patch connectors are used to adjust the system for the aircraft concerned. The nine-term compensator is used by the RAF/BAe Nimrods and the US Navy/Lockheed P-3C Orion and S-3A Viking sub-hunters.

Figure of merit: less than 1 gamma
Max compensation field: 50 gamma on each side of aircraft
Mtbf: >1800 h
Dimensions: (Control indicator) 9 × 5¾ × 6½ inches (229 × 146 × 165 mm)
(Electronic control amplifier) 7¾ × 5⅞ × 13⅝ inches (197 × 149 × 346 mm)
(Magnetometer assembly) 6 inches (152 mm) cube
(Coil assembly) 3½ inches (89 mm) cube
Total weight: 29·5 lb (13·4 kg)
Power requirement: 115 V 100 W plus 10 W at 28 V dc or ac for panel lamps
Specification: MIL STD

STATUS: in service on Nimrod, P-3C and S-3A.

AN/ASA-65 mad compensator group adapter

CAE has devised this add-on device for the AN/ASA-65(V) mad compensator to automate the data-gathering and thus reduce the time required for the compensation exercise. The compensator group adapter requires no modification to the original compensator and is operational with RAF/BAe Nimrod MR.2s and US Navy/Lockheed S-3A Vikings and P-3C Orions. The compensator group adapter comprises an indicator and a micro-computer, which calculates the changes required in each of the nine terms to provide optimum compensation. The compensation update requires only six minutes flying, during which time low amplitude manoeuvres are conducted on four different headings. The operator then has to adjust the compensation terms manually, as shown on the indicator. This update, which has to be done at least once a week, can take 90 minutes when manual compensation is used, and 30 minutes without the adapter. The compensator group adapter also compensates for four terms after weapon drop. This takes two minutes, and requires small random manoeuvres to be conducted on two or four headings. As well as automatic data acquisition, the device includes built-in test equipment, which reduces the time taken in pre-flight checks and troubleshooting. Aircraft availability is thereby enhanced considerably.

Dimensions: (Indicator) 3¾ × 5¾ × 5 inches (95 × 146 × 127 mm)
(Computer) 8⁷/₁₀ × 9 × 16½ inches (221 × 229 × 418 mm)
Weight: (Indicator) 3 lb (1·35 kg)
(Computer) 22 lb (10 kg)
Power requirement: 115 V 65 W, plus 10 W at 5 V ac or dc for panel lighting and 5 W at 28 V dc for annunciator lamps

STATUS: in service and in production for P-3C retrofit.

AN/ASQ-502 magnetic anomaly detector

CAE claims that the AN/ASQ-502 is the most sensitive airborne magnetic anomaly detector (mad) available. The quoted sensitivity is 0·01 gamma, or a change in the earth's magnetic field of one in five million. The AN/ASQ-502 is in service on the CP-140 Auroras of the Canadian Armed Forces, and CAE has received interest from several other countries, including Britain, where it has been evaluated by the RAF. A towed version for helicopters is under consideration.

CAE has abandoned the normal multiple-cell detection method in favour of a single caesium cell mounted on gymbals to allow for dead zones and heading errors, which are normally taken care of by the use of multiple cells. The device also features a stable self-oscillator caesium sensor loop, and a low-noise frequency-to-voltage converter which is claimed to have virtually infinite dynamic range.

Sensitivity: 0·01 gamma (1 × 10⁻⁷ Oersted)
Detecting head figure-of-merit: (uncompensated) 1·5 gamma
(compensated) 0·3 gamma

CAE AN/ASA-65 compensator group adapter

Larmor frequency of detector: 3·5 Hz/gamma
Operating range: 0·2 – 0·75 Oersted
Larmor output (total field): 14 Hz/gamma
Specification: MIL-E-5400
Operating temperature range: –54 to +55° C
Mtbf: 1500 h
Power required: 115 V 120 W, plus 10 W at 5 V ac or dc for panel lighting

STATUS: in service.

OA-5154/ASQ fully automatic compensation system (facs)

The 16-term fully-automatic compensation system is the next step after the nine-term semi-automatic AN/ASA-65(V). It is in service with the Canadian Armed Forces CP-140 Auroras, and has been delivered to the West German and Netherlands navies for their Atlantic sub-hunters. Facs conditions the raw magneto-meter data, using its mini-computer for dis-plays and other aircraft systems. The other necessary input is the orthogonal vector magnetometer signals which provide heading, manoeuvre, and total earth-field data. The use of completely electronic compensation means that there is no need for output coils to generate opposing fields and no operator input is needed.

The compensation flight-programme lasts for no more than six minutes and comprises four one-minute low-amplitude manoeuvres on headings approximately at right angles to each other. Additional trapping circles and clover-leaf manoeuvres are optional. As well as automating the recompensation exercise, the facs allows the operator to update the system with minor magnetic variations at the touch of a button. Built-in test equipment is included.

Figure of merit: <0·4 gamma
Residual interference signals: 0·03 gamma average, or magnetometer internal noise if that is greater
Input: Minimally filtered analogue magneto-meter detector signal
Power required: 115 V 150 W plus 5 V ac/dc for panel lighting, plus 10 W at 28 V dc for annunciator lights
Mtbf: 1900 h
Specification: MIL-E-5400
Dimensions: (control indicator) $7^1/_5 \times 5^4/_5 \times 6^1/_2$ inches (182 × 147 × 166 mm)
(electronic amplifier) 9 × 9 × $21^1/_2$ inches (229 × 229 × 546 mm)
(vector magnetometer) 6 inches (152 mm) cube
Weight: 31·5 lb (14·2 kg)

STATUS: in service on CP-140 and in produc-tion for Franco – German Atlantic.

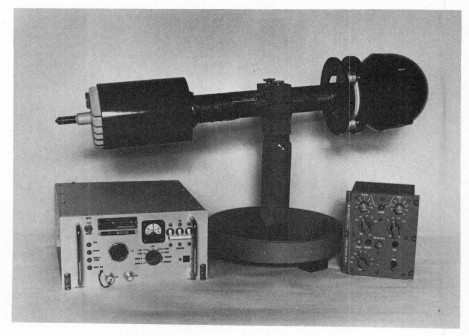

CAE AN/ASQ-502 magnetic anomaly detector

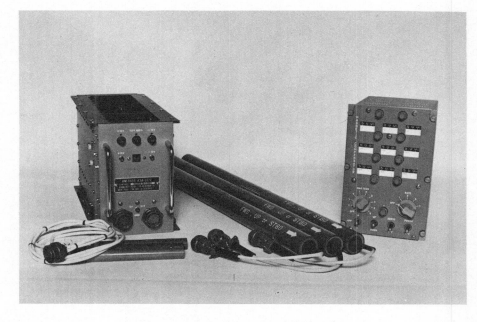

CAE AN/ASA-65(V) nine-term compensator

Canadian Armed Forces' Lockheed CP-140 Aurora sub-hunter, with CAE AN/ASQ-502 magnetic anomaly detection system in rear-fuselage

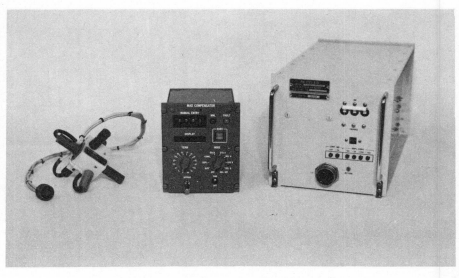

CAE OA-5154/ASQ fully automatic compensation system

Computing Devices

Computing Devices Company, Division of Control Data Corporation, PO Box 8508, Ottawa, Ontario K1G 3M9

AQS-901 acoustic processing system fast Fourier transform analyser

Computing Devices provides the fast Fourier transform analyser of the AQS-901 acoustic processing system, which is built by Marconi Avionics. The analyser can perform 2048 complex transform in 11·25 ms, and it forms an important part of the system. Computing Devices is also involved in the Marconi advanced signal processing system, probably with a similar role.

STATUS: in production.

FRANCE

Crouzet

Crouzet SA, Division Aérospatiale, 25 rue Jules Védrines, 26027 Valence cedex

Magnetic anomaly detector for Atlantic NG

Crouzet is a leading French magnetometer company, having been involved in the field since 1974. It is currently building the improved magnetic anomaly detector (mad) for the Atlantic Nouvelle Génération (ANG) for Aéronavale, the French Navy. This comprises twin magnetic sensors mounted longitudinally in the aircraft's mad boom. Since the two sensors are in different positions with respect to the mass of the aircraft, any change in the aircraft's magnetic field due to the dropping of stores or the performing of manoeuvres will be detected by the combination. The use of sum and difference calculations from the two sensors obviates the necessity for separate compensation equipment.

STATUS: in production.

Towed magnetometer Mk 3 for helicopters

This is the latest version of Crouzet's mad equipment and is intended for ASW helicopters; it detects the presence of a submersible by measuring the resulting disturbance to the earth's magnetic field. In order to eliminate disturbances created by the carrier helicopter, the detector probe is placed in a streamlined 'bird' which is towed at the end of a 230-foot (70 metre) cable. The digital computer measures the signals from the sensor and transforms them into suitable formats for the graphic recorder at the operator's station.

The measurement probe operates according to the nuclear magnetic resonance principle with electronic pumping. It offers the following advantages: it is ready to use as soon as it is switched on, even down to –40° C; it is easy to implement; the sensor lifetime is virtually infinite; weight and size are low; electrical power consumption is low; and sensitivity is very high. The Mk 3 mad comprises a detection element, computer, graphic recorder, and the winch and cradle assembly. Maintainability is assisted by in-flight checking of the 'bird' (the towed magnetometer), computer, control unit, and recorder. The ground-test points are easily accessible, and the sub-assemblies are plug-in units.

The 'bird' assembly includes the detection probe and an electronic unit, combining to produce a nuclear oscillator whose Larmor frequency is a function of the magnetic field exerted on the probe. The geometry of the probe is chosen so that its position with respect to the magnetic-field vector can be ignored.

The computer processes the signals from the probe and delivers the measurement to the recorder. Two processors are included in the computer. The first is a high-precision frequency meter which performs the frequency/voltage conversion of the signal from the probe. The second performs signal formatting, filtering, and amplification so that the operator sees pre-processed data. The computer also controls the power supply to the 'bird' and control unit, performs the cyclic and triggered tests,

Electronics set, including chart recorder, for Atlantic NG mad system

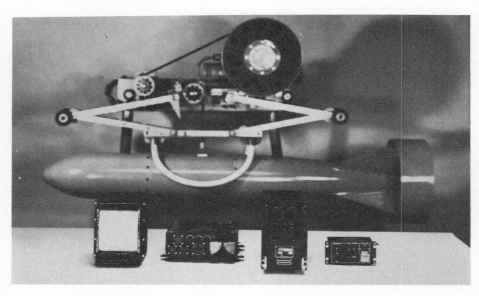

Crouzet towed magnetometer Mk 3 for helicopters

and interfaces with the control unit. The latter allows the operator to switch on and off, choose the sensitivity, and operate the built-in test equipment. A third, digital processor can be accommodated in the computer if required. It improves the signal filtering by adapting the filter characteristics to the ambient noise spectrum, and analyses mathematically the signals received as a further aid to target detection. This computed data is displayed to the operator on one of the two graphic recorder traces, the other remaining in its original state.

The graphic recorder sub-assembly is a potentiometric device which displays simultaneously two signals, recorded by red and black styluses on paper which unrolls at a constant speed. A light-emitting plate gives a permanent recording display. The operating controls are on/off, speed selection, test, and lighting.

The winch and cradle assembly comprises a

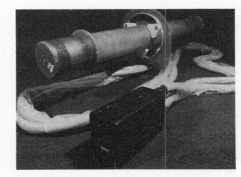

Crouzet mad probe for Atlantic NG sub-hunter

winch, gear system, drum, cradle with locking device, power unit and control unit. The probe is connected to the computer by three co-axial cables, and a cutter is provided for emergency jettisonning.

Background noise: typical deviation 0·006 gamma
Pass band: adaptable to carrier
Measurement range: 25 000–70 000 gamma
Sensitivity for relative field output: 1, 2, 5, or 10 gamma for 10 cm stylus deviation
Paper speed: 6, 75, or 300 mm/minute

Streamlined body
Length: 1300 mm
Weight: 16 kg
Diameter: 160 mm

Computer
Dimensions: 346 × 124 × 194 mm
Weight: 7·5 kg
Power supply: 100 W at 200 V ac 400 Hz, three-phase, plus 5 W at 28 V dc

Control unit
Dimensions: 165 × 150 × 57 mm
Weight: 1 kg

Recorder
Dimensions: 190 × 150 × 190 mm
Weight: 4 kg
Power supply: 50 W at 115 V ac 400 Hz, plus 20 A at 28 V dc

Winch and cradle
Dimensions: 1330 × 350 × 755 mm
Weight: 44 kg
Power supply: 40 A at 27 V dc

STATUS: in development.

Thomson-CSF
Thomson-CSF, Division des Activités Sous-Marines, BP 53, 06802 Cagnes-sur-Mer

TSM 8200 & TSM 8210 Sadang acoustic processor
Thomson-CSF is the project leader for the acoustic systems on board the Franco–German Atlantic Nouvelle Génération (ANG) sub-hunter, for which it developed the Système Acoustique d'ANG (Sadang) acoustic processing system. The Atlantic NG is fitted with the TSM 8200 system, comprising two identical modules called TSM 8210. The TSM 8210 is available in single units, in which form it is called the Mini-Sadang. This configuration is suitable for smaller aircraft or helicopters. The TSM 8210 has a single processor capable of monitoring up to eight omni-directional passive or four omni-directional active sonobuoys, and it can process data from directional buoys. Thus it can handle most types of sonobuoy which exist now or are under development.

The sub-units which make up the TSM 8210 are described in the following paragraphs; the ANG system is built up of two such systems.
Vhf receiver The TSM 8210 has a 99-channel vhf receiver capable of receiving the sonobuoy transmissions from eight buoys simultaneously. A pre-amplifier is located close to the antenna, and a remote control unit is integrated into the operator's console. For testing the system, an acoustic signal source generator is incorporated. This simulates the transmissions from sonobuoys, and can be used at any time during a mission.
Signal processing The demodulated signals are processed successively in analogue and then digital form, depending on the types of sonobuoys being monitored and the types of processing selected by the operator. For position-fixing with passive sonobuoys, the processor uses various methods based on launching several waves of buoys from different positions, and comparing their returns. Discrepancies in amplitude, frequency, time, or phase between the signals received from two or more sonobuoys are used to calculate the position of the target submarine. Closer to the attack, active sonobuoys are used to provide range–only or range-plus-azimuth (as with the Dicass buoy). The Sadang processor can deal with these types as well.

Intelligibility of the noises sensed by the hydrophones in the sonobuoys is improved by the possibility of hearing selected frequency bands or listening in selected directions. Sadang can arrange for this, and the operator can also listen simultaneously to transmissions from two different sonobuoys, one in each ear.

Displays A special processing unit is devoted to the man-machine interface for maximum efficiency and flexibility of communication with the system. Inputs are made via two physically identical keyboards, one alphanumeric and one functional, plus a rolling ball to control the marker on the cathode ray tube display. The output is shown on a cathode ray tube in raster-scan mode, and on a graphic recorder. The cathode ray tube is used to display a wide variety of pictures selected by the operator according to the tactical situation and the various types of sonobuoys in use.

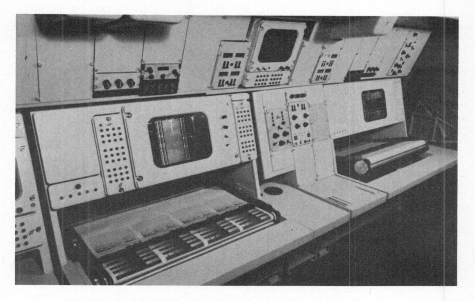
Thomson-CSF TSM 8200 Sadang acoustic processing system for Atlantic NG

Aérospatiale Dauphin 365 helicopter, candidate for Lamparo system

Atlantic NG prototype

Several pages are updated simultaneously in the mass memory, each page being available for display at any time. Each page displays a spectral analysis of the acoustic signals detected from the sonobuoys, within various frequency bands and integration times. Alphanumeric information is also available. A graphic recorder keeps a permanent trace of the processed data, printing out four simultaneous channels with a resolution of 750 points each.

Successive analyses of the frequency lines are memorised and displayed on the cathode ray tube or on paper, thus benefiting from visual correlation by enhanced sensitivity and resolution. The various detectable frequency lines can be gathered into families with the help of the electronic facilities at the operator's disposal. Fundamental frequencies can be compared with signatures stored in a library, and the noise emitter can be identified or at least classified as belonging to a category of vessels. The frequency lines in the lower part of the spectrum (under 100 Hz) are especially typical, and modulate the amplitude of the higher part of the frequency spectrum. In order to detect these lines, a processing method called demon (demodulation of noise) is available.

Active sonobuoy remote control The operation of active sonobuoys must be as quick and efficient as possible, since they can be detected by the target submarine. For this reason, some active sonobuoys are equipped with a remote-control capability by which a single pulse, or a few pulses, may be triggered where strictly necessary with pre-selected transmission characteristics (pulse duration, continuous wave or modulated frequency, etc). This function is performed by a sonobuoy remote control unit connected to a uhf transmitter tuned to the receiving frequency of the sonobuoy. The transmitter may be either a dedicated transmitter or the centralised communication system's transmitter.

On-top position indicator (otpi) Target fixes are given relative to the sonobuoy positions, but the buoys drift with winds and sea currents, so their positions must be updated from time to time. This updating is done by flying over the various buoys in succession. Flying towards these buoys is made possible by a radio direction-finder connected to a detection and display system, the on-top position indicator (otpi).

Navtac The sonobuoy data is transmitted via a digital data-bus to the tactical computer, and thence to the tactical co-ordinator. The tactical computer, called Navtac, correlates the data from sonobuoys, radar, forward-looking infrared, electronic support measures, magnetometer, and navigation systems. The tactical co-ordinator sees all this information on his two-colour display, and can thus make decisions such as choice of sonobuoy launch patterns, and order the update of sonobuoy positions, leading to the attack.

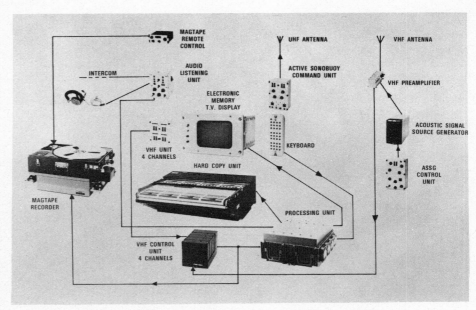

Thomson-CSF TSM-8220 Lamparo acoustic detection system

Options The TSM 8210 has optional sub-units for the recording of acoustic signals for later examination, and a sonobuoy launcher remote-control system.

STATUS: in production for Atlantic NG and on offer for various other types, such as the Franco-German C-160S Transall, British Aerospace HS748 Coastguarder, and Fokker F27 Maritime.

TSM 8220 Lamparo acoustic processor

Thomson-CSF derived the Lamparo as a lightweight, compact version of the TSM 8210, suitable for smaller aircraft and light helicopters. The aim was to provide the performance level of the basic system, but with fewer functions. For example, the processing of directional sonobuoys of the difar and dicass type has been deleted. The Lamparo furthermore cannot handle as many sonobuoys, but it can still process data from four omni-directional passive or two omni-directional active buoys simultaneously.

Lamparo has all the analogue, digital and management processing functions gathered within a single unit and the audio processing is simplified. One keyboard and the rolling-ball marker have been dropped, leaving a single keyboard with one cathode ray tube display. The hard-copy recorder is optional, its deletion reducing the space requirement considerably. The vhf receiver can handle only four channels, as it is much smaller than that of the Sadang. Lamparo is fully compatible with Navtac (see TSM 8200 above).

Despite the reduction in size and functions, Lamparo can later be upgraded if the user has

Display and controls for Thomson-CSF TSM 8210 Mini-Sadang

the space available and decides that he needs the extra capability. More extensive audio processing can be added, as can the remote control and processing of active sonobuoys such as difar and dicass. Likewise, a graphics recorder and a magnetic-type recorder can be added.

STATUS: under development. Suitable for smaller applications such as ASW versions of the Embraer Bandeirante, and the Aérospatiale Dauphin 2 and Super Puma.

ITALY

Elettronica

Elettronica SpA, Via Tiburtina km 13.7, 00131 Rome

ELT-810 sonar performance prediction system

Elettronica has developed the ELT-810 sonar performance and propagation prediction system in conjunction with its Elettronica Ingegneria Sistemi (EIS) subsidiary. It enables the sonar propagation characteristics to be fully exploited for both defensive and offensive purposes. The system comprises a single compact unit, which houses the computer, its memory, a graphic cathode ray tube display, and a dedicated command and control panel. Installation is foreseen in both fixed-wing aircraft and helicopters, as well as on submarines and surface vessels.

STATUS: in development.

Selenia

Selenia Industrie Elettroniche Associate SpA, Via Tiburtina km 12400, 00131 Rome

Falco submarine locating system

Few details have been revealed of Selenia's anti-submarine warfare activities, but the Falco system is known to be an airborne equipment designed to increase the capability for helicopters and fixed-wing aircraft to detect, classify and locate submarines. The system operates on the low-frequency noise radiated by targets and gathered by directional low-frequency passive sonobuoys.

Noise spectral analysis, target data processing and display are performed by a real-time digital computer. It is claimed that the Falco system can determine target position with an accuracy sufficient to carry out an attack with automatic homing torpedoes, and with errors due to sonobuoys drifting automatically cancelled out. The equipment is produced in several versions.

STATUS: in production and service, probably on Italian Navy Atlantic fixed-wing aircraft and SH-3 helicopters.

UNITED KINGDOM

Marconi Avionics

Marconi Avionics Limited, Military Aircraft Systems Division, Airport Works, Rochester, Kent ME1 2XX

ACT-1 airborne crew-trainer

The airborne crew-trainer is being used by crews of RAF Nimrod MR2 sub-hunters to simulate sonobuoy returns without real submarines, without wasting sonobuoys, and independent of other air traffic. ACT-1 comprises a portable instructor station, which is plugged into the aircraft's AQS-901 processing system (described below), and a reel of magnetic tape which contains the computer program.

The crew-member acting as instructor keys into the exercise control unit the position, course, and speed of target submarines, along with sonobuoy data such as type of buoy, active frequency, and last-known location. In this way a simulated threat situation is set up, and the trainees see activity corresponding to the simulation on the five screens which make up the AQS-901. Different submarine manoeuvres can be injected at any time to evade location.

The ACT-1 is used mainly when the aircraft is in transit between its base and a real area to be patrolled. Instantaneous transfer to 'live' configuration is available so that the Nimrod is always available if required. Marconi stresses that ACT-1, which it claims to be the first airborne training aid in squadron service, complements rather than replaces the ground simulator.

STATUS: in production for the RAF.

AQS-901 acoustic processing system

The Marconi AQS-901 acoustic processing system is being retrofitted into RAF/British Aerospace Nimrods as part of the conversion from MR1 to MR2 standard. It is also in service with Royal Australian Air Force/Lockheed P-3C Orion anti-submarine patrol aircraft. Computing Devices of Canada is the principal sub-contractor on the AQS-901, which is claimed to be the only processing system able to cope with all the types of sonobuoy in the NATO inventory. This includes the modern types such as barra and cambs, which are intended to combat the movement of quieter submarines in deeper water. The AQS-901 uses modern digital processing techniques, and presents information to the crew on five cathode ray tube displays as well as chart recorders.

The AQS-901 is based on the 920 ATC digital computer, which has an expandable memory up to a total of 256 K words of directly addressable memory. An important feature is the Computing Devices fast Fourier transform analyser, which can perform a 2048 complex transform in 11·25 ms. Micro-programs in the analyser are initiated by the 920 ATC, providing the flexibility and adaptability needed to maximise performance. The processing equipment which makes up the AQS-901 is modular, allowing a choice of configuration to suit different aircraft types. The operating, processing, and displaying modes are easily changed by the system operator for increased effectiveness. As well as handling all the NATO

Marconi Avionics ACT-1 airborne crew trainer

Marconi Avionics AQS-901 equip the Nimrod MR2

sonobuoys, the AQS-901 can accommodate magnetic anomaly detection functions.

The AQS-901 receives sonobuoy data and controls the buoys via an advanced eight-channel receiver and command system. Each channel meets the latest international standards and is tunable to any of the 99 NATO sonobuoy frequencies. The command subsystem allows the deployed buoys to be controlled in terms of pulse length, type and repetition frequency. The receiver is supplied by McMichael Limited.

Beam-forming, spectral analysis, and broadband power analysis are used by the AQS-901 to process in real time the data received from multi-hydrophone array buoys. Three passive location processing techniques are employed: hyperbolic and Doppler fix-processing for omni-directional buoys, and direct-bearing processing for barra buoys. When these techniques are not sufficient, and in the attack phase when very accurate data is required, active buoy processing is used. The directional-array active buoys provide bearing, range, and relative velocity of the submarine, and the AQS-901 uses frequency domain beam forming and correlation analysis to process this data. Detection and tracking of the target are performed automatically. The displays are optimised for long-range detection, and to separate slow-moving targets from reverberation and shipwrecks.

The AQS-901 has a target-oriented auto alert feature, which tells the operator when a target has been spotted. The operator sets the relevant thresholds and parameters so that real targets may be distinguished from spurious ones such as surface ships and friendly submarines. The AQS-901 can still carry on with other processing tasks while looking for certain target features, as demanded by the operator. Those features can be changed at will.

STATUS: in production and service. RAF Nimrods receiving AQS-901 in MR2 update. RAAF/Lockheed P-3C update handled by Commonwealth Aircraft Corporation, Amalgamated Wireless Australia, EMI and Computer Sciences Australia.

AQS-902 lapads lightweight acoustic processing/display system

Marconi Avionics has developed the AQS-902 lightweight acoustic processing and display system (lapads) as a range of acoustic processing devices for helicopters and smaller maritime patrol aircraft, to complement the large AQS-901 system described above. Lapads is fitted aboard the Royal Navy's Sea King HAS.5 ASW helicopters, and the British Aerospace Coastguarder and Fokker F27 Maritime are examples of possible fixed-wing applications. The modular design makes lapads suitable for most types of ASW aircraft, including those without a current sonobuoy capability. The system can process data from either sonobuoys or dipping-sonar or both, so it is one factor which helps to remove the dipping-sonar constraint from ASW helicopters. Marconi cites the following advantages and capabilities for a lapads/ASW helicopter combination: increased flexibility, endurance, search area, mission range and success probability; continuous onboard processing; ability to work with sonobuoy fields already sown; covert search capability; and autonomous operation.

Marconi Avionics AQS-901 installed in RAF Nimrod MR2

Other variations are possible depending on customer requirements. Much of the control is normally accomplished by a control/display unit which has an alpha-numeric keyboard with a small cathode ray tube. The data is presented on hard copy or cathode ray tube or both in a fully annotated form, and cathode ray tube installations have a cursor and rolling ball for the operator to interrogate the data. Frequency dividers, markers, and data readout by interrogation are provided. The operator has fixing aids for correlation detection and recording, manual Doppler fix, manual hyperbolic fix and Lloyds-mirror depth analysis.

As far as the sonar operator is concerned, he enjoys enhanced performance with earlier detection, a choice of buoy processing modes, a versatile data display, extensive detection, classification and fixing aids, and better tethered sonar processing. As well as showing submarine positions, lapads can calculate and display attack profiles.

Lapads is said to be inexpensive to buy and operate. The equipment is up-to-date, and much of it has been proven in the Nimrod's AQS-901 system. It is compatible with existing sonobuoy receivers, and can process the following sonobuoy types: AN/SSQ-36, 41B, 47B, 53, 57, 62, 77, 801, 904, 905, 937 and 963.

Lapads modules are interchangeable without calibration. Built-in test equipment provides interruptive and non-interruptive confidence testing, indication of system readiness, go/no-go testing and failure isolation to a high confidence level. First-line fault-diagnosis is accomplished by the built-in test equipment without special equipment, while second-line testing down to module level is performed by hot-rig or second-line test equipment.

AQS-901 in Royal Australian Air Force/ Lockheed P-3C Orion

Advanced signal-processing systems

Marconi is developing signal-processing equipment with multiple applications mainly for the next generation anti-submarine helicopters and fixed-wing aircraft. A major target for Marconi is the projected Westland/Agusta EH101 Sea King replacement. The system under development, in collaboration with Computing Devices (which builds part of the AQS-901 and 902), is understood to embody a universal multiplexer which will be able to process data from different types of sonobuoy by calling up the appropriate part of the software. A single processor should thus be able to handle data from all sonobuoy types in the same mission, while showing a reduced weight and volume over the AQS-902.

STATUS: in development.

Marconi Avionics lapads in Royal Navy/ Westland Sea King HAS.5 helicopter. Note canister of sonobuoys on floor next to seat

Marconi quotes five examples of lapads configurations:
AQS-902A passive omni-directional buoy processing with hard-copy display
AQS-902B passive omni-directional buoy processing with cathode ray tube display
AQS-902C passive omni-directional and directional buoy processing with hard-copy display
AQS-902D passive omni-directional and directional buoy processing with cathode ray tube display
AQS-902D-DS AQS-902D with dipping-sonar processing and display

AQS-902A
Buoys processed: 4 Jezebel, 1 bathy, or 4 difar
Weight: 112 lb (51 kg)
Volume: 5⁵/₈ ft³ (0·16 m³)
Power: 555 W

AQS-902D
Buoys processsed: 4 Jezebel, 1 bathy, 4 Ranger, 4 difar, 2 dicass, or 4 vlad
Weight: 130 lb (59 kg)
Volume: 3¹/₂ ft³ (0·1 m³)
Power: 567 W

STATUS: in production and service.

UNITED STATES OF AMERICA

Cubic

Cubic Corporation, Defence Systems Division, 9333 Balboa Avenue, San Diego, California 92123

AN/ARS-2 sonobuoy reference system

Cubic developed the AN/ARS-2 sonobuoy reference system to determine automatically sonobuoy locations and the system is fitted on the US Navy/Lockheed S-3A Viking. Ten fuselage and wing-mounted antennas receive radio signals from the sonobuoy, from which the sonobuoy reference system computer determines the angles and slant ranges to the detector. This data is fed into a Sperry Univac 1832 mission computer for further processing so that sonobuoy positions can be displayed at the aircraft's tactical station. The aircraft may be shown at the centre of the screen with the sonobuoys surrounding it, or a tactical reference point can be placed in the centre, so the operator sees his aircraft's location with respect to the detectors.

The AN/ARS-2 can work in either active or passive modes. In its passive role, it receives the buoys' vhf signals on the antennas, which are arranged in longitudinal and transverse arrays. The sonobuoy reference system tunes automatically into any of the 31 standard sonobuoy channels, so no modification to the passive buoy is necessary. The phase difference between the various antenna pairs is used to calculate the buoy's bearing, and several measurements are taken as the aircraft flies close to the buoy. To measure its range, an active method must be used, but one reply fixes the buoy's position. A modulated signal is sent to the buoy on the uhf command link, and this triggers a reply, which is returned on the normal vhf link. The round-trip phase delay is a measure of the slant range.

The Univac computer uses aircraft position as measured by the inertial navigation system to determine the spatial position of the buoys. About 4000 words of computer memory are used. The distribution of the antennas is important in determining accuracy: wide antenna spacings are best for accuracy, but pairs with shorter spacings help to resolve ambiguities. The sonobuoy reference system has no external controller other than the activation button, and it automatically calibrates itself and sets itself to zero on instructions from the mission computer.

The sonobuoy reference system replaces the former method of sonobuoy location, which involved a visual sighting while flying directly overhead the buoy. This was difficult because the buoy's position had to be estimated from the position of the aircraft when it was dropped, with attendant uncertainties arising from wind, sea state, and aircraft navigation drift. So the sonobuoy reference system fixes the buoy network almost immediately and allows operation at higher altitude or longer stand-off range.

Dimensions: 8¾ × 9 × 19 inches (222 × 229 × 483 mm)

Cubic AN/ARS-2 sonobuoy reference system is installed in US Navy/Lockheed S-3A Viking sub-hunters

Weight: 56 lb (25·4 kg) (whole system, including antennas)
Power required: 115 V 150 W
Temperature range: −48 to +22° C
Cooling air: 0·54 lb/minute (0·24 kg/minute)
Mtbf: 2650 h
Specification: MIL-E-5400

STATUS: in service on US Navy S-3A Vikings. Cubic has received a US $3.5 million contract to produce five prototypes of a 99-channel version of the AN/ARS-2. This would be fitted in the S-3B Viking update.

AN/ARS-3 sonobuoy reference system

The AN/ARS-3 sonobuoy reference system is being retrofitted into the US Navy/Lockheed P-3C Orions as part of the Lockheed Update II programme, and Japan and the Netherlands are other customers for the system. This sonobuoy reference system can measure the locations of up to 31 sonobuoys using a passive method similar to that used by the AN/ARS-2, described above. This particular system relies on several measurements of slant angle from the buoy to the aircraft to ascertain its position. There is capacity to provide slant range measurement using an active method, again similar to that of the AN/ARS-2. The data is given to the Sperry Univac CP-901 mission computer. This uses the aircraft's inertial navigation system to determine the geographical position of the buoys, which need no modification to be used by the sonobuoy reference system. Eight of the ten vhf blade antennas are fuselage mounted, while the other

Cubic AN/ARS-3 sonobuoy reference system

two are on the tailplane. The AN/ARS-3 has built-in test equipment which can locate a faulty module 95 per cent of the time. Unlike the AN/ARS-2, the system needs no cooling air.

Dimensions: 11 × 21½ × 12¼ inches (279 × 546 × 311 mm)
Weight: 79 lb (35·8 kg)
Power required: 115 V 150 W
Temperature range: −54 to +71° C
Cooling air: none
Mtbf: 1000 h
Specification: MIL-E-5400

STATUS: in production for P-3C, with deliveries to start in April 1983.

Edmac

Edmac Associates, PO Box 391, 333 West Commercial Street, East Rochester, New York 14445

AN/AKT-22(V)4 telemetry data transmitting set

This system relays up to eight channels of sonobuoy data from an asw helicopter to a ship. It comprises the T-1220B transmitter-multiplexer, the C-8988A control indicator, an AS-3033 antenna, and a TG-229 actuator. The sonobuoy signals are received by dual AN/ARR-75 radio receiving sets, and passed into the transmitter-multiplexer. The control indicator has four trigger switches, each of which disables two data channels. Composite trigger tones are brought into the multiplexer separately via the control indicator, and combined with the sonic data channels and the single voice channel. The resulting fm signal is used to modulate the transmitter.

The data-transmitting set has a difar operating mode, in which the extra voice channel is inoperative, and the composite fm modulating signal is disconnected from the transmitter input. Two difar sonobuoy transmissions enter the transmitter-multiplexer on dedicated channels. After conditioning in an amplifier-adapter, the difar signal is split into two components, difar A and difar B. The A signal is conditioned in a low-pass filter, while the B signal drives a variable-cycle oscillator centred at 70 kHz, and then passes through a bandpass filter. The two filter outputs are combined linearly, and the resulting composite modulation signal drives the transmitter. A switch on the controller-indicator controls whether the normal sonic or composite difar signals are transmitted.

The AS-3033 antenna has two sections: a vhf element for receiving the sonobuoy signals, and a uhf part which sends the multiplexed data down to the ship. The ship receives the information on an AN/SKR telemetric data receiving set.

Primary power input: 115 V ac, three-phase, at 0·85 A, 1·5 A, and 1 A respectively

Warm-up time: <1 minute in standby mode; < 15 minutes under environmental extremes

Operating stability: >100 h for continuous or intermittent operation

Output frequency: 2200–2290 MHz, one of 20 switch-selectable S-band channels

Multiplexer inputs: (a) 8 sonar data channels (7 with 10 – 2000 Hz bandwidth; 1 with 10–2800 Hz), at 0·16–16 V

(b) 4 sonar trigger channels, 26–38 kHz, at 1–3 V

(c) 1 voice channel, 300–2000 Hz bandwidth, at 0–0·25 V

(d) Two composite difar channels, 10–2000 Hz bandwidth at 3–6 V or 10·6 V

Channel phase correlation: difference in phase delay between any two passive data channels <1° (10–500 Hz)

Modules of Edmac AN/ARR-72 sonobuoy receiver (from left): pre-amplifier, assg, receiver, audio assembly and control indicator

Edmac AN/AKT-22(V)4 telemetry data transmitting set

Specifications: (transmitter set) MIL-T-81695A(AS)
(environmental) MIL-E-5400, MIL-STD-461
(reliability) MIL-STD-781 and 785
(communications) MIL-STD-188
(service conditions) MIL-E-5400 class 1A
Operating temperature: -54 to +71° C
Vibration: 2 g to 2 kHz (MIL-E-5400)
Shock: 15 g for 11 ms

STATUS: in service with US Navy.

AN/ARR-72 sonobuoy receiving system

This system is used on the Lockheed P-3C Orions of the US Navy and Japanese Maritime Self-Defence Force to receive, amplify, and demodulate the vhf fm signals from sonobuoys deployed by those aircraft. The AN/ARR-72 is included in the P-3C Updates I and III, and in the US Navy Sikorsky SH-60 Lamps III helicopter system. It is compatible with the AN/SSQ-36, 41, 47, 50, 53, and 62 sonobuoys and receives on 31 channels. The receiver system is in five parts: AM-4966 pre-amplifier, CH-619 receiver, SA-1605 audio assembly, C-7617 control indicator, and SG-791 acoustic sensor signal generator (assg), which performs diagnostic functions.

The radio signals from the sonobuoys contain data on the frequency and directional pattern of the sound pattern received by the buoy's hydrophone array. These signals are amplified by the AM-4966 before passing into the radio receiver unit, where a multi-coupler distributes them to 31 fixed-tune receivers according to their frequency. The receivers further amplify and demodulate the signals to provide baseband audio and radio level signals. Each receiver channel contains two plug-in converters, a discriminator/amplifier, and a

filter; all parts are identical for each channel except for the crystal oscillator. The receiver power supply, designated PP-5000, uses integrated circuits, as does the SA-1605 audio assembly. This unit contains 19 audio channels, each of which has a 31 by 1 switching matrix to select the output of a given receiver. The signal is amplified a final time before transmission to the data processing equipment. The desired receiver channel may be selected by the mission computer, or manually on the C-7617 dual-channel control indicator.

Built-in test equipment is contained within the assg, which generates test radio signals for the pre-amplifier, the multicoupler, or a radiating antenna. Internal circuits generate simulated signals to test processing equipment for sonobuoys such as lofar, extended lofar, and range-only. End-to-end checks of sophisticated equipment such as the AN/AQA-7(V) difar sonobuoy indicator are performed by accepting modulation from their target generators. The 31 radio frequencies are generated by a frequency synthesiser comprising a multi-frequency oscillator which reduces the channelling time to less than a second.

Weight: (pre-amplifier) 2 lb (0·9 kg)
(receiver) 5·5 lb (26·3 kg)
(audio) 39 lb (17·55 kg)
(assg) 19·2 lb (8·64 kg)
(8 control indicators) 32 lb (14·4 kg)
Total weight: 150·7 lb (67·8 kg)
Power: 300 W max at 115 V ac, 400 Hz single-phase, plus 280 mA at 28 V dc for panel lighting of each control indicator and assg, plus 250 mA at 18 V dc for assg annunciators
Frequency range: 31 channels between 162·25 and 173·5 kHz
Noise figure: 5 dB max (3·5 dB is typical)
IF rejection: 66 dB minimum (more than 100 dB is typical)
Image rejection: 66 dB minimum (more than 100 dB is typical)
High audio level: 16 V at ±75 Hz deviation
Standard audio level: 2 V at ±75 Hz deviation
Crosstalk: 54 dB minimum
Output isolation: 60 dB minimum
Audio frequency response: ±1 dB from 20 Hz to 20 kHz; ±6 dB from 5 Hz to 40 kHz
Specifications: MIL-E-5400, MIL-STD-781, AR-5, AR-10, AR-34, MIL-I-6181
Warm-up time: 30 s max
Temperature: -20 to +55° C (pre-amp can go to -54° C)
Mtbf: 500 h including assg and ten dual-channel control-indicators
Operational stability: 500 h
Operating life: 20 000 h

STATUS: in production.

AN/ARR-75 sonobuoy receiving set

The AN/ARR-75 is a 31-channel sonobuoy receiving set intended mainly for helicopters. It is currently specified for the Kaman SH-2 Lamps I, Sikorsky SH-3, and Sikorsky SH-60 Lamps III helicopters. Independent receiver modules provide four simultaneous demodulated audio outputs each capable of selecting one of the 31 standard frequencies. Its volume is about a third of its predecessor, the AN/ARR-52A, which also weighs twice as much. The reliability of the new system is ten times that of the earlier one, through the use of solid-state electronics and fixed-tuned circuits which remove tuning adjustments.

The AN/ARR-75 comprises two units, the OR-69 receiver group assembly and the C-8658 (or C-10429) radio set control. The former unit contains most of the electronics: the power supply, four receiver modules and an electrical equipment chassis. The power supply contains the transformer and rectifier assembly and four switching regulator boards which each supply a very pure (low-noise) voltage. Each receiver module contains a suppression filter, local oscillator, converter, amplifier-discriminator, and the audio frequency output circuit-board. The electrical equipment chassis contains the antenna relay and filter, pre-amp-multicoupler, reference oscillator, built-in test reference divider, suppression filter cavity and interface connections.

Edmac AN/ARR-75 sonobuoy receiver

The control unit provides independent selection and meter monitoring of any one of the 31 channels for each of the four modules. A push-button built-in test function is included. Maintainability is improved by the modularity of the electronics. Each circuit board can be replaced without soldering, and built-in test functions down to plug-in module level are provided by the optional Mark 1634 Active Remote Module (arm) support equipment.

Edmac says that the main features of the AN/ARR-75 are a large dynamic range and high sensitivity over a broad base-wavelength. A wide, linear-phase audio output is provided, with close uniformity between the four audio channels.

Construction: modular, with 36 quick-replacement assemblies
Dimensions: (receiver) 8 × 7³/10 × 12 inches (203 × 186 × 305 mm)
(controller) 4⁴/5 × 5⁷/10 × 2⁴/5 inches (122 × 145 × 71 mm)
Weight: (receiver) 21·5 lb (9·8 kg)
(controller) 2·5 lb (1·1 kg)
Input power: 75 W at 115 V ac, plus 5·6 W for lighting at either 27 V (C-8658) or 5 V (C-10429)
Frequency range: 162·25 to 173·5 MHz (31 channels)
Noise figure: 5 dB at 50 ohms
Audio outputs (four simultaneous): (standard) 2 V at ±75 kHz deviation
(high) 16 V at ±75 kHz
Specifications: MIL-STD-461, 462 463, 781; MIL-E-5400; MIL-R-81681; AR-5, 8, 10, 34
Mtbf: 1500 h
Life: 20 000 h
Temperature: –54 to +55° C (+71° C for 30 minutes)

STATUS: in production.

AN/ARR-75(?) sonobuoy receiving set

Edmac has designed this replacement for the AN/ARR-75, but it has not yet been purchased and its designation remains AN/ARR-75(?). In physical size and audio characteristics, the system is the same as its predecessor, but channel capability goes up from 31 to 99. Four AN/ARR-75(?) sets may be used to replace the AN/ARR-72 in a large anti-submarine aircraft. This multiple installation improves reliability through redundancy, as well as by modernising the electronics. No changes are necessary to the aircraft's electrical harness for such a retrofit.

This improved receiver uses the same electronic design as the AN/ARR-75, with a few improvements such as higher data output capability, digital output at radio frequency level for remote indication, better components, and computer controlled built-in test. The same active remote module is used for diagnostic testing.

Edmac AN/ARR-75(?)

Specifications: as AN/ARR-75 except:
Weight: (receiver) 22·5 lb (10·2 kg)
(controller) 2·5 lb (1·1 kg)
Power: 70 W

STATUS: ready for production.

R-1651/ARA on-top position indicator

The US Navy/Lockheed P-3C Orion Updates I and III and the Sikorsky SH-60 Lamps III helicopter all have the R-1651 on board. This unit is a sonobuoy on-top position indicator (otpi). It is used in locating sonobuoys, and as a receiving converter when used with the aircraft's automatic direction finder equipment. The AN/ARA-25 automatic direction finder control box contains a switch which com-

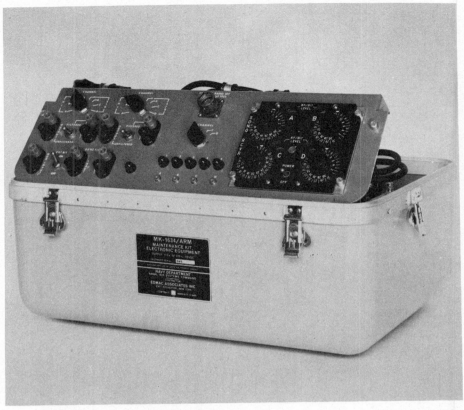

Edmac Mark 1634 is an active remote module used for ground test of AN/ARR-75

mands the R-1651 to switch the direction finder from a uhf automatic direction finder to a vhf sonobuoy. To establish the position of a sonobuoy, which moves with wind and sea currents, the R-1651 uses the vhf signals from the buoy. The system has the standard 31 channels, the correct one being chosen by binary code.

Dimensions: 3 × 4⁹/10 × 5³/10 inches (76 × 124 × 135 mm)
Weight: 3·5 lb (1·6 kg)
Input power: 15 W at 115 V ac, plus 20 W at 28 V dc for lighting
Output: 30 mA at 225 V dc
Frequency range: 162·25 to 173·5 MHz (31 channels)
Sensitivity: 10 dB minimum
Noise: 6 dB nominal
Audio output: 0·5 V at 100 Hz
Audio frequency response: ±2 dB, 75–500 Hz
Specifications: MIL-R-81680, MIL-E-5400, MIL-T-5422, MIL-STD-461, 704, 781, 785, MIL-I-6181
Temperature: –54 to +55° C
Mtbf: 1000 h
Operational stability: 1000 h
Operating life: 5000 h

STATUS: in production.

SG-1196/S acoustic test signal generator

Edmac developed the SG-1196 acoustic test signal generator for the US Navy/Lockheed P-3C Orion Update III programme. The unit provides test patterns representing the formats of most active and passive sonobuoys for in-flight testing of AN/ARR-78 sonobuoy receivers and Proteus acoustic signal processors. It acts on all 99 standard sonobuoy vhf channels, and it has been designed for calibration duties as well as functional tests. The SG-1196 is interchangeable mechanically and electrically with the SG-791, the acoustic sensor signal generator used in conjunction with the AN/ARR-72 on the P-3C. The acoustic test signal generator can thus be retrofitted easily, and will then provide a greater number of available test signals, which are more accurate and more stable. A single 23 MHz crystal oscillator synthesises all the radio frequencies

in the SG-1196, while coherent base-band signals are generated by a 7–68 MHz crystal oscillator. A pseudo-random digital source provides simulated sea noise.

Edmac R-1651/ARA sonobuoy on-top position indicator

Weight: 21 lb (9·5 kg)
Frequency accuracy: ±5 kHz
Output level accuracy: ±0·5 dB
Number of rf outputs: 4
Number of pre-established levels per output: 4
Spurious levels (referred to carrier): –48 dBc second harmonic, –70 dBc all others
Channelling time for an on-channel signal: 100 ms typical, 250 ms max
Max fm deviation: ±150 kHz
Distortion and noise: 1% for deviations to 75 kHz and modulations to 50 kHz
1·5% for deviations to 150 kHz and modulations to 100 kHz

Difar and lofar modes
Number of targets: 8
Target frequency accuracy: ±0·01%
Signal-to-noise ratio accuracy: ±0·5 dB

Range-only mode
Number of targets: 84, in two patterns
Number of Doppler frequencies: 11 per pattern; each accurate to ±0·1 Hz
Number of signal levels: 4
Signal-to-noise accuracy: ±0·5 dB
Timing accuracy: ±1 ms
Repeat time: 12·8 s

Cass mode
Number of targets: 8, in 2 patterns
Number of Doppler frequencies: 4 per pattern, each accurate to ±0·1 Hz
Number of signal levels: 4
Signal-to-noise ratio accuracy: 0·5 dB
Timing accuracy: ±1 ms
Repeat time: 12·8 s

Dicass mode
Number of targets: 8, in 2 patterns
Number of bearings: 4, accurate to ±1°
Doppler frequencies as in cass mode

B/T mode
B/T frequency: 1700 Hz
Number of levels: 4, accurate to ±3%

STATUS: in production.

Emerson Electric

Emerson Electric Company, Electronics and Space Division, 8100 West Florissant Avenue, St Louis, Missouri 63136

Emerson's anti-submarine warfare activities include acoustic processing and magnetic-anomaly detection. The company is responsible for the sonobuoy processing of the West German Navy Atlantics, under the project leadership of Dornier. Emerson is supplying a modification kit for the Atlantic sonobuoy processing system, as part of the Atlantic improvement programme which Dornier is performing during 1982-83. The new equipment will be able to handle a larger frequency spectrum and will analyse eight buoys simultaneously. Better directional accuracy and an improved signal-to-noise ratio will extend detection to quieter submarines, and they will be located with more accuracy. Digital technology and the use of electronic rather than mechanical switching will improve maintainability and reliability.

Emerson also produces the AN/ASQ-10A magnetic anomaly detector, which is used by ·the Nimrod MR2s of the Royal Air Force.

Emerson Electric is providing a modification kit for sonobuoy processing system in West German Atlantic improvement programme

General Electric

General Electric Company, Aircraft Equipment Division, French Road, Utica, New York 13503

AN/AYA-8B data-processing system

General Electric has been producing the data processing system (dps) for the US Navy/Lockheed P-3C Orion anti-submarine warfare aircraft since 1968. It provides the interface between the CP-901 central computer and the aircraft systems and so constitutes a major part of the P-3C's mission avionics.

The dps is connected to all of the crew stations on the aircraft: tactical co-ordinator (tacco), non-acoustic sensor, navigation/communications, acoustic sensor, and flight-deck. In addition, it communicates directly with the radar interface unit, armament/ordnance system, navigation systems, sonar receiver, submarine anomaly detector, and Omega. A few systems such as forward-looking infra-red sensor and the data-link go directly into the CP-901.

The P-3C's data-processing activities are divided into four logic units, which form separate boxes in which all electronic operations are conducted in analogue format. Various key-sets and panels complete the dps hardware.

Logic unit 1 interfaces between the central computer and four types of peripheral information systems: manual data entry, system status, sonobuoy receiver, and auxiliary readout display.

The manual entry sub-system provides the communication between the various operator stations and the central computer. Each operator has a panel of illuminated switches and indicators by which he communicates with the central computer. System status for the navigation, sonobuoy receiver, and submarine anomaly detector is received and stored by part of logic unit 1, and transmitted to the central computer. The computer receives the status information on demand, or when any status changes. The sonobuoy receiver logic allows

US Navy/Lockheed P-3C Orion has General Electric's AN/AYA-8B ASW data-processing system

the central computer or operator to tune each of the 20 receiver processor channels to one of the 31 sonobuoy frequencies. Finally, the auxiliary readout display logic interfaces between the central computer and the auxiliary displays at the tacco and nav/com stations. The radar and electronic support measures interfaces are also achieved by logic unit 1.

Logic unit 2 is concerned with communicating between the central computer and the navigation, armament/ordnance, and magnetic tape transport systems. This logic unit transmits to the computer Doppler and inertial

navigation data, and instructions to launch search-and-kill stores and read and write digital information on magnetic tapes.

The dps' logic unit 3 controls the cathode ray tube displays provided for the tacco, sensor operators, and pilot. The tacco and pilot displays can generate characters, vectors and conics, while the sensor displays use characters and vectors only.

The final logic unit, number 4, is mainly an expansion unit, comprising two items: the data multiplexer sub-unit (dms) and the drum auxiliary memory sub-unit (dams). These provide extra input/output capacity and memory capacity respectively. The dms can service up to eight input and output peripherals, as selected by the central computer. One output channel presents characters, vectors, and conics for the auxiliary sensor display, and one input channel is used for the aircraft's Omega navigation system. An input and an output channel are used for built-in test equipment. The remaining channels were left free in the original design, to allow for future expansion. The P-3C Update II Orions carry a command launch system for the Harpoon anti-ship missile, and a sonobuoy reference system, both of which interface with the dms. The

dams was incorporated to give an additional 393 216 words of memory to the computer, so that the operational program could be expanded to accommodate extra functions and equipment. At least some of this capacity has probably been taken up by the Update II expansions, which also include forward-looking infra-red sensing and fleet satellite communications.

Various keysets and control panels allow access to the central computer via the dps. Three universal keysets allow the transfer of information between the computer and the navigation/communications operator and the two acoustic sensor operators. The pilot uses

his own keyset for controlling the information presented on his cathode ray tube display, entering navigation stabilisation data, dropping weapons and flares, and entering information on visual contacts. The ordnance panel displays the commands which the computer has given to the ordnance operator concerning status and position of the search stores, such as sonobuoys, which are available for deployment. Finally, there is an armament/ordnance test panel, which monitors the output from the dps logic unit 2 to those systems.

STATUS: in production for P-3C Orion.

Hazeltine

Hazeltine Corporation, Greenlawn, New York 11740

AN/ARR-78 & AN/ALQ-158 advanced sonobuoy communication link (ascl)

US Navy/Lockheed P-3C Orions are receiving the Hazeltine ascl as part of their Update III improvement, and the system is proposed for the forthcoming S-3B Viking II. Hazeltine describes the ascl as a versatile and effective high-performance system designed to satisfy the most demanding current and future ASW requirements. The standard configuration provides 20 receiver channels, each of which can operate on any of the 99 standard sonobuoy vhf frequencies. In the P-3C, a single AN/ALQ-158 antenna processor feeds two AN/ARR-78 receivers, which thus send 40 channels of data into the aircraft's Proteus advanced signal processor.

The ALQ-158 comprises a vhf blade antenna, radio frequency pre-amplifier, processor unit, and operator's control box. The system is computer controlled, with a micro-processor-based control unit. Components, including radio frequency amplifiers and mixers, have a high dynamic range to avoid third-order intermodulation effects, and surface-acoustic-wave filters and delay lines are used because of their linear phase characteristics, low signal distortion, and lack of field alignment requirement.

Hazeltine says that the ascl receiver, the AN/ARR-78, is the first such sonobuoy device to be computer controlled, and the first to have 99 channels. It has five units: the AM-6875 pre-amplifier, R-2033 receiver assembly unit, C-10126 indicator control unit, ID-2086 receiver status indicator, and C-10127 automatic direction finder (adf) receiver control. The incorporation of an adf function is another 'first' claimed for the ascl. The adf transmitter is used as a sonobuoy on-top position indicator, telling the crew when the aircraft is directly over a sonobuoy. This means that a separate unit is not required to monitor sonobuoy positions, which change due to current and wind conditions.

The receiver unit has two separate front-ends. One is used as an amplifier/filter for the adf signal and drives one of the 20 receiver modules, which is dedicated to the adf function and which provides the adf antenna drive error signal. The second receiver front-end amplifies and filters the sonobuoy signals before splitting them 19 ways for the remaining receiver modules. Sixteen of these modules are identical and provide the fm demodulated

Five modules which make up Hazeltine AN/ARR-78 advanced sonobuoy communication link

acoustic outputs to the Proteus processor. The other three are auxiliary receiver modules, which provide miscellaneous functions and contain am and fm demodulators. Two of the auxiliary modules provide audio outputs to the operator's headphones, and the other monitors radio frequency signal strength, sending an indication to the Proteus.

The common receiver modules each measure 10 × 7½ × ⅔ inches (254 × 191 × 17 mm), and make extensive use of custom hybrid circuits in the synthesiser, intermediate frequency and demodulation areas. Only slight differences in the latter distinguish between the normal and auxiliary acoustic modules. Surface-acoustic-wave filters are used to allow a low-profile mechanical design and confer high selectivity. Each module is divided into five compartments: dc regulator, high-level mixer, synthesiser, amplifier, and fm demodulator.

The receiver assembly unit contains an input/output module which acts as the interface with the three control and status units. This input/output unit contains a 8-bit micro-processor which communicates with the 32-bit Proteus via a module that slows down the 32-bit signals into sets of four 8-bit words. The Proteus can command all operational and test modes and request detailed receiver status information such as sonobuoy type, radio frequency channel, and power level. All commands that can be sent from the Proteus can also be initiated by the operator on his panel, using switches and a 12-digit keyboard on the indicator control unit (icu). This unit contains lighted status and test message indicators and a row of light-emitting diodes to indicate which commands are being processed. The icu has its own micro-processor to format commands and decode messages from the receiver assembly.

Hazeltine has built extensive self-test into the AN/ARR-78. The built-in test equipment can be initiated automatically by the Proteus, or manually by the operator. This system includes a radio frequency-modulated signal-generator capable of 99-channel operation, special baseband test validation circuitry, and test-related micro-processor routines. A number of tests are run automatically by the micro-processor before each flight. There are also loop tests of the digital circuits, to verify that all the data transfer lines are operational. A check sum routine in the two micro-processor memories makes sure that they are both working. Hazeltine claims to have demonstrated that built-in test equipment can detect 96 per cent of all faults, and isolate 99 per cent of these to the defective module.

In the P-3C Update III, the 20 receiver modules are inserted vertically into the box, along with five other modules such as the computer input/output unit. The power supply and reference oscillator modules are placed above this part of the receiver. Hazeltine has also repackaged the system into a smaller, uniform box which places all the modules alongside each other in boxes with the same form factor. Fewer receiver channels (about ten) are provided, but this is suitable for applications in smaller aircraft such as the S-3B Viking, or in helicopters.

Power: 115 V ac, 380 – 440 Hz, 3-phase, 500 W, plus 7 W at 18 – 32 V dc plus 50 W at 26·5 V ac, 400 Hz 3-phase
Audio output: (analogue) 2 V rms balanced (monitor) 50 mW, 300 ohms
Environmental standard: MIL-E-5400 Class 1A (modified)
Temperature: –40 to +55° C (up to +71° C for 30 minutes)
Altitude: 30 000 ft (9100 m)
Shock: 15 g
Vibration: ±2 g, 9 – 500 Hz
Electro-magnetic interference: MIL-STD-461A
Dimensions: (AM-6875) 3 × 5¾ × 4¼ inches (76 × 146 × 108 mm)
(R-2033) 12 × 21⅓ × 15⅓ inches (309 × 541 × 389 mm)
(C-10126) 9 × 5¾ × 6½ inches (229 × 146 × 165 mm)
(ID-2086) 10½ × 5¾ × 5 inches (267 × 146 × 127 mm)
(C-10127) 2¼ × 5¾ × 3¼ inches (57 × 146 × 83 mm)
Weight: (AM-6875) 2·1 lb (0·95 kg)
(R-2033) 101 lb (45·9 kg)
(C-10126) 6·6 lb (3 kg)
(ID-2086) 4·1 lb (1·9 kg)
(C-10127) 1·4 lb (0·6 kg)

STATUS: in production.

Sparton

Sparton Corporation, 2400 East Ganson Street, Jackson, Michigan 49202

Airborne sonar processing equipment

Sparton's modular airborne processing systems and components provide a flexible, lightweight, relatively inexpensive means to process and display data from sonobuoys and similar submarine tracking devices. The modular approach allows individual units to be used separately or combined to form a compact system capable of processing and displaying signals from most of the sonobuoy types currently in use or being developed. The following descriptions apply to typical equipment.

Active processing system

The active processing system comprises five modules: command signal transmitter, command signal generator, active processor, sonar display, and power supply. It can simultaneously process and display data from two AN/SSQ-47 or AN/SSQ-522 active sonobuoys. The system can be modified to provide two-channel processing and display of a single AN/SSQ-50 or AN/SSQ-553 command active sonobuoy. The modifications provide the matched filter processing required for the additional continuous-wave pulse lengths, and a fm processor which uses replica correlation. The system can be further expanded to process and display the AN/SSQ-62 or proposed dicancass directional command-active sonobuoys by the addition of a sixth module. This module contains a digital arc-tangent computer which calculates and displays bearing with a direct digital readout in degrees.

The command signal-generator is the active sonobuoy's information source. It is a dual-channel control unit which generates the sonar pulse information for two sonobuoys. Selected channels from the sonobuoy receiver are routed to the active processor. The command signal-generator also provides the active processor with trigger pulses and other digital control information.

This unit controls operating channel selection (A or B), automatic or manual ping cycle mode and characteristic, transducer depth commands, and sonobuoy scuttling command. Six different command signals (one for each channel) are selectable at the front control panel. The sonic continuous-wave signals trigger sonar transmission in the sonobuoy at selected vhf/uhf channel frequencies.

Indicator lights are installed on the control panel: a 'uhf on' light (sonobuoys A and B) to signify that the transmitter is operating; a 'ready' light (sonobuoys A and B) to signify that a manual sonic command will be transmitted if initiated; and a 'channel select error' light to indicate that sonobuoys A and B are erroneously being operated on the same channel. A 'notch in/out' switch prevents reverberations from saturating the A-scan trace when operating with this back-up display for continuous-wave Doppler detection. A two-position switch and a head-phone jack permit audio-monitoring of sonar from sonobuoy A or B.

The active processor is a lightweight, dual-channel unit capable of processing simultaneously two command-active sonobuoys. It analyses automatically information for the detection and localisation of submarines, and

is completely solid-state, with integrated circuits used extensively in the analyser portion of the unit. The processed signals from two sonobuoys can be displayed in a split-screen mode on the sonar display.

The active processor receives signals from the command signal-generator, where it is processed in the continuous wave or lfm mode. When operating in the various continuous wave modes, the active processor performs a spectral analysis transform. When operated in the lfm mode, it performs replica correlation. The processor also provides the horizontal, vertical, z-axis and other control outputs necessary to drive the sonar display.

When the unit is operating in continuous wave mode, spectral analysis of the target is Doppler velocity versus range. In the lfm mode, target echoes or wideband-frequency sweeps are correlated with a recirculating replica of the transmitted pulse. The portion of the echo with coherent phase and frequency results in a detection envelope which is presented on the A-scan sonar display as amplitude versus range. As a result of multiple-bit digital quantisation, high sample rates, and a large memory capability, the fine details of received echoes are retained in amplitude, frequency, and time.

The processor also uses matched filter processing for optimum detection. Target correlation envelopes are developed linearly with low loss. This allows programmable thresholding of preset clear targets to the display unit, while minimising false alarms and reverberation clutter. Linear processing is also necessary for active directional systems because the amplitudes of the correlation envelopes must be preserved in the direction process so that accurate bearing estimates may be computed.

The sonar display unit can be used with the active processor to provide target range and Doppler information in the continuous wave mode, and target detection and range data in the fm mode. The Sparton sonar display uses a variable-persistence cathode ray tube to give bright high-quality readouts. The screen is divided vertically into two displays, where a combination of A- or B-scans can be displayed for one of two sonobuoys simultaneously. Persistence time is adjustable from zero to full storage. Stored spot resolution is equivalent to 50 lines/inch (20 lines/cm).

Directional sonar processor

The directional sonar processor accepts a composite difar signal from any sonobuoy receiver and demultiplexes it to provide north-south, east-west, and omni-directional outputs. This equipment features precise target bearing resolution, immunity to phase pilot noise sidebands, built-in test, outputs for computer analysis, selectable lock retention-time constants, and manual or remote gain control.

Frequency range: broadband difar audio
Input impedance: 10 k ohms minimum
Output impedance: 10 k ohms max
Variable gain (manual or remote): 48 dB
Time constants: 0·1, 1 and 10 s
Phase pilot bandwidth: 1 Hz max
Processor analysis bandwidth: 1 Hz max
Mtbf: 1500 h minimum
Demultiplexer outputs: n-s, e-w and omni-directional
Processor outputs: n-s, e-w and omni-directional
Mechanical configuration: ATR format

Spectrum analyser

The Sparton spectrum analyser is a low-cost, special-purpose digital processor which performs real-time spectral analysis. The signature of the target being analysed is displayed simultaneously on both channels of a Sparton sonar display. Discrete Fourier transform technology is used with circuits operating at optimally low bit-rates. Features include linear integration of spectral estimates, manual or automatic modes, internal cursor for accurate frequency identification, and total spectrum coverage.

Frequency range: 14–912 Hz
Frequency bands: 14–114 Hz (0·25 Hz resolution)
112–912 Hz (2 Hz resolution)
Number of spectral estimates per band: 400
Input amplitude dynamic range: 80 dB

OL-5003 (–)/ARR signal processor group

The Sparton OL-5003(–)/ARR sonobuoy signal processor is a lightweight, versatile, active processor and display system. Designed to make data from either the AN/SSQ-47 or AN/SSQ-522 sonobuoys available instantly with minimum operator involvement, the system automatically synchronises with sonobuoy range trigger, analyses signal returns for echo level and Doppler, and displays the result on a variable persistance storage cathode ray tube. The single display tube presents simultaneously, in a split-screen mode, the data from two sonobuoys. The B-scan format displays range on the horizontal axis, Doppler on the vertical axis, and echo level on the intensity axis.

This processor is solid-state except for the storage cathode ray tube. Integrated circuits are used extensively in the analyser portion of the unit. The processor is also compatible with the AN/SSQ-523 sonobuoy system in the short-pulse mode.

Dimensions: 9¼ × 7⁴/₅ × 19³/₅ inches (235 × 198 × 498 mm)
Weight: 27·9 lb (12·7 kg)
Operating temperature: –20 to +55° C
Humidity: 95% at 40° C
Type of input: AN/SSQ-47 signals are supplied by the standard output of the AN/ARR-52, AN/ARR-72, or AN/ARR-75 sonobuoy receiver
Number of processing and display channels: 2, simultaneous
Analysis bandwidth: 10 Hz
Doppler coverage: compatible with AN/SSQ-47B or AN/SSQ-522 sonobuoy
Doppler accuracy: ±1 knot
Range scale: selectable, 3600 or 7200 yards (3292 or 6584 m)
Power: 120 W (115 V, 50–400 Hz, single-phase)

Dicass command-signal monitor

The Sparton dicass (directional command active sonobuoy system) command signal monitor is used to monitor valid command signals during testing of the AN/SSQ-50. The unit is self-contained and requires a uhf antenna and 115 V ac for operation. It provides a permanent printed record of the vhf channel command, identification of command function, radio-frequency signal level during command, time of signal transmission, length of command, and radio frequency signal level prior to command. Although designed originally for the AN/SSQ-50 cass sonobuoy, the command signal monitor can be used with dicass buoys.

Texas Instruments

Texas Instruments Inc, PO Box 6015, Dallas, Texas 75222

AN/ASQ-81(V) magnetic anomaly detector system

The AN/ASQ-81(V) magnetic anomaly detector (mad) system was developed by Texas Instruments for the US Navy for use in the detection of submarines from an airborne platform. The system operates on the atomic properties of optically pumped metastable helium atoms to detect variations of intensity in the local magnetic field. The Larmor frequency of the sensing elements is converted to an analogue voltage which is processed by band-pass filters before it is displayed to the operator.

Two configurations of the AN/ASQ-81(V) are available, one for installation within an airframe, and one for towing behind an aircraft. The US Navy uses the AN/ASQ-81(V)-1 inboard installation on the carrier-based Lockheed S-3A Viking aircraft, where it is extended on a boom. This version is housed in a tail 'sting' on the land-based Lockheed P-3C Orion aircraft. The towing configuration is designated AN/ASQ-81(V)-2, and this is employed by US Navy Sikorsky SH-3H and Kaman SH-2D helicopters. This version has also been selected by the US Navy for its Sikorsky SH-60 Lamps III helicopter. In addition, the towed version is produced for the Netherlands for use on the

US Navy's P-3C Orion uses Texas Instruments AN/ASQ-81(V)-1 mad system

Westland WG.13 Lynx, and for Japan, which uses it on the Mitsubishi HSS-2 helicopter.

Both versions of the AN/ASQ-81(V) have the same C-6983 detecting set control and AM-4535 amplifier and power supply unit. The AN/ASQ-81(V)-1 uses a DT-323 magnetic detector, while the AN/ASQ-81(V)-2 has a TB-623 magnetic detecting towed body. The towed version is controlled by the C-6984 reel control, which works the RL-305 magnetic-detector launching and reeling machine.

STATUS: in production and service.

Flight Control Systems

FRANCE

Sfena

Société Française d'Equipements pour la Navigation Aérienne, Controls and Systems Division, Aérodrome de Villacoublay, BP 59, 78140 Vélizy-Villacoublay

Automatic flight-control system for Airbus A300

Although collaboratively developed with other nations involved in the European Airbus consortium, Sfena leads the afcs development programme and has sub-contracted system components to Smiths Industries (UK) and Bodenseewerk (West Germany). Sfena was appointed prime contractor in November 1969.

The predominantly analogue, duo-duplex, fail-operational automatic flight-control system developed by the three participant companies has been installed in all A300 airliners built, except those newer models which have a largely digital version of the same afcs, described separately below.

The A300 afcs is designed to provide automatic control of all stages of flight from initial climb to touchdown. It is certificated for operations to Category IIIA weather minima, and approval may be extended to Category IIIB. Sfena points out that the afcs is closely linked to the aircraft flight controls and uses sensors which are available in other systems in addition to dedicated sensors. External references are provided by heading-reference units, air-data computers, ils/vor receivers and radio altimeters. The afcs is divided into six independent sub-systems:

Autopilot/flight director. All processing functions in this sub-system are duplicated, and processors are used which have two computational paths, so that there is an output for both a real and an 'imaginary' servo. The system can operate as a simplex self-monitoring system in non-critical operating modes, and as a duo-duplex (or quadruplex self-monitoring) fail-operational system in safety-critical modes. In cruise only one set of computers is used, but the second set is continuously synchronised and can be selected as a replacement if there is a malfunction. Both sets operate in unison during automatic-landings.

Each computer set comprises a longitudinal (pitch) and lateral (roll and yaw) computer, plus a logic computer. The latter generates operating logic and engagements, including indications on the cockpit control unit and mode indicators. This is a digital electronic unit which uses a 4K 8-bit word read-only memory. Basic autopilot modes provided by the other computers, and sequenced by the logic computer, include pitch and roll/heading hold.

Other autopilot modes from the pitch computer are altitude, speed and vertical-speed hold, plus altitude acquisition with arm and capture-phase facilities. The lateral computer provides heading pre-select, vor/loc arm, capture and track, and navigation mode facilities. The latter include back-beam approach using flight-director guidance. Facilities jointly provided by the two computers are turbulence, auto-approach, automatic-landing (flare and roll-out), overshoot, and control-wheel steering modes.

When cruise modes are used the computers automatically limit the amplitude and speed of commands, compare the amplifier outputs in both lanes of each computer, and compare the actual servo output with the output of a pseudo-servo, or electronic 'image'. Mechanical limiting is also incorporated in the control-surface servos.

In the automatic-landing mode, both sets of computers are engaged and the processor configuration produces two direct servo-motor drives, and two 'imaginary' servo drives. A majority-vote assures continued operation of the system after an initial failure, and passive disconnection of the system after a second failure.

Top-selling A300 Airbus airliner has Sfena afcs

In addition to the processors a control unit in the flight-deck glareshield provides all the controls for engagement of the autopilot/flight-director modes and autothrottle. Each pilot also has a failure and performance indicator which shows autopilot and autothrottle warnings, the system modes in use, and the redundancy level available during approach and landing. Two sets of dynamometric rods, comprising two rods per control column, are used to sense crew inputs during control-wheel steering phases.

Pitch trim. This also uses a duo-duplex, fail-operational processor configuration. The system is designed to minimise residual longitudinal control forces during periods of control activity, and this is achieved by varying the setting of the variable-incidence tailplane, when the autopilot is in use. During manual flight the crew can place the trim datum at any desired position. The system also incorporates Mach-trim compensation. Each active lane drives an independent trim actuator, and if a defective lane is detected the better lane will take over. All the above operations are handled by two identical pitch-trim computers. Two additional computers, which are for pitch compensation, produce additional pitch-trim demands, using triple angle-of-attack data. When necessary, these units disengage the trim computers. A small control unit installed on the flight deck permits selection of either or both lanes of the overall trim system.

Yaw Damper. This sub-system is also based on a duo-duplex processor layout, so that it will survive an initial failure, and fail passively after a second failure. A control-panel provides for selection of either system lane, and each lane has an identical set of lateral accelerometer and yaw-rate gyro units, in each of which are two similar sensors. There is a connection from the lateral processors to each yaw-damper processor so that roll angle can be used to coordinate yawing manoeuvres.

Autothrottle. This is a single-lane, self-monitoring system, analogous to one dual-lane set of the duo-duplex processors. It should disengage automatically on detection of a failure. The system has inputs appropriate to drive speed select, N_1 limit, go-around and automatic-landing flare modes. There is a single actuator on each engine throttle, and dynamometric rods assure easy pilot-override when necessary. Autothrottle mode controls are incorporated in the glareshield control panel.

Test. This is a sub-system dedicated to fulfilling the double function of fault location during maintenance, and monitoring of safety circuits on the ground. The detection of a defective unit is ensured by a continuous reading, in-flight and on the ground, of the 'go/no-go' signals exchanged between the afcs units and peripherals. This detection function is completed by what are known as complementary tests, initiated on the ground. The safety-monitoring function consists of automatically testing the afcs sub-system comparator during a ground test.

The test system is duplicated, to ensure no loss of integrity and to maintain complete segregation between the sub-systems. In flight, the system is continuously checked by a self-test programme which shares processor time with the monitoring programme.

All test facilities are included in two identical computers, each with a 2K 16-bit word core memory. A test control and display panel is on the flight deck.

Speed Reference System. This simplex, self-monitoring facility generates the pitch commands required during take-off and go-around (after autoland disconnect) manoeuvres, and commands the autothrottles in N_1-limiting mode during abnormal angle-of-attack manoeuvres, or if windshear is encountered during approach. Take-off references are to hold V_2+10 knot or 18 degrees pitch attitude and on go-around it will hold V_{ref}+10 knots or 18 degrees pitch attitude. In the event of an engine failure it will hold V_{ref}.

Computer units:
2 longitudinal computers each ¾ ATR Long
2 lateral computers each ¾ ATR Long
2 logic computers each ¾ ATR Long
2 pitch trim computers each ½ ATR Long
2 pitch compensation computers each ⅜ ATR Short
2 yaw damper computers each ½ ATR Long
1 autothrottle computer ½ ATR Short
2 test computers each ½ ATR Short
1 speed reference system computer ⅜ ATR Short

Control units:
1 glareshield control panel
2 mode indicators
1 trim engagement unit
1 yaw-damper engagement unit
1 test control and display panel

Other units:
2 pitch dynamometric rods (on control columns)
2 roll dynamometric rods (on control columns)
2 trim actuators
2 lateral accelerometer units
2 yaw-rate gyro units
1 autothrottle actuator assembly
2 autothrottle coupling assemblies
2 autothrottle dynamometric rods
2 two-axis accelerometer units

STATUS: in service.

Airbus Industrie's new A310 has Sfena digital automatic flight-control system

Digital automatic flight-control system for Airbus A300/A310

New digital processors which replace the larger analogue elements of the existing automatic flight control system in the Airbus Industrie A300 and A310 wide-body airliners have been developed and put into service, so bringing the standard of this afcs up to the standard of an almost wholly digital system. This will be available as an alternative to the existing analogue system in all future A300 production. All A310s have the new digital automatic control system as standard.

The new system was first flown in the A300 in December 1980, and entered service with Garuda Indonesian Airways in January 1982.

The digital automatic control system provides flight augmentation functions (pitch trim in all modes of flight; yaw-damping, including automatic engine-failure compensation when the autopilot is engaged, and flight-envelope protection); has a comprehensive complement of autopilot and flight-director modes which permit automatic operations from take-off to landing; a thrust-control system which operates throughout the flight envelope, has a derate capability, contains protection features against excessive angle-of-attack; and a fault-isolation and detection system for line maintenance.

Design has been in accordance with ARINC 600 and 700 characteristics, and has led to the adoption of ARINC 429 standard links between the automatic control system processors and sensors. Four or five processors are used, these comprising two flight-augmentation computers, one or two flight-control computers (the second unit being necessary only if Category III automatic-landing capability is required), together with a thrust-control computer. A second thrust-control option is available, in which the last-named item is replaced by a flight-control unit providing pilot interface with the autopilot/ flight-director and autothrottle functions, plus a thrust-rating panel which allows crew access to thrust-limit computations.

A further new item of equipment is an engagement unit, with pitch-trim and yaw-damper engage levers, autothrottle-arm and engine-trim controls, and which is mounted in the flight-deck roof panel. Two pitch and roll dynamometric rods are also used as control-wheel steering sensors, there are two new trim actuators, a new autothrottle actuator, two coupling units (one on each engine), and two further dynamometric rods connected to the throttle-control linkage.

Flight-deck controller equipment has been revised, and a new autopilot/flight-director and autothrottle mode selector is installed in the centre glareshield. Variable data can be entered by rotating selector knobs, and shown by illuminated readouts. The various modes are engaged by push-buttons. Modes available are altitude capture and hold, heading select, profile (capture and maintain vertical profiles and thrust commands from flight-management system), localiser, landing and speed reference. Autothrottle modes include delayed-flap approach, speed/Mach-number select, and engine N_1 or engine pressure ratio selection.

There is also a thrust-rating panel, mounted above the centre pedestal. This has comprehensive controls permitting thrust-levels to be selected, depending on operating mode, and providing for selection of such facilities as derated-thrust take-off.

The fault isolation and detection system has a dedicated maintenance-test panel which, on a 2 line × 16 character display, provides written alert messages based on fault information from automatic-testing activities. This is conducted in all line-replaceable units, and includes fault-isolation, tests to check for correct operation after maintenance action and on-ground automatic-landing availability checks. Up to 30 faults, from six flights, can be stored and retrieved.

Production of the new system is being jointly conducted by Sfena (as prime contractor) and Smiths Industries (UK) with Bodenseewerk (West Germany).

Computer units:
two flight control computers (10 MCU size)
thrust-control computer (8 MCU size)
flight augmentation computer (8 MCU size)
Control units:
flight-control unit (glareshield)
thrust-rating panel (centre panel)
fac/ats engagement unit (roof panel)
Other units:
two pitch dynamometric rods
two roll dynamometric rods

two trim actuators
autothrottle actuator
two engine coupling units
two engine dynamometric rods

STATUS: in service.

Widely selling Dassault Mirage F1 fighter has Sfena AP 505 autopilot

AP 205 autopilot

This system was designed at the request of the French Air Force for the French version of the Anglo-French Jaguar tactical-support aircraft. It is a simple and rugged system allowing pilots to fly hands-off while conducting other operational tasks.

Autopilot facilities include, in the pitch axis, pitch attitude- and altitude-hold modes, and in the roll axis, heading-hold mode. Trim functions provide automatic pitch trim and manual trim-assist in roll. Further modes for navigation or guidance purposes can be added. Modular construction is claimed to minimise maintenance actions, and the system is designed to be retrofitted into aircraft not equipped with autopilot. The first autopilot was delivered to the French Air Force in November 1981, under the terms of a production contract worth FFr 50 million.

Unit: 123·5 × 210 × 253·2 mm
Weight: 5·4 kg
Power requirements: <40 VA 200 V ac 400 Hz three-phase
<1 VA 26 V ac 400 Hz three-phase
<15 W 28 V dc

STATUS: in service.

AP 305 autopilot/flight-director system

This system is derived from an earlier Sfena system, the Tapir, which equipped civil-operated Fokker F27 and Armée de l'Air Nord 262 light transports. For Aéronavale Nord 262s a radio-altimeter low-altitude alert facility has been added as a safety feature for use during low-flying operations over the sea. It is an analogue system with sideslip suppression, turn co-ordination and Category II auto-approach capability. Flight-director facilities include altitude capture and hold, selected-heading capture and hold, vor/ils capture and tracking, attitude-hold, and appropriate operation indications and alerts.

System computer: 421 × 125 × 194 mm
Control panel: 180 × 146 × 57 mm
Mode selector: 135 × 146 × 27 mm
There is also a sideslip detector, and four servo actuators.
Weight: (excluding actuators) 8·6 kg (including actuators) 21·6 kg
Power requirements: <80 VA 115 V ac 400 Hz single-phase
<150 W 28 V dc

STATUS: in service.

AP 405 autopilot and autothrottle system

The AP 405 has been specially designed to meet the requirements of the French Navy for

the low-altitude and high-speed operations of its carrier-borne Super Etendards, and is able to hold and monitor radio height above sea level. Precise attitude and speed-holding is provided by the autothrottle, including during carrier approaches when precise control is a deciding factor in achieving consistency and safety of operations.

Autopilot functions are pitch-hold, altitude acquisition and hold, actual heading hold (with control-wheel steering override), semi-automatic pitch trim and the provision for instinctive disconnect to make rapid flight-path changes. At very low altitudes there is also a radio-altitude hold with continuous flight-path monitoring. The autothrottle system can capture and hold a selected angle-of-attack.

System computer: 409 × 198 × 94 mm
Control/indicator panel: 115 × 26 × 150 mm
Autothrottle actuator: 120 × 120 × 200 mm
Angle-of-attack selector panel: 37 × 45 × 50 mm
Total weight: 7·9 kg
Power requirements: <60 VA 115 V ac 400 Hz single-phase
<1 VA 26 V ac 400 Hz single-phase
<60 W 28 V dc

STATUS: in service.

AP 505 autopilot

The AP 505 is constructed entirely with integrated circuits, and has been designed for Mach 2+ combat aircraft. Pilots of the Mirage F1, which has the system, are claimed to have expressed every satisfaction with its reliability and ease of use. The system can be switched on before take-off, and engaged or disengaged by means of a hand-grip trigger in the control column. In basic mode it maintains the longitudinal attitude held when the pilot releases the stick trigger, and either the heading or bank angle, depending on whether re-engagement is effected at a bank angle of less or more than 10 degrees. Autopilot modes provide for automatic flight at a pre-selected altitude, heading or vor/Tacan/ils radio-aid bearing. Limit on attitude hold facilities are plus or minus 40 degrees in pitch, and plus or minus 60 degrees in roll.

System Computer: 190 × 202 × 522˙mm
Function selector unit: 132 × 142 × 27 mm
Heading selector unit: 99 × 35 × 80 mm
Total weight: 14·9 kg
Power requirements: <100 VA 200 V ac 400 Hz three-phase
<6 VA 26 V ac 400 Hz single-phase
<15 W 28 V dc

STATUS: in service. More than 600 sets have been built and supplied in eight countries.

Sfena's AP 605 digital flight-control system was developed for new Dassault Mirage 2000 multi-role fighter

AP 605 autopilot

The AP 605 is a digital flight-control system developed for the new Dassault Mirage 2000 supersonic combat aircraft. The high-capacity Sfena Series 7000 computer provides for future extensions to the basic autopilot, for instance to provide new modes suited to the particular missions flown by the aircraft. Flight-control signals which pass between the autopilot and the Mirage 2000 fly-by-wire flight control system use a digibus serial data-transmission system.

The AP 605 provides semi-transparent control, that means it is engaged or disengaged simply by releasing or taking hold of the control stick, which has a trigger-switch in the handgrip. There is a high degree of internal monitoring, and computer design and organisation have been configured to reduce on-board maintenance. In basic mode the autopilot will hold pitch angle to any value within the range plus or minus 40 degrees, or bank angle in the range plus or minus 60 degrees. Additionally there are altitude capture and hold modes, preset altitude acquisition and fully-automatic-approach capability down to 200 feet.

System Computer: 194 × 124 × 496 mm
Control unit: 24 × 146 × 115 mm
Total weight: 12·7 kg
Power requirements: <100 VA 200 V ac 400 Hz three-phase
<1 VA 26 V ac 400 Hz single-phase
<30 W 28 V dc

STATUS: entering service.

AP 705 autopilot

The AP 705, equipping the Franco-German Atlantic Nouvelle Generation ASW aircraft is designed to provide precise flight-path control and high levels of safety at very low altitudes above the sea. The system can hold a course to within half a degree while maintaining altitude with an accuracy in the order of 2 metres. These levels of performance are made possible by the quality of inertial data available on the aircraft, and by the versatility of the micro-processor-based computer. Built-in automatic testing enables the operation of the system and its safety devices to be checked before take-off, and also facilitates on-board maintenance.

The three-axis autopilot includes pitch trim and can drive a flight-director system. Autopilot modes permit pressure-altitude hold, glide-slope-beam tracking (in Category I weather minima) and radio altitude hold over the sea down to 150 feet in good visibility, or down to 300 feet in poor visibility. Lateral modes provide for holding the heading or course at the time of engagement, heading hold and homing or tracking on radio navaids, or navigation way-points.

System computer: 384 × 256 × 194 mm
Control unit: 200 × 164 × 67 mm
Servo-actuator: 185 × 183 × 101 mm
Total weight: 23 kg
Power requirements: <50 VA 200 V ac 400 Hz three-phase
<150 VA 28 V ac

STATUS: in production.

B 39 auto-command autopilot

Introduced into French Armée de l'Air Mirage III interceptors since 1975 to replace older autopilots, the B 39 is now available for retrofit to other versions of the same aircraft type. It is a relatively simple system, but uses modern integrated-circuit techniques to confer benefits in terms of performance, safety and maintenance standards which are appropriate to modern equipment. The new autopilot computer is physically interchangeable with the original equipment.

Autopilot functions are confined to attitude and altitude hold modes. Attitude hold includes short-term capability, stability augmentation, and uniform artificial-feel load against load factor irrespective of flight conditions.

Computer: 264 × 200 × 140 mm
Weight: 6·5 kg
Power requirements: <25 VA 200 V ac 400 Hz three-phase
<1 A 28 V dc

STATUS: in service.

FDS-90 helicopter flight-director system

The FDS-90 system comprises an attitude director indicator and a nav coupler/computer unit.

The attitude director indicator is a 4-inch unit which can operate autonomously using self-contained gyros and power inverters. It uses a ball-type real-world display and has a three-cue command capability: bank pointer, pitch and roll command bars, and aircraft symbol. A collective command indicator and inclinometer are built in, and there are annunciators for go-around, decision height and flight-director mode monitor. The gyro can be caged.

The B152 nav coupler/computer is a small, panel-mounted unit which interfaces the attitude director indicator to navigation equipment. It has eight buttons which permit selection of different flight-director operating modes. These are:
hdg: captures and tracks the heading selected on the horizontal-situation indicator.
v/l: captures and tracks vor and ils localiser beam.
bc: tracks the back-course localiser.
alt: maintains the altitude existing at the time of selection.
gs: captures and tracks an ils glideslope beam.
vs: maintains the vertical speed that exists at the time of engagement.
ias: maintains the airspeed that exists at the time of engagement.
fd: removes the flight-director bars from view.
There is also a button on the collective-pitch

control to initiate the go-around mode and establish the correct pitch angle for climb-out. An additional safety feature is that irrespective of the mode selected, any relevant deviation outside preset limits for more than ten seconds causes the flight-director annunciation on the automatic director indicator to flash.

H140 HSI
Size: 4 ATI standard case
Weight: 2·5 kg

B152 Unit
Size: 3 ATI standard case
Weight: 1·3 kg
Power requirements: <1 A 28 V dc
<0·25 A 26 V ac 400 Hz
Lighting power: <0·5 A 28 V dc
Equipment is fully TSO'd, and meets environmental category D0160

STATUS: in production.

Ministab helicopter stability-augmentation system

This is the most basic of three related helicopter stability/flight-control systems produced by Sfena. It consists of three independent channels for roll, pitch and yaw. Each incorporates a computer with integral rate-gyro and an electro-mechanical or electro-hydraulic actuator in series with the control linkage and having approximately ten per cent control authority.

Upstream of the actuator a magnetic-brake/force-gradient assembly is used for stability augmentation system (sas) control activity isolation, artificial feel and stick trim, and detection of pilot control inputs. In manoeuvring flight the pilot's control inputs are detected and integration terms removed to eliminate any resistance from the stability augmentation system. Dynamic stability is still ensured by pure rate terms, and integration is restored when the manoeuvre is complete. This technique provides good damping and positive static stability, but response to control inputs is as though the stabilisation were inherent to the airframe.

If the helicopter has a transmitting directional-gyro, Ministab can maintain the heading being flown on engagement. An inexpensive static transducer can be added also to give long-term altitude hold.

Ministab installation weight and performance is specific to the helicopter type.

STATUS: in service.

Duplex Ministab helicopter stability-augmentation system

In duplex Ministab there are two cyclic channels which are monitored to detect any potential hardover commands. This feature extends the attitude-hold qualities of the basic Ministab system, so permitting true 'hands-off' operation, a considerable reduction in pilot workload, and fulfilling all requirements for single-pilot ifr operation.

Monitoring in the cyclic channels (pitch and roll) is achieved by a separate computer which generates an 'image' of the real actuator position, and inhibits demands if a discrepancy is detected. Hardover commands are eliminated in this way, and any system failure causes reversion to natural helicopter stability.

The installation consists of five Ministab computers driving three actuators (the anti-torque channel is unmonitored because hardover faults are not critical). Altitude hold through the pitch channel is an important part of the system, and heading hold through the roll channel is optional. 'Beep' trims are provided in pitch and roll.

The duplex Ministab has been demonstrated to meet all the requirements for CAA certification of single-pilot ifr operations.

STATUS: in production.

Sfena Ministab stability-augmentation system was designed for such helicopters as the Aérospatiale SA 340-series, known in Britain as Gazelle

Helistab helicopter stability-augmentation system

Busy missions which include safety-critical operations require two pilots. Although sophisticated flight-control systems can be used, fail-operational requirements are not essential, so Sfena has developed a system which offers substantial automatic capability with relatively low cost.

Helistab is a modular system consisting of a basic three-axis Ministab system, plus an autopilot computer and associated equipment. In normal operation sufficient Ministab functions are in use that this constitutes an adequate reversionary mode in the event of an autopilot failure.

The autopilot replaces the rate-gyro integrated terms by vertical and directional-gyro inputs, providing references for automatic control relative to attitude or heading. Basic autopilot functions are long-term pitch and roll attitude hold, long-term heading hold in cruise and hover, and auto-trim in pitch. Functions which can be added include turn co-ordination and heading capture in cruise, automatic hands-off recovery from unusual attitudes, barometric altitude or airspeed hold in cruise.

Doppler-based longitudinal and lateral speed control in hover, and collective-to-yaw coupling.

Helistab is totally transparent, ie the pilot's control inputs are detected and the attitude-hold terms removed to keep the autopilot from resisting pilot control in manoeuvring flight. The peripheral equipment used by Helistab makes the installation helicopter dependent.

STATUS: in production.

Sfim

Société de Fabrication d'Instruments de Mesure, 13 Avenue Marcel Ramolfo-Garnier, 91301 Massy

AFCS 85 helicopter autopilot/flight-director system

The AFCS 85 system is a simplex, two-axis autopilot with hands-off vfr capability, including heading hold, and to which can be added flight-director facilities such as automatic lateral and vertical guidance and control.

Further additional equipment provides an off-airway, ifr system, two-pilot ifr or complete single-pilot ifr helicopter flight-control system. Three-axis versions of all basic configurations are also available.

Basic two-axis system. This is a simplex autopilot. There are two basic modes: 'damp' for angular-rate damping in the pitch and roll axes, and 'ap' for attitude-hold in the pitch and roll axes. In the autopilot mode, heading-select facilities provide indication of roll channel demands on the horizontal-situation indicator.

Basic three-axis system. This additionally provides heading stabilisation in the yaw axis and heading-hold capability.

Off-airways ifr and two-pilot ifr system. A barometric sensor is added to the basic system and adds airspeed hold, baro-altitude hold, and basic attitude hold to the available autopilot functions. A lighted trim-demand indicator simplifies trim adjustment, since declutching the magnetic brake and moving the stick as demanded by the indicator allows the actuator to be recentred, without losing the reference-attitudes previously set. Two- and three-axis versions are available.

Single-pilot ifr. To the last-named system is added a flight-director/coupler and a command-bar equipped gyro-horizon. This adds area-nav route intercept and track, vor route intercept and track, localiser (normal and back course) intercept and track, and glideslope intercept and track facilities. Two- and three-axis versions are available.

Basic two-axis system:
1 Computer (4 ATI case, panel-mounted); 2·3 kg
2 Linear actuators; 0·9 kg each
1 Panel-mounted gyro-horizon (HDV 78 recommended)
1 Remote vertical-gyro (GV 76 recommended)
1 Gyro-magnetic compass system
Off-airway ifr and two-pilot ifr system: as basic, plus
Baran 85 baro-sensor: 134 × 64 × 106 mm; 0·85 kg
1 set of trim-demand indicators
Single-pilot ifr system: as above, plus
FDC85 Flight-director/coupler: 106 × 106 × 225 mm; 1·5 kg
Gyro-horizon with command bars

Three-axis systems use an extra actuator and a pedal-movement sensor. To the two more complex systems are also added a yaw-actuator position sensor and a transverse accelerometer.

Pedal-movement sensor: 115 × 42 × 42 mm; 0·2 kg
Transverse accelerometer: 64 × 8 × 39 mm; 0·2 kg

STATUS: in production.

AFCS 155 helicopter autopilot/flight-director system

The AFCS 155D is a duplex, fail-operational autopilot, optimised for both single and dual pilot ifr operation, covering the whole flight envelope from hover to VNE. It is available for Super-Puma, Super Frelon, Dauphin and Twin-Star helicopters and can be integrated with the FDC85 flight-director/coupler also produced by Sfim.

In this duplex system each channel has its own computer, power supplies, sensors and interconnections. The components are installed in a common case, physically separated by a partition, allowing a small, light and inexpensive design and installation.

Basic functions include long-term attitude

Sfim FDC155 flight-director system for helicopters has many operating modes, and can be used for asw work

and heading hold, duplex (fail-operational) stabilisation and auto-trims. The system also features transparent, fly-through, handling. All configurations satisfy the needs of single-pilot ifr flight, with three basic upper modes: heading select, altitude hold and air-speed hold. Adding flight-director coupler facilities additionally provides ifr automatic nav, radio-nav and approach capabilities, plus special modes suitable for anti-submarine warfare and off-shore or search and rescue operations. These include Doppler/radio height or cable hover, down and up or translational low-altitude manoeuvring, microwave landing system radio-beacon or area-nav automatic approach to hover at low speed.

The AFCS 155 can be used with two dissimilar vertical gyros, eg one panel-mounted and one remote. There is no need for twin gyro-compasses, and each channel or axis can be engaged and monitored independently.

Sfim PA85 autopilot is used in Aérospatiale Gazelle light helicopter

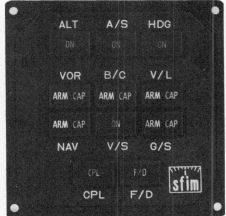

Sfim PA155 autopilot has engagement buttons for each axis of the dual-lane system, autotrim controls and test facilities

Controller unit for Sfim FDC85 flight-director has lighted push-buttons serving as controls and annunciators

Fly-through handling characteristics allow the pilot to make quick attitude and heading changes while benefiting from dynamic damping. On releasing the controls the autopilot returns to stabilised flight at the previously-set attitudes. If new attitude settings are required the pilot can either use the 'stick-release' button on the cyclic-pitch grip, or he can direct the autopilot to change the settings by using the 'beep-trim' button on the cyclic grip. Extensive periods of manoeuvring can be flown with simple stabilisation modes, and there is a turbulence mode which provides smoother flight in turbulence.

All versions of the AFCS 155 include automatic trim, which keeps the stick centred at all times, so that the autopilot always has full control authority. The basic stabilisation functions are duplex, and in the event of a failure, simplex auto-stabilisation is available. Control and aircraft movements due to a failure are mild, and a clear indication of the system working conditions is provided to the crew.

Weight: (autopilot computer) 8 kg
(servo amplifier) 4·2 kg
(autopilot control box) 1·6 kg
(4 actuators) 1 kg each
(3 trim servos) 1·3 kg each
(baro-sensor) 1·5 kg
(flight-director computer) 2 kg
(flight-director coupler box) 2 kg
(collective pitch motor) 2 kg

STATUS: in production.

GERMANY, FEDERAL REPUBLIC

Bodenseewerk

Bodenseewerk Gerätetechnik, Postfach 1120, D-7770 Überlingen

Yaw-damper for PAH-1 helicopter

In close co-operation with MBB, Bodenseewerk has developed a yaw-damper for the PAH-1 anti-tank version of the MBB BO 105 helicopter under collaborative development by Germany and Italy. During attacks the unit is responsible for maintaining target alignment, and during manoeuvring or cruising flight it improves basic flight characteristics.

STATUS: in production.

Automatic flight-control system for Airbus A300

The full description of this system appears in the French section under Sfena, which is the prime contractor responsible to Airbus Industrie for the A300 afcs. A considerable proportion of design, development and production of sub-system equipment is sub-contracted to Bodenseewerk.

IAI

Israel Aircraft Industries Limited, Ben-Gurion Airport, Israel 70100

Autocommand Mabat flight-control system

This is an analogue flight-control system used in the IAI Kfir fighter and developed from a Sfena unit used in the Dassault Mirage III fighter. Autocommand is a pitch-only stabilisation system which alleviates undesirable transient handling characteristics (including during the transonic region of the flight envelope); improves dynamic stability with consequent benefits in target pursuit, aiming, firing and weapon-release accuracy; provides stick-force/g scheduling; and provides attitude hold for a limited period after stick release, and automatic barometric height hold (except during transonic flight).

Autocommand Mabat is a single line-replaceable unit, comprising three interconnected modules; an analogue processor, air-data system and power supplies. The main unit is the processor, which uses dc operational amplifier technology. Integrated circuit ac/dc and dc/ac conversion units, and self-test circuits are incorporated, and the unit uses both multi-layer printed circuit boards and a motherboard to reduce internal wiring to a minimum.

Pitch autostabilisation is provided to requirements set out in Mil F-8785, with a stick-force/g characteristic set by the manufacturer, and pitch-attitude hold in the plus or minus 20 degrees pitch range. After releasing the stick maximum pitch angle change is less than 5 degrees per minute. The maximum change of altitude due to a malfunction cannot exceed ten metres during the first second following the failure, and any normal acceleration produced is within a specified envelope. Self-test can be initiated by the pilot and if the test-button lamp does not light within ten seconds this is taken as an indication of failure. System engagement/disconnection can take place anywhere within the flight envelope. Automatic disconnection

Israel's Kfir adaptation of French Mirage III fighter has IAI Autocommand Mabat analogue flight-control system

will occur only if a malfunction occurs in the compensator system, the autocommand safety system, or if the no 1 hydraulic system disconnects and the preliminary servo changes over to the no 2. Manual disconnect is achieved via several separate override or disconnect buttons, or by exerting a high stick-force.

Altitude hold is available only if pitch stabilisation has been engaged. It is usable throughout the flight envelope, except in the transonic region (Mach 0·95 to 1·15). In level flight altitude errors are within plus or minus 30 feet or 0·1 per cent of altitude, or double these values at 30 degrees roll angle, and not outside a linear extension to 150 feet error at 60 degrees roll angle. Aircraft oscillations are limited to periods in excess of 20 seconds, and maximum normal acceleration is less than 0·1 g. The hold mode can only be engaged if normal acceleration is less than 0·5 g. Automatic disconnect will occur if rate of height

change exceeds 2000 feet/minute, when the speed is between Mach 0·95 and 1·15, or if the undercarriage is extended.

Dimensions: 306 × 198 × 195 mm
Weight: 10 kg
Power requirements: 100 mA 115 V ac 400 Hz single-phase
1 A 28 V dc
Mtbf: >1400 h
Sensors and output devices associated with the unit include a rate gyro unit, accelerometer unit, left- and right-hand compensation servos, pitch pre-servo, stick dynameter and trim motor
A range of dedicated cockpit controls and annunciators are also installed
System performance conforms to MIL 18224C and MIL F-8785, and electrical characteristics are to MIL-STD-704A

STATUS: in service.

UNITED KINGDOM

Louis Newmark

Louis Newmark, Aircraft and Instrument Division, 80 Gloucester Road, Croydon, Surrey CR9 2LD

FN31 helicopter flight-control system

The Type FN31 is a versatile autopilot for the Royal Navy/Westland Sea King helicopter with the following facilities: stability augmentation, heading hold, barometric height hold, radio height hold, automatic transitions, hover control, and plan-position control. It is a simplex system derived from the company's FN30 duplex system, which is currently also in service with the Royal Navy.

FN31 flight-control system facilities are engaged on the centrally-mounted pilot's controller unit. This provides indications of the modes selected and disengagement or engagement of any of the system's four channels can be effected. Servo-amplifier output selectors actuate variable-orifice valves in the auxiliary servo unit, and these act in series with the pilot's own demands, but with limited authority. Extended control authority is by spring mechanisms in the auxiliary servo-unit and by automatic inching of the cyclic trim during automatic transitions.

Selection of the stabiliser facility gives a combination of attitude and heading hold, with three-axis damping of aircraft motion. The flight attitude of the helicopter becomes stabilised relative to the cyclic column position, and heading hold takes effect whenever the pilot completes any heading manoeuvre. A heading trim control is provided for controlling the helicopter during flat turns.

In cross-country flying barometric altitude and stabilisation facilities may be engaged together or separately. By pressing a manoeuvre button, situated on the hand-grip of the collective lever, the pilot can disengage altitude hold temporarily, and fly the helicopter to a different altitude.

Radio altitude hold can be engaged separately. When this mode is engaged the helicopter is stabilised relative to an inertial height, which is derived from radio altitude and vertical acceleration data. This stabilises the helicopter when flying over undulating terrain. With this mode engaged the pilot can make controlled changes in altitude by turning the 'set radio height' control knob to other settings.

When the pilot engages 'trans down', the helicopter begins a controlled descent to hovering altitude, at the same time decelerating to zero ground-speed. Forward speed information is obtained from a Doppler-radar sensor. The helicopter hovers at an altitude which is preset by the pilot on a 'set hover height' control. When 'trans up' is engaged the helicopter climbs from the hover altitude to the altitude set on 'set radio height', and accelerates to the ground-speed set on the 'set exit speed' control. Both transition manoeuvres are programmed for completion in minimum time.

In asw operations, at the completion of a transition to hover, a sonar can be lowered, and as it enters the water the pilot can select 'cable hover', after which plan-position and height are controlled relative to the submerged sonar. On retrieving the sonar the pilot reselects 'Doppler hover' in preparation for an up-transition. In rescue operations the pilot can set the hover switch at 'Doppler' and engage an auxiliary hover-trim control, which allows the winch-operator to command small increments of fore-aft and lateral ground-speed.

Four attitude-indicators are included in the flight control system. Attitude signals for the indicators, and for other systems in the aircraft, are obtained electrically from two vertical-gyro sets and a repeater platform. Each vertical-gyro unit transmits synchro signals to the main

Pilot's controller for Louis Newmark LN400 flight-control system has engagement buttons for each axis of two-lane autopilot, and other facilities

indicator of one pilot and the standby indicator of the other pilot, ensuring that neither pilot suffers complete loss of attitude information through a single system failure. As the accuracy of transmitted information depends on having accurate alignment between the repeater platform-unit and the associated vertical-gyro unit, these are mounted on a common base-plate.

Dimensions and weight: (pilot's controller) 240 × 147 × 117 mm, 6 lb (2·7 kg)
(channel selector) 147 × 150 × 144 mm, 3½ lb (1·5 kg)
(stick canceller) 90 × 58 × 51 mm, ½ lb (0·2 kg)
(altitude controller) 230 × 141 × 135 mm, 2½ lb (1·2 kg)
(amplifier) 531 × 286 × 104 mm, 22 lb (10 kg)
(2 vertical-gyro units) 230 × 160 × 134 mm, 8½ lb (3·8 kg) each
(repeater platform) 410 × 205 mm dia, 19½ lb (8·9 kg)
(2 5-inch attitude ind) 141 × 141 × 230 mm, 4⅘ lb (2·2 kg) each
(2 3-inch attitude ind) 83 × 83 × 250 mm, 2⅘ lb (1·2 kg) each
(rate gyro unit) 103 × 103 × 51 mm, 1 lb (0·5 kg)
Power requirements: 220 VA 200 V ac 400 Hz three-phase
100 W 28 V dc

STATUS: in service.

LN400 helicopter flight control system

The LN400, in production for the Westland WG30 Civil Lynx, is designed for single-pilot ifr operation and provides duplex stability-augmentation for pitch, yaw, roll and collective axes. Automatic trim in pitch and roll ensures mid-point operation of the series actuators, so ensuring full authority at all times. The system incorporates collective-acceleration control, heading hold and yaw-trim facilities. A radio or barometric-height hold can be provided as an option. The system comprises a computer, controller and sensor.

Computer unit (NDN 8919-01) This is contained in a standard ½ ATR Short case with front plug connectors. The computing circuits are of modular form using plug-in circuit boards for each lane. Simple manipulation of control-law parameters are effected by appropriate linkages on the printed circuit boards. Duplex integrity is assured throughout, each lane of each channel containing an independent stabilised power supply. The lane isolation is maintained by the embodiment of separate front plug connectors.

The duplex stabilisation boards of each channel are identical, each board containing the necessary electronics for interfacing with the appropriate sensor, processing the control-law signals and providing the series-actuator drive circuitry and control logic. In addition, simplex boards contain the heading-hold circuitry and provide the parallel-actuator drive for automatic trim in the pitch and roll axes.

Growth capability in the computer unit allows for an optional barometric or radio-height hold facility.

Pilot's controller (NDN 8921-01) This unit provides push-buttons to engage and disengage both lanes of pitch, yaw, roll or collective channels. There are two dual-purpose meters that may be used to monitor each of the four channels during flight or alternatively (as part of the built-in test equipment) to diagnose channel or lane faults.

The auto-stabilisation functions engage as soon as power is available, and illuminate the collective channel push-buttons. On achieving remote-gyro verticality, pitch, roll and yaw channels of both lanes are engaged by depressing the 'ase engage' button. When both lanes of all channels are engaged, feedback signals from the series actuators are compared in each channel. If a predetermined level of disparity is detected between a pair of channel lanes, the relevant fault light illuminates, and the central-warning system is informed. The pilot then takes appropriate action, based on built-in test indications, to disengage one or both channels.

The selected heading may be fine trimmed by using a 'beep' switch on the controller, and there is space for an optional height-hold control.

Sensor unit (NDN 8923-01) This contains duplex yaw-rate gyros, duplex vertical accelerometers and has growth capability for additional sensors. Each sensor is excited from separate power supplies and electrical connections are via two connectors, thus maintaining lane isolation. An external three-wire heading signal is required to complete the total system.

Dimensions and weight: (computer unit) 125 × 194 × 319 mm; 12 lb (5·5 kg)
(pilot's controller) 137 × 95 × 165 mm; 5³⁄₁₀ lb (2·4 kg)
(sensor unit) 130 × 65 × 212 mm; 4¹⁄₁₀ lb (1·9 kg)
Power requirements: 48 VA 115 V ac 400 Hz single-phase
210 W 28 V dc

STATUS: in service.

Royal Navy Sea King HAS5 has FN31 flight control system

LN450 digital helicopter flight-path control unit

The LN450 provides a digital interface between sensors and the flight-control system and it offers a wide range of flight-path control options. It especially offers facilities tailored to offshore and high-density commuter traffic operations. Simplex and duplex installations are available to suit customer requirements.

Pilot's Controller (NDN 9642) This incorporates mode-dedicated push-button switches, each of which illuminates when the appropriate logic circuitry is selected. Digital light-emitting diode displays are provided for 'set hover height', 'radio height' and 'indicated airspeed hold' inputs. The system incorporates high-integrity self-testing and continual amplitude- and rate-monitoring, with fault indicators on the controller.

Duplex or simplex inputs from standard aircraft sensors are acceptable. The following coupled/uncoupled modes are available: survivor overfly, winch-man hover-trim, transition up/down, radio or barometric height hold, vertical speed hold, indicated airspeed hold and heading hold. There are navigation mode selections for ils/loc, vor, area-nav, microwave landing system, back-course approach and go-around. Except for go-around, all navigation modes include capture and track logic. There is growth capability to incorporate an asw mode.

Computer unit (NDN 9634) Each computer unit contains a signal conditioner, microprocessor and power supplies. In each unit the sensor analogue input signals are conditioned, multiplexed, and fed into the micro-processor, where autopilot and flight-director information

Westland 30 Civil Lynx has Louis Newmark LN400 afcs

for all four axes is computed in accordance with the programmed control laws. The autopilot outputs control the helicopter through the existing stability-augmentation system, while the flight-director outputs drive command bars on the cockpit attitude-director indicator units. To maintain maximum integrity the system is arranged so that only two-channel outputs for either autopilot or flight-director are active from each computer unit. All outputs are fed back for signal self-monitoring, and should a discrepancy, in excess of a pre-determined threshold occur, the computer unit automatically disconnects the system then reverting to simplex. If the micro-processor cannot

establish in which unit the fault has occurred, the pilot must ascertain the corrective action.

Dimensions and weight (duplex system):
(2 computer units) 91 × 193 × 320 mm; 7·7 lb (3·5 kg) each
(Pilot's controller) 146 × 162 × 165 mm; 6·1 lb (2·8 kg)
(A simplex installation has only one computer unit, and the pilot's controller weighs only 3·3 lb (1·5 kg))
Power requirements: 0·4 VA 115 V ac 400 Hz single-phase
63 W 28 V dc

STATUS: in development.

Marconi Avionics

Marconi Avionics, Airport Works, Rochester, Kent ME1 2XX

Automatic flight control system for Tornado

This is a tri-national venture which Marconi shares with Bodenseewerk of West Germany and Aeritalia of Italy. Design and development of the system started in 1971 and covers three major components; the command and stability augmentation system (csas), autopilot and flight-director system (afds) and spin and incidence limiting system (spils). Work on the csas is shared with Bodenseewerk, and on the afds with Aeritalia. Spils will be fitted only to Tornados operated by the RAF.

Command and stability augmentation system The csas is an analogue, triple-redundant, fly-by-wire manoeuvre-demand system. It provides electrically-signalled pitch, roll and yaw control, and automatic stabilisation of the aircraft response to either pilot commands or turbulence. Gain scheduling enhances handling qualities and control stability over the full flight envelope, and operates in conjunction with the spin prevention and incidence limiting system. 'Carefree manoeuvring' enables the aircraft's full lift capability to be exploited at all flight conditions and irrespective of aircraft configuration without risk of structural damage as a result of the pilot inadvertently exceeding the design load factors.

Autopilot and flight director system The digital autopilot and flight-director system automatically controls the flight-path in all modes, including terrain-following, and provides signals to the director instruments enabling the crew to monitor autopilot performance or to fly the aircraft manually; pitch autotrim is also incorporated. The duplex, self-monitoring processor configuration provides high-integrity automatic control, permitting low-altitude cruise with appropriate safety margins. The flight-director remains available after most single failures. Autopilot manoeuvre-demand signals are routed to the control actuators via the csas. A 12-bit processor is used, with 6 K words of stored program and 1 K words of data

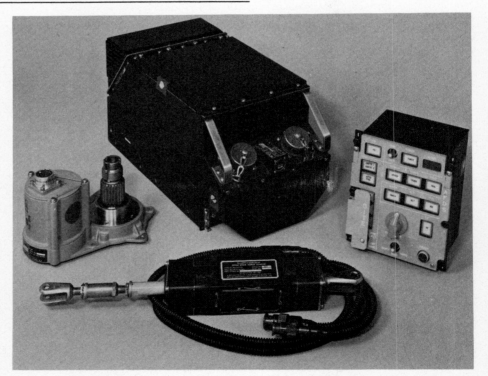

Tornado equipment by Marconi Avionics includes control panel, computer, pitch stick-force sensor and drive motors

store. Typical computing speed is around 160 KOPS and program cycle time is 32 ms.

Spin and incidence limiting system Spils is a duplex analogue system which operates in conjunction with the csas to enable the aircraft's full lift capability to be exploited at low-level. It limits incidence, irrespective of the pilot's demands, when maximum angles of attack are reached.

Individual line-replaceable units for the control and stability augmentation system are: csas pitch computer, csas lateral computer, csas control unit, pitch, roll and yaw-rate gyros, and pitch, roll and yaw-position transmitters.

For the autopilot and flight director system they are: afds computer no 1, afds computer no 2, afds control unit, and autothrottle actuator, and pitch and roll stick-force sensors.

The spin and incidence limiting system has two line-replaceable units: the spils computer, and the spils control unit.

STATUS: in service.

Auto-stabilisation system for Harrier

Designed and developed specifically for V/STOL operations, this system provides three-axis stability augmentation during vertical take-off, transition, wing-borne flight, hover and landing. Over 400 systems have been produced for all UK, US and Spanish armed forces customers. There is no autopilot in this aircraft. Rate gyros and self-test facilities are included in each processor.

Individual line-replaceable units are: pitch/roll auto-stabilisation computer, yaw auto-stabilisation computer and lateral accelerometer unit.

STATUS: in service.

Tornado spils computer opened up to show four-board, front-connector layout used in this Marconi Avionics/Bodenseewerk equipment

Autopilot for Sea Harrier
The Sea Harrier is the first Harrier variant to have an autopilot. Designed to work in conjunction with the Harrier auto-stabilisation system, the autopilot currently provides roll and pitch attitude hold, heading hold, barometric height hold, and self-test. The autopilot is standard equipment for all Sea Harriers.

STATUS: in service.

Fly-by-wire flight control system for Jaguar
The integrated flight control system for the BAe Jaguar fly-by-wire (fbw) technology-demonstrator programme commenced development in 1977 and achieved first flight in October 1981. This project is the world's first complete and all-digital fly-by-wire system, operating without mechanical or electrical reversionary controls. The system has been manufactured to production standards and provides a ready basis for future fbw applications.

Individual line-replaceable units are: four flight control computers, two actuator drive and monitor computers, pilot's control panel, five control surface position transmitters, a lateral accelerometer, and three quadruplex gyro packs.

A target loss-probability of 10⁻⁷ per hour for the complete system led to adoption of a quadruplex system configuration. Where essential data is measured, quadruplex sensors are used also, and their outputs are routed via optically-isolated lanes for consolidation in each flight-control computer. Processing is accomplished using 16-bit words and operating with a 28 K word program and 6 K words of stored data. Typical processor operating speed is 800 KOPS and computing takes place in cyclic frames of between 5 and 80 ms duration. Failure-survival relies on majority-voting of the quadruplex components.

Flight-control computer outputs to the tail-plane (pitch/roll), rudder (yaw) and spoilers (roll) drive two-stage hydraulic actuators. Two hydraulic supplies are connected to either two or three sub-actuator units at each control-

surface location, thus providing the degree of redundancy necessary for the actuator to survive a single hydraulic power-supply failure without generating catastrophically large actuator movements. Each flight-control computer is supplied from up to four independent electrical sources, including a battery-backed supply.

Marconi Avionics supplies the quadruplex rate-gyro packs and duplex lateral acceleration units which are used as sensors. A diagnostic display unit stores failure information for ground maintenance personnel.

STATUS: technology-demonstrator in preparation for the next generation of combat aircraft.

General-purpose three-axis auto-stabiliser
Design and development by Marconi Avionics of a general-purpose three-axis auto-stabilisation system, with integral gyro pack and self-test, was started in 1978 and the system is now in quantity production for an undisclosed customer. The entire system is enclosed by a single ½ ATR box.

STATUS: in production.

Flight control system for Lynx helicopter
Marconi Avionics has produced three variants of the basic Lynx afcs. Facilities common to all

variants include pitch and roll attitude stabilisation with yaw-rate stabilisation in heading hold. Army versions of the Lynx additionally have a barometric height hold, using collective and lateral acceleration control. Two naval versions provide radio-altitude height hold using collective control, or radio-altitude acquire and hold with automatic transition to hover for dunking sonar in asw operations.

STATUS: in service.

Quadruplex rate-gyro sensor packs have been developed by Marconi Avionics for the Jaguar fly-by-wire demonstrator

Marconi Avionics/Bodenseewerk csas equipment for Tornado includes pitch/yaw rate-gyro pack, computer units, bite panel and drive units shown here

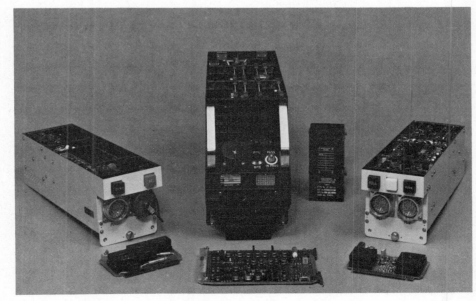

For Royal Navy Sea Harrier V/STOL fighters Marconi Avionics autopilot facilities have been added to basic RAF Harrier auto-stabilisation system. Computer and sensor units are shown here

Automatic flight control system for Concorde

Marconi Avionics jointly developed with Sfena (France) the Concorde afcs, which comprises eight separate systems with no fewer than 38 individual units, and provides fully-automatic control from climb-out to automatic landing. The system is duplex-monitored and fail-operational, to meet the operational requirements of Icao Category IIIA, and is cleared for automatic landing with 15 feet (4·5 metres) decision height and 200 metres runway visual range. It was the first afcs to make use of a push-button controller fitting into the flight-deck glareshield.

STATUS: in service.

Flight-control system for Westland/Aérospatiale Lynx helicopter includes this computer, controller and drive units

Jaguar fbw installation contains two actuator drive and monitoring computers (left) and four flight-control computers (right)

Automatic flight control system for BAC One-Eleven

Several afcs variants are in service, ranging from three-axis auto-stabilisation units with autopilot approach facilities meeting Icao Category I requirements, to a fail-passive automatic-landing system which includes automatic-throttle control and a minimum decision height of 60 feet.

STATUS: in service.

Flap/slat control computer for A310

The A310 flap/slat control computer is part of a fully-monitored, fail-operative, duplex system designed for this new short/medium-range derivative of the A300 wide-body airliner. It is a digital system using dissimilar redundancy to satisfy high-integrity requirements. Built-in safety features include protection following detection of a mechanical jam. This could be caused by a runaway resulting from a mechanical breakage.

STATUS: in production.

BAe Jaguar fly-by-wire quadruplex digital fcs demonstrator

Marconi Avionics Machan RPV and flight-control system

Control systems for rpvs and drones

Marconi Avionics currently has two rpv systems and one drone system in production, providing either comprehensive or partial control functions. Full afcs is provided for the Jindivik drone/target vehicle, with automatic stabilisation. Radio-link ground-control demands are added to on-board computed auto-stabilisation demands. The original Mark L5 system comprised six units (rate gyro, attitude gyro, amplifier unit, air-data unit, accelerometer, and radio altitude computer) and has been superseded in production by the Mark 4 system, comprising only a flight-control computer, accelerometer and three-axis rate-gyro unit.

For drone versions of the early-1950s Hawker Siddeley Sea Vixen naval fighter, Marconi Avionics supplies a universal pack comprising

Drone conversions of Sea Vixen deck-landing fighter have two sensor packs, computer, and programming unit. Supplier is Marconi Avionics

a drone computer, accelerometer, controller and three-axis rate-gyro unit, which can be fitted in place of the pilot's seat. The auto-stabilisation and control facilities enable the aircraft to be flown as an unmanned target from the ground via a command radio-link.

Towed-target systems produced by the UK company, Ruston, can be equipped with a Marconi-developed height-control computer, so enabling the target to be flown at a preset altitude, independent of aircraft attitude. Sea-skimming targets are simulated in this way.

STATUS: in service.

Fly-by-light demonstration rig

Fly-by-light is a development of the fly-by-wire principle enabling the flight-control system to operate even in the presence of high levels of electro-magnetic interference, and virtually eliminating the risk of lightning damage to the flight control system in aircraft constructed of composite materials. A demonstration rig representative of one-lane of a triplex flight control system and containing a force-stick control column, rate gyro, flight-control computer actuator and actuator drive unit has been developed. It uses fibre-optic data transmission from the flight control computer to the actuator, a hydraulic-to-electric generator to provide electrical power at the actuator, and monitoring equipment in the actuator which sends data, via a fibre-optic link, to a monitor processor within the flight control computer.

STATUS: in development.

Australian-designed Jindivik drones have latest Marconi Avionics flight-control equipment shown here

Smiths Industries

Smiths Industries, Aerospace and Defence Systems Company, Cricklewood, London NW2 6JN

SEP10 automatic flight-control system

The SEP10 has been selected for the British Aerospace 146 feeder-liner. It provides three-axis control or stabilisation and incorporates a two-axis (pitch and roll) autopilot, elevator trim, flight director and yaw-damping facilities. It uses simple, well-proven control laws and the minimum of sensors. There is also a 'transparent' control facility which allows the pilot to temporarily disengage the autopilot clutches and sensor chasers, and to manually manoeuvre the aircraft, so adjusting the datum of the basic and manometric autopilot modes.

The autopilot is based on rate-type control laws. Pitch and roll-rate signals are derived from ARINC three-wire attitude references, thus eliminating the need for rate gyros. Other ARINC standard interfaces accept a wide range of sensor inputs, including those from barometric and radio-navaid sensors, and allow systems to be tailored to suit particular operator's needs. The autopilot computer uses digital computing techniques to provide outer-loop control and to organise the mode logic, and has capacity to accommodate optional facilities. Analogue computing is used for the inner-loop stabilisation computing, servo-drive amplifiers and safety monitors.

The system can be supplied with either a parallel-acting yaw-damper, which uses a rotary servomotor to drive the rudder and rudder pedals, or a series yaw-damper which drives a linear actuator in series with the rudder control run. In each case, the yaw-damping system is self-contained and consists of an analogue yaw computer, sensor and the relevant actuator or servo motor.

Flight-director computations are performed within the digital section of the autopilot computer, which can supply commands to V-bar or split-axis flight-directors. The flight-director and autopilot share common mode-selection and outer-loop guidance, but if desired they can be operated independently.

Emphasis has been placed on maintainability

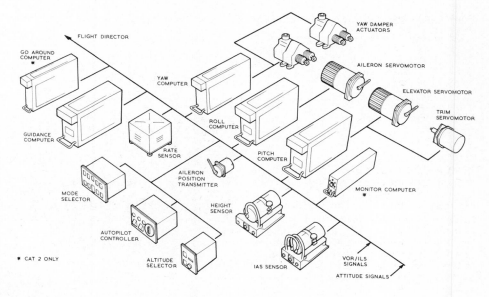

Components of Smiths Industries SEP10 automatic flight-control system as installed in the British Aerospace 146 feederliner

and ease of testing, both for the installed system and for individual units in the workshop. Routine testing is designed to confirm correct functioning of safety devices, the tests being performed by operating a test button in conjunction with buttons on the mode selector. Alternatively, a stored digital test program, which checks digital computer operations, is activated by pressing a separate button on the autopilot computer. Modular construction has been used extensively to ensure that faulty equipment can be corrected and recertified easily and quickly.

The following descriptions of individual line-replaceable units outline the operation of a full SEP10 system.

Autopilot Controller The autopilot controller, in addition to providing autopilot and yaw damper engage or disengage controls, also includes pitch-rate and roll-angle selectors, and the elevator and rudder trim-position indicators. Engagement of the autopilot and yaw damper is confirmed by the illumination of

a legend within each selector, and by the illumination of engage monitor lights on the mode selector. Pitch control uses a spring-centred lever which has a non-linear feel so that minor adjustments can be made instinctively. Roll control is accomplished by rotation of the control knob, which remains offset by a displacement proportional to the roll angle demanded in the basic mode, but returns automatically to the central position on selection of an alternative mode. Each elevator and rudder trim indicator displays the trim force held by the control servos.

Mode-selector There are 11 push-button switches, each illuminating as mode-indicators, for the selection of both autopilot and flight-director functions. Control mode engagement is confirmed by the illumination of a green triangle on the appropriate button, and for modes that include both arm and engage facilities, an amber triangle is illuminated during the arm phase, changing to a green triangle when the mode is engaged. A turbulence facility is included to soften flight disturbances in turbulent air. This reverts the autopilot to the basic stabilisation mode and at the same time reduces the overall gain of the system. Autopilot and yaw damper engagement lights are provided so that the full engagement state of the system can be seen on the mode selector. There is provision for remote mode indication also.

Autopilot computer The autopilot computer receives both analogue and logic information from sensors, controllers and selectors, and processes them to formulate the pitch and roll axis demands and the flight-director commands. The majority of autopilot computing is performed digitally, although analogue techniques are used to provide pitch and roll-stabilisation and authority limitations. Correct functioning of the computer safety circuits is verified by a test facility that can be located on the front face of the computer, or at a convenient remote station.

Yaw computer This unit takes short-term damping information from a yaw-rate gyro, while the lateral accelerometer senses acceleration along the lateral axis for slip-skid prevention. A suitable series yaw computer is available in the event of an operator electing to fit a series yaw-damper system.

Autopilot engage panel As the 'transparent' control facility (provided as standard) permits datum adjustments, an autopilot engage panel without alternative switches for these facilities is available. Adoption of this unit reduces the panel space requirement and total equipment weight.

Altitude alerting unit The addition of an altitude-alerting unit provides facilities for altitude pre-select and vertical-speed hold modes. Altitude information is obtained from either a servo-altimeter or an air-data source. Selected altitude is displayed on a counter readout. A warning flag obscures this display in the event of a power failure or absence of

BAe 146 feeder-liners have Smiths Industries SEP10 autopilot

altitude valid signal, and a test facility allows checking of associated audio-visual signals and altitude pre-select function.

Air data unit Where there is a requirement for a Mach-hold facility to secure better fuel economy, the basic airspeed sensor can be replaced with an air-data unit providing the necessary extra outputs.

Monitor computer For operation to Category II weather minima, this unit completes the performance monitor functions necessary to provide a fail-safe pitch channel. It independently monitors autopilot pitch, localiser and glideslope deviation and provides outputs that can be used to disconnect the autopilot and provide warnings to the pilot. The computer is completely independent of the autpilot, and a self-test facility allows a check to be made on the correct operation of all the monitoring functions.

Flight-director computer FAA clearance for Category II approaches may require that flight-director information be provided from an independent source. Smith Industries therefore offers a flight-director computer whose outputs can control the deflection of pitch and roll directors on the attitude-director indicator,

Altitude selector for Smiths Industries SEP10 afcs

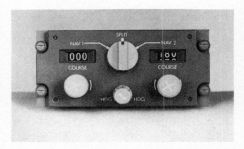
Heading/course selector panel for Smiths Industries SEP10 afcs

during the approach phase. It can be used to provide independent flight-director signals for heading, altitude and vor as well as the localiser and glideslope. A flight-director mode selector is available for use in conjunction with the flight-director computer.

STATUS: in production.

UNITED STATES OF AMERICA

Astronautics

Astronautics Corporation of America, 2416 Amsler, Torrance, California 90505

Helicopter flight-director system

The company has developed a three-cue system which gives full ifr capability to helicopters. It combines a versatile flight-director computer with a three-command-bar attitude director indicator, dual-bearing pointer horizontal-situation indicator and a multi-mode computer controller. In ifr conditions, the system provides instrument landing system, vor, and automatic direction finder approach capability, including the option to include steep-angle instrument landing system approaches.

The three-cue system adds a collective-command steering bar to the pitch and lateral command steering bars used in two-cue systems. Continuous altitude, airspeed, vertical speed, automatic direction finder, vor/ils, and (optional) Doppler and altitude-alert inputs permit the flight-director computer to respond to pilot-selected flight modes. A pilot can execute automatic vor/ils intercepts, glideslope intercepts, vertical and airspeed holds, deceleration rates and altitude holds. The computer provides automatic intercept and tracking of vor and automatic direction finder courses, glideslope and localiser, and automatic initiation of deceleration instrument landing system approaches. Pilot and co-pilot displays and controls are effected by optional slaved horizontal situation indicators and transfer controls.

The indicators are 5-inch units, hermetically sealed with a dry nitrogen-helium atmosphere. Direct-current servos used in these units are claimed to result in considerably less heat

Astronautics three-axis helicopter flight-director system

dissipation, less power drain, higher torque and greater reliability than typical ac-servoed units.

The mode controller has four rotary switches for mode, nav select, vertical speed and airspeed selections. An optional remote dual-course selector can be added to supplement the basic controller.

The flight-director computer accepts inputs from the mode controller and an array of sensors. It computes the pitch, lateral and collective commands required to adhere to selected and/or scheduled flight parameters, and displays the commands on the attitude director and horizontal situation indicators.

Dimensions and weight: (adi) 127 × 133 × 194 mm; 7 lb (3·2 kg)
(hsi) 127 × 108 × 174 mm; 6 lb (2·7 kg)
(flight director computer) 59 × 194 × 319 mm; 6·5 lb (2·9 kg)
(mode controller) 146 × 105 × 127 mm; 3·5 lb (1·6 kg)
(remote course selector) 146 × 32 × 165 mm; 1·5 lb (0·7 kg)

Power requirements: 45 VA 115 V ac 400 Hz
0·5 A 28 V dc
8 VA 5 V ac (for lighting)

STATUS: in production.

Three-axis autopilot for Hughes 500 helicopter

Astronautics has developed an autopilot for light helicopters, which has been FAA certificated, and is available to Hughes 500D users. The autopilot provides full three-axis control to greatly reduce pilot fatigue. There are seven basic operating modes, plus hands-off stabilisation. In basic attitude-retention mode, the helicopter can be flown hands-off not only in straight-level flight, but also during climbs, descents and turns. The desired attitude will be held even in autorotation, and all turns are automatically co-ordinated. In altitude-hold mode the desired altitude is maintained to within plus or minus 20 feet. The altitude can be captured with vertical velocities as high as 1000 feet per minute. Following high vertical velocity climb/descent the helicopter will smoothly change pitch attitude and capture the selected altitude. In heading-hold mode the pilot may select the desired heading either before or after the mode is engaged. All turns to the new heading are automatically co-ordinated at a bank angle of 20 degrees. Hands-off hover capability is provided either in or out of ground effect. Heading hold may be engaged any time during hover. Any new desired heading is entered by moving the 'bug' on the heading gyro, and yaw-damping is provided throughout the autopilot flight envelope.

STATUS: in service.

Astronautics 3-inch adi used in Northrop F-5 fighter

Pathfinder P3B general-aviation autopilot

Astronautics produces a range of general-aviation autopilots under the trade-name Pathfinder, of which the P3B is one embodying almost all of the components available in this family of modular autopilots. The P3B system is all-electric, and is a three-axis system which provides roll, yaw and pitch stabilisation, heading selection and control, vor/loc tracking with automatic cross-wind correction, automatic unlimited-angle vor/loc intercepts, manually-selected standard-rate turns, pitch stabilisation in all attitudes, altitude hold with automatic pitch-trim, automatic or manual glideslope intercept, glideslope tracking with automatic pitch trim and manually-operated electric pitch trim.

STATUS: in service.

Autopilot for Boeing Mini-rpv

Astronautics is currently flight-testing and evaluating an autopilot for a Boeing-manufactured mini-rpv system. No details of this vehicle or system performance have been released.

STATUS: under development.

Flight-director instruments and computers

Astronautics manufactures a wide range of 5-, 4-, and 3-inch attitude-director indicator and horizontal-situation indicator units, and associated computers, for use in military and civil aircraft world-wide. The following is a list of recent contributions to current aircraft programmes. Equipment is specified by the customer, but is not necessarily composed of standard-fit items.

Equipment
5-inch attitude director indicator: three-cue
Application
Sikorsky S-61 and S-76,
Bell 212, 412 and 214ST,
Aérospatiale Super Puma,
Agusta-Bell AB212 and AB412

Equipment
5-inch attitude director indicator: two-cue
Application
General Dynamic F-111,
Vought A-7 Corsair,
Lockheed C-130 Hercules,
Aermacchi MB.326 and MB.339,
BAe Hawk
Equipment
4-inch attitude director indicator: two-cue
Application
McDonnell Douglas F-15 Eagle
Equipment
3-inch attitude director indicator: two-cue
Application
General Dynamics F-16,
Northrop F-5
Equipment
Flight-director computers
Application
Vought A-7 Corsair,
Lockheed C-130 Hercules,
Aermacchi MB.326 and MB.339,
Lockheed P-3C Orion

STATUS: wide range of models in production.

Astronautics 5-inch adi used in Bell 212 helicopter

Bendix

Bendix Avionics Division, 2100 NW 62nd Street, Fort Lauderdale, Florida 33310

FCS-870 automatic flight-control system

Bendix has combined integrated-circuit technology with several new automatic flight-control features to produce a system which offers optimum performance across a wide performance spectrum. The FCS-870 is a basic autopilot, with flight-director and independent yaw-damper options which form the basis of many options. It is designed for installation in a broad range of aircraft types, from heavy singles through most turboprop powered types. The system meets or exceeds the TSOs for these classes of aircraft.

The complete FCS-870 consists of the cockpit instruments (including flight controller, attitude-director indicator, horizontal-situation indicator and mode annunciator) and a remote-mounted computer amplifier and servos. The yaw-damper option adds a slip-slip sensor, a panel-mounted turn-and-slip indicator, and a remote yaw servo.

Flight controller This is a small panel-mounted unit, used to select the desired operational modes of the system. All nomenclature on the panel is back-lighted for easy night-viewing.
Mode annunciator This can be mounted in any convenient 'head-up' panel location so that the pilot can monitor the autopilot or flight-director functions in use. The unit also alerts the pilot to some fault and armed-system conditions.
Computer amplifier This is the main flight control system unit and is an all-solid-state device which houses all of the lateral and longitudinal computational circuitry, power supplies and altitude transducer. It also contains the calibration circuits which ensure compatibility with specific aircraft sensor and output requirements, plus an additional circuit board providing input signals for flight-director operations. Lateral and longitudinal data circuits are segregated on opposite sides of the unit, so reducing the amount of inter-wiring, and augmenting reliability and serviceability. Relays have been eliminated and all heat-generating components are near the outside of the unit to improve cooling.
Servos Advanced design servos provide the greatest torque required for the highest-performance aircraft likely to use the system. Three similar units are used for pitch, roll and trim control.
Yaw-damper The independent yaw-damper provides positive turn co-ordination and rudder control in all flight conditions. In twin-engined types the yaw-damper is claimed to provide substantial assistance in maintaining directional control during an engine-out sequence.
Instrumentation The company recommends integration with the following Bendix flight instruments:
DH-886A director horizon indicator (4-inch model) or
DH-841V director horizon indicator (3-inch model), plus
HSD-880 horizontal-situation indicator (4-inch) or
HSD-830 horizontal-situation indicator (3-inch).
Specific features of the flight control system are: command-turn (half- or full-rate co-ordinated turns can be initiated simply by rotating a knob), full-pitch integration (provides smooth capture of desired altitude, and eliminates stand-off errors), automatic altitude pre-select, pitch synchronisation (keeps elevator surfaces aligned with trim to eliminate disengagement disturbances), control-wheel steering (manoeuvre to desired attitude with button depressed, and then release to leave flight control system maintaining pilot's demand), coupled go-around, automatic cross-wind correction, all-angle intercepts, pitch-rate command and manual or automatic glideslope capture.

Weight: (basic autopilot) 19·7 lb (8·94 kg)
(with 3-inch FD/HSI) 31·1 lb (14·11 kg)
(with 4-inch FD/HSI) 37·65 lb (17·08 kg)
(optional yaw-damper) 7·35 lb (3·33 kg)
TSO compliance: C9c, C52a, DO160.
STATUS: in service.

Fly-by-wire flight control system for F-16 AFTI

Bendix has developed a triplex digital fly-by-wire (FBW) flight control system for the advanced fighter technology integration (afti) programme, which is directed by the Air Force System Command Flight Dynamics Laboratory at Wright-Patterson AFB, and which is based on a substantially modified General Dynamics F-16 fighter. In addition to conventional F-16 control surfaces the aircraft has two large manoeuvring canards on the forward fuselage for direct-force control. Much of the equipment is in a dorsal fairing above the aircraft fuselage. Flight-testing commenced at Edwards AFB in 1982.

STATUS: research programme.

Collins

Collins Avionics Divisions, Rockwell International, 400 Cedar Road NE, Cedar Rapids, Iowa 52406.

AP-105 autopilot (ProLine)

Available for several aircraft types in the large turboprop and executive-jet category, the AP-105 offers three-axis automatic flight control and can be associated with the Collins FDS-84 or FD-112V flight-directors. The system offers a full complement of navigation and vertical mode options, including altitude pre-select, and can present all computed steering data on flight-director displays for manual control guidance, or to monitor the autopilot performance.

The AP-105 is designed to exceed automatic and manual Category II approach requirements. Glideslope scheduling using radio-altimeter information is incorporated, and an all-angle vor/loc capture capability is included, plus automatic back-course operation. The glidescope function can automatically capture from above or below, either before or after localiser capture. Vertical speed hold, airspeed hold, Mach hold and altitude hold modes can be integrated to fly departure, en-route and arrival segments. Automatic mode changing is standard and automatic elevator trim permits ripple-free transitions between modes, and autopilot disconnects.

A control-wheel synchronisation facility is engaged by depressing a button on the yoke, and the autopilot then follows pilot control inputs, taking up the desired attitude when the wheel is released. The rudder channel can be operated in manual flying modes to enhance turn co-ordination, or engaged for all flying modes. In established attitudes the aircraft is gyro-mode stabilised, but displacement of pitch or roll rate manoeuvre controls permits manual override. Attitude rates are used as a basis for all control inputs and automatic lift-compensation is provided in the basic control laws.

Dimensions and weights:
Autopilot/flight-guidance system
(servos (3)) 86 × 122 × 205 mm, 7·4 lb (3·4 kg) each
(autopilot amplifier) 57 × 194 × 311 mm, 5·7 lb (2·6 kg)
(yaw-damper) 57 × 88 × 325 mm, 2·7 lb (1·2 kg)
(airspeed sensor) 64 × 64 × 67 mm, 0·7 lb (0·3 kg)
(autopilot controller) 101 × 57 × 133 mm, 1·3 lb (0·6 kg)
(mode coupler) 57 × 83 × 325 mm, 3·3 lb (1·4 kg)
(flight computer) 57 × 194 × 311 mm, 7·7 lb (3·5 kg)
(altitude controller) 91 × 89 × 319 mm, 4·6 lb (2·1 kg)
(mode selector) 146 × 67 × 165 mm, 2·9 lb (1·3 kg)
Displays
(adi (329B-8Y)) 129 × 129 × 216 mm, 8 lb (3·6 kg)
with (hsi (331A-9G)) 127 × 108 × 229 mm, 6·7 lb (3 kg)
or (adi (ADI-84)) 106 × 106 × 238 mm, 5 lb (2·3 kg)
with (hsi (HSI-84)) 106 × 106 × 224 mm, 5·3 lb (2·4 kg)
or (FD-112V) 106 × 187 × 191 mm, 7·4 lb (3·3 kg)

STATUS: in production.

AP-106A autopilot (ProLine)

This three-axis automatic flight control system can be used in a wide range of small twin-turboprop aircraft types, and integrated with the Collins FD-112 flight-director. Facilities are similar to those provided by the AP-105 autopilot and include: Category II glide-slope steering, all-angle course capture, coupled go-around, airspeed hold, linear vor coupling,

Flight panels of Mitsubishi Marquise with comprehensive avionics suite by Rockwell Collins, including FCS-80 flight-control system and FDS-84 flight-director equipment

RNav-type linear vor-deviation steering, and control wheel synchronisation.

Dimensions and weights:
Autopilot
(selector/computer) 83 × 83 × 229 mm, 2·3 lb (1·05 kg)
(pitch/roll control) 64 × 64 × 76 mm, 0·3 lb (0·14 kg)
(programmer) 89 × 89 × 356 mm, 2·5 lb (1·14 kg)
(primary servo) 102 × 102 × 178 mm, 3·7 lb (1·72 kg)
(trim servo) 102 × 102 × 178 mm, 3·7 lb (1·72 kg)
(altitude controller) 76 × 76 × 102 mm, 2 lb (0·91 kg)
(airspeed sensor) 64 × 64 × 64 mm, 0·5 lb (0·23 kg)
Flight-director
(adi & hsi (FD-112V/C)) 106 × 187 × 191 mm, 7·5 lb (3·4 kg)
(vertical gyro) 127 × 127 × 152 mm, 6·9 lb (3·13 kg)
(directional gyro) 133 × 197 × 133 mm, 5 lb (2·27 kg)
(flux detector) 124 × 124 × 73 mm, 1·2 lb (0·53 kg)
(slaving amplifier) 60 × 88 × 356 mm, 3·5 lb (1·59 kg)
Optional flight instruments
(course situation display) 83 × 83 × 191 mm, 3·2 lb (1·45 kg)
(turn & slip indicator) 83 × 83 × 165 mm, 3 lb (1·36 kg)
(adi-70) 106 × 106 × 210 mm, 4·6 lb (2·1 kg)
(hsi-70) 106 × 106 × 229 mm, 4·9 lb (2·2 kg)

STATUS: in production.

APS-80 autopilot (ProLine)

Designed for high-performance executive and corporate jets, the APS-80 autopilot provides three-axis flight-control from initial climb to final approach. It has been developed from experience accumulated with the FCS-110 all-weather landing/automatic flight-control system. A fully-integrated installation would use a Collins FDS-85 (or FDS-84) flight-director system. The APS-80 incorporates many integrated-circuit components and has automatic test functions. Smooth autopilot control during aircraft configuration changes is claimed through the use of torque-rate limiting servos. Each servo is designed to fail-passive on

detection of a 0·6 g pitch disturbance caused by servo failure, or if preset values of roll angle and roll rate are exceeded.

Category II approach performance is available, including radio altimeter-based flight-path scheduling to extend performance further if necessary. An internal comparator replaces external monitoring systems to maintain acceptable roll and pitch attitude, localiser and glideslope deviation, and heading and radio altimeter data during the approach.

Climb, en-route and descent profiles can be flown using autopilot modes. These incorporate gain-programming control laws, automatic turn co-ordination, lift compensation, command smoothing, all-angle adaptive vor/loc capture, turbulence mode, vertical synchronisation and half-bank angle limiting for high-altitude manoeuvring. The autopilot is compatible with a wide variety of navigation systems, including vor/dme, inertial navigation system and Omega/vlf.

Weight:
Autopilot
(autopilot amplifier) 5·3 lb (2·4 kg)
(autopilot computer) 4·9 lb (2·22 kg)
(autopilot panel) 1·5 lb (0·68 kg)
(2 primary servos) 10·8 lb (4·9 kg)
(2 servo mounts) 4·8 lb (2·18 kg)
Yaw-damper system
(primary servo) 5·4 lb (2·45 kg)
(servo mount) 2·4 lb (1·09 kg)
(computer) 2·8 lb (1·27 kg)
Dual flight-director system
(2 flight-guidance computers) 11·4 lb (5·18 kg)
(2 flight-guidance panels) 1·7 lb (0·77 kg)
(2 normal accelerometers) 1·4 lb (0·64 kg)
(2 vertical references) 14·8 lb (6·72 kg)
(2 ADI-85) 13·5 lb (6·12 kg)
(2 HSI-85) 14·5 lb (6·6 kg)
(remote hdg/crs selector) 1 lb (0·63 kg)
(rate-of-turn sensor) 2¹/₅ lb (1 kg)

STATUS: in production.

APS-841H autopilot (MicroLine)

This pitch/roll autopilot system is available for light helicopters such as the Aérospatiale AStar, Bell JetRanger variants, and Hughes 500D. It integrates with radio and navigation system inputs to provide hands-off operation during climb, en-route and approach phases of flight. Full-time attitude stabilisation can be selected alone or integrated with coupled-navigation modes. Automatic-trim capability is available and permits manual flight-path adjustments without system disengagements.

Uninstalled two-axis autopilot and guidance control system mass, excluding flight instrumentation, is in the order of 10·5 kg. The panel-mounted computer-controller measures 3¼ × 3¼ × 9 inches (83 × 83 × 229 mm). Both 115 V ac and 28 V dc power supplies are required.

System TSOs:
Computer-controller C9, C52a
Servos C9c
Aircraft altitude monitor C9c
Vertical gyro C9c, C4c, C52a
Altitude controller C9c, C52a

STATUS: in production.

FCS-80II flight-control system (ProLine)
This comprises the Collins APS-80 autopilot and FDS-85 flight-director system. Each system is described separately.

FCS-110 flight-control system
This analogue flight-control system has been installed in the majority of Lockheed L-1011 TriStar wide-body airliners produced. A comprehensive set of autopilot modes, including Category III automatic-landing capability is incorporated. It is installed in conjunction with the Collins FD-110 flight-director system.

STATUS: production has ceased.

Boeing's new 757 airliner has Rockwell Collins FCS-700 digital flight control system

FCS-700 flight-control system
This is a fully-digital triplex autopilot and flight-director system which features fail-operational automatic-landing capability.

Development has proceeded from the FCS-111X experimental flight-control system which Collins flew in Boeing development aircraft during the late-1970s. The fully-digital architecture permits integration of comprehensive self-test and failure-protection monitoring, improved system performance monitoring, built-in growth and flexibility, and improved equipment test and serviceability

TriStar 500 has Rockwell Collins FCS-240 flight-control system

FCS-240 flight-control system
This advanced digital flight-control system has replaced the FCS-110. It is a dual-dual integrated autopilot/flight-director system and is initially tailored to the Lockheed TriStar-500 wide-body transport. Compared to its analogue predecessor the number of computing units has diminished, from seven to two, and benefits are said to extend across mass savings, improved reliability, and lower cost of ownership. Category IIIA automatic-landing capability is included with a comprehensive set of conventional autopilot modes.

STATUS: in production. Service entry was represented by the delivery of Lockheed TriStar-500s to Air Canada in February 1981.

ACS-240 active control system
The Collins ACS-240 digital active flight-control system is designed for the Lockheed TriStar, and can be integrated with the FCS-110 or FCS-240 flight control systems. It is a lateral flight control system which uses aileron deflection scheduling to alleviate gust-loads and to improve ride-quality in turbulence. The system is in service with British Airways, Delta Airlines and LTU.

features. Large-scale integrated circuit technology is used to confer a 3:1 mass reduction compared with the experimental FCS-111X.

Autopilot modes include control-wheel steering, automatic cruise hold/select modes for heading, altitude, vertical speed and air-speed/Mach number, approach modes with back-course capability, and automatic approach and landing. The latter facility includes automatic flare control, roll-out guidance and coupled go-around. Computed flight-director steering is available in non-coupled flight, including take-off. Turn co-ordination and dutch-roll damping is provided. Dual or triplex control computer configurations can be used.

Steering commands from a navigation computer, which will permit completely automatic control of preplanned vertical and lateral flight profiles, is possible using a wide variety of navigation sensors. Large-scale integrated circuit technology is used in ARINC 429 bus interface drives and digital multiplexers, and in several areas of each flight-control computer, which is based on the Collins CAP-6 processor configuration. Computer operating speed is 256 KOPS, and high-order language programming is used throughout the system.

Maintainability improvements are claimed

from the maintenance central display panel which uses a micro-processor to perform control, display and data-management tasks. In-flight system failures are indicated by the maintenance control display panel and fault data is stored for up to ten flights.

STATUS: in development and production for the new generation, all-digital Boeing 767 and 757 airliners.

FD-108/109 flight director system
Essentially similar systems, the FD-108 uses 4-inch instruments and the FD-109 attitude director and horizontal situation indicator 5-inch instruments. Two categories of each are available, suffixed Y and Z, the latter incorporating air-data and other additional features. Either flight-director system can be integrated with many types of modern autopilots, and both versions are used on current production airliners.

The attitude director indicator has a V-bar display and a flat-tape background which provides linear pitch information through plus or minus 90 degrees. Outputs are provided for annunciations of all submodes and approach progress indications. A conventional compass-rose horizontal situation indicator is provided with an electronic digital display of distance-to-go and a mechanical course readout.

STATUS: in production.

FD-110 flight director system
The FD-110 system uses 5-inch attitude director and horizontal situation indicator instruments. Several configurations are available, making them suitable for individual airliner, company and autopilot combinations. Configurations can use cross-pointer or V-bar attitude director indications, various ARINC system interface compatibilities, or RNav and inertial navigation system integration. The equipment is used in many current airliner types.

STATUS: in production.

FD-112 flight director system (ProLine)
Developed specifically for high-performance piston and light turboprop twins to permit more efficient single-pilot operations. The system is compatible with the AP-106A autopilot, and comprises a single package of 4-inch attitude director and horizontal situation indicator displays. Cross-pointer or V-bar steering displays are available, and all systems are able to provide a pitch-hold function and have built-in self-test, initiated by a panel-mounted button.

FD-112V combined adi/hsi characteristics
Dimensions: 4⅕ × 7⅓ × 7½ inches (106 × 187 × 191 mm)
Weight: 8 lb (3·62 kg)
Power: 12 VA 26 V ac 400 Hz
0·85 A 27·5 V dc
plus lighting (28 V ac/dc)
TSOs C4c, C6c, C52a

Operating temperature ranges: –15 to +71° C
Max operating humidity: 95%
Max operating altitude: 25000 ft

STATUS: in production.

FDS-84 flight director system (ProLine)

Featuring separate 4-inch attitude director and horizontal situation indicator displays, the FDS-84 system is compatible with Collins FCS-80 and FCS-105 autopilot/flight-control systems. It is a comprehensive system suitable for high-performance business and executive aircraft, and commuter airliners.

The ADI uses a flat-tape attitude display background and has V-bar steering command symbology. A radio-altimeter readout is optional in this unit. The horizontal situation indicator uses a compass-rose presentation with an electronic readout which can show distance, time-to-go or ground-speed information. The equipment is fully-compatible with ARINC standard vor, distance measuring equipment, inertial navigation system and Omega/vlf sensors.

ADI-84 attitude-director indicator (4-inch)
Dimensions: 4$\frac{1}{5}$ × 4$\frac{1}{5}$ × 9$\frac{2}{5}$ inches (106 × 106 × 238 mm)
Weight: 5 lb (2·27 kg)
Power: 1 A 26 V ac 400 Hz
Temperature range: –20 to +70° C
Altitude range: –1000 to +50 000 ft
TSOs: C3b, C4c, C52a
HSI-84 horizontal-situation indicator (4-inch)
Dimensions: 4$\frac{1}{5}$ × 4$\frac{1}{5}$ × 8$\frac{4}{5}$ inches (106 × 106 × 224 mm)
Weight: 5·3 lb (2·4 kg)
Power: 165 mA 26 V ac 400 Hz
200 mA 28 V dc
Temperature range: –20 to +70° C
Altitude range: 0 to 35 000 ft
TSOs: C6c, C52a, C66a
REU-84 remote electronic unit
Dimensions: 3$\frac{1}{4}$ × 4$\frac{2}{3}$ × 5$\frac{1}{5}$ inches (83 × 118 × 131 mm)
Weight: 1·1 lb (0·5 kg)
Power: 110 mA 28 V dc
Temperature range: –40 to +70° C
Altitude range: 0 to 50 000 ft
TSOs: C6c, C52a, C66a

STATUS: in production.

FDS-85 flight director system (ProLine)

Based on 5-inch attitude director and hori-

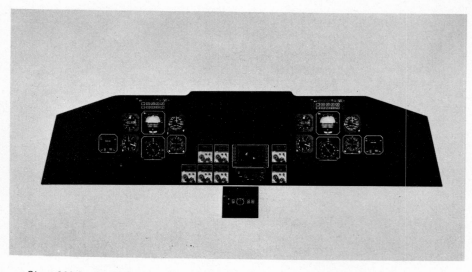

Short 360 feeder-liner with FDS-84 flight-director system and Rockwell Collins avionics

zontal situation indicator instruments, and compatible with the Collins APS-80 autopilot, this system is aimed at the high-performance business aircraft and commercial airliner markets. It is the most comprehensive mechanical flight-director system offered by the company. Flat-tape attitude indication, V-bar steering commands and electronic readout of distance, time-to-go, speed or elapsed time are standard. A separate heading/course control panel is used.

Attitude director indicator ADI-85
Dimensions: 5$\frac{1}{5}$ × 5$\frac{1}{5}$ × 8$\frac{1}{10}$ inches (131 × 131 × 206 mm)
Weight: 7 lb (3·2 kg)
Horizontal situation indicator HSI-85
Dimensions: 5$\frac{1}{5}$ × 4$\frac{2}{5}$ × 9$\frac{1}{10}$ inches (131 × 112 × 229 mm)
Weight: 7·2 lb (3·3 kg)
Heading/course control panel HCP-86
Dimensions: 5$\frac{3}{4}$ × 1$\frac{1}{2}$ × 6 inches (146 × 38 × 152 mm)
Weight: 1·4 lb (0·63 lb)
Overall system
Temperature range: –20 to +70° C
Altitude range: Up to 35 000 ft
Cooling: convection
Power: 44 VA 26 V ac 400 Hz
260 mA 28 V dc
10 W 5 V ac/dc for lighting

STATUS: in production.

FIS-70 flight instrumentation system (ProLine)

Compatible with the Collins APS-80 or AP-106A autopilots, the FIS-70 comprises 4-inch attitude director and horizontal situation indicator instruments and is suitable for high-performance turboprop and business jet aircraft. It is essentially a low-cost version of the FDS-84 system, and excludes the digital distance/course readouts, and radio-altimeter displays options. All other FDS-84 features are incorporated.

Attitude-director indicator ADI-70
Dimensions: 4$\frac{1}{5}$ × 4$\frac{1}{5}$ × 8$\frac{1}{4}$ inches (106 × 106 × 210 mm)
Weight: 4·6 lb (2·1 kg)
Temperature range: –15 to +71° C
Altitude range: –1000 to +50 000 ft
TSOs: C3b, C4c, C52a
Horizontal-situation indicator HSI-70
Dimensions: 4$\frac{1}{5}$ × 4$\frac{1}{5}$ × 9 inches (106 × 106 × 229 mm)
Weight: 4·9 lb (2·2 kg)
Temperature range: –20 to +70° C
Altitude range: 0 to 35 000 ft
TSOs: C6c, C52a
Total power:
1·04 A 26 V ac 400 Hz
0·2 A 28 V dc
0·34 A 28 V for lighting.

STATUS: in production.

Lear Siegler

Lear Siegler Inc, Astronics Division, 3171 South Bundy Drive, Santa Monica, California 90406

Fly-by-wire flight-control system for F-16

This system, originally developed for the US Air Force LWF lightweight fighter technology/demonstrator programme launched in the early 1970s, uses a quadruplex-redundant flight-control computer configuration, and a unique side-stick control column. There is corresponding redundancy in the arrangement of sensors, also manufactured by Lear Siegler. No mechanical back-up is provided.

Flight-control computers Each analogue computer receives information from air-data and aircraft-motion sensors and processes the information to provide electrical signals to the servo-actuators that move the flying controls. The system provides artificial stability so that the aircraft can be flown under conditions which would conventionally be regarded as aerodynamically unacceptable in view of the reduction of natural stability. By accepting much reduced stability, great airframe weight-savings can be achieved. The servo-actuators

World's first production fly-by-wire flight-control system is Lear Siegler equipment on F-16

operate with full authority on the ailerons, elevators, rudder, leading-edge and trailing-edge flaps.

Side-stick control column The F-16 pilot has a side-stick on the right-hand side of the cockpit, instead of the conventional centre-stick control. It is a fixed unit, but transducers within the stick determine fore/aft or sideways forces exerted by the pilot, and produce independent quadruplex pitch and roll demand signals which are proportional to the applied force in each direction. These are fed directly to the flight-control computers. The non-moving stick, again a unique feature, is claimed to allow precise control even during high *g* manoeuvring flight.

STATUS: in service.

Target-drone autopilot for BQM-34 RPV

The Lear Siegler VDA versatile drone autopilot has been most extensively used in the Teledyne-Ryan BQM-34 Firebee RPV, one of the very few remote-piloted vehicles to have seen operational service. Examples have also been installed in QT-33 drone versions of the Lockheed T-33 trainer, and in a Nasa-developed ⅜-scale F-15 model which was used to assess some aspects of full-size aircraft behaviour, notably spinning.

The VDA is a modular, single-box, autopilot which contains all sensors, processing and output control circuits. Sensors included in the system are a vertical gyro, altitude transducer, airspeed transducer or Mach computer, and a yaw-rate gyro. There is a basic electronic module to which can be added customised modules to suit different vehicles. The extra modules permit subsonic, supersonic, low-altitude, advanced manoeuvrability, proportional command, evasive flying and pre-programmed mission manoeuvres. It is claimed to be suitable or adaptable for a wide-range of vehicles and missions.

The self-contained air-data equipment uses solid-state electronics and is supplemented by digital synchronisers and manoeuvre programming devices. The VDA can hold *g*-levels to within ±0·25 *g* up to airframe limits, and within 3 seconds of command, hold high-*g* while climbing, diving or holding altitude, maintain low-level flight (50 feet above sea level) and accommodate pre-programmed manoeuvres, plus proportional or augmented-proportional control.

Dimensions: 6½ × 8 × 18 inches (165 × 203 × 457 mm)
Weight: 28 lb (13 kg)
Power: 150 VA 115 V ac
30 W 28 V dc

STATUS: in service.

Update RPV avionics system

This is a proposed autopilot development for future RPV applications. It would initially accommodate mission programmes for reconnaissance and sensor-seeding tasks, but has considerable growth capability to undertake delivery of Maverick/Hobo weapons, laser-target designation, night-sensor reconnaissance, electronic countermeasures/Wild Weasel, and electronic reconnaissance missions. Automatic take-off and landing capability can be incorporated too.

The digital system is initially designed to be compatible with loran, Doppler radar, radar altimeter and tercom sensors and processors. It can also accommodate inertial navigation, Tacan or Navsat sensor inputs. Basic autopilot sensors are an air-data computer, magnetic compass, and attitude and heading reference system. Communication for remote operation is by a microwave link. Automatic pre-flight check-out should take less than one hour, interactive software mission-planning facilities are proposed, and a supporting modular fault-isolation facility is incorporated.

STATUS: in development.

Sperry

Sperry Flight Systems, PO Box 21111, Phoenix, Arizona 85036

Automatic flight-control system for AS-350/355 Écureuil helicopter

Tailored to the Aérospatiale AS-350/355 helicopter family, this is a combined autopilot/flight-director system comprising independent 'building blocks' so that major functions can be installed either separately or together. Single-pilot ifr operation is possible with this equipment.

The autopilot (helipilot) is a three-axis stabilisation system with sufficient authority to control the helicopter throughout the entire flight regime. Autotrim is also available to enhance performance and to reduce pilot workload. The system comprises the helipilot computer, a trim computer, three series actuators, two trim actuators, three actuator-position indicators, a controller, C-14 gyro-compass and GH-14 gyro-horizon/attitude-director indicator.

Autopilot operating modes are divided into three categories. The stability-augmentation system modes provide transient motion-rate damping, and pilot cyclic inputs determine aircraft attitude-rate responses. Attitude hold modes permit gyro-stabilisation to any pre-selected attitude, and pilot cyclic inputs are used to determine aircraft attitude response. Flight-director coupling modes permit hands-off flying and flight-director display on the attitude-director indicator.

The flight-director system performs the computation necessary to bracket and hold the flightpath selected by the pilot. Facilities effectively anticipate manoeuvre requirements and cause automatic flight-path transfer between modes. All command information is presented on the attitude-director indicator. The system comprises a flight-director computer, airspeed sensor unit, altitude computer, navigation receiver, flux-valve and compensation for a C-14 gyro-compass, mode selector, and RD-550 horizontal-situation indicator. A radio-altimeter can also be included as a flight-director sensor. The flight-director system shares with the autopilot the CH-14 gyro-horizon/attitude-director indicator to display information to the pilot.

Flight-director operating modes permit selection of heading, vertical speed, barometric

Flight panel of Aérospatiale AS-350D AStar helicopter, with Sperry flight-director/flight-control system

altitude, airspeed and go-around procedures. External references which can be coupled to the system include vor, ils, mls, Omega/vlf, area-navigation systems, loran or Tacan.

STATUS: in service.

SA-365N Digital flight-control system for Dauphin helicopter

This is an advanced digital system which combines autopilot and flight-director functions and caters for dual gyro-compass sensors and displays, with both attitude-director indicators and horizontal-situation indicators for each crew member.

Instrument-panel units consist of flight-computer controllers, mode selectors and altitude pre-select/air-data displays. The flight computer modes cover all conventional stability-augmentation, attitude-hold and flight-director coupling capabilities, including hover

augmentation and radar-altitude hold. Flight-director modes cater for on-board sensor coupling to heading, vertical speed, barometric altitude, airspeed, pre-selected altitude, go-around and deceleration scheduling. External sensor data can be accommodated from vor, ils, mls, Omega/vlf, area-navigation systems, loran and Tacan. The altitude pre-select panel permits the pilot to enter the desired altitude, and can be used as a momentary display of air-data information.

The system is a complete four-axis stability-augmentation/autopilot, with dual flight computers, dual flight-director mode selectors, three-cue attitude-director indicators, and incorporating automatic test and system diagnostics and monitoring features. It is microprocessor-based and fail-passive. After first failure there is no aircraft reaction, and in the case of a second failure a 3-second delay can be accommodated.

Equipment weight:
Flight control system:
(digital computer) 7 lb (3·2 kg)
(2 vertical gyros) 18·8 lb (8·6 kg)
(2 gyro/synchroniser assemblies) 9·4 lb (4·2 kg)
(2 flux valves) 2·6 lb (1·2 kg)
(dual remote compensator) 1 lb (0·5 kg)
(flight-computer controller) 1·6 lb (0·7 kg)
(actuator-position indicator) 1·4 lb (0·6 kg)
(pitch actuator) 4·7 lb (2·1 kg)
(roll actuator) 5·6 lb (2·5 kg)
(yaw actuator) 4 lb (1·8 kg)
(trim actuator) 4 lb (1·8 kg)
Flight-director coupler:
(altitude/airspeed sensor) 1·4 lb (0·6 kg)
(2 mode selectors) 2·4 lb (1·1 kg)
(radio altimeter unit) 5 lb (2·3 kg)
(2 radio altimeter antennas) 1·6 lb (0·8 kg)
Instruments:
(2 adis (AD-650 units)) 14 lb (6·4 kg)
(2 hsis (RD-650A units)) 16 lb (7·2 kg)
(instrument remote controller) 0·8 lb (0·4 kg)
(2 radio-altimeter indicators) 1·5 lb (0·7 kg)
(2 rmis (RH-404 units)) 6·4 lb (2·9 kg)

STATUS: in service.

Automatic flight-control system for Bell 412 helicopter

The Bell 412 system has been designed to minimise pilot workload, improve mission reliability and contribute to flight safety, without incurring high costs. An installation can comprise one or two autopilot (helipilot) systems (combined in series for redundancy) and may be used with or without a flight-director system.

A single helipilot system consists of one directional and one vertical gyro, controller panel, computer unit and three series actuators, one each for pitch, roll and yaw control. A second helipilot system, if fitted, comprises a second vertical gyro unit, extra computer unit, and pitch and roll actuators installed in series with the first unit. A trim computer and two trim actuators, for pitch and roll adjustments, can be added also.

If a flight-director facility is added it requires the addition of a mode controller, navigation receiver inputs, plus attitude-director and horizontal-situation indicators. Flight-director functions include heading, vertical speed, barometric altitude and airspeed select, go-around procedure selection, and vor, ils, mls, Omega/vlf, area-navigation system, and loran or Tacan coupling.

STATUS: in service.

Digital flight-control system for Sikorsky S-76 helicopter

This digital system uses components identical to those described above for the Aérospatiale SA-365N Dauphin helicopter. The configuration and facilities are identical, and much of the equipment is common. The digital computers used are however programmed to correspond with the flying qualities of the particular type of helicopter.

Equipment weights:
Flight Control system
(2 digital computers) 14 lb (6·35 kg)
(2 vertical gyros) 18·8 lb (8·53 kg)
(2 gyros and synchronisers) 9·4 lb (4·26 kg)
(2 flux valves) 2⅗ lb (1·18 kg)
(dual remote compensator) 1 lb (0·45 kg)
(flight computer controller) 1·6 lb (0·73 kg)
(actuator position indicator) 1·4 lb (0·64 kg)
(rate monitor) 1·5 lb (0·68 kg)
(forward pitch actuator) 2·5 lb (1·13 kg)
(aft pitch actuator) 2·6 lb (1·18 kg)
(2 roll actuators) 4·8 lb (2·18 kg)
(dual yaw actuator) 4 lb (1·81 kg)
(4 trim actuators) 16 lb (7·26 kg)

Sperry digital flight-control system and associated flight-director equipment installed in Aérospatiale SA-365C Twin Dauphin helicopter

Sperry flight-directors and other avionics in a Bell 212 helicopter

Flight-director coupler
(altitude/airspeed sensor) 1·4 lb (0·64 kg)
(2 mode selectors) 2·4 lb (1·09 kg)
(radio altimeter unit) 5 lb (2·27 kg)
(radio altimeter antennas) 1·6 lb (0·73 kg)
(air-data pre-select panel) 1·4 lb (0·64 kg)
Instruments
(2 adis (AD-650 units)) 14 lb (6·35 kg)
(2 hsis (RD-650A)) 16 lb (7·26 kg)
(instrument remote controller) 0·8 lb (0·36 kg)
(radio altimeter indicator) 3 lb (1·36 kg)

STATUS: in service.

Sikorsky S-76 executive helicopter has Sperry flight-control system

SP-150 automatic flight-control system

This automatic flight-control system embodies an automatic-landing function, and is used extensively in Boeing 727 airliners. It superseded the earlier SP-50 automatic-landing capable system, and employs more solid-state components for rate-sensing and integration. The system has a dual-channel configuration which meets FAA Category IIIA requirements (50-foot (15·24 metres) decision height, 700 foot (213·36 metres) runway visual range).

A wide-range of stabilisation, attitude hold and external-sensor steering modes are available in the basic autopilot. Additionally there is provision for area-navigation steering and for a radio-altimeter input so that control law gains can be varied during the approach. Fail-operational performance is assured by the dual-computer configuration and a dual-channel yaw-damper is also part of the overall flight control system.

Computational equipment used in the system includes two pitch computers (each ½ ATR size), two yaw-damper couplers (each ⅜ ATR size) and a single roll-control computer (¾ ATR size).

STATUS: in service.

Many examples of Boeing's 727 airliner have Sperry SP-150 AFCS

SPZ-500 automatic flight-control system

This is an integrated autopilot/flight-director system, suitable for corporate aircraft, that provides all of the pilot interface controls, air-data computation and control servos necessary to automatically fly a selected flight profile. It integrates with companion flight-director and air-data systems.

The autopilot is a full-time system which provides continuous control through all phases of climb, cruise and descent, and with a full complement of lateral and vertical modes. These may be flown automatically by engaging the SPZ-500, or manually by following computed steering commands presented on the flight-director instruments. The latter also provide the pilot with a means to monitor autopilot performance.

To provide air-data information over a wide range of flight profiles the SPZ-500 contains an air-data system.

The SPZ-500 is a full three-axis autopilot with a yaw damper. It has both acceleration and rate-limiting circuits to provide smooth autopilot performance without compromising positive control action. Turn entry and exit is smooth, and by programming the roll-rate limit as a function of selected mode, rates are matched to the required manoeuvres.

Control of the autopilot for basic stabilisation and attitude-command is provided through the autopilot controller. Engaging the system with no flight-director mode selected causes the aircraft to maintain the existing pitch attitude, roll to wings level and then hold the existing heading. With a navigation or vertical-path mode selected, engaging the autopilot auto-

Flight-director displays (HSIs and ADIs), computers and controllers shown here are part of the Sperry SPZ-600 AFCS

matically couples the selected mode. When the autopilot is engaged the yaw-damper is automatically engaged to provide yaw stabilisation through control of the rudder. When the autopilot is not engaged, the yaw-damper may be used separately to assist the pilot during manual flight.

The autopilot controller includes the turn knob and pitch wheel, allowing the pilot to insert pitch and roll commands manually. The amount of bank or pitch change is proportional to the command selected. The soft-ride engage button reduces autopilot gains for smoother operation in turbulence, and the automatic elevator trim annunciators show any out-of-trim condition. The autopilot may be pre-flight checked with the test button.

Touch control steering allows the pilot to take control of the aircraft momentarily without disengaging the system. The pilot can push the button on the control wheel and manually change the aircraft flight-path. While the touch control steering button is pressed, the autopilot synchronises on the existing aircraft attitude. On releasing the button the system holds the new attitude and resumes the coupled-flight mode.

Several flight-director instrumentation sets, either 4- or 5-inch units, can be integrated with the flight-control system. If the Sperry ADZ-241/242 air-data system is installed this includes Sperry air-data instrumentation.

Dimensions and weights:
Autopilot:
(controller) 67 × 146 × 114 mm, 1·5 lb (0·68 kg)
(computer) 194 × 71 × 321 mm, 6 lb (2·72 kg)
(3 servos & mounts) 100 × 129 × 224 mm, 16·3 lb (7·35 kg)
(normal accelerometer) 51 × 25 × 61 mm, 0·3 lb (0·14 kg)
(trim servo) 56 × 84 × 175 mm, 5·5 lb (2·5 kg)
Air-data system:
(computer) 193 × 124 × 361 mm, 11·5 lb (5·22 kg)
(altimeter) 83 × 83 × 159 mm, 2·5 lb (1·13 kg)
(VNav computer/controller) 38 × 83 × 272 mm, 2 lb (0·91 kg)
(vertical speed indicator) 83 × 83 × 140 mm, 2 lb (0·91 kg)
(Mach/airspeed indicator) 83 × 83 × 186 mm, 2·8 lb (1·27 kg)

Flight-director system
(adi (AD-650B unit)) 129 × 129 × 223 mm, 7 lb (3·18 kg)
(hsi (RD-650B unit)) 103 × 129 × 208 mm, 7·3 lb (3·31 kg)
(remote controller) 38 × 146 × 66 mm, 0·8 lb (0·36 kg)
(computer) 194 × 71 × 321 mm, 5 lb (2·27 kg)
(mode selector) 48 × 146 × 114 mm, 1·3 lb (0·59 kg)
(rate gyro) 46 × 52 × 95 mm, 1 lb (0·45 kg)

STATUS: in service.

SPZ-600 automatic flight-control system

The SPZ-600 AFCS is designed specifically for the new generation of long-range, high-performance, business-jet aircraft and is claimed to be business aviation's first and only completely automatic fail-passive/fail-operational flight-control system.

It is a complete dual-channel system from the sensors to the servos. All channels (roll, pitch and optional yaw) are fully-operational at all times and the performance of each is continuously monitored and compared. If a failure that would result in a hardover manoeuvre occurs in any channel, it is immediately shutdown, resulting in single-channel operation in that axis only. Performance status is continuously displayed on the autopilot status panel, as well as the autopilot master warning annunciator. Sperry says that the unique monitoring system and duplex servo design, have been well tried in air-transport aircraft. Each of the duplex servos incorporates two independent servo motors that operate from signals applied by their own autopilot channels. The common tie to a single control surface is accomplished through a mechanical differential gear mechanism.

In the event of a system fault, the master warning annunciator will flash amber, and may be cancelled by pressing the annunciator. The status panel then indicates that the system has automatically disconnected the malfunctioning channel, and is single-channel in that axis only. The status panel also provides means to manually select single-channel operation in any axis, and has provisions for testing the dual autopilot channels and monitoring circuits prior to flight.

Control of the autopilot for basic stabilisation and attitude command is provided through the autopilot controller. Engaging the system with no flight-director mode selected causes the aircraft to maintain the existing pitch attitude, roll to wings-level, and then hold the existing heading. With a navigation or vertical path mode selected, engaging the autopilot automatically couples the selected mode. When the autopilot is engaged, the yaw-damper is automatically in use. When the autopilot is not engaged the yaw-damper may be selected separately to assist the pilot during manual flight. The autopilot controller also has a turn knob and pitch wheel, which allow the pilot to insert pitch and roll commands manually. The amount of bank or pitch change is proportional to the command selected. A soft-ride engage button reduces autopilot gains for smoother operation in turbulence, and a couple button selects which flight-director is driving the autopilot.

Touch control steering allows the pilot to take control of the aircraft momentarily without disengaging the system. He simply pushes the touch control steering button on the control wheel and manually changes the flight path as desired. While the button is pressed, the autopilot synchronises to the existing aircraft attitude, and on releasing the button the system holds the new attitude, or resumes the coupled-flight mode.

The flight-director system uses standard 5-inch instruments and there is a choice of displays, with either split-cue or V-bar directors, and vertical-scale differences. The ADZ-242 air-data system includes a full set of appropriate instruments, including altitude-alert controller and true airspeed/temperature indicator.

Dimensions and weight:
Autopilot:
(controller) 69 × 146 × 114 mm, 1·5 lb (0·63 kg)
(status/switching panel) 48 × 146 × 114 mm, 1·6 lb (0·73 kg)
(2 duplex servos & brackets) 106 × 205 × 298 mm, 20·8 lb (9·44 kg)
(computer) 194 × 71 × 321 mm, 5 lb (2·27 kg)
(yaw actuator) 54 dia × 232 mm, 2 lb (0·91 kg)
(normal accelerometer) 31 × 25 × 61 mm, 0·3 lb (0·14 kg)

Flight-director system
(adi (AD-650B unit)) 129 × 129 × 224 mm, 7 lb (3·18 kg)
(hsi (RD-650B unit)) 103 × 129 × 203 mm, 7·3 lb (3·31 kg)
(remote controller) 38 × 146 × 66 mm, 0·8 lb (0·36 kg)
(computer) 194 × 71 × 321 mm, 5 lb (2·27 kg)
(mode select) 47 × 146 × 114 mm, 1·3 lb (0·59 kg)
(rate gyro) 46 × 52 × 65 mm, 1 lb (0·45 kg)

Air-data system
(computer) 193 × 125 × 361 mm, 11·5 lb (5·22 kg)
(altimeter) 83 × 83 × 159 mm, 2·5 lb (1·13 kg)
(vertical speed indicator) 83 × 83 × 146 mm, 2 lb (0·91 kg)
(Mach/airspeed indicator) 83 × 83 × 186 mm, 2·8 lb (1·27 kg)
(VNav computer/controller) 38 × 87 × 500 mm, 2 lb (0·91 kg)

Gyro references
(vertical gyro) 157 × 165 × 238 mm, 7·3 lb (3·31 kg)
(directional gyro) 191 × 154 × 229 mm, 4·7 lb (2·13 kg)
(fluxvalve and compensator) 121 dia × 73 mm, 1·4 lb (0·64 kg)

STATUS: in service.

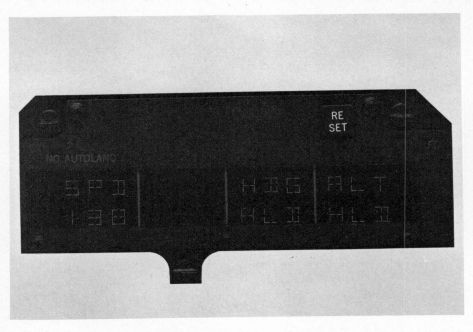

Annunciator panel for Sperry SFS-980 digital flight-guidance system showing speed selected, together with heading and height holds engaged

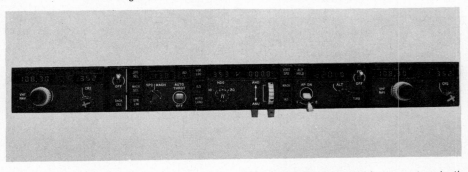

Glareshield-mounted controller for the Sperry SFS-980 digital flight-guidance system in the McDonnell Douglas DC-9 Super 80 airliner

SFS-980 digital flight-guidance system

Used exclusively on the McDonnell Douglas DC-9 Super 80 transport, which has been described as the first digital airliner, the dfgs is a totally-new automatic flight-guidance system based on relatively few line-replaceable units, especially in respect of the range of functions performed. In addition to a comprehensive selection of conventional autopilot/flight-director operating modes the system is FAA-cleared for Category IIIA automatic-landing operations (50-foot (15·24 metres) decision height, 700-foot (213·36 metres) runway visual range) and has a full-time autothrottle. The large-scale use of digital computing has led to the installation of a comprehensive built-in self-monitoring and maintenance diagnosis capability. Major line-replaceable units and functions are described in more detail below.

Digital flight-guidance computer (dfgc) There are two identical 1 ATR Long (29 lb, 13·15 kg) dfgc units, each autonomous with respect to the other, and capable by itself of handling all system functions, including fail-passive automatic landing. Within each unit analogue/digital conversions take place, and the complete system occupies 31 board places, against a maximum capacity of 51 boards. Each digital processor has 30 K words of read-only memory and 4 K words of random-access memory. In addition to all autopilot/flight-director processing, each unit also generates thrust-rating indication signals, plus maintenance-data storage and status test-panel data. The aircraft can have (optionally) a head-up display for take-off

and go-around data presentation, and guidance signals for this unit are also generated in each dfgc.

Flight-guidance control panel (fgcp) This unit fits in the centre glareshield and contains mode selection and control functions for full-time autothrottle, both flight-directors, the autopilot and altitude alerting. Pilots may select which dfgc controls all functions. Autothrottle speed/ Mach, selected heading, vertical speed and selected altitude readouts are shown on seven-segment incandescent lamps which may be dimmed by a control knob on the bottom of the panel. The spd/Mach knob is a three-position control, while the heading selection is one of two concentric knobs. The outer knob provides selection of maximum bank angle for all autopilot/flight-director lateral modes except loc. The inner knob is a four-position device providing heading and autopilot/flight director heading mode selection. The three-position alt knob provides altitude selection and autopilot/ flight director altitude pre-select mode arming.

Vhf/nav control panels Two panels on either side of the fgcp allow selection of vortac station frequencies and courses. Displays are incandescent-lamp readouts.

Flight mode annunciator (fma) One fma on each pilot's panel provides instrument failure warning for ils, attitude, heading, automatic-landing, auto-trim or instrument-monitor functions. They indicate which autopilot or flight-director system is engaged, and warn of autothrottle or autopilot disconnects. The units also annunciate which autothrottle, flight-director and/or autopilot modes are armed, and in which mode the system is currently controlling.

Autopilot duplex servo drive Three duplex servos drive the ailerons, elevators and rudder. The duplex servo functions only during ils, land, or go-around mode. A linear actuator provides normal yaw damping functions in other flight regimes.

Each servo has two separate dc electric motors whose outputs are summed in a differential gear-train with a single output. Each servo sends position and rate-feedback signals to the dfgc, where servo 'models' monitor actual servo operation. The duplex servo design provides fail-passive protection against hardover manoeuvres. A fault in one servo channel is cancelled by the mechanical velocity summing conducted in the dual servo differential.

McDonnell Douglas DC-9-80 has Sperry's SFS-980 flight-guidance system

Autothrottle/speed control system Conducted in the dfgc, this function provides fast/slow attitude-director indicator commands throughout the entire flight regime. During take-off and go-around it provides pitch guidance for the flight-director, autopilot and optional head-up display. The speed control system provides a speed margin above stall (alpha speed) for all autothrottle modes, in addition to the autopilot stall-protection feature.

The autothrottle system automatically prevents excursions beyond maximum operating airspeed and Mach values, slat and flap placard speeds, and engine epr limits. It also keeps the aircraft at or above alpha speed for prevailing flap/slat position and angle-of-attack, using limit data stored in the dfgc solid-state memory.

Automatic reserve thrust enhances safety, operational economy and noise reduction. In the event of an engine failure, as indicated by a difference of more than 30 per cent between engine N_1 (fan) speeds, simultaneously with slats-extended indication, the good engine thrust is automatically boosted by about 4 per cent. Automatic reserve thrust also allows use of less-than-maximum certificated thrust for take-offs and go-arounds without corresponding reduction in take-off gross weight, with benefit to fuel and maintenance costs.

STATUS: in service.

Avionics system for QF-100 drone

Sperry Flight Systems is developing a set of drone avionics for installation in US Air Force/North American F-100 Super Sabres and so transform them into full-scale aerial targets under the designation QF-100. Initial operational clearance is set for June 1983.

The targets will be used for air-to-air and ground-to-air missile evaluation and combat-crew training and to meet the demands of this range of operations each will have a digital flight control system which interfaces with much existing ground-control and test equipment.

The digital processor in the aircraft is based on Sperry SDP 175 equipment with air-data sensors, analogue/digital and digital/analogue converters, power supplies and interface electronics. All this equipment is installed in a single line-replaceable unit, compared to four units in a broadly equivalent analogue system. The digital signalling techniques have facilitated the incorporation of a large amount of system testing capability, aimed at meeting pre-mission test-stand and airborne system test-set requirements. These reduce the time needed to complete tests and improve the ability to find faults quickly.

STATUS: in development. Current work is involved with a nine-aircraft full-scale equipment development programme, and production is in prospect for 72 aircraft sets.

Altitude	Effective radius against radar type (nm)	
	L-band (8–10 GHz)	F-band (3–4 GHz)
300 ft	26	34
1000 ft	41	50
2000 ft	47	63

STATUS: in production.

ELT/263 receiver system

Suitable for light maritime patrol aircraft such as the Beechcraft King Air and Gates Learjet 35. It is virtually the same receiver system as the Colibri (ELT/161 plus ELT/261) but modified for fixed-wing operations. The system is capable of the following instantaneous surveillance coverage:

Altitude	Effective radius against radar type (nm)	
	L-band (8–10 GHz)	F-band (3–4 GHz)
2000 ft	42	65
6000 ft	53	100

STATUS: in production.

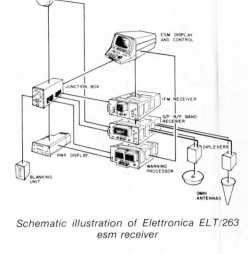

Schematic illustration of Elettronica ELT/263 esm receiver

ELT/457-460 supersonic noise jammer pod

A set of four related noise jammer pods is available for use on any high-performance strike/fighter or light-attack aircraft, mounted on an underwing pylon. All units are self-contained with a ram-air turbine-driven generator in the nose section and each is dedicated to a particular waveband. A heat-exchanger is situated behind the turbine, and there are fore-and-aft antennas in all pods. The ELT/459 and ELT/460 versions have additional antennas beneath the body of the pod. Processing within the system allocates threat-jamming power on proportional bases against any type of pulsed-radar threat, and includes built-in test equipment and control of blanking with other aircraft systems. All units are claimed to be highly resistant to electronic counter-countermeasures, including any type of frequency-agility, and to incorporate a large threat library. Maximum operating speeds are Mach 1·1 at sea level, and Mach 1·5 at 40 000 feet.

Pod length: 3120 mm
Diameter: 340 mm
Weight: 145 kg
Interfacing: Stanag 3726 and MIL-A-8591D

STATUS: in development.

ELT/555 supersonic deceptive jammer pod

Similar configuration to ELT/457 system, with the same ram-air turbine on the nose of the pod. Internal equipment includes fore-and-aft facing antennas which receive pulsed and continuous wave signals and transmit pulsed responses, plus separate fore-and-aft facing continuous wave transmitter antennas on the undersurface. The system is sensitive to H to J band (6 to 20 GHz) threats, and is designed to operate in dense electro-magnetic environments. It has multiple-target engagement capability and features built-in test equipment. Operating limitations and interface specifications are the same as those of the ELT/457. The cockpit display can be tailored to customer's requirements.

Pod length: 3000 mm
Diameter: 340 mm
Weight: 140 kg

STATUS: in development.

EL-70 deception repeater jammer

Developed jointly by Elettronica and AEG-Telefunken. Between 200 and 400 systems have been produced for West German- and Italian-operated NATO/Lockheed F-104 fighter-bombers. The system operates in the 2·6 to 5·1 GHz and 8 to 10 GHz bands and combines radar warning with travelling-wave tube countermeasure system.

STATUS: in service, production completed.

Elettronica ELT/156 radar-warning receiver

Elettronica ELT/460 supersonic ecm pod

Aermacchi MB-339K light strike aircraft carrying underwing an ELT/555 ecm pod

EL/73 deception jammer

Under development between Elettronica, AEG-Telefunken and Marconi Space and Defence Systems for the Panavia Tornado. The system is a development of the EL/70 and embodies technology used in the ELT/555 and similar systems in the Elettronica range. It is likely that the EL/73 system will become standard active ew equipment for West German- and Italian-operated Tornados.

STATUS: in development.

Mechanical check-out of Elettronica ELT/460 ecm pod ram-air turbine

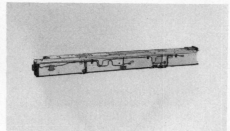
Electronic module for Elettronica ELT/555 ecm pod

Selenia

Selenia, Special Equipment and Systems Division, via dei Castelli Romani 2, 00040 Pomezia, Rome

IHS-6 esm/ecm system

Designed for tactical electronic surveillance measures missions such as stand-off jamming, air-strike support or fleet protection, the system has two major components; RQH-5/2 esm system and TQN-2 jamming system. The esm component can detect, analyse and identify threats in the 1 to 18 GHz band and has a wide-open receiver with instantaneous frequency measuring capability. It provides monopulse direction finding and can accommodate a library of 2000 threats. At least 50 targets can be tracked simultaneously. Data is displayed on a cathode ray tube with alpha-numeric or graphic presentation.

The TQN-2 jammer can operate on up to four bands, using steerable I to J band (8 to 20 GHz) antennas and fixed installations for lower frequencies. Either spot, barrage or hybrid jamming is possible, with automatic release of chaff countermeasures.

Systems using two antennas and operating in I to J band have been shown for the Agusta 109A helicopter, and a full four-band system has been installed in Egyptian-operated Westland Commando medium assault helicopters.

STATUS: in production.

SL/ALQ-34 ecm pod

Available as a self-defence aid for high-performance aircraft, although no customer has been clearly identified with the system. The SL/ALQ-34 provides a radar-warning facility and can jam anti-aircraft artillery or surface-to-air missile radars. Current versions operate across a 6 GHz bandwidth in the H to J band (6 to 20 GHz) but lower frequency systems (C, D, E and F bands), effective against air-defence search radars, are in prospect.

The receiver processor conducts threat assessment and ranking, using a stored library of threat data, and performs jamming power-management tasks. Travelling-wave tube transmitters (probably two) are used, and cooled by a closed-loop liquid cooling system. Supersonic, high-altitude clearance has been obtained.

STATUS: in production.

JAPAN

Tokyo Keiki

Tokyo Keiki, 16, 2-Chrome, Minami-Kamata, Ohta-ku, Tokyo

This company leads Japanese airborne ew technical effort and has conducted most of the systems production to date. Other contributors to the Japanese ew field are Mitsubishi, Nikon, Toshiba and Fujitsu. The following brief descriptions outline most known facts about Japanese ew equipment, and to which all the above companies contributed.

APR-1 radar warning system

Used on Lockheed F-104J Starfighters and developed from the US AN/APR-25/26 system. Production has long since been completed, and the equipment about to be retired from service.

APR-2 radar warning system

Used on McDonnell Douglas F-4EJ Phantoms and developed from the US AN/APR-36/37 system. Production is complete and equipment is likely to have been withdrawn in favour of APR-4.

APR-3 radar warning system

Used on Mitsubishi F1 attack aircraft, but origins and capability unknown. May still be in limited production.

APR-4 radar warning system

Used in McDonnell Douglas F-4EJ Phantoms and F-15EJ Eagles. Its specification calls for the ability to process multiple inputs simultaneously in a dense electro-magnetic environment. The system incorporates a digital processor with reprogrammable software which permits reconfiguration to meet developing threats. A tactical situation cathode ray tube display presents multiple-threat data in alpha-numeric and graphic form. Interfaces with other on-board electronic weapons systems can be accommodated.

STATUS: in production.

ALQ-2 esm system

Used in Lockheed T-33 operational trainers believed to be withdrawn from service now.

ALQ-3 esm system

Installed initially on Curtiss C-46 Commando and later on esm variants of the NAMC YS-11 transport aircraft. It is able to receive and jam fixed and mobile ground-radars. Production has probably ceased.

ALQ-4 esm system

Used in Lockheed F-104EJ Starfighters, and probably withdrawn from service.

ALQ-5 esm system

Used on esm variants of the Kawasaki C-1 medium transport and in development since 1978, the system receives and jams surface-to-air missile radars and is complementary to the ALQ-3 system.

ALQ-6 jamming system

In development to provide an airborne-radar jamming capability for F-15EJ, F-4EJ and Mitsubishi F1 aircraft and to replace the APR-4 radar-warning system on these types.

ALQ-8 jamming system

In development for F-4EJ, and may be a variant of the ALQ-6 set. It is understood to provide countermeasures in the 1 to 4, 4 to 8 and 7·5 to 18 GHz bands.

SWEDEN

Philips

Philips Elektronikindustrier AB, Defence Electronics, S-175 88 Jarafalla

Manufacturer of the **BOZ-100** high-performance aircraft chaff and infra-red flare dispenser, and **BOZ-3** training chaff dispenser. The BOZ-100 system has been supplied to several export customers.

STATUS: in production.

Philips BOZ-100 chaff/flare dispenser for high-performance aircraft

Philips BOZ-3 training chaff dispenser

Satt

Satt Elektronik AB, PO Box 32006, S-126 11 Stockholm

Radar warning system

No designation appears to have been allocated, but the system is used on the Saab 105G and Viggen ground-attack aircraft. The system has four detector heads aligned to provide 360-degree azimuth coverage, and can determine quadrant of I to J band pulse and continuous-wave radar threats. E to F and G to H band options are available.

The Saab 105G has detectors in the tailplane bullet fairing, while the Viggen uses two forward-facing antennas in wing-tip installations and two rearward-facing antennas in the fuselage. A circular display together with aural warning is provided in the cockpit. Equipment facilities include non-threat data rejection, provision for radar blanking and built-in test equipment.

Detector heads: 430 × 105 × 105 mm; weight 3 kg each
Indicator: 79 mm dia × 69 mm
Frequency band: 8 to 20 GHz (4 to 8, 2 to 4 GHz options)
Power requirement: 300 mA 28 V dc (excluding indicator lamps)

AR-765 radar warning receiver

This is a lightweight, low-cost system developed for helicopters, and particularly suitable for the Agusta-Bell JetRanger, where the installation replaces the battery access panel and the radar warning receiver takes the place of the battery cover plate. Dual forward-facing circularly-polarised spiral antennas provide full forward hemisphere coverage with plus or

minus 45 degrees' elevation. A visual indication of the threat quadrant is provided, together with aural warning.

STATUS: in production.

AQ-31 deception jamming pod

Tailored to the Saab Viggen, the pod uses digital electronics, operates over 2 to 20 GHz waveband, providing am, fm noise/sawtooth jamming, and has its own ram-air cooling system. A single cockpit panel controls two pods simultaneously, and operation is compatible with the AR-753 jammer/set-on receiver pod.

STATUS: in production.

AR-753 jammer/set-on receiver pod

Although precise operating details have not been disclosed, this appears to be a home-on-jam pod, and to have no active elements within it. It is claimed to offer high resolution over a wide operating band, and it has its own cockpit control panel, containing the display, antenna selection switch and gain control.

Pod: 140 mm dia × 1600 mm length
Weight: 25 kg

STATUS: in production.

AR-777 airborne microwave receiver

Able to receive and locate threats in the 1 to 8 GHz waveband, this system comprises a single radio frequency head and a receiver unit. A cockpit display unit shows threat characteristics and has a keyboard for processor control inputs. Data can be transmitted to an on-board store for subsequent analysis.

Viggen carries Satt Elektronik AQ-31 deception jamming system

Frequency range: 1-8 GHz
Sweep width: 100/300 MHz
Sweep rate: 10 sweeps/s
Frequency accuracy: 5 MHz
Frequency resolution: 5 MHz
Prf measurement: 200 – 2000 Hz
Prf accuracy: 1 Hz
Power requirement: 300 VA, 200 V ac

STATUS: in production.

SRA

SRA Communications AB, Box 1, S-16300 Spanga-Stockholm

Loke repeater jammer

Believed to be a 8 to 12 GHz system employing travelling-wave tube technology for noise or deception jamming of airborne threats, Loke is

designed for compatibility with the Saab Viggen. Within the pod are three receiving antennas which provide 3 × 120 degree all-round coverage, and there are two independent jammers. Deception jamming falsifies range and bearing data, and additionally both spot and barrage jamming is possible.

STATUS: in development and production.

UNITED KINGDOM

Marconi Space and Defence Systems Limited

Marconi Space and Defence Systems Limited, The Grove, Warren Lane, Stanmore, Middlesex HA7 4LY

ARI 18223 radar-warning receiver

This is a self-contained radar-warning receiver system suitable for single-seat aircraft, and it is installed in RAF Jaguar and Harrier close-support aircraft. Systems have also been supplied in some overseas deliveries of the two types. Frequency coverage is 2 to 20 GHz (E to J band), and a lamp-type display is used to show the quadrant in which the highest priority threat is determined. Antennas are on the vertical fins of the aircraft and an audio-alarm is also triggered when a threat is displayed.

STATUS: in service.

ARI 18228 radar-warning receiver

This system uses the same antennas and has the same frequency-band coverage as the ARI 18223 system. It is designed for two-seat aircraft such as the Buccaneer and Phantom bombers. It is believed that the Vulcan also used this system. The display is a small circular cathode ray tube on which the direction of several threats from the aircraft can be shown simultaneously.

STATUS: in service.

Radar-warning receiver for Tornado interceptor

Based on ARI 18223/18228 experience, Marconi Space and Defence Systems has begun development of a more capable radar warning receiver system, suitable for the Tornado F2 air-defence variant of Europe's multi-role combat aircraft. This will be a new-technology system covering a similar frequency-range to its predecessor systems (2 to 20 GHz), which can determine the direction of threats more accurately. A 'modular' threat-processor capability is claimed, and probably refers to a digital unit into which various levels of threat-priority software can be loaded, so that all mission cases can be individually embraced.

STATUS: in development.

ARI 23246/1 ecm pod

Initially developed for use with RAF Tornados and Jaguars, but recently redesignated as a Tornado-only system, this electronic-countermeasures pod is sometimes referred to as Sky Shadow, and was formerly known as Ajax. Marconi Space and Defence Systems is prime contractor and assembles the system from components developed by Racal-Decca, British Aerospace, Marconi Avionics and Plessey. Marconi Space and Defence Systems is responsible for overall design and development and subsequent testing.

It has been suggested that a high-power

Royal Air Force Jaguars are equipped with Marconi Space and Defence Systems ARI 18223 radar-warning receiver units near fin tip

English Electric travelling-wave tube amplifier is used with a dual-mode capability for deceptive and continuous-wave jamming. A voltage-controlled oscillator uses a varactor-tuned Gunn diode to cover the full frequency band. The set-on receiver is of Decca design, and signal processing uses Marconi Avionics hardware.

The pod is known to incorporate both active and passive electronic warfare systems, and to include an integral transmitter/receiver, processor and cooling system. It has radomes at both ends and is stated to be capable of countering multiple ground and air threats, including surveillance, missile and airborne radars. Some automatic power management can be assumed, and modular construction probably allows the system to be 'tuned' to differing operational missions.

Equipment interfaces are probably similar to those of the Elettronica EL/73 deception jamming pod which is to be used on German and Italian Tornados. Inital reports suggested that all three countries would use the same electronic countermeasures pod, and although this has not been officially denied, the prospect now seems unlikely.

Sky Shadow flight-tests were completed in 1980 and a production contract has been let to Marconi Space and Defence Systems.

Approximate pod dimensions: 380 mm dia × 3350 mm length

STATUS: in production.

Internal ecm for Jaguar and Harrier

It is known that an electronic-warfare retrofit programme for RAF Jaguar and Harrier close-support aircraft has been considered by the Ministry of Defence, and that tenders have been submitted for internally-mounted active elec-

Royal Air Force Tornado carrying Marconi Space and Defence Systems ARI 23246/1 ecm pod on each outer pylon

tronic countermeasures systems. These could either replace or add to the ARI 18223 radar warning receiver system, and probably would compare in capability with the Loral Rapport III system specified for Belgian Air Force/General Dynamics F-16 fighters. A date for contract award has not been revealed either, but may be delayed if the RAF aims to achieve commonality with the US/UK AV-8B V/stol close-support aircraft.

STATUS: in development.

MEL

MEL, Manor Royal, Crawley, West Sussex RH10 2PZ

Airborne Matilda threat-warning system

Following many years of protracted development, in late 1981 MEL finally revealed the definitive version of the naval electronic warfare system, Matilda, a radar-intercept device capable of providing warning of a locked-on fire-control radar or the target-seeking radar beam of a guided weapon. Coincident with the announcement, the company also revealed that the technology used in the development of the

Matilda unit for small ships was also being applied to the development of a miniaturised, lightweight system for service with helicopters and light aircraft.

The surface-vessel version of Matilda is designed to react to either pulsed or continuous wave illuminating radar signals. The criteria used to initiate an alarm are either more than 250 pulses of similar bearing, pulse width and amplitude in a 500 ms period (in the case of a pulsed radar), or continuous wave emission detection for more than 100 ms.

Outputs take the form of an aural warning and a visual display and, if desired, an automatic countermeasures initiating signal

which activates chaff dispensing equipment. In the surface-vessel case, a voice module is used for verbal indication of the sector bearing of the threat. Simultaneously, the visual display provides a pictorial presentation of the same data. The airborne unit will probably dispense with the voice alarm and will use a warning tone in its place, although sector bearing will be displayed on a panel instrument.

The current Matilda unit uses four wide-beam antennas which cover 360 degrees and are deployed around the vessel's masthead and MEL is presently investigating the substitution of suitable replacements for airborne uses. Direction sensing of the received signal is carried out by means of amplitude comparison of the signal received through different antennas.

A detector diode in the receiver section produces a dc-to-10 MHz video signal which is amplified for processing and for simplification of bearing measurement. A video processor is used to compare channel outputs and the signal output from this is digitised for trans-mission to a data processor. Digitised values of bearing, amplitude and pulse-width are assembled to form an address to an 8 K × 8 K random-access memory. Each time a particular location is accessed by pulse data, its contents are increased and threat-warning is initiated when a preset level is reached within any one integration period of 0·5 second. A test facility which simulates a locked-on radar is incor-porated for confidence checking.

The present display system takes the form of an eight sector lamp indicator presentation which is likely to be reduced from its present size to fit a standard aircraft display for panel mounting.

Dimensions: processor unit ¾ ATR

STATUS: under development.

MS Instruments

MS Instruments Ltd, Rowden Road, Becken-ham, Kent BR3 4NA

Hofin hostile-fire indicator

The hostile-fire indicator (Hofin) uses shock-wave sensors to detect gun-fire aimed at the host aircraft, and is a passive-warning system suitable for helicopter use. It uses a five-armed sensor (four arranged in one plane, at 90 degrees to each other, and one perpendicu-larly) which is mounted beneath the helicopter.

Shock-waves generated by projectiles are converted to electrical signals, and then pro-cessed in an electronic unit which occupies a ⅜ ATR Short box. This produces outputs which can be used to generate audio or visual warning of nearby arms fire. An audio warning lasts for about one second after a shock-wave has been detected, and a visual warning is presented on a 4-ATI-size unit which has eight 45-degree wide segments. Four segments which approximately indicate the direction from which the shock is detected will illuminate for about five seconds.

Dimensions and weight: (sensor array) 305 × 305 × 195 mm; 4·25 lb (1·93 kg) (computer) 94 × 418 × 228 mm; 6 lb (2·72 kg) (indicator) 106 × 106 × 125 mm; 2·2 lb (1 kg)
Power: 60 W at 22 to 28·5 V dc
Sensitivity: responsive to supersonic projec-tiles at 20 m
Operating temperature range: –20 to +50° C
Humidity limit: 95% non-condensing

STATUS: in service.

Racal-Decca

Racal-Decca Defence Systems (Radar) Lim-ited, Lyon Road, Walton-on-Thames, Surrey KT12 3PS

HWR-2 radar-warning receiver

This is a miniature hand-held radar-warning receiver. It is suitable for use in light helicopters, and other small craft, where weight and volume restrictions might preclude a permanent panel-mounted installation. How-ever, the system is self-contained and can be hard-mounted if necessary.

The unit is sensitive to radar frequencies from 2 to 11 GHz and is scanned manually, providing a bearing indication to within approximately plus or minus 10 degrees. Antenna polarisation, when the unit is held with the handle vertical, is 45 degrees, so vertical, horizontal and circularly-polarised signals are all received. An audio tone indicates reception of a signal, and by rotating the receiver about the axis pointing towards the indicated target, polarisation is shown by an illuminated arrow on the back of the unit. Knowledge of polarisation character-istics is a valuable guide to deducing the type of radar illuminating the aircraft. The audio signal-frequency is representative of the radar pulse repetition frequency detected, and to deter-mine frequency more accurately either E/F or I/J band suppression can be selected. Audio signal intensity is representative of range and power, and if required a 15 dB audio reduction can be obtained by operating a sensitivity switch.

Production deliveries have been made to unspecified customers.

STATUS: in service.

MIR-2 esm system

Production of over 100 MIR-2 sets has been

Some Royal Navy Lynx are equipped with Racal-Decca MIR-2 warning receivers

reported for Royal Navy/Westland asw Lynx helicopters, since mid-1978. The system is suitable for small rotating- or fixed-wing aircraft, as well as small naval craft, and uses an advanced digital receiver served by a fully solid-state wide-band antenna system. The cockpit display is a compact light emitting diode unit.

STATUS: in service.

UNITED STATES OF AMERICA

AEL

American Electronic Laboratories Inc, PO Box 552, Lansdale, Philadelphia 19446

AN/APR-44 radar warning receiver

With the company designation CMR-500B, this low-cost unit is based on the AN/APR-42, with which it is electronically interchangeable and compatible. It has been adopted by the US Army and Navy.

The system consists of three units. The AEL Model A05-1543 antenna is a monopole unit which provides omni-directional, vertically-polarised, coverage within a 50-degree eleva-tion sector and is designed to be mounted on a flat horizontal surface. The receiver is a radio frequency-chopped, crystal-video type with a bandpass filter followed by a radio frequency switch/detector module, a linear video ampli-fier and processing circuitry. The video amplifier output is routed to a comparator before being compared with a radio-frequency-chopping signal in the processing logic. This procedure detects any continuous-wave signal and triggers a 3 kHz audio output, a logic output, an alert lamp drive and a logic blink signal (2 Hz). The third unit is a control panel which includes the alert light indicator.

Receiver frequency and bandwidth: selectable
Sensitivity: –45 dBm (minimum)
Max rf input level: 1 W cw (integral limiter)
Receiver dimensions: 91 × 91 × 33 mm
Receiver weight: 0·77 lb (0·35 kg)

STATUS: in production.

AIL

Eaton Corporation, AIL Division, Commack Road, Deer Park, NY 11729

AN/ALQ-61 elint system

A major programme, completed in 1967, by which time 500 systems had been delivered to the US Navy at a total cost of $65 million. The system is still operational in some US Navy RA-5C Vigilante and RF-4B/C Phantoms. It combines radar, infra-red, electronic and photographic sensor data for military intelligence assessment.

STATUS: in service.

AN/ALQ-99 jamming system

AIL (Cutler-Hammer) provides the tactical jamming system to the AN/ALQ-99 electronic warfare system which is currently installed in Grumman EA-6B Prowler and General Dynamics EF-111A aircraft. Grumman is prime contractor for the whole system, and a complete description of features is presented in that company's entry.

STATUS: in production.

AN/ALQ-112 expendable chaff-size ecm jammers

This was a research programme in 1974-75 which aimed to provide a low-cost, expendable electronic countermeasures jammer. Funds totalling $443,000 were earmarked, but the project was shelved. It is noteworthy that the Goodyear AN/ALE-39 chaff dispenser is designed to handle such payload.

AN/ALQ-130 communications jamming system

Delivery of this system commenced in 1974 and it is now used on US Navy A-4 Skyhawk, A-6 Intruder, EA-6B Prowler, A-7 Corsair, and F-4 Phantom aircraft. The system's principal function is the disruption of enemy defence communication links and surface-to-air missile radar operations. Jamming may include broadband and/or acoustic noise or spot-frequency jamming, but the former is more likely. Development was intiated as an AN/ALQ-92 update programme. It is internally-mounted on EA-6B, but may be externally carried on other types.

STATUS: in service.

AN/ALQ-161 ecm system

AIL/Cutler-Hammer has been appointed prime contractor to provide an on-board electronic countermeasures system for installation on the Rockwell B-1B strategic bomber. The AN/

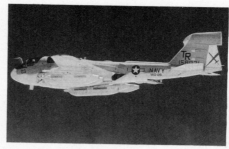

US Navy/Grumman EA-6B Prowler carries the AIL AN/ALQ-130 communications jamming system internally

ALQ-161 system is claimed to provide airborne electronic countermeasures management of emitted signals in direction, frequency, time and amplitude. Trials have been conducted since 1979 using the No 4 B-1A prototype at Edwards AFB, California.

Operating frequency range is approximately 0·5 to 10 GHz, chosen to cover Soviet early-warning and ground-controlled interception radar, surface-to-air missile and interceptor radar frequencies. Jamming signals in the higher regions of the electronic countermeasures spectrum are emitted from three electronically-steerable, phased-array antennas; one in each wing-glove leading-edge, and the other in the fuselage tail-cone. Each antenna provides 120-degree azimuth and 90-degree elevation coverage. Lower frequency signals are emitted from quadrantal-horn antennas mounted alongside the high-frequency equipment.

Major sub-contractors in the AN/ALQ-161 programme will be Northrop, Litton Industries and Sedco Systems. All companies serve as sub-system managers and have responsibilities for receivers, data-processing and jamming techniques. Litton supplies the LC-5416D mini-computer which handles the majority of system processing. Basic capacity is 48 K 16-bit words, with a rise to 64 K words likely before production. It is likely that Northrop will provide the low-frequency jamming antennas, and Sedco is supplying the phased-array antennas. AIL/Cutler-Hammer is responsible for system integration. There is a direct link between the Litton processor and the Singer-Kearfott defence management computer on the B-1B. Infra-red and pulse-Doppler tail-warning sensors, at least one of them located in the rear of the tailplane bullet fairing, will be integrated when sub-contractors have been chosen.

AIL AN/ALQ-161 is being developed for resuscitated US Air Force/Rockwell B-1 bomber

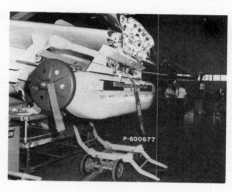

AIL (Cutler-Hammer) AN/ALQ-99 jammer pod installed on US Navy/Grumman EA-6B Prowler

Some indication of the magnitude of the system is provided by the observation that during early design it was found that although the preferred sub-system configuration involved an increase in uninstalled weight of 125 lb (56·7 kg), it eliminated 750 lb (304·3 kg) of cabling.

Development contracts worth in excess of $70 million have been let, and the target price per ship-set, set in 1972, was $1·1 million. Ceiling price was set at $1·4 million. It now seems unlikely that this system will be retrofitted to the US Air Force/Boeing B-52 Stratofortress strategic bombers.

STATUS: in development.

Argo

Argo Systems, 1069 East Meadow Circle, Palo Alto, California 94303

AN/ALR-52 ecm receiver

Used by the US Navy, this is a multiband instantaneous frequency measuring receiver, which is also used as a shipborne system under the designation AW/WLR-11. Typical coverage is 0·5 to 18 GHz using sets of receiver modules which cover octave bandwidths. It is a large system, providing for up to two operator positions, each with a control unit and display, although a single-operator version with direct link to digital hardware is also available. Capabilities are extensive and permit measurement of radar parameters, analysis of continuous wave and pulse signals, and direction finding measurements. The shipborne system is one of the major electronic countermeasures warning sets used by the US Navy, and airborne systems can only be installed in relatively large aircraft, eg the US Navy/Lockheed P-3 Orion.

STATUS: in service.

Bunker-Ramo

Bunker-Ramo Corporation, Electronic Systems Division, 3171 La Tienda Drive, Box 5009, Westlake Village, California 91359

AN/ALQ-86 esm system

Bunker-Ramo developed this set and shared production work with Loral and Hewlett-Packard. It is used in the US Navy/Grumman EA-6A Intruder carrier-borne electronic-warfare aircraft, in conjunction with the Loral AN/ALQ-53 electronic reconnaissance system, and the AN/ALH-6 recorder, which is also made by Bunker-Ramo.

STATUS: in service.

Cincinnati Electronics

Cincinnati Electronics, 2630 Glendale-Milford Road, Cincinnati, Ohio 45241

AN/ALR-23 and AN/AAR-34 infra-red warning receivers

An estimated 700 sets of these two sensors have been fitted on US Air Force/General Dynamics F-111 fighter-bombers. They are receiver/missile warning sets which detect infra-red emissions characteristic of burning hydrocarbon exhausts.

STATUS: in service.

Dalmo-Victor

Dalmo-Victor, Bell Aerospace Textron, 1515 Industrial Way, Belmont, California 94002

DSA-20 processor

This is a digital computer which can be used as a direct replacement for analogue, hardwired units used in earlier equipment such as AN/ALR-45, AN/APR-25, AN/APR-36 and AN/APR-107. It is claimed to present information to the crew more clearly than an analogue system, as it augments a strobe-line display with alphanumeric symbols which indicate the class of contact, and permits distinction between overlapping threats. The digital processor also incorporates a reprogrammable threat library which can be readily updated as new intelligence data is gathered, and is a major component of the AN/ALR-46A (Compass Tie) system.

STATUS: in production.

E-Systems

E-Systems Inc, Memcor Division, PO Box 540, Huntington, Indiana 46750

AN/APR-39(V)1 radar warning receiver

This system is used in US Navy helicopters and can determine the frequency, pulse repetition frequency, pulsewidth, persistence and threshold power level of surface-to-air missile and anti-aircraft gun-laying radars. Coverage is continuous across the E/I band, and there is some additional coverage in C, D and part of J band. An under-fuselage blade antenna and four spiral antennas are used. Two dual-receiver units are associated with each pair of spiral antennas. The latest version, AN/APR-39(V)2, uses a Loral CM-480 processor which offers 19 K words prom/ram storage in a single 4·9 kg unit. Warning information is presented to the crew as strobe lines on a 3-inch diameter cathode ray tube and is accompanied with audio tones. Helicopters equipped with the system include the Boeing-Vertol CH-47 and Bell AH-1, OH-58 and UH-1H.

STATUS: in production. The first production order, for 359 sets, was placed in October 1975, and deliveries commenced in 1980.

AN/AAR-44 infra-red warning receiver

This is an advanced warning system, the specification for which calls for continuous search of the lower hemisphere while tracking and verifying missile launches. Aircrew is warned of missile position, and automatic countermeasure action can be initiated. The system includes continuous track-while-search processing, multiple missile threat capability, and is able to discriminate missiles against solar radiation, terrain, and countermeasure backgrounds.

AN/ALR-62 radar warning receiver

Developed for the US Air Force/General Dynamics F-111, FB-111 and EF-111A tactical fighters and jamming aircraft, this is the major sensor element in a recently-established US Air Force programme to update the electronic warfare equipment on these types. Other elements are the AN/ALQ-137 internal jammer and AN/AAR-44 infra-red warning receiver. At least 400 aircraft sets were expected to be delivered in the late-1970s for F-111/FB-111 operations, and a further 40 sets for EF-111A use, with last delivery due in July 1982.

STATUS: in service.

AN/APS-109 radar homing and warning system

Dalmo-Victor/Textron supplied about 500 examples of this system for initial production of the US Air Force/General Dynamics F-111 and FB-111 swing-wing tactical fighters and strate-

gic bombers. It has been phased out as AN/ALR-62 equipment is received and installed.

STATUS: being withdrawn from service.

Mk III miniature digital radar warning system

Specifically designed for helicopters, light aircraft and fast patrol boats. A lightweight digital processor is used to process sensor data, and presentation is on an eight-segment liquid-crystal display unit. The processor threat-data store employs ultraviolet-erasable prom and ram memories and is readily updated.

STATUS: in production.

Dimensions and weight: (sensor) 376 mm dia × 366 mm; 34·3 lb (15·6 kg)
(processor) 191 × 168 × 259 mm; 9·3 lb (4·22 kg)
(control/display unit) 104 × 145 × 79 mm; 1·42 lb (0·64 kg)

STATUS: in development and limited production.

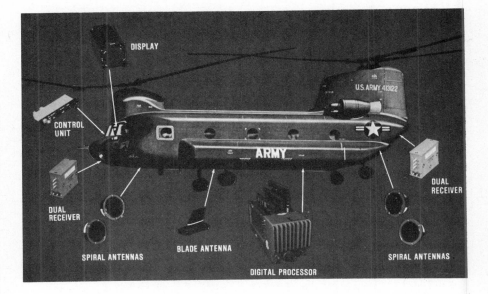

AN/APR-39(V)2 radar-warning equipment by E-Systems on US Army Chinook

Electronic Specialty

Electronic Specialty Corporation, Datron Systems Inc, PO Box 20055, Portland, Oregon 97220

AN/ALR-32 esm receiver

Developed to monitor airborne and ground-based radar transmissions, this system was installed in E-66, RB-66 and B-52 Stratofortress. Total production reached about 700 aircraft sets, and was completed several years ago. It can cover 360 degrees in azimuth and can control jamming equipment.

STATUS: in service.

ESL

ESL Inc, 495 Java Drive, Sunnyvale, California 94086

AN/ALQ-151 airborne jammer

Although little is known about this system produced by a TRW subsidiary, it is understood to be currently installed in US Army/Bell EH-1H electronic countermeasures helicopters. Design commenced in 1975 and is believed to have reached the full-scale production phase. In view of its recent development this is probably an advanced deceptive jammer.

STATUS: in production.

General Electric

General Electric, Aerospace Electronic Systems Dept, French Road, Utica, New York 13503

AN/ALQ-71/72 jammer pod

The ALQ-71 and ALQ-72 pods were both developed by General Electric and production was carried out by Hughes Aircraft in the early 1970s. About 1600 sets were built for use on LTV A-7 Corsair, Republic F-105 Thunderchief and McDonnell Douglas F-4 Phantom variants.

ALQ-71 operates in the 1 to 8 GHz band and has two noise jammers which can interfere with surface-to-air missile guidance and tracking radars. The ALQ-72 is a higher frequency-band counterpart. Both units are self-contained and use a Garrett AiResearch ram-air turbine to provide electrical power and cooling. Some units are thought to have been supplied to Israel.

STATUS: in service, possibly being withdrawn gradually.

AN/ALQ-87 barrage jammer pod

Operates in the 1 to 8 GHz band and provides frequency-modulated barrage jamming from a self-contained pod. About 350 sets were built and supplied for use on US Air Force F-111, FB-111, F-4 Phantom and F-105 Thunderchief attack aircraft.

STATUS: in service.

General Instruments

General Instrument Corporation, Electronic Systems Division, 600 West John Street, Hicksville, New York 11802

AN/ALR-66 radar warning receiver

A fully-programmable digital radar warning receiver system which uses four independent spiral antennas to provide data on up to 15 simultaneous hostile transmitters. The central processor is a high-speed, 8 K word ram/prom unit with provision for up to 68 K words capacity. The warning system will interface with chaff/flare dispensers, telemetry links, recorders or jammers. Production is current for the US Navy, including a version designated AN/ALR-66(VE) which is interchangeable with older systems, such as AN/APR-25, AN/APR-36/37, AN/ALR-48 and AN/ALR-46.

Frequency coverage: continuous over 4 bands
Azimuth coverage: 360°

Direction-finding accuracy: better than 15°
Shadow time: 2 μs
Weight: 59 lb (27 kg) for basic AN/ALR-66
66 lb (29 kg) for AN/ALR-66(VE)
Power requirement: 400 W, 200 V ac

STATUS: in production.

AN/ALQ-98 and AN/ALQ-102 ecm systems

Developed for use on US Navy helicopters as either internally-mounted (ALQ-98) or pod-mounted (ALQ-102) systems. They were proposed as part of an anti-cruise missile system, but development has not been acknowledged since the late 1970s.

AN/ALR-606 radar warning receiver

An improved version of the AN/ALR-66(VE) with identical installation characteristics, but sensor sensitivity has been extended to cover E to J band (2 to 20 GHz). A blinking data display which provides alpha-numeric indications of threat pulsewidth, pulse repetition frequency, and frequency band is provided. Customer is unknown, but is not one of US armed forces.

STATUS: in production.

AN/ALR-646 esm system

Supplements the AN/ALR-66 radar warning system and can receive the E to J bands (2 to 20 GHz) using six cavity-backed planar spiral antennas. Improved processing and data presentation provide broad-band, high-sensitivity, signal detection, with instantaneous frequency measurement and accurate direction-finding. Dual-mode cathode ray tube display shows polar plot of total signal activity and a detailed radar parameter readout of active radars.

STATUS: in development, but US armed forces procurement not expected.

Goodyear

Goodyear Aerospace Corporation, Lichfield Park, Arizona 85340

AN/ALE-39 chaff dispenser

One of the leading in-service chaff-dispenser types, the AN/ALE-39 replaces the ALE-29 on many US Navy aircraft. It is already in operation on A-4 Skyhawk, A-6 Intruder, EA-6B Prowler, A-7 Corsair, F-4 and RF-4 Phantoms, and F-14 Tomcat. It will be used on the new US Navy/McDonnell Douglas F-18 Hornet, possibly on AV-8B, and is about to be adopted for US Marine Corps AV-8As. Contracts for almost 1500 sets had been placed by 1976, and it seems certain that all of them, plus perhaps

several more batches, have now been delivered.

The system accommodates three types of payload; chaff, infra-red flares or expendable jammers. Cartridges are loaded in batches of 10 up to a maximum of 60. Each cartridge is 1·42 inches (3·61 cm) diameter and 5·8 inches (14·73 cm) long. Typical payload would be RR-129 chaff cartridges or Mk46 infra-red flares. No manufacturer of expendable jammers is acknowledged but research programmes such as the AN/ALQ-112 were directed towards studying the feasibility of such devices.

STATUS: in production.

US Marine Corps McDonnell Douglas AV-8B V/Stol fighter is a candidate for Goodyear AN/ALE-39 chaff dispenser

Grumman

Grumman Aerospace Corporation, Bethpage, New York 11714

AN/ALQ-99 ecm system

This tactical noise-jammer, operated on US Navy/Grumman EA-6B Prowler and US Air Force/General Dynamics EF-111A electronic warfare aircraft is one of the most sophisticated electronic countermeasures systems currently operated by any of the US armed forces. Funding for project development and initial

production exceeded $250 million by 1979 and system management is a Grumman responsibility. There are several manufacturing subcontractors.

Installation on the EA-6B relies on up to five externally-mounted pods, each containing jamming equipment. This has the advantage that jamming can be tailored to mission requirements without having to carry redundant equipment. All processing and receiver equipment is within the airframe, several antennas being in a large fin-tip fairing. The

main processor is an IBM 4 Pi general-purpose machine. When jamming is initiated it can be directed into any combination of 30-degree azimuth sectors. There are three basic operating modes (automatic, semi-automatic and manual), each combining different proportions of automatic and operator-conducted threat selection and jamming.

The EF-111A installation is similar to the EA-6B, but probably incorporates more receiver antennas, either to improve direction finding resolution or to extend waveband coverage. Several processing improvements, including more rapid threat response, have been included in this later system and will be retrofitted to the carrier-based EA-6B.

Major components and suppliers are as follows:

AN/ALQ-99 tactical jamming system (Eaton Corp/Cutler-Hammer)
AN/ALQ-92 communications jamming system (Sanders)
AN/ALQ-126 radar deception system (Sanders)
AN/ALQ-42 radar warning receivers
AN/ALE-29 chaff/flare dispensers (Tracor)
AN/ALQ-41 track-breaker system (Sanders)
IBM: 4 Pi general-purpose central computer
Raytheon: transmitters and exciters
AEL: transmitters
Hughes: travelling-wave tubes
Microwave Associates: travelling-wave tubes
Teledyne: travelling-wave tubes
McDonnell Douglas: pods
Garrett AiResearch: ram-air turbines
Astronautics Corp of America: displays

STATUS: in production.

US Air Force/General Dynamics EF-111 incorporates Grumman AN/ALQ-99 tactical noise jammer

Hughes Aircraft
Hughes Aircraft, Radar Systems Group, PO Box 92426, Los Angeles, California 90009

AN/AAR-37 infra-red warning receiver
This unit is installed in US Navy/Lockheed P-3 Orion asw aircraft, but no technical details are available.

STATUS: in service.

AN/ALQ-71/72 jamming pods
Production carried out in the early 1970s, based on a General Electric design. Over 1600 units were delivered to the US Air Force. Details given in General Electric section.

STATUS: in service, possibly becoming obsolete.

AN/ALQ-119 ecm pod
Developed and produced in quantity, more than 1600 units, by Westinghouse. Hughes Aircraft was selected by the US Air Force in 1978 as a second-source supplier of this important electronic warfare system.

IBM
IBM, Federal Systems Division, 10215 Fernwood Road, Bethesda, Maryland 20034

AN/ALR-47 radar homing and warning system
Developed for the US Navy/Lockheed S-3A Viking asw carrier-borne aircraft, this is a comprehensive passive electronic warfare system which uses four cavity-backed planar spiral antennas in each wing-tip. The aerials are orthogonally directed to enhance monopulse direction-finding, and therefore ensuring that threat direction is measured very accurately. Associated with them are twin, highly-sensitive, narrow-band receivers, and a comprehensive processor. Manual or automatic system operation is possible, control being exercised over frequency-band limits, speed of tuning and signal selection. The processor indicates frequency-scanning limits, scan speed, pulse-length, pulse repetition frequency and bearing limits of any detected radar transmissions.

STATUS: in production.

F-4G Wild Weasel, with IBM AN/APR-38 warning equipment

AN/APR-38 radar warning receiver
IBM supplies radar warning receiver equipment used in the AN/APR-38 electronic surveillance measures system installed in US Air Force F-4G 'Wild Weasel' aircraft. Details of the system, as much as has been revealed, are given in the Loral entry.

STATUS: in production.

US Navy/Lockheed S-3A Viking carries IBM's AN/ALR-47 radar homing and warning system

Itek
Itek Corporation, Applied Technology Division, 645 Almanor Avenue, Sunnyvale, California 94086

AN/ALR-45 radar warning receiver
Perhaps the most significant US Navy radar warning receiver in service, and used in A-6A Intruder, A-7E Corsair, F-4J Phantom, RA-5C Vigilante and F-14 Tomcat. Extension for use in virtually all US Navy types is planned in the immediate future. Production commenced in mid-1970s and is likely to extend over a decade, although the improved performance of AN/ALR-67/68 sets could curtail this dominant programme. Several thousand sets are believed to have been delivered, some of them for Denmark and Canada.

The AN/ALR-45 is basically a digital version of the analogue AN/ALR-25 system, incorporating wider frequency-band characteristics (2 to 14 GHz). Four cavity-backed planar spiral antennas, disposed to give 360-degree azimuth coverage, are used, each with a pre-amplifier and interface unit. There is a central digital processor, and a strobe-line indicator display unit. The processor contains a reprogrammable threat library.

STATUS: in production.

AN/ALR-46 radar warning receiver
A very significant US Air Force system, used in all front-line types except the F-15 Eagle, F-111 and FB-111 types. It replaced APR-25/26 and APR-36/37 systems, and in its AN/ALR-46(V)3 version it is virtually a digital version of the latter. Replacement is under way already however, as the AN/ALR-69 radar warning receiver enters service.

The AN/ALR-46 is a 2 to 18 GHz system. It uses a wide-open, front-end, crystal video type receiver, with a digital processor (see Dalmo-Victor DSA-20 processor entry), which can be reprogrammed to accept new threat data. In addition to providing cockpit visual and audio warning cues it can directly control jamming systems. Export examples have been supplied to West Germany, Iran, Korea and Saudi Arabia.

STATUS: in service.

AN/ALR-67/68 advanced radar warning receiver
Developed in the late 1970s and entering production in 1979, this is the latest US Navy radar warning receiver. It is claimed to provide 1 to 16 GHz frequency coverage, and will be used initially on Grumman F-14 Tomcat and EA-6B Prowler, and McDonnell Douglas F-18 Hornet.

It may also be installed in the AV-8B. Many of the system features are similar to AN/APR-36/37 functions. There are four spiral antennas, and an Itek Victor V processor is used for all threat assessments. Conventional cockpit displays and aural warnings are provided, and integration with active electronic warfare systems is possible.

STATUS: in production.

AN/ALR-69 advanced radar warning receiver
Basically an updated AN/ALR-46, and when used in conjunction with the AN/ALQ-119 electronic countermeasures pod together form the Compass Tie system. In addition to ALR-46 functions the new system can also detect continuous-wave signals and has direction-finding capability in the 0·5 to 2 GHz range. A Dalmo-Victor processor is retained, and the unit will be used in A-10, F-16, A-7D Corsair, F-4 and RF-4 Phantom.

STATUS: entering service.

ITT Avionics
ITT Avionics, Avionics Division, Nutley, New Jersey 07110

AN/ALQ-117 jamming system
Over 400 examples of this system have been produced for the US Air Force/Boeing B-52 strategic bomber fleet. An initial production contract was awarded in 1975, and installation took place during Rivet Ace programme lay-ups, in which all B-52G and H variants were fitted with electro-optical viewing equipment in chin blister installations. AN/ALQ-117 is an I-band (8 to 10 GHz) system and may include both noise and deception jamming elements. The system was initially regarded as an interim fit, to be replaced by the AN/ALQ-161 (currently under development for the US Air Force/Rockwell B-1B bomber) but costs may rule out this possible substitution.

STATUS: in service.

AN/ALQ-129 advanced mini-jammer system
Although not committed to production, AN/ALQ-129 equipment represented a significant US Navy electronic warfare system development. The system was funded in 1972, and flight-tested around 1977 in a US Navy/LTV A-7E Corsair. It was an internally-mounted jammer unit of relatively small size, but with limited bandwidth capability. Experience with this unit led to the AN/ALQ-136 research system and has been an important technology-demonstrator for the advanced self-protective jammer programme.

STATUS: research only.

AN/ALQ-136 advanced jammer system
Developed from the AN/ALQ-129 research

US Air Force/Boeing B-52G and -H strategic bombers carry ITT AN/ALQ-117 jamming system

system, and funded by the US Army in the late 1970s as a possible helicopter electronic countermeasures system. Engineering development models were produced in 1977, but no production contracts have been let. Although technical data is not available, it is widely thought that a system combining AN/ALQ-136 and advanced self-protective jammer features may evolve to meet future US Army requirements.

STATUS: research only.

AN/ALQ-165 advanced self-protective jammer
ITT Avionics and Westinghouse were awarded contracts in late 1981 to jointly produce the AN/ALQ-165 aspj system. A competitive design had been funded also, and was proposed by Northrop Corporation and Sanders Associates.

STATUS: in development (see ITT/Westinghouse entry).

ITT/Westinghouse
ITT Avionics Division, Nutley, New Jersey 07110 and Westinghouse Electric Corporation, Aerospace Division, Baltimore, Maryland 21203

AN/ALQ-165 advanced self-protective jammer
In terms of quantity production, this is likely to be the most significant US electronic warfare system in the current decade. Production is expected to exceed 3000 sets, and contract

value, at 1978 prices, will exceed $1500 million. It is a joint US Navy/US Air Force programme aiming to provide a common active electronic countermeasures capability in F-14 Tomcat, F-18 Hornet, EA-6B Prowler, AV-8B, F-111 and F-16. Other current aircraft remaining in front-line service after the above types have been equipped also stand to have the system installed.

The joint-services specification for advanced self-protective jammers was developed around 1976. It was evident then that an internal

electronic countermeasures system, small enough to fit into modern combat aircraft, and with wideband jamming capability, was a logical future development. Even the US electronics industry, where electronic warfare capability far exceeds anywhere else in the Western World, recognised that the system would be a challenge to the resources of individual companies. Accordingly, with direct encouragement from the services, industrial consortia were formed to bid for this contract. Major development contracts were let eventually to ITT/Westinghouse and Northrop Corporation/Sanders Associates.

In its original proposals the US Navy asked for 'off-the-shelf' hardware, and considered that innovative application of available technology with miniaturised jammer techniques (based on dual-mode travelling-wave tubes) would meet the requirements. It now seems unlikely that these components will be used, but that new technology will be incorporated instead, probably as GaAs fet amplifiers and very high speed integrated circuit components. The technical risk in the programme could therefore be very high, but the aim is to achieve hitherto unavailable capability in very small volume. Criticisms have been reported in US Navy circles which claim that the degree of miniaturisation demanded by the US Air Force is too much of a technology and cost driver. There is little doubt, however, that if the joint-service goals can be met, aspj will set a new standard in internal electronic countermeasures technology.

The aspj specification is summarised below. It also calls for all the equipment to fit in 2 ft³ (0·065 m³) volume and to expand up to 35 GHz capability before 1990. Eventual expansion up to 140 GHz is a long-term goal. Some Soviet fire-control and interception radars are already reported to be operating up to 40 GHz.

Frequency coverage: 0·7 – 18 GHz
System response time: 0·1 – 0·25 s

Receiver-processor
(dynamic range) 50 dB
(sensitivity) –71 dBm
(resolution) 5 MHz
(instantaneous bandwidth) 1·44 GHz
(pulsewidth) 0·1 μs (minimum)
(input pulse amplifier) 20 dBm (max)
(false alarm rate) 5/h (max)
(signal detection capability) pulse, cw, pd, agile

Jammer
(peak power) 58 – 63 dBm
(pulse-up capability) 5 – 7 dB
(set-on accuracy) ±0·5 to ±20 MHz
(duty cycle) 5 to 10%
(jamming capacity) 16 – 32 signals
(modes) noise, deception

STATUS: in development.

US Marine Corps/McDonnell Douglas F/A-18 Hornet is one of the combat aircraft types that will receive the ITT/Westinghouse AN/ALQ-165 system

Comprehensive internal active ew specification for McDonnell Douglas F/A-18 fighter led to US Navy aspj contract, awarded as AN/ALQ-165

Litton

Litton Systems Inc, Amecon Division, 5115 Calvert Road, College Park, Maryland 20740

AN/ALR-59 esm system

Developed to meet US Navy requirements for an effective passive detection system to supplement the early-warning radar used in Grumman E-2C Hawkeye aircraft, this highly-capable electronic surveillance measures system is tailored to the role and configuration of the E-2C and its systems. It has an extensive array of antennas, considerable on-board processing, and its main purpose is to pass any received data to the E-2C central processor for correlation with radar data.

The system uses 16 antennas, one set of four for each of four wavebands, and with each complete set positioned to look at 90-degree sectors, and thus provide 360-degree azimuth coverage. The forward and aft antennas are in the fuselage extremities, and the sideways-looking aerials are in the tailplane tips. All receiver sets are under separate control, so wavebands are scanned independently and simultaneously in all sectors. Each antenna has

Litton AN/ALR-59 esm system is installed aboard US Navy/Grumman E-2C Hawkeyes

dual processing channels and uses digital closed-loop rapid-tuned local oscillators. The latter provides instantaneous frequency measurement with fast time response and high-accuracy frequency determination.

Receiver outputs are collected at a signal preprocessor unit which performs pulse-train separation, direction-finding correlation, band-tuning and timing, and built-in test equipment tasks. Data is then in a form suitable for the general-purpose digital computer, which has overall control of electronic surveillance measures operations, and will vary frequency coverage, dwell time and processing time according to prescribed procedures. Control of these parameters is aimed at maximising the probability of intercepting signals in particular missions. Other on-board sensor data and crew inputs will determine the technique adopted. Data such as signal direction of arrival, frequency, pulsewidth, pulse repetition frequency, pulse amplitude and special tags is sorted by the computer and transmitted to the E-2C central processor.

STATUS: in production.

AN/ALQ-125 ecm system

Sometimes referred to as Terec (tactical electronic reconnaissance) this passive electronic countermeasures system is designed to locate and identify radar emitters, and is installed in a limited number of US Air Force/McDonnell Douglas RF-4C Phantoms. All types of fixed and mobile ground-based radars are included in the system threat library, and data can be recorded for replay, or transmitted on data-link. Eighteen systems are believed to have been procured by the US Air Force, and it is likely that RF-4C operations with high-speed anti-radiation missile will be associated with Terec-equipped aircraft.

STATUS: in production.

Loral

Loral Electronics Systems, 999 Central Park Avenue, Yonkers, New York 10704

AN/ALR-31 esm system

An early electronic surveillance measures system which was supplied to the US Air Force and installed in F-105 Thunderchief and F-4 Phantom variants. Some equipment may remain in service, but in diminishing numbers.

STATUS: being withdrawn from service.

AN/ALR-39/41 homing receiver

Used in initial production US Air Force/General Dynamics F-111 and FB-111 aircraft, and built under sub-contract to the airframe company. Little is known about the system, which is being replaced in service by the AN/ALR-62.

STATUS: being withdrawn from service.

AN/ALR-56 radar warning receiver

Together with the AN/ALQ-135 internal electronic surveillance measures set, this radar warning receiver system forms the tactical electronic warfare system (TEWS) which is installed in all models of the US Air Force F-15 Eagle.

The radar warning receiver system comprises five antennas, a high-band tuner unit, low-band receiver/processor, power supply unit and cockpit control and display. There are four circularly-polarised planar spiral aerials (two facing forward, one in each wing-tip, and one rearward-facing at the tip of each fin) to provide 360-degree azimuth high-band coverage. Additionally there is a blade antenna under the fuselage for lower hemisphere low-band coverage.

The system uses separate high and low-band receiver units. The high-band tuner unit provides digital control of the antennas, and is a dual-system capable of scanning a very wide portion of the electro-magnetic spectrum. The single-conversion superheterodyne design provides high sensitivity, and the use of dual YIG radio frequency pre-selection affords excellent sensitivity and spurious signal rejection.

The low-band receiver unit is integrated with the system processor, which is a 32 K 16-bit word unit with a cycle time of 1·5 μs. With the comprehensive antenna/receiver-derived data the processor is able to conduct direction-finding operations, and low-band electronic tuning is carried out directly. Pre-processor facilities, which convert intercepted signals into digital data for the central processor element, are included in the processor unit. A re-programmable threat library is held in memory and can be easily changed.

A self-contained precision signal source is provided within the system for calibration and/or built in testing over the entire tuning range, and there is provision for multiplexing important receiver functions. The latter are controlled by serially-generated NRZ Manchester-coded data. The precise data rate also serves as the reference clock for the receiver frequency synthesiser.

Information provided in the cockpit includes data for a circular threat-evaluation display and audio signals. The display unit shows threat

Loral AN/ALR-56 radar warning receiver is installed on US Air Force's F-15 Eagle fleet

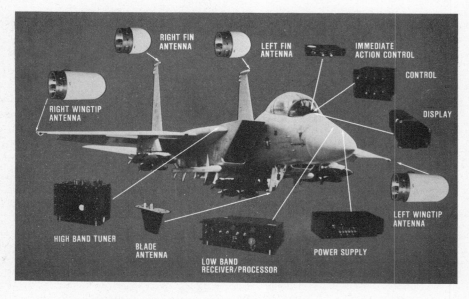

Loral AN/ALR-56 radar-warning equipment on US Air Force F-15 Eagle

bearing and distance, and threat characteristics are presented by alpha-numeric symbology. The display is designed to be legible in high ambient lighting conditions and has pilot-selectable clutter-elimination programs. There are two cockpit control units. One unit combines basic radar warning receiver and tactical electronic warfare systems and TEWS control functions, and the other, smaller, unit is specifically for radar warning receiver use, with interrogated data readout facilities.

Dimensions and weight:
(power supply unit) 445 × 267 × 156 mm; 26·4 lb (12 kg)
(wing-tip antennas) 89 dia × 144 mm; 1·75 lb (0·8 kg) each
(fin antennas) 89 dia × 100 mm; 0·94 lb (0·43 kg) each
(low-band antenna) 133 × 84 × 45 mm; 0·4 lb (0·2 kg)
(high-band receiver) 241 × 187 × 202 mm; 22 lb (10 kg)

(low-band receiver/processor) 451 × 404 × 165 mm; 53 lb (24 kg)

(display unit) 406 × 149 × 137 mm; 13·3 lb (6 kg)

(control unit) 165 × 111 × 114 mm; 2·37 lb (1·1 kg)

(immediate action controls) 29 × 76 × 127 mm; 0·66 lb (0·3 kg)

Total system weight: 138 lb (63 kg)

Total system volume: 2·41 ft³ (0·068 m³)

Installation: 13 lb (5·9 kg) for RF transmission lines

Power requirement: 550 VA, 115 V ac 3 to 7 A, 28 V dc

STATUS: in production.

AN/APR-38 (Wild Weasel) esm system

This significant US Air Force programme was sub-contracted to IBM (radar warning receiver), Texas Instruments (radar warning computer) and Loral (displays). The important tactical nature of the tasks associated with this system has prevented release of any detailed information, except in respect of the display equipment (see Loral entry in Displays section).

The 'Wild Weasel' programme was initiated in 1965, following introduction of Soviet SA-2 radar-guided surface-to-air missiles in North Viet-nam in answer to US bombing. It is a self-contained electronic surveillance measures system, and was the first ever to provide both bearing and range data on missile acquisition and target radar locations. The aircraft can neutralise threats with conventional weapons, high-speed anti-radiation missiles, or electro-optically guided Maverick missiles.

Wild Weasel equipment is installed in F-4D Phantoms, substantially modified and reclassified as F-4Gs. In total 116 aircraft are involved, of which 96 are used by combat units. Deliveries commenced in April 1978, and were due to be completed in 1982. It is claimed that each Wild Weasel aircraft has 52 additional antennas, some of which are attributed to McDonnell-Douglas supplied items, although the acknowledged radar warning receiver sub-contractor for the programme was IBM (AN/APR-38 rwr system).

At the centre of the system is a Texas Instruments processor, which sifts all received data to identify threats, using information contained in a reprogrammable threat library.

The Loral-furnished control/indicator set comprises seven items in the front and rear cockpits housing the pilot and system-operator. Data is assigned priorities by the central processor before transmission to the cockpit items. All displays are controlled from a special electronics unit: they comprise a plan-position indicator for each crewman, and a panoramic analysis and homing indicator/control unit for the system operator. The remaining units are a lighting controller, system and missile control panel and aircraft commander's warning panel. Alpha-numeric displays are used to designate signals, eg 'A' corresponds to anti-aircraft gun radar, '2' to Soviet SA-2 tracking radar and so on. The 15 highest priority threats are assigned and labelled, and the most significant threat is placed within a triangular symbol. The operator can override this with his diamond-shaped 'Bear' symbol.

Automatic weapon-release is also handled by the system, blind attack being possible by sighting a green aircraft cursor on the gunsight display over a red reticule positioned by the threat processor.

STATUS: in production and service.

AN/ALQ-78 esm system

Used in US Navy/Lockheed P-3C Orion subhunter, the AN/ALQ-78 is a radar-warning system designed to detect anti-submarine and conventional electronic warfare threats. A high-speed rotating antenna in the aircraft belly operates in an omni-directional search mode, and using threat-processor control it initiates

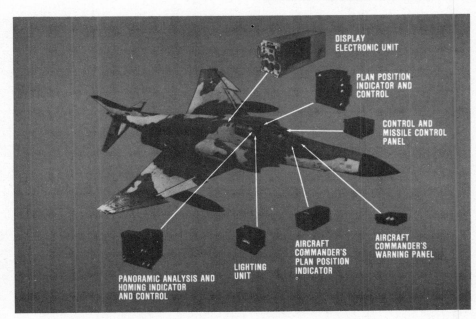

AN/APR-38 control indicator sub-system is a part of overall F-4G Wild Weasel ecm system

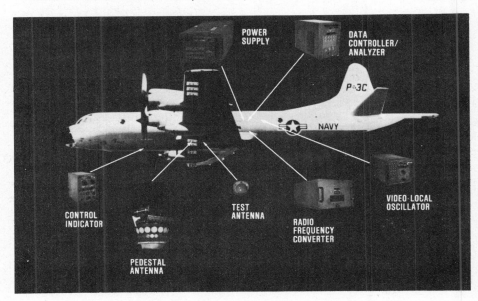

Loral AN/ALQ-78 esm equipment Loral on US Navy P-3C Orion

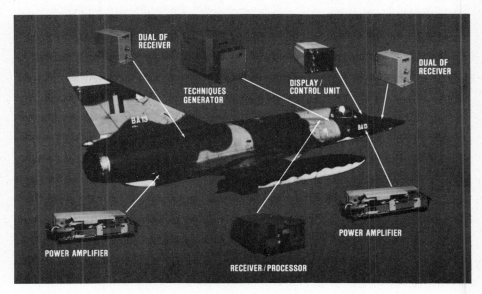

Loral Rapport II ecm equipment on Belgian Air Force Mirage V

automatic direction-finding as signals are acquired. Threat analysis is carried out within the equipment and data is displayed to system operators in the aircraft.

STATUS: in production, at a rate of some 12 sets a year since the mid-1970s.

Rapport II ecm system

Rapport II is designed to protect aircraft against airborne and ground-based radar-directed weapon systems. The system continuously analyses, identifies and determines the bearing of each detected threat radar. It can rank up to

14 such radars in priority order according to reprogrammable data held in the threat-processor memory. An alpha-numeric display presents data on threat nature, direction and status to the pilot, and audio alert tones are provided. Automatic noise or continuous-wave/pulse countermeasure action can be initiated.

The system has been developed and installed in Belgian Air Force Mirage Vs.

It comprises seven line-replaceable units, plus antennas. The units are: two amplifiers, two dual DF receivers, a techniques generator, receiver/processor, and a cockpit display/control unit.

Frequency range: E to J band (2 to 20 GHz)
System weight: 282 lb (128 kg)
System volume: 5·1 ft³ (0·145 m³)
Data-bus compatibility: growth to MIL-STD-1553

STATUS: in production and service.

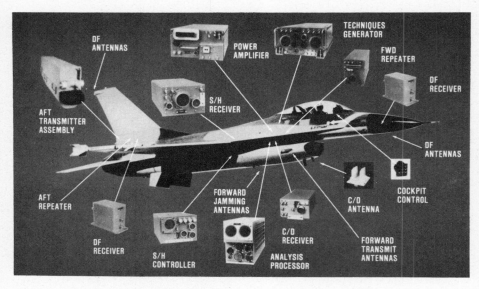

Loral Rapport III ecm equipment on Belgian Air Force F-16

Rapport III ecm system

Basically a Rapport II development, the Rapport III system is tailored to the General Dynamics F-16 fighter, and will be delivered to the Belgian Air Force. It can use the same receiver equipment as Rapport II, or is compatible with the AN/ALR-69 radar warning receiver system which is a standard fit on US Air Force F-16s. It is currently the only internally-mounted electronic countermeasures system for the F-16, and is available ahead of the US Air Force planned installation of AN/ALQ-165 equipment.

Rapport III offers E to J band (2 to 20 GHz) detection and jamming, with simultaneous fore-and-aft jamming capability. Equipment modes include a 'look-through' facility, which permits rapid assessment of jammer effectiveness. Threat data derived by the system includes direction-finding, pulse repetition frequency, pulsewidth and scan rate. Information is presented to the pilot on a circular cathode ray tube indicator with alpha-numeric detail.

System weight: 368 lb (166 kg)
Installed volume, in avionics bays: 2·8 ft³ (0·079 m³)
 in tail compartment: 5·5 ft³ (0·156 m³)

STATUS: in production.

ARI-18240/1 esm system

Loral is supplying a new esm system to the Royal Air Force for installation on the British Aerospace Nimrod AEW3 and (possibly) Nimrod MR2 aircraft. Sets of fore-and-aft facing cavity-backed planar spiral antennas are in wing-tip pods for a total of 16 aerial units. Also in each pod is a receiver unit, which extracts threat data from the radar warning receiver sensors for onward transmission to an intermediate-frequency processor unit in the centre fuselage. Loral also supplies a power supply unit, and the absence of any further hardware suggests that all subsequent use of threat data is integrated into the basic aircraft systems.

Few technical details are available, although it seems probable that each antenna set covers a 90-degree azimuth sector and thus provides all-round electronic surveillance measure coverage. Both high- and low-band antennas are used and frequency coverage is probably E to J band (2 to 20 GHz).

STATUS: in production.

EW-1017 esm system

Loral is marketing a new electronic surveillance system under this designation, several units of which are apparently similar to those in production for the ARI-18240/1 system. A

RAF's new Nimrod AEW3 early warning aircraft carry Loral ARI-18240 esm system

Loral supplies passive receiver equipment for use on RAF Nimrod MR2 and AEW Mk.3 maritime patrol and early warning aircraft

similar capability to that system can be assumed. No orders are known to have been received.

Multibeam transmitter assembly

Loral is developing, as part of a company-funded activity, a distributed miniature travelling-wave tube transmitter system, which it is claimed will improve considerably the efficiency and directional-jamming capability of an active electronic warfare system.

Demonstrations were conducted in early 1981 to show the feasibility of a multibeam transmitter system which would use a ten-element forward-looking array to provide a field-of-view measuring 120 degrees in azimuth and 60 degrees in elevation. This would use five transmitter elements on either side of the fuselage. A nine-element aft-looking array is proposed to complement this unit. Several miniature 40-watt travelling-wave tubes (Varian Associates VRT-6110-A3 devices in prototype units) are used to feed each array element. As a result of this redundancy, the system is claimed to be less likely to fail than a conventional system using only a single unit, and also to provide graceful degradation in the event of a failure, as the contribution of any one unit is relatively small. Operating range is 4·8 to 18 GHz.

In a total electronic surveillance measures system Loral is proposing that a superheterodyne receiver with crystal-video receivers should be used for direction-finding. Demonstration of the effectiveness of such a system against monopulse and simultaneous targets in different azimuth locations has been completed. Total system power requirements are 3·5 kVA forward and 2 kVA aft in the type of installation described above, which would be suitable for the General Dynamics F-16 fighter.

A configuration that could be installed in the F-16 has been outlined by Loral, and exemplifies the most compact internally-mounted electronic surveillance measures system the firm has so far contemplated. The two sets of forward-looking arrays and their associated crystal-video receivers would be installed in blisters on either side of the chin engine air-intake. The forward transmitter box would be located in space already reserved for possible electronic warfare use and connected to the aerials by co-axial cables about 7 feet (2·2 metres) in length. Loral has also stated that it will consider the applicability of this new technology for retrofit to aircraft such as the F-15 Eagle, F-14 Tomcat or FB-111.

STATUS: in development.

Lundy Electronics

Lundy Electronics and Systems Inc, 3901 NE 12th Avenue, Pompano Beach, Florida 33064

Lundy Electronics specialises in all types of expendable countermeasure systems, including aircraft chaff dispensers and missile penetration-aid systems. Several separately designated units (although many of them are variants of a basic design) are in service.

AN/ALE-24 chaff dispenser

Chaff dispensing system for US Air Force/ Boeing B-52G and -H Stratofortress strategic bomber.

STATUS: in service.

AN/ALE-29 chaff dispenser

Widely used on a variety of US Navy tactical aircraft. Each unit can accommodate up to 30 RR-129 or RR-144 chaff cartridges or Mk 46/ Mk 47 infra-red decoy flares. The normal aircraft installation is usually two units. A cockpit-mounted programmer unit allows crew selection of countermeasure deployment, or automatic operation can be initiated by a direct link from threat-warning devices.

STATUS: in service.

AN/ALE-32 chaff dispenser

Chaff-dispensing system for US Navy/Grumman EA-6A Intruder carrier-borne electronic countermeasures aircraft.

STATUS: in service.

AN/ALE-33 chaff dispenser

Chaff-dispenser system for Teledyne Ryan BQM-34 Firebee target drone and rpv.

STATUS: in service.

AN/ALE-43 chaff cutter/dispenser pod

This is a high-capacity chaff system which holds rolls of chaff material and cuts it to the appropriate dipole length during operation. Lundy claims to have overcome the difficulties associated with past chaff-cutters by developing a unique chaff-roving supply system.

The pod has a ram-air intake in the nose and a chaff-roving hopper in the centre section. Behind this is the chaff-cutter system, and it can be presumed that the processor controlling operation of the cutter is in the nose section,

ahead of the hopper. An existing pod profile, that used for the AN/ALE-2 system, is used.

Chaff is drawn simultaneously from up to nine chaff-roving packages in the hopper. Each roving passes through a guide-tube which terminates at draw rollers and a cutting roller. As dipole lengths are cut, they are discharged into a turbulent airflow produced by ram-air tubes from the pod nose, and so distributed efficiently behind the pod. Each cutter assembly consists of a drive motor, clutch/brake unit and a cutting roller, the latter embodying blades which yield specific combinations of dipole lengths.

Pod length: 13⅚ ft (4·22 m)
Pod max diameter: 1³/₇ ft (0·48 m)
Pod empty weight: 185 lb (83·9 kg)
Pod loaded weight: 535 lb (242·6 kg)
Dispensing rate: up to 1 lb/s (0·45 kg/s)
On-time: select 1-9 s in 1 s steps, or continuous
Off-time: select 1-9 s, in 1 s steps
Pod attachments: for NATO 14 or 30 standard pylons

STATUS: in production.

AN/ALE-44 chaff dispenser pod

This is a chaff or flare dispenser system suitable for supersonic aircraft, usually installed as a two-pod system with a control unit in the cockpit. Each pod houses two dispenser modules and a sequencer. The pods are lightweight units which can be installed on wing-tip, underwing or under-fuselage store locations. They have dual-channel dispensing capability and can be quickly reloaded with RR-129 chaff or Mk 46 infra-red flare cartridges. The cockpit unit permits selection of burst rates, burst interval, and units per burst. Flares and chaff can be dispensed simultaneously.

Pod length: 6³/₄ ft (2060mm)
Pod section: 4³/₈ × 7⅛ inches (110 × 180 mm)
Pod empty weight: 30 lb (13·6 kg)
Pod weight with 32 chaff cartridges: 44 lb (19·9 kg)
Pod weight with 32 flare cartridges: 50 lb (22·6 kg)
Control unit weight: 2½ lb (1·1 kg)
Modes: 1 or 2 units per burst
Programs: 1, 2, 4, 8 or continuous bursts per program
Rate: 4, 2, 1 or ½ bursts/s

STATUS: in production.

Lundy supersonic countermeasure pod

Developed by Lundy Technical Centre, this new system is suitable for a wide range of supersonic strike aircraft. It comprises a control unit and two dispensing pods, each pod holding two dispensing modules and a sequencer. Total system capacity is 64 RR-129 chaff dispensers or Mk 46 infra-red flare cartridges. Flight-qualified for supersonic flight, and with a low frontal area, the system is marketed as being suitable for installation in a pod or on a pylon. The control unit permits selection of burst rate, burst interval and units per burst. Specification is identical to AN/ ALE-44 unit, also built by Lundy.

STATUS: in development.

FAC countermeasure dispenser

This is a lightweight system designed for the US Air-Force and successfully flight-tested on Cessna O-2 and Bell UH-1 aircraft. It carries up to 20 Mk 50 infra-red decoy flares or chaff cartridges, and primary use is in forward air controller (FAC) aircraft.

STATUS: in development.

Lundy new countermeasure system

A recent development has been a 14-cartridge chaff/flare dispenser which can be accommodated on virtually any surface by using a contoured mounting plate. Aircraft countermeasure loads can be tailored to missions by using different numbers of dispensers, and chaff cartridge capacity is claimed to be adequate to protect aircraft such as F-16 and F-4 Phantom within 2 seconds of ejection, against radars in E to J band.

Dispenser
Dimensions: 710 × 130 × 150 mm
Weight: 31 lb (14·2 kg) with flares

Cartridge
Dimensions: 40 mm dia × 127 mm
Weight: (I/R) ⁷/₁₀ lb (0·32 kg) (Chaff) ³/₅ lb (0·3 kg)
Flare: 15 kW output, 2–5 μm range
Chaff: RCS of 20 m² within 1-2 s of ejection

MB Associates

MB Associates, PO Box 196, San Ramon, California 94583

AN/ALE-38/41 bulk chaff dispenser

This bulk chaff system carries up to 300 lb (136 kg) of expendable countermeasure material in rolls (RR171-6). Metallised glass, aluminium chaff or aluminium-backed Mylar

tape can be loaded. The wavelength of hostile emissions detected by any electronic surveillance measures systems will cause the unit to dispense payload cut to the appropriate dipole wavelengths.

The unit is contained in a podded installation and was initially developed for US Air Force/ Republic F-105 Thunderchief and McDonnell Douglas F-4 fighter-bombers. It was also adopted later for the Teledyne-Ryan AQM-

34G/H Firebee rpv, which in operations as a chaff-seeding drone is known as Combat Angel. Examples have also been supplied to Israel. Production status is uncertain, all known contracts having been completed.

A version of the US Air Force unit was supplied to the US Navy in the late 1970s under the designation AN/ALE-41.

STATUS: in service.

McDonnell Douglas

McDonnell Douglas Electronic Company, 2600 N, Third Street, Box 426, St Charles, Missouri 63301

AN/ALQ-76 noise jamming pod

This unit was derived from the earlier AN/

ALQ-31 and is part of the AN/ALQ-99 system. It is a podded unit, which is superseding older types of noise and deception jamming systems. McDonnell Douglas undertook design for the US Navy, and was awarded prime contractor status, although development and production

is with Raytheon. Operating frequencies are between 2 and 8 GHz, and 650 pod units will be supplied for use on EA-6A Intruder, EA-6B Prowler and A-4E Skyhawk.

STATUS: in service.

Magnavox

Magnavox, Government and Industrial Electronics Company, 1313 Production Road, Fort Wayne, Indiana 46808

AN/ALR-50 radar warning receiver

A very substantial US Navy programme which has passed production phase but units are still

widely used in the service. The system was a development of the early AN/APR-27 system, also a Magnavox-produced unit, and it in turn now seems likely to be superseded by the AN/APR-67. Operating frequencies were probably in the range 4 to 20 GHz (G to J band), and production was almost continuous throughout the early 1970s, with as many as 800 units,

worth $22 million, being ordered in 1972. At least 1300 sets were delivered to the US Navy and used on A-4 Skyhawk, EA-6A Intruder, EA-6B Prowler, A-7 Corsair, F-8J/RF-8G Crusader, RF-4B/F-4N Phantom, F-14 Tomcat and RA-5G Vigilante.

STATUS: in service.

AN/ALQ-91 communications jammer

Developed for the US Navy as a counter-measure system in the vhf/uhf and hf communications waveband. It forms part of the tactical homing and warning system for F-14 Tomcat fighters and A-4M Skyhawk light bombers. An order for 98 sets was made in 1971, and although it remains in service, more advanced sets are now believed to be replacing it.

STATUS: being withdrawn from service.

AN/ALQ-108 iff deception set

Representing a little-publicised sector of the electronic warfare market, this system is used in US Navy E-2C Hawkeye, EP-3E Orion and S-3A Viking types to improve survivability in anti-submarine warfare and elint operations. Production of about 300 sets is reported, most systems having been delivered in the early 1970s. It seems likely that this is still an important system.

STATUS: in service.

US Navy/Grumman F-14 Tomcat fighters have Magnavox AN/ALR-50 radar-warning receivers

Motorola

Motorola, Government Electronics Division, PO Box 1417, Scottsdale, Arizona 85252

AN/ALQ-122 noise jamming system

Production is in hand for 274 sets to equip all US Air Force/Boeing B-52G and H Stratofortress strategic bombers. System development began in 1975 and was initially referred to as the Soviet noise operation equipment (Snoe). It is a high-power noise-jammer capable of generating multiple false targets on Soviet SA-3 and Tu-126 Awacs-type radars, and therefore operating probably only at the lower end of the electro-magnetic spectrum.

STATUS: in production.

Northrop

Northrop Corporation, Defence Systems Division, 600 Hicks Road, Rolling Meadows, Illinois 60008

AN/AAQ-8 infra-red countermeasures pod

This podded system is a self-contained unit with a ram-air turbine-powered generator in the nose, and a rearward-facing infra-red counter-measure aperture which provides full rear hemisphere protection. Units have been supplied to the US Air Force for use on tactical aircraft.

STATUS: in service.

AN/ALQ-135 noise/deception jamming system

This is an internal countermeasure set developed as a component of the tactical electronic warfare system for the US Air Force/McDonnell Douglas F-15 Eagle fighter. It operates with the AN/ALR-56 radar-warning system.

The AN/ALQ-135 is a noise/deception jamming system which uses dual-mode, pulsed/continuous wave, travelling-wave tube transmitters. All equipment is mounted internally and jamming system management is provided by the AN/ALR-56 radar warning receiver processor. No details of operating frequency range or installation data has been revealed, but it is thought to use a four-antenna array providing 360-degree azimuth jamming coverage.

Initial development funding commenced in August 1974, and led to an initial $25 million production contract in September 1975. This has been followed by further production contracts worth $350 million by 1981. By 1980 more than 300 sets had been delivered and production is expected to continue beyond 1982.

STATUS: in production.

F-15 Eagle Tactical Electronic Warfare System (TEWS) includes a Northrop AN/ALQ-135 noise/deception jamming system carried internally

AN/ALQ-136 esm jammer

One of two compact internal radar-jammer systems (the other is AN/ALQ-162) currently available from Northrop. No technical data has been released and, as yet, it is uncertain whether this system will be committed to production. It is intended for use on aircraft too small to accommodate the new AN/ALQ-165 advanced self-protective jammer system.

STATUS: in development.

AN/ALQ-155 ecm system

This is a powerful system, almost certainly combining noise and deception jamming techniques, and currently being installed in US Air Force/Boeing B-52G and -H Stratofortress strategic bombers. It is broad-band, complementary to both the AN/ALQ-122 unit which operates at the low-frequency end of the electronic warfare spectrum and the AN/ALQ-117 unit which is an 8 to 10 GHz jamming system. Existing AN/ALT-28 noise-jammer equipment is believed to be integrated into the new system, which uses eight antennas (as against six in previously installed systems).

An initial production order was let in June 1978 at a value of $27·9 million for 27 aircraft sets. A further 64 sets were ordered in the following year, and additional orders sufficient to equip all 269 B-52G and -H variants currently in service, are expected. Production will continue until 1983.

STATUS: in production.

AN/ALQ-162 esm jammer

Funded by the US Navy, and expected to be given a production go-ahead in late 1982, the AN/ALQ-162 is a small radar-jamming system which can be supplied with its own receiver/esm management processor, or made compatible with many existing types of radar warning receiver processor systems. Internal or podded installations are proposed. It is virtually complementary to the AN/ALQ-165 advanced self-protective jammer, and is therefore suitable for US Navy A-4M Skyhawk, F-4J and RF-4B Phantom, A-7E Corsair and other types not eligible for advanced self-protective jammer fitment on account of age.

It is claimed to use advanced jamming techniques, and probably has dual-channel, pulsed/continuous wave, transmitter elements. A reprogrammable threat library/techniques generator system can be provided. The receiver/processor and power supply/transmitter units are built as two separate items but they may be installed as a single unit. Proposed future applications of the system are in NATO tactical aircraft, probably F-16, and US Army fixed-wing and helicopter types.

STATUS: in development.

AN/ALQ-171 ecm system

Possibly using AN/ALQ-162 technology, the AN/ALQ-171 ecm system is a lightweight self-contained receiver/processor/transmitter system proposed for conformal installation in a slender pod beneath any Northrop F-5 variant, including the new F-5G Tigershark single-engined project. The unit extends along about one-third of the fuselage centre-line length, and is claimed to permit good aerodynamic performance, and to be compatible with the use of any weapon or stores combination. A reprogrammable threat-processor is included in the total system and modular system design features have been embodied. No development or production plans have been revealed, but a demonstration system seems likely to be flown concurrently with the F-5G aircraft development programme, in 1984.

STATUS: in development.

Raytheon

Raytheon, Electro-magnetic Systems Division, 6380 Hollister Avenue, Goleta, California 93117

Raytheon handles many major electronic warfare sub-contracts and in 1980 was ranked eighth largest contractor in the US market, with an estimated $85 million turnover that year. The company does not, however, hold prime-

RCA

RCA, Government Systems Division, Building 206-1, Route 38, Cherry Hill, New Jersey 08358

AN/ALQ-127 tail warning radar

Initial concepts for US Air Force Phase IV ecm

Sanders

Sanders Associates Inc, Daniel Webster Highway South, Nashua, New Hampshire 03060

AN/ALQ-41 track-breaking system

This designation has been quoted as the track-breaking element of the US Navy AN/ALQ-99 tactical noise-jamming system installed in Grumman EA-6B Prowler and General Dynamics EF-111A electronic-warfare aircraft.

STATUS: in service.

AN/ALQ-81 jamming pod

In production between 1971 and 1974 and supplied to the US Navy for use on LTV A-7B and A-7E Corsair variants. Total funding exceeded $20 million.

STATUS: in service.

AN/ALQ-92 communications jamming system

This designation has been quoted as the communications jamming element of the US Navy AN/ALQ-99 tactical noise-jamming system. This particular unit may be installed only in the Grumman EA-6B Prowler carrier-borne electronic warfare aircraft, and is understood to use an antenna below the forward fuselage.

STATUS: in service.

AN/ALQ-94 noise deception jammer

This is an internally-mounted system used in US Air Force/General Dynamics F-111 fighters and FB-111 strategic bombers. Production of about 500 sets has been reported, and although still widely used it is being replaced by the AN/ALQ-137 system in the FB-111.

STATUS: in service.

AN/ALQ-100 deception jamming pod

The AN/ALQ-100 is being superseded by the AN/ALQ-126 system, but since the late 1960s it has been used by the US Navy on A-4 Skyhawk, A-6 Intruder, EA-6B Prowler, A-7 Corsair and F-14 Tomcat. The system operates in the 2 to

AN/ALT-13 noise jammer

About 1000 examples of this unit were supplied to the US Air Force for use in the Boeing B-52 Stratofortress. The unit covers 2 to 8 GHz waveband and is now being withdrawn from service.

AN/ALT-28 noise jammer

Produced for B-52D, -G and -H variants of the US Air Force/Boeing Stratofortress strategic bomber, and as these represent the only models which remain in front-line use (in operational and training squadrons), this is likely to be the sole noise-only jammer type still in service on B-52s. Installation into the 340 or so B-52s still operational was due to be

contractor status on any current large airborne electronic warfare programmes.

The most significant airborne electronic warfare system to which Raytheon contributes, at present, is the AN/ALQ-76 noise jamming pod (designed by McDonnell Douglas, but developed by Raytheon which is now producing 680 production sets). This unit is used in

update of US Air Force/Boeing B-52G and -H Stratofortress bombers called for an active tail-warning radar. RCA was contracted to produce five sets in 1974, and these prototype items were delivered for evaluation, although production did not follow. The unit was a pulse-Doppler radar, and in view of the subsequent

completed in 1978. Frequency range capability is said to exceed that of the 2 to 8 GHz AN/ALT-13 unit, and may therefore have been extended into higher frequencies.

STATUS: in service.

the US Navy AN/ALQ-99 programme managed by Grumman.

A high proportion of airborne jamming systems made in the USA uses transmitters and antennas produced by Raytheon, and the company has a substantial involvement in ground-based and seaborne electronic warfare systems.

decision to install Westinghouse AN/ALQ-153 tail-warning radar in the current B-52 Phase VII electronic countermeasures fit this project is now concluded.

STATUS: development only.

US Navy/LTV A-7E Corsair IIs were among aircraft that had Sanders AN/ALQ-81 jamming system

8 GHz band and has been the major US Navy tactical-aircraft jamming system for many years. Funding exceeded $130 million before effort was switched to the AN/ALQ-126, which is also a Sanders-produced system. Initial systems were followed by an updated design, called Pride, and the role of these units has been stated to be the jamming of Soviet air-to-air missile radiation frequencies, using either noise or break-track jamming techniques.

STATUS: in service.

AN/ALQ-126 deception jamming system

Replacing the AN/ALQ-100, and initially referred to as the AN/ALQ-100X system, it will be used on US Navy A-6 Intruder, A-7 Corsair, F-4 Phantom and RA-5C Vigilante. It is now the most important jammer in US Navy service, with some 1500 sets either on order or delivered. Full-scale production was preceded by a programme in 1974, worth $53.2 million, to convert 464 ALQ-100 sets to ALQ-126 configuration, and although peak production rate was passed in the late 1970s, the AN/ALQ-126 is still being manufactured. It is internally-mounted, and examples have been supplied to the Royal Netherlands Air Force for use on its F-104 Starfighters.

Frequency range: thought to be 2 to 18 GHz
System weight: 190 lb (86 kg)
System volume: 2·3 ft³ (0·065 m³)

STATUS: in production.

AN/ALQ-132 infra-red countermeasure set

One of several countermeasure, airborne infra-red (CAIR) projects designated in the early 1970s. This unit was developed as the CAIR II system, and was in simultaneous production with the CAIR I (AN/ALQ-123), which was a Sanders/Xerox venture. AN/ALQ-132 uses propane fuel to heat a ceramic block which then produces an intense source of infra-red radiation. The output is modulated at frequencies designed to break the track-lock of infra-red homing missiles. It is a podded system and was to be compatible with A-4 Skyhawk, A-6 Intruder, A-7 Corsair, A-10 and OV-10 aircraft. The system is reported to have been carried by P-3C Orion, but production rate and operational uses have not been given by official sources.

STATUS: in service.

AN/ALQ-134 expendable chaff-size ecm jammers

Funding for the programme was about $1 million in 1975-76, but no further contracts have been announced. During 1974-75 Eaton Corporation/Cutler-Hammer had conducted similar studies (resulting in the AN/ALQ-112). The AN/ALE-37 and AN/ALE-39 chaff dispensers are designed to accept such stores. It is assumed that research did not indicate a feasible solution.

AN/ALQ-137 ecm system

Forming, with the AN/ALR-62 radar warning receiver, a comprehensive internally-mounted electronic warfare system for the US Air Force/General Dynamics F-111 tactical fighter and FB-111A strategic bomber, this system is in full-scale production. It supersedes the AN/ALQ-94 noise/deception jammer, and is referred to as an advanced development, so probably retains a similar dual-channel, continuous wave/deception, jamming capability, but with more advanced processing. Trials were conducted in 1974 and the set was first ordered in 1977 under a $24 million contract. The equipment may also be installed in US Air Force/General Dynamics EF-111A electronic warfare aircraft.

STATUS: in production.

AN/ALQ-140 infra-red countermeasure set

Derived from the AN/ALQ-132, and initially called CAIR III (countermeasure, airborne infra-red), the system is specifically designed for installation in McDonnell Douglas F-4 Phantom fighter-bombers. It comprises a heated ceramic block, using an electrical-heating element, and the unit fits in place of the tail-parachute doors. Funding has been sporadic and it is not clear whether the set has been committed to production.

STATUS: in development.

AN/ALQ-144 infra-red countermeasure set

Similar to other Sanders-produced infra-red countermeasure sets, this system uses a heated ceramic block. It has been designed for helicopters and was tested on US Army/Bell AH-1 and UH-1 attack and utility types before production commenced in 1978. The set weighs 31 lb (14 kg).

STATUS: in service.

AN/ALQ-146 infra-red countermeasure set

Funded by the US Navy, and in production for use on Boeing CH-46D assault helicopters. The infra-red source is an electrically-heated ceramic block. Flight tests were conducted in 1976 and production orders were placed in 1977, initially for 15 test sets only. Up to 275 sets were due to be ordered for operational use. Two units are used on each CH-46D, one above the cockpit and one above the tail-loading doors. Crews can select the modulating frequencies used to break missile track-lock.

STATUS: in production.

AN/ALQ-147 infra-red countermesure set

A variant of other heated ceramic block infra-red countermeasure systems developed by Sanders, the AN/ALQ-147 is designed for aircraft which cannot support the large power-generation load associated with electrical heating of the ceramic block, and instead JP fuel is burned in a ram-air filled duct. The AN/ALQ-147(V)1 unit for the OV-10 Bronco is mounted on the rear of a 150-gallon fuel tank, and the infra-red unit is directable. Installation includes a filter which reduces visible emissions, and therefore makes the system suitable for night operations.

STATUS: in production.

AN/ALQ-156 missile detection system

This is a small pulse-Doppler radar which is claimed to be an omni-directional detector of missiles approaching any aircraft on which it is installed. Initial applications are US Army/Boeing CH-47C assault helicopter and RU-21 liaison types. On these aircraft the system responds to infra-red seeking missile threats by triggering the release of infra-red decoy flares. Pulse-Doppler techniques are used to provide good coverage when flying at low altitudes.

There are four units; a receiver/transmitter, a control unit, and two antennas, weighing in total about 44 lb (20 kg).

The processing element is a reprogrammable digital unit which can be developed to operate in helicopter or fixed-wing aircraft, and correlated with laser or radar-warning receiver units.

STATUS: in production.

Sperry

Sperry Corporation, Great Neck, New York 10020

Jampac(V) noise jamming system

Developed for light strike/attack aircraft and helicopter applications, and especially for export versions of the US Air Force/Northrop F-5, this is a compact and relatively simple jamming system. Jampac units can be attached to aircraft singly or in sets of up to three (typically) or four (maximum). They provide self-protection during action, or can be used in peace-time for combat-system evaluation or electronic counter-countermeasures training. Within each unit is a transmitter operating in the range 1 to 15·5 GHz, using 30 per cent bandwidth voltage-tuneable magnetrons (vtms). Two vtm units can be accommodated in one Jampac. Noise-power output per jammer is 150 to 400 watts (continuous wave), and power-management receivers control jammer-frequency tuning. Where a radar warning receiver system is already installed in an aircraft, the power-management receivers can be deleted.

Jampac units are carried underwing or fuselage, and should not interfere with stores carriage. Aircraft performance deterioration with Jampacs attached is said to be negligible.

Frequency range: 1-15·5 GHz with 30% bandwidth vtms

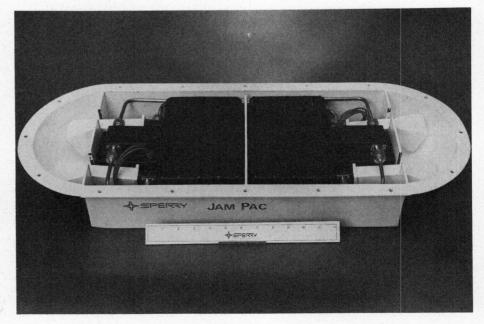

Sperry Jampac(V) unit fitted under aircraft fuselage or wing

Power output: 150-400 W (cw)
Jammers/Jampac: 1 or 2

Unit weight: 60 lb (27 kg)
Unit volume: 0.84 ft³ (0.027 m³)

Support Pac(V) noise jamming pod

Designed to provide high-power jamming capability for aircraft such as the Northrop F-5 light fighter. Two versions of Support Pac(V) are available; one being an aircraft-powered unit and the other a self-contained unit using a ram-air turbine. Either one or two pods can be used simultaneously. The pods are claimed to be useful for tactical electronic counter-measures support, stand-off jamming or electronic counter-countermeasures training, and have standard weapon-station attachments. Two voltage-tuned magnetron (vtm) jammers are installed in each pod, and overall frequency-range in which jamming can take place is 1 to 15·5 GHz. Solid-state jammers can be added to provide up to 150 watts additional power at lower frequencies.

Sperry Support Pac(V) ecm pod

Frequency range: 1-15·5 GHz using vtms, below 1 GHz using solid-state transmitters
Power output: 150-400 W (cw) per vtm jammer, 150 W per solid-stater jammer
Pod length: (3 canisters) 8½ ft (2·59 m) (2 canisters) 6½ ft (1·99 m)
Pod max diameter: 10 inches (0·25 m)

Pod weight: (3 canisters) with ram-air turbine 267 lb (121 kg), without ram-air turbine 255 lb (114 kg)
(2 canisters) with ram air turbine 197 lb (89 kg), without ram air turbine 185 lb (84 kg)
Vtm dimensions: 290 × 184 × 115 mm

Vtm weight: 25 lb (11 kg) incl antenna
Solid-state jammer dimensions: 150 × 105 × 46 mm
Solid-state jammer weight: 5 lb (2·25 kg)

STATUS: development completed.

Texas Instruments

Texas Instruments Inc, PO Box 225474, MS 240, Dallas, Texas 75265

AN/ALR-21 infra-red warning receiver

Developed for the US Air Force as a tail-mounted system suitable for General Dynamics F-111 and Boeing B-52 types. The specification called for detection of approaching aircraft or missile launch events in the rear hemisphere. Only development funding has been recorded, around 1970, and the system was not installed in operational aircraft.

STATUS: development only.

AN/APR-38 radar warning computer

An unidentified model of the Texas Instrument airborne computer range is used as the radar-warning processor in the AN/APR-38 Wild Weasel system. The unit has a reprogrammable threat library.

STATUS: in service.

AN/APR-41 advanced radar warning receiver

This unit was developed in the mid-1970s for US Army helicopters, but beyond approximately $1 million funding sanctioned in 1972-73, it does not appear to have been placed in production.

STATUS: development only.

Tracor

Tracor Inc, 6500 Tracor Lane, Austin, Texas 78721

AN/ALE-29 chaff dispenser

Described in Lundy entry. Both companies are suppliers of this widely-used chaff/flare dispenser unit.

STATUS: in service.

AN/ALE-40 chaff dispenser

This system was developed to meet a US Air Force requirement calling for a high counter-measure payload capability, simple handling and loading procedures, improved decoy effectiveness over current systems, high reliability, and a modular design which could be adapted easily to different aircraft types. Design commenced in 1974, and concentrated initially on a RR-170 chaff-cartridge-compatible unit for the McDonnell Douglas F-4 Phantom fighter-bomber. In 1977 the system was tested with MJU-7/B infra-red flares.

The F-4 installation units are now designated AN/ALE-40(V)1, 2, 3. There are two small cock-pit units, and four similar dispensers which can be attached on either side of the two inboard armament pylons. Each dispenser can accommodate 30 RR-170 chaff cartridges, or the outer modules can carry 15 MJU-7/B flares. With each rechargeable cartridge module is a

AN/ALE-40 chaff/flare dispenser by Tracor for US Air Force/General Dynamics F-16 fighter

streamlined nose-cap. Aerodynamic drag is said to be similar to that of a Sidewinder missile and launcher.

A smaller system, AN/ALE-40(N) was developed for the Royal Netherlands Air Force/Northrop NF-5 fighter. In this variant there are two skin-mounted dispensers which attach to the rear fuselage, each of which carries 30 RR-170 chaff cartridges, or 15 MJU-7/B flares. Loaded weight is less than 32 kg. A similar unit has been flown on a Lockheed F-104 Starfighter.

The AN/ALE-40(V)4, 5, 6 is a variant for internal-installation within the US Air Force/General Dynamics F-16 aft fuselage. Capacity is the same as that of the AN/ALE-40(N). A similar-size system, designated AN/ALE-40 (V)7, 8, 9 is available for internal mounting within the F-5E/F left wing root.

An extremely large system, AN/ALE-40(V)10 in an internal installation for the US Air Force/Fairchild A-10 anti-tank aircraft. It comprises up to 16 dispensers holding as many as 30 cartridges each; 480 RR-170 chaff cartridges or 480 M-206 infra-red flares constitute the maximum capacity. Four dispensers are mounted in each wing-tip and each wheel-well.

Other variants of the AN/ALE-40 are operationally deployed on Hunter, G-91 and Mirage. All systems have cockpit control-units tailored to individual requirements, and which allow pilot selection of salvo/burst rate, ripple rate, and so on. All units show the number of cartridges remaining in each dispenser.

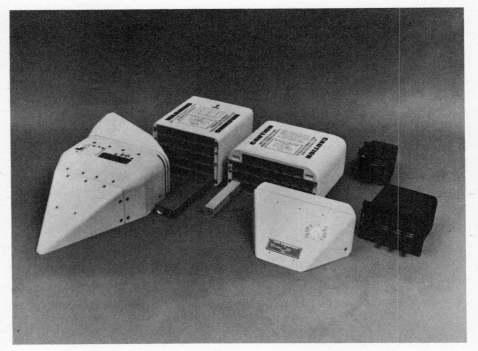

Tracor AN/ALE-40 chaff/flare dispenser for McDonnell Douglas F-4

	F-4 pylon	**F-16 internal**	**A-10 internal**
Dispenser length	outb'd 528 mm inb'd 348 mm	216 mm	193 mm
Dispenser depth	123 mm	269 mm	260 mm
Dispenser width	241 mm	170 mm	147 mm
Load weight	(120 chaff) 48·1 kg	(60 chaff) 22·5 kg	(480 chaff) 158 kg
	(60 chaff + 30 flare) 58 kg	(30 chaff + 15 flare) 25·2 kg	(240 chaff + 240 flares) 168 kg
		(30 flares) 27·9 kg	(480 flares) 177 kg

Power (standby): 0·1A 28 V dc
(dispensing at 100 ms intervals) 6A 28 V dc

STATUS: in production.

M-130 general-purpose countermeaure dispenser

Developed from AN/ALE-40 experience, and incorporating a greater proportion of modular design features, this system bears a strong family resemblance to its predecessor. Each payload module can accommodate 30 cartridges; either M-1 countermeasure chaff or M-206 flare types, and a double-dispenser

module is also available with capacity for 60 cartridges. The cartridges are 25 mm square and 210 mm long.

In addition to the dispenser there is a remote control box and a cockpit-mounted control unit on which the crew can select single-shot or ripple deployment of countermeasures. The controller also indicates the number of chaff/flare cartridges remaining.

Dispenser dimensions: 323 × 114 × 226 mm
Capacity: 60 chaff, 30 chaff/30 flare, or 60 flare cartridges
Total weight: with three payloads above, either 24·6, 26·2 or 27·9 kg
Power requirement: 6 A 28 V dc (when dispensing at 100 ms intervals)

United Technology Laboratories

United Technology Laboratories, 410 Kirby, Garland, Texas 75042

AN/ALQ-133 elint system

This is a highly-accurate direction-finding electronic warfare system which can operate over wavelengths up to 18 GHz. It is installed in Grumman EV-1 Mohawks operated by the US Army, and a variant system is produced for the US Air Force/Fairchild A10 tank destroyer.

An interferometer antenna configuration is used to provide search over a 90-degree sector abeam of the aircraft, and it can locate emitters

to a typical accuracy of 0·5 degree. A 120-degree sector coverage is possible with reduced direction-finding accuracy. Currently the antenna is installed in a pod, and the associated data-processing equipment is in another pod. The processor analyses intercepted signals to determine which are hostile radars. A third pod may be added to the A-10 to transmit processor output data to the ground via a data-link.

The project was initially funded in 1974, and in 1977 about $63 million was made available, of which $52 million was for production.

STATUS: in production.

US Air Force/Fairchild A-10 tank-destroyer has United Technology AN/ALQ-133 elint system

Watkins-Johnson

Watkins-Johnson Company, 3333 Hillview Avenue, Palo Alto, California 94304

AN/APR-34 elint system

This is a self-contained microwave system which operates between 20 and 60 GHz, and is a development of an early surveillance radar, designated WJ-10007. System development started in 1969, and production commenced two years later. About 225 systems were

delivered to the US Air Force before 1974, and were installed in Lockheed EC-121 Super Constellation and Boeing EC-135/RC-135 reconnaissance aircraft.

STATUS: in service.

Westinghouse

Westinghouse Electric Corporation, Aerospace Division, Baltimore, Maryland 21203

AN/ALQ-101 noise jamming pod

This is one of the US Air Force's most widely-used jamming pods, and it has also been supplied abroad. The project began in 1966 and was referred to as Sesame Seed. Five examples of a development pod, designated QRC-335A/101-1, were developed, each with fore-and-aft-facing antennas and operating in the 2 to 8 GHz frequency range.

The first production units, AN/ALQ-101(V)3, were more powerful versions of the prototype systems, but were soon followed into production by the AN/ALQ-101(V)4, which covered the 2 to 20 GHz frequency range. A further frequency band extension was added in Model AN/ALQ-101(V)6.

The final production version, and the most numerous unit in service, was the AN/ALQ-101(V)8. This was distinctive in being the first electronic countermeasures pod to use a gondola layout, where a trough compartment underneath the main body considerably increases the available volume in the pod, without significantly increasing the cross-sectional area. A further frequency-range extension was incorporated in the system, and many earlier production pods were later updated to this standard.

AN/ALQ-101 pods are also in service in the UK, Israel and West Germany.

Production was as follows:

QRC-335A/101-1	5 prototypes
ALQ-101(V)3	71 pods
ALQ-101(V)4	324 pods
ALQ-101(V)6	58 pods
ALQ-101(V)8	400-plus pods

Pod length: (QRC-335A/101-1) 7¹⁄₂ ft (2·3 m) (all later models) 12³⁄₄ ft (3·9 m)
Pod width: 9⁴⁄₅ inches (250 mm)

STATUS: in service.

AN/ALQ-105 noise jammer pod

This was a repackaged AN/ALQ-101 system which effectively split the original pod in half and located the two resulting units on either side of an F-105 Thunderchief fuselage. About 90 examples of this system were produced.

STATUS: in service.

AN/ALQ-119 noise/deception jamming pod

Initiated as project QRC-522 in 1970 and now one of the most numerous jamming pods in service with the US Air Force. AN/ALQ-119 is a dual-mode jamming pod initially used on the McDonnell Douglas F-4 Phantom, but now used also by Fairchild A-10 and General Dynamics F-16 squadrons. The system has a three-band frequency-range transmitter which almost certainly covers the 2 to 10 GHz range, and probably higher frequencies too. Both noise and deception jamming modes can be employed. Each pod has dual-mode travelling-wave tube emitter elements and represent the first dual-mode jammer to enter service. Used by the US Air Force in 1972, and still in limited production, the pod has the gondola cross-section introduced by Westinghouse on the AN/ALQ-101 pod.

Recent producton has been of the AN/ALQ-119(V)15 version, which is recognisable alongside earlier versions of the pod by the addition of a radome below the front end of the gondola portion. Earlier versions of the pod in US Air Force service are being upgraded to this standard. Features of the (V)15 version include

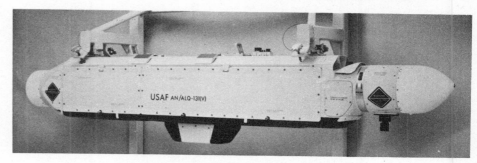

Westinghouse AN/ALQ-131(V) ECM pod

Westinghouse AN/ALQ-101(V)8 ecm pod

automatic control of power radiated, frequency selection and signal type. A shorter-body version, the AN/ALQ-119(V)17, is also operated by the US Air Force.

Hughes Aircraft has been designated second-source supplier for this equipment, the total production run for which exceeds 1600 units, at production rates of between 10 and 30 sets per month. Sets are operated by Israel and West Germany. It will be superseded, in US Air Force service, by the AN/ALQ-131 system.

STATUS: in service.

AN/ALQ-131 noise/deception jamming pod

The first development contract for this system, let by the US Air Force in 1972, called for an advanced noise/deception jamming pod covering five wavebands, and incorporating a digital processor for direct control of the emitter elements. The specification also required a modular design suitable for extension or traction to meet specific application requirements. The standard pod appears to comprise a single processor and two independent, fore-and-aft, jamming sets. Motorola was sub-contracted to provide the two lower band jamming units, and Loral developed the processor. Total waveband capability may exceed the range 2 to 20 GHz.

Operationally the pod can be tailored to meet the waveband requirements of differing missions, and the processor is reprogrammable on the flight-line to take account of revised threat priorities, or new threats as they appear. The processor is the power-management unit, and it ranks all received threats to assign priorities and correlate jammer actions.

After one year in development the AN/ALQ-131 was shelved temporarily, but work soon recommenced and flight-test and development got under way in 1976. Initial production contracts were let in that year, and by 1980 production was being funded at $91 million annually, with at least $400 million worth of production still outstanding. Aircraft using the AN/ALQ-131 include F-16, F-4 Phantom, F-105 Thunderchief, A-7 Corsair, A-10 and AC-130 Hercules.

Eventual replacement will be the AN/ALQ-

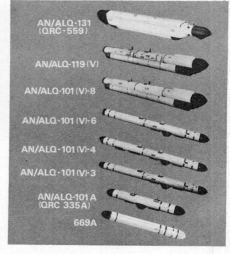

Several Westinghouse noise/deception jammers in US Air Force service

165 internally-mounted advanced self-protective jammer system.

Pod length: 112¹⁄₅ inches (2850 mm)
Pod depth: 24⁴⁄₅ inches (630 mm)
Pod width: 11⁴⁄₅ inches (300 mm)
Weight: 629 lb (286 kg)

STATUS: in production.

AN/ALQ-153 tail warning radar

Development of this system began in 1975, ground-tests were completed in 1976 and a competitive fly-off against a similar set took place in 1976 and 1977. The AN/ALQ-153 was selected in 1978 and was followed by full-scale production.

This is a pulse-Doppler radar that increases protection from rear-closing threats, either aircraft or missiles, and will be a prime sensor for countermeasure deployment and crew warning in the US Air Force/Boeing B-52G and -H Stratofortress strategic bombers. Installation in US Air Force/General Dynamics F-111 and McDonnel Douglas F-15 are also proposed, and Westinghouse claims that the set is small enough for the F-14 Tomcat, F-16, F-18 Hornet and A-10.

STATUS: in production.

AN/ALQ-165 advanced self-protective jammer

Westinghouse and ITT Avionics were awarded contracts to jointly produce the AN/ALQ-165 aspj system in late 1981. A competitive design has been funded also, and was proposed by Northrop Corporation and Sanders Associates.

STATUS: in development (see ITT/Westinghouse entry).

Xerox Corporation

Xerox Corporation, Electro-optical Systems, 300N Halstead Street, Pasadena, California 91107

AN/ALQ-123 infra-red countermeasure system

This infra-red jamming system is a pod-mounted unit with a ram-air turbine powered generator. It is designed to provide protection against SA-7 surface-to-air missiles and is suitable for A-6 Intruder and A-7 Corsair types. Tail installations in other US Navy types have been proposed.

Jamming is achieved by modulating an infra-red source in a manner which will break the lock-on of an infra-red seeking missile. Development was completed in March 1973, and production commenced a few months later. The system is now used by the US Navy and several export customers.

STATUS: in service.

Television and Thermal Imaging Systems

DENMARK

Jørgen Andersen Ingeniørfirma

Jørgen Andersen Ingeniørfirma a-s, 1 Produktionsvej, DK-2600 Glostrup, Copenhagen

SIT/ISIT low-light level television system

Jørgen Andersen's SIT/ISIT system is fundamentally a man-portable low-light level television system designed for use as an electronic news gathering equipment by broadcasting organisations. It has however been applied to a number of airborne military, para-military and civil roles, particularly aboard helicopters. These include reconnaissance, fisheries patrol and inspection, police night surveillance activities, mountain rescue and, of course, night electronic news gathering work.

In such applications, the camera may be shoulder-mounted for use from a helicopter cabin footstep or used with a Ronford fluid 15S mounting head combined with a helicopter shock mount to counteract the effects of rotor-induced vibration.

The system operates over a wide range of lighting conditions down to moonlight and starlight illumination levels. For helicopter work, the system is supplied with a 4½-inch (114 mm) viewfinder and a double pan bar for use with the 15S head. A full range of accessories is available and special equipment obtainable for outside broadcast applications includes a video transmitter-receiver operating in the 2 GHz band, a real-time video image processing system and a man-portable video tape recording unit. The company also supplies a range of motorised zoom lens assemblies which meet the special optical requirements of low light level television systems. With these lenses, says the manufacturer, low-light level television systems can operate over a range of high or low ambient lighting conditions.

Weight: variable, according to lens system fitted, but typically totalling approximately 26·5 lb (12·04 kg) including camera, lens system and nickel cadmium belt-mounted battery power supply.

STATUS: in production and service.

Type 771 low-light level television system

Developed originally for naval applications in conjunction with director radar weapon control systems, the Jørgen Andersen Type 771 camera is a ruggedised low-light level television sensor particularly suited to harsh operational environments. It operates over an ambient lighting range from sunlight to starlight at extremes of temperature and is well-adapted to aircraft applications. Among possible applications are target identification, weapon-aiming, and use with other sensors such as laser rangefinders. Of primary importance to aerospace users is the special attention paid to the Type 771's reliability, including electronic circuit-protection measures. A mean time between failures of over 8000 hours is claimed for this system.

Dimensions: 29½ × 13¾ × 12⁹/₁₀ inches (750 × 350 × 330 mm)

STATUS: in production and service.

FRANCE

TRT

Télécommunications Radioélectriques et Téléphoniques (TRT), Defence and Avionics Commercial Division, 88 Rue Brillat-Savarin, 75460 Cedex 13, Paris

TRT has been concerned with development of thermal-imaging technology since 1964 and has worked under a number of French Government contracts in developing a range of experimental equipment designed to check the validity of the technical approaches, and to explore their operational possibilities. A number of operational systems have been produced under this development programme. Further development work, in collaboration with Société Anonyme de Télécommunications, SAT, has been conducted with the aim of developing a range of common modules for the French Ministry of Defence.

Helicopter-borne camera

In 1973, TRT developed a helicopter-borne thermal camera system under contract from the Services Techniques des Télécommunications de l'Air. The system is characterised by a 36 × 13 degree field of view and has been installed on gyrostabilised platforms aboard Aérospatiale Puma helicopters.

STATUS: in service.

Thermidor thermal camera

Thermidor thermal cameras, developed in 1976 under contract from the French Section d'Etudes et Fabrications des Télécommunications have accumulated several thousand hours of operation.

STATUS: in service.

Hector thermal imager

TRT's Hector system is an airborne thermal imager designed for Aérospatiale's Dauphin helicopter for night and poor-visibility firing of the Franco/German HOT anti-tank missile, operating in the 8 to 13 micron band. The sensor has array of fewer than 60 detector elements and a series-parallel scanning method is employed, providing an output suitable for either television or light-emitting

Elements of TRT's SMT common modules

diode displays. In the Dauphin application, information is presented on a micro-monitor of the missile operator's day-firing eyepiece. The Hector camera, which provides a choice of wide or narrow fields of view, is integrated in a gyro-stabilised platform for helicopter mounting.

STATUS: in production and service.

Cantor thermal imager

Initiated under a contract from the Service Technique des Télécommunications de l'Air, Cantor was built for flight-trials as a component of a weapons system designed to permit helicopter launching of HOT anti-tank missiles by day or night. As in the case of Hector (see preceding entry) the Cantor system operates in the 8 to 13 micron range, has fewer than 60 detector elements and uses series-parallel scanning to provide an output compatible with either television or light-emitting diode display systems.

The system is to be used with an automatic tracking or a direction-finding device on a missile tracker.

STATUS: in development.

Modular thermal system (SMT)

Under a common modules development programme for the French Defence Ministry, TRT, in collaboration with SAT, has developed a range of thermal-imaging modules to meet the requirements of the three armed services. Under an inter-company agreement, each is responsible for the manufacture of specific modules and either one may be the prime contractor, dependent on particular project requirements and current workload, for any system application. Design co-ordination authority however is vested in TRT.

Initiation of the common modules programme followed a study conducted by TRT and SAT on behalf of the French Ministry of Armed Forces, to define a thermal imaging system to meet national specifications and to establish the technical design. The resulting parameters were:

eight to twelve micron spectral bandwidth with series-parallel scanning; full television display compatibility on a standard CCIR 625-line cathode ray tube; a single, double or triple field of view optical system requiring a minimum number of lenses; a limited number of detector elements; possibility of application to light-emitting diode displays; ease of coupling with image-processing systems; open or closed-circuit cryogenic cooling; highly stabilised optic axis; reliable behaviour in the presence of heat sources such as missile traces or fires.

The design aims included a number of requirements which are generally similar to those of the common modules programmes of other nations. These include:

easy adaptation to a variety of applications; an increase in the number of systems able to be produced due to standardisation of the camera (ie sensor) units; easy incorporation of new developments; and facilitation of fault-finding and qualification testing of camera systems through the use of already approved modules. Additional advantages from the programme include reduction in study, development and production costs for a given application, simplification of maintenance, enhanced reliability and reductions in production timescales and training.

The basic module range comprises:

a scanner module, designed to explore the field of view at the appropriate frame and line frequencies in order to feed the detectors with a suitable signal, using series-parallel scanning techniques; a detector module comprising a cadmium mercury telluride, two-dimen-sional mosaic detector cooled to 77° K and associated pre-amplifiers; a linear electronics module designed to amplify, rephase for each line, and shape the signals from the various detector module channels; a signal-processing module for producing a television-compatible, standard CCIR 625-line video output from simultaneous signals produced by the linear electronics unit; display modules for direct viewing of the field of view by the operator through a light-emitting diode viewing lens, or for indirect viewing on a remote cathode ray tube monitor screen, depending on the application; a cryogenic module designed to cool the detector array; and a test and diagnosis module which runs internal self-tests for fault-detection and isolation.

The modules may be assembled in various configurations depending on particular requirements. They fulfil the basic functions of a thermal imager while specific elements are added to tailor the unit to particular applications; for example, a larger or smaller field of view would require a new optical train, and the television monitor display would be added when remote viewing of the image is desired.

Typical performance and characteristics quoted for a long-range system are diagonal fields of view of 2·6 degrees and 7·7 degrees with a detection range of 14·3 miles (23 km) against a small maritime target (eg fast patrol boat), and a so-called reconnaissance range of 9·6 miles (15·5 km) against a land target such as a tank. For missile-guidance purposes a system would offer fields of view of 3·6 degrees and 10·8 degrees with detection and reconnaissance ranges of 2·6 miles (4·2 km) and 1·92 miles (3·1 km) respectively, against land targets such as tanks.

TRT flir sensor package

TRT modular systems have been applied to both fixed-wing aircraft and helicopters for a number of surveillance roles including target detection, identification and designation, and for reconnaissance and navigational purposes.

Weight: (experimental version) 20·9 lb (9·5 kg) (military version) 15·4 lb (7 kg)

STATUS: in production and service.

UNITED KINGDOM

British Aerospace

British Aerospace Dynamics Group, Manor Road, Hatfield, Hertfordshire AL10 9LL

The Dynamics Group of British Aerospace is broadly responsible for all of BAe's non-aircraft projects. Its principal products are tactical guided weapons and space satellites and their associated systems, but the group's activities also cover the design and manufacture of gyroscopes, radomes, air-conditioning systems, hybrid and integrated micro-circuits, propellers, sightline stabilisation systems, and a wide variety of aerospace products including infra-red equipment.

Linescan 401 infra-red reconnaissance/surveillance system

The British Aerospace 401 infra-red linescan system is designed for low-level, high-speed air reconnaissance and can be installed in a wide variety of aircraft types or equipment pods. The system operates in the 8 to 14 micron band and employs a five-element mercury-cadmium-telluride detector array which is cooled to liquid-air temperature by a Joule Thomson high-pressure air minicooler. Typical endurance times from a standard air cylinder are 120 minutes at plus 20° C soak temperature and 45 minutes at plus 70° C soak temperature. Detector cool-down time is less than one minute.

Imagery is recorded on standard 70 mm photographic film which is contained in a large-capacity, quick-change magazine. Automatic velocity/height ratio control is provided and the flight-navigation data is recorded on the film edge, reducing both aircrew and photo interpreter workloads.

The transverse field of view covers 120 degrees and can be offset left or right to provide horizon/horizon coverage. Roll stabilisation to plus or minus 55 degrees is provided and allows high angles of aircraft bank without loss of continuity of the scanned area or the resulting imagery.

Cockpit control items include an event-

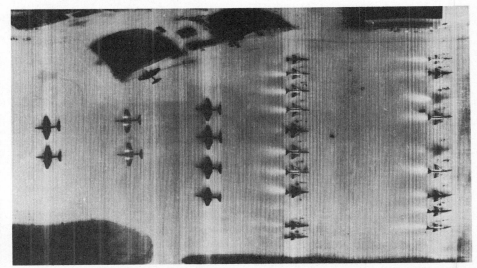

Combat airfield pictures by British Aerospace Linescan system. White areas denote heat; note hot engine casings on the two Canberras

marker and a film-footage indicator. Built-in self test circuitry activates a cockpit warning lamp in the event of equipment unserviceability.

Details of spatial resolution, thermal sensitivity and velocity/height ratio are classified. The equipment is said to be compatible with a number of high performance strike aircraft. Currently, British Aerospace Linescan systems are offered as pod installations by several manufacturers including Vinten and Aeronautical General Instruments in the United Kingdom, MBB and Dornier in West Germany, Pacific Aero Systems in the USA, and Hawker de Havilland in Australia.

Dimensions: (linescan) 23³/₄ × 12³/₅ × 11 inches (604 × 320 × 280 mm)
(roll swept dia) 14²/₅ inches (366 mm)
(cooling pack) 15¹/₃ × 9 × 4³/₄ inches (390 × 230 × 120 mm)

Weight: (linescan) 74·8 lb (34 kg) (cooling pack) 19·8 lb (9 kg)

STATUS: in production and service.

Linescan 214 infra-red reconnaissance/surveillance system

The British Aerospace 214 Linescan is a member of the company's range of 200 Series equipment, claimed to be the first lightweight production infra-red surveillance systems in the world. The 200 Series systems were initially developed for the UK Government for military application and several hundred 201 systems have been delivered and are in service in a number of countries. The 201 system is, for example, the standard equipment aboard the Canadair CL89 surveillance rpv. Another member of this series is the Linescan 212 system, designed for use in light aircraft and helicopters. Likewise, a number of these systems are also in service throughout the world.

Both the 201 and 212 systems record their imagery on standard 70 mm photographic film, but the 214 Linescan has been developed to meet a requirement for real-time imaging, although facilities for recording the output have been incorporated.

The 214 system operates in the 8 to 14 micrometre band. It has an instantaneous field of view of 1·5 milliradians with an across-track scan of 120 degrees. Thermal resolution (rms noise equivalent temperature) is 0·25° C at 22° C. The system is cooled by liquid nitrogen and will operate up to five hours from a single reservoir charge. An integral film recorder, using standard aerial photographic film, is incorporated and the six-metre capacity cassette provides a typical coverage of 65 × 1 km at 300 metres altitude (40 × 0·6 miles at 1000 feet). Velocity/height range is within the band 0·1 to 1 radians per second depending on the recording method in use. Selection of velocity/height ratio is made at the cockpit control unit in a series of eight steps. The unit also contains an automatic gain-control switch which provides overland/oversea selection in four steps. A video inversion control is also incorporated which permits hot targets to be displayed in either black or white polarity.

The mixed video and synchronisation pulse output is compatible with a wide range of recording and data transmission systems and equipment. Real-time oscillographic recorders suitable for use with Linescan 214 are available from a number of manufacturers, including Honeywell and Medelec, and there is a choice of recording paper and image development processes according to the particular application. Tape-recording systems for recording and playback, either in the air or on the ground, are also available from such manufacturers as Racal or Sony. Suitable data-transmission equipment is manufactured by Meteor and Pacific Aerosystems of Italy and the United States respectively.

With these two latter organisations, British Aerospace markets the sensor with the Meteor 'Alamak' air-to-ground data-link and ground station in a combined system entitled 'Rigel', of which a simplified variant known as 'Rigel Sim' is also available. Both systems permit infra-red sensor information to be transmitted directly to the ground for immediate viewing. The data-link can also send commands to the aircraft directing the pilot to fly over a selected area. The ground station also includes a map display which indicates the aircraft's position and flight path. Further details of Rigel are given in a subsequent entry.

Linescan 214 may be either directly installed in the aircraft's fuselage or within an underwing or underbelly pod which may also contain conventional photographic reconnaissance cameras. Suitable pod systems are manufactured by Vinten and Aeronautical and General Instrument of the UK, MBB and Dornier in West Germany, Pacific Aerosystems in the USA and by Hawker de Havilland in Australia.

Dimensions: (linescan sensor) 10¹/₆ × 17¹/₅ × 12¹/₂ inches (258 × 437 × 318 mm)
(cooling pack) 10³/₄ × 5¹/₂ × 4¹/₈ inches (273 × 140 × 105 mm)
(control unit) 4²/₅ × 5³/₄ × 4²/₅ inches (113 × 146 × 113 mm)
Weight: (linescan sensor) 26·4 lb (12 kg)
(cooling pack) 3·74 lb (1·7 kg)
(control unit) 2·42 lb (1·1 kg)
STATUS: in production and service.

Rigel infra-red linescan/television reconnaissance system

The Rigel system comprises a British Aerospace Linescan 214 sensor, a low-light level television camera, and the Italian Meteor air-to-ground data-link. This assembly is normally pod-contained for either underwing or fuselage

DAY

NIGHT

British Aerospace Linescan 401 day and night pictures of yacht marina and oil storage tanks

British Aerospace Linescan 401

British Aerospace Linescan with print-out aboard aircraft

centre-line mounting but may be installed inside the fuselage if space permits. The system's low-light level television camera is mounted in the nose of the pod and has a motorised zoom/focus facility for ground control. The Linescan 214 infra-red sensor is centrally mounted, together with its cooling-pack and control unit, and views the ground through an open window in the underside of the pod with an unobstructed arc of 120 degrees across track. All Linescan functions are controllable from the ground, including on/off selection of photographic-film recording in addition to real-time air-to-ground imagery transmission.

Also contained within the pod are electronic units including the command receiver/decoder and power amplifier, the tv/tm transmitter, altitude and airspeed transducers, antennas for

tv/tm transmission and command signal reception and the electrical supplies. A small battery, for ground-checkout purposes, is mounted at the rear of the pod.

In normal use, the ground-based operator can remotely select all major functions without reference to the pilot in the aircraft, but facilities to display sensor information can be installed aboard the aircraft if required. During surveillance, the ground operator can scan the terrain with the low-light level television camera and select individual targets for the infra-red Linescan unit. Typically, infra-red imagery is available at the ground station within ten seconds of the target being overflown. The infra-red sensor may also be operated in a continuous-scan mode.

The pod itself has NATO standard (STANAG 3726) attachments at 14 inch (355 mm) centres.

The Rigel ground station consists of three mobile units mounted on military-pattern, four-wheel-drive truck chassis each carrying a 250 volt 50 Hz 250 kVA power supply. One of the trucks contains a fully air-conditioned ground station on which is mounted the main antenna. In the station is a television monitor screen to display the low-light level television camera picture, and a trace recorder for infra-red imagery. The station also contains a map-plotter display which continuously records the position of the aircraft. A control panel is provided whereby the operator can remotely select the various sensor parameters, including high-quality photographic film recording for subsequent processing on the ground. The second truck contains the film processing and interpretation facilities, while the third vehicle is used to accommodate calibration and maintenance equipment for the aircraft pod and the ground station.

The aircraft infra-red sensor specification is the same as that of the standard Linescan 214 equipment (see preceding entry). The low-light level television camera can be either a 625-line, 50 Hz or 525-line 60 Hz unit with a dynamic light range of $5 \times 10^6{:}1$ and automatic gain control. Signals from both the thermal imaging and television sections are transmitted to the ground over L-band links with a choice of five fm frequencies. The command link likewise is a five-frequency unit but can be either L-band or uhf. The airborne vehicle can be tracked at ranges of up to 78 miles (125 km).

In the ground station, the aircraft position is displayed on a 29·9 × 29·9 inch (760 × 760 mm) plotting board with presentation on a map. Altitude and airspeed are displayed by analogue and digital means, range and azimuth position in digital form only. Television images are shown on a 12-inch (304·8 mm) monitor. Infra-red linescan recording is oscillographic with a paper development time of between five

British Aerospace Linescan 204 is similar to 214, but has no external cooling

and fifteen seconds. Grey scales are from 6 to 16 and resolution of the system is eight lines/mm. A magnetic tape signal recording system is optionally available.

A simplified variant of Rigel, known as Rigel Sim, is similar to the major system but dispenses with the low-light television element. In the Rigel Sim airborne pod, the low-light television camera and its gimbal mountings are replaced by ballast weights and the altitude and airspeed sensors are also deleted. The Rigel Sim ground station is correspondingly simplified and contains no television monitor or aircraft plotting facilities.

Rigel has been jointly designed and developed by British Aerospace Dynamics Division in the UK, by Meteor SpA in Italy and Pacific Aerosystems in the USA. The Linescan

sensor is produced in Britain and the remaining parts of the Rigel system in Italy but for the US market, over 90 per cent of the total system will be produced by Pacific Aerosystems in San Diego, California.

Dimensions: (pod) 57^1/$_{10}$ × 15 inches dia (1450 × 380 mm dia)
(infra-red sensor) 10^1/$_6$ × 17^1/$_5$ × 13^1/$_2$ inches (258 × 437 × 343 mm)
(cooling pack) 10^3/$_4$ × 5^1/$_2$ × 4^1/$_{10}$ inches (273 × 140 × 105 mm)
Weight: (pod) 132 lb (60 kg);
(infra-red sensor) 26·4 lb (12 kg)
(cooling pack) 4·4 lb (2 kg)

STATUS: in production and service.

Ferranti
Ferranti plc, Robertson Avenue, Edinburgh EH11 1PX

Type 221 thermal-imaging surveillance system

The Type 221 thermal-imaging surveillance system is designed for service with military helicopters. Developed by Ferranti in conjunction with Barr & Stroud Limited, the system incorporates an IR18 thermal imager and telescope by the latter company, with sightline stabilisation steering provided by a Ferranti stabilised mirror. The assembly is contained in a pod which can be mounted either beneath the nose of a helicopter or can project through an aperture in the aircraft's floor.

The Barr & Stroud IR18 imager unit provides a normal field of view of 38 degrees in azimuth and 25·5 degrees in elevation. In the Type 221 application users can choose a telescope magnification of either ×2·5 or ×9, with corresponding wide or narrow fields of view. The wider field of view (15·2 degrees azimuth and 10·2 degrees elevation) would be used for general surveillance, target acquisition or navigation purposes. The narrow field of view permits detailed observation for target identification or engagement of targets detected in the wide field of view mode. In the narrow field of view case, the angles are 4·2 degrees in azimuth and 2·6 degrees in elevation. With the Ferranti sighting mirror in the pod installation, the system has fields of regard of plus 15 to minus 30 degrees in elevation and plus or minus 178 degrees in azimuth. The entire sensor system is vertically mounted above the mirror which is angled, periscope fashion, at 45 degrees to the horizontal, to provide views in the horizontal plane.

The sensor system employs Mullard Sprite detector units cooled by means of a Joule Thomson minicooler supplied with high-

Ferranti/Barr & Stroud Type 221 surveillance system

pressure compressed air. The air source is a bottle, mounted on the equipment and charged immediately prior to flying. This has a capacity of one litre and provides a system operation time of approximately 2·5 hours. If greater endurance is required, then other cooling options, involving the use of mini-compressors permanently connected to the equipment, are optionally available. The system operates in the 8 to 13 micron band and has a sensitivity of between 17 and 35° C to target background and surroundings.

The Ferranti-produced mirror has an aluminium reflective element which is diamond machined to a flatness of two fringes at 550 nanometres. This mirror has a reflectivity of greater than 97·5 per cent at 45-degree incidence, averaged over 8 to 12 micrometres. Its stabilisation system comprises a two-axis device with integrating rate gyros as rate sensors. The mechanism is driven in each axis by a direct-drive dc torque motor while steering is obtained by torquing the integrating rate gyro. Angular information is derived from a resolver fitted to each axis. The mirror sightline is controlled by means of signals provided by an electronics unit which also provides power and signals to the main turret azimuth drive on the pod. System control may be exercised either through a digital computer or by a hand controller unit. According to Ferranti, the use of the mirror system provides a higher degree of stabilisation than is normally attainable by other methods and claims that the resultant image is blur-free under typical aircraft vibration conditions. Infra-red spectrum vision is obtained via a germanium window on the front of the mirror turret assembly.

The display may be presented on either 525- or 625-line television monitors. The output is either in CCIR or EIA composite video formats, as required. This television-compatible output may be displayed on one or more monitors situated at various locations throughout the aircraft, and Barr & Stroud emphasises the desirability of having monitors located at both the pilot and winch operator positions of a helicopter in order to co-ordinate crewmembers for better hover control during night and bad visibility winching operations.

The Type 221 system has been developed to a UK Ministry of Defence specification and has undergone both Royal Navy and Royal Air Force trials. It is intended for service with medium-to-large helicopters and roles envisaged include maritime reconnaissance, search and rescue, and integration with on-board weapon systems to improve their all-weather capability.

Dimensions: (pod unit) 34 inches (865 mm) × 16½ inches (420 mm) max dia
(electronics unit) 5 × 17 × 13 inches (127 × 432 × 330 mm) max.
Weight: (scan head unit) 8·36 lb (3·8 kg)
(electronic processor unit) 5·28 lb (2·4 kg)
(power supply unit) 2·64 lb (1·2 kg)
(total IR18 thermal imager) 16·28 lb (7·4 kg)
(×2·5/×9 telescope) 18·48 lb (8·4 kg)
(pod complete) 165 lb (75 kg)
(electronics unit) 17·5 lb (8 kg)
(total system) 183 lb (83 kg)

STATUS: under development.

Lasergage

Lasergage Limited, Newtown Road, Hove, East Sussex BN3 7DL

Hand-held thermal imager

The Lasergage hand-held thermal imaging system was originally developed by the Philips Electronics MEL Division. However, following an enthusiastic reception for the system in the civil helicopter market (particularly in the United States) the unit was not built in quantities sufficient for the market, and MEL disposed of its opto-electronics business in early 1982, withdrawing from this sector of the industry. The further development, manufacturing and marketing of the thermal imager has since been undertaken by Lasergage, a small but successful opto-electronics manufacturer best known for lightweight laser ranging and target-designation equipment.

Lasergage has been acquiring certain marketing rights for the Flir Systems Inc 100A thermal imaging system. The opportunity to obtain the assets of MEL's thermal imaging development and production facilities, together with the services of key scientists and technicians, enabled the company to form the base of development, manufacturing and support capability to promote the sales of both systems.

The hand-held system is compact, lightweight and self-contained, being powered from a small battery pack which may be either belt-mounted or carried over the operator's shoulder by a sling strap. Operating in the three to five micron band, the imager is unusual in that it requires no external cooling supplies, the detector cooling being achieved by thermo-electric methods. This factor makes the system particularly suitable for use in aircraft or in any confined space where the presence of pressurised gas bottles could constitute a physical hazard. The freedom from dependence on a cooling gas supply also extends the operational duration of the system, a single battery charge being capable of maintaining operation for periods of up to eight hours or more. While the battery powered method of operation is ideal for complete portable use, the system may however derive its power supply from standard 12- or 24-volt vehicle batteries.

In standard form, the unit has a field of view of ten degrees in azimuth and seven degrees in elevation with a lens providing a ×1 magnification. Optionally, a ×0·44 magnification lens is available offering a field of view of 22·5 degrees azimuth and 16 degrees elevation. Other lenses, offering longer focal lengths for long ranges and greater angular resolutions, and shorter focal lengths for wider fields of view, are understood to be in development.

The Lasergage imager employs a mechanical system for scanning an array of 12 detectors placed at the focal point of the objective lens. Sequential scanning is carried out by an eight-faceted rotating mirror and the detector output, following processing, is fed to a bank of light-emitting diodes which are again scanned by the mirror at a point diametrically opposite the detector bank. Light output from the light-emitting diodes is fed to a small eyepiece screen and appears to the operator as a picture similar to that produced by a 96-line television screen which, despite the relatively small number of lines, produces a high-quality image of good resolution.

The system is extremely easy to operate, possessing only four controls, an on/off switch and knobs which control focus, brightness and contrast. It is normally held to the user's eye by grip straps on the side of the unit's casing but a removable pistol grip is also provided. For prolonged operations from the footstep of a helicopter, a chest harness may be used to relieve the operator of the need to support the imager but it is in fact small and light enough to be supported by bungee strips suspended from the top of a cabin or cockpit door aperture, thereby not only removing the need to support the unit but also eliminating the problem of aircraft vibration, which tends to affect the use of some portable, direct viewing eyepiece systems.

The thermal imager's eyepiece may be fitted with adaptors to permit attachment of still or cinematography cameras to obtain permanent records. A television camera interface may also be used for remote display on a monitor or for videotape recording. When not in use, the entire system, including the NiCd rechargeable battery pack, can be stored or transported in a briefcase-sized carrying case.

Lasergage hand-held thermal imager aboard JetRanger helicopter

The system is operable over a temperature range varying from minus 20 to plus 40° C and can detect targets exhibiting a temperature differential of less than 1° C from their background or surroundings. This degree of sensitivity led to the system being instrumental in the discovery in a JetRanger helicopter, after less than one hour's flying, of a dead body which had defied all attempts at discovery by 100 police officials, aided by dog teams, over a two week period during a search in southern England.

Dimensions: (hand-held imager (excluding pistol grip)) 12⅕ × 6³⁄₁₀ × 4¾ inches (310 × 160 × 120 mm)
(shoulder battery pack) 5¼ × 2⁹⁄₁₀ × 2¾ inches (133 × 75 × 70 mm)
(carrying case) 18 × 13³⁄₅ × 6½ inches (457 × 345 × 165 mm)
Weight: (hand-held imager) 6·6 lb (3 kg)
(battery pack) 4·4 lb (2 kg)
(carrying case (empty)) 2·2 lb (1 kg)

STATUS: in production and service.

Marconi Avionics

Marconi Avionics Limited, Airport Works, Rochester, Kent ME1 2XX

Lantirn head-up display

The Marconi Avionics Lantirn head-up display (low altitude navigation, targeting infra-red for night) is a new technology system which combines the wide angle capability of diffractive (holographic) optical systems with a Marconi Avionics developed method of combining raster and cursive scan information in a single display with the objective of reducing the hardware requirements of day/night head-up display systems. The equipment is further described in the Displays section.

Heli-Tele television system for helicopters

Marconi's Heli-Tele is an airborne television system combined with a microwave link and designed for helicopter applications. The system, developed by the company's Electro-Optical Surveillance Division, provides long-range, real-time surveillance of the ground to meet the requirements of police and military forces and other security authorities.

Heli-Tele comprises a colour camera with a 20:1 zoom lens giving a narrow, 1-degree field of view, mounted on a gyro-stabilised platform aboard the helicopter. The operator can control the pointing angle of the camera and adjust the field of view to display any selected area on the

Marconi Avionics Heli-Tele system on Bell 222 helicopter

ground. Video information thereby obtained is transmitted to any number of ground stations via an omni-directional microwave link. Receiving stations may be located at distances up to approximately 50 miles (80 km) from the helicopter.

The airborne equipment, which has been certificated by the United Kingdom's Civil Aviation Authority, has been fitted to nine different types of helicopter and is in use with a number of police forces.

STATUS: in production and service.

V325 low-light television camera (LLTV)

The Marconi V325 low-light television camera has been designed by the company's Electro-Optical Surveillance Division for fixed-wing aircraft and helicopters to meet pilots' night-vision requirements. Operating from 28 volts dc, the intensified isocon camera system has full air clearance and provides high resolution video pictures over a range of conditions from bright sunshine to largely overcast starlight.

The V325 low-light television camera provides a 24-hour operational capability, allowing pilots to fly with confidence at low level, using the television picture displayed through a head-up display with a 1:1 overlay on the outside-world view.

STATUS: in production and service with the Royal Air Force.

Helicopter night-vision aid

This helicopter night-vision aid has been developed to meet a requirement to provide all-round night vision for helicopter pilots. Development of the Marconi unit was carried out under a contract from the UK Ministry of Defence (Procurement Executive) and the system is due for evaluation at the Royal Aircraft Establishment, Farnborough using a Sea King helicopter as the test vehicle.

The system is produced by Marconi's Electro-Optical Surveillance Division at Basildon, Essex. It comprises a highly stabilised platform, mounted beneath the nose of the helicopter, and contains a night vision sensor. In the case of the Sea King trial installation, the sensor is a Marconi Avionics V325 television camera (see preceding entry). This intensified isocon camera, which is already in production and in service with the Royal Air Force, can produce a usable picture at light levels down to overcast starlight conditions.

The stabilised platform has also been designed to carry interchangeable payloads. These include thermal imagers from the United Kingdom's thermal-imaging common modules programme, to which Marconi is a contributor.

STATUS: under development.

Thermal-imaging common modules

Marconi Avionics' Electro-Optical Surveillance Division and the company's major sub-contractor Rank Taylor Hobson are producing a range of thermal imaging common modules under contract from the UK Ministry of Defence (Procurement Executive). These modules, which provide indirect view, television-compatible thermal images, can be used as the building blocks to assemble flir systems that will meet a variety of requirements from the three armed services.

The imagers operate in the 8 to 13 micron band, permitting smoke and haze penetration. They operate from 28 volts dc and are designed to meet a comprehensive range of applications, including a number of airborne roles. Initial limited production has commenced in order to support future UK thermal imaging system requirements.

STATUS: in production.

Units of Marconi Avionics common modules

Marconi Avionics V325 camera system

Lltv picture seen through Marconi Avionics head-up display

Marconi Avionics helicopter night vision aid mounted on Sea King nose. Note also helmet-mounted display on pilot

This thermal imager is assembled from Marconi Avionics common modules

Rpv sensor payloads and command ground station

Marconi Avionics is known to have manufactured a number of lightweight air-vehicle payloads suitable for remotely piloted vehicles. Sensors employed are believed to include both conventional and image-intensified low-light-level television cameras and thermal imaging systems, and would undoubtedly suit the company's own Machan RPV, which first flew in February 1981.

A typical rpv payload would accommodate an attitude-stabilised sensor and command receiver and a video transmitter, decoder telemetry, and power supply. Payloads have been successfully flown on trials conducted for the UK Ministry of Defence and have undergone stringent environmental testing. The ground station used to control these payloads is housed in a CB300 container box body which contains the main command computer, trials analysis equipment, and the pilot and image interpreter.

STATUS: in development and limited production.

Marconi Avionics ground-control station for rpvs

Downward-looking sensor in a Marconi Avionics rpv payload

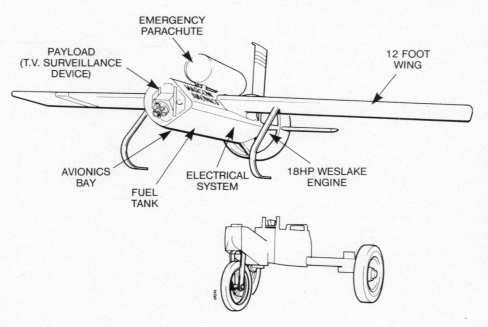

Payload installation in Marconi Avionics Machan rpv. Rear-mounted engine permits unobstructed view by payload sensors

Thorn EMI

Thorn EMI Electronics Limited, Victoria Road, Feltham, Middlesex TW13 7DZ

Multi-role thermal-imagers

Thorn EMI has developed a number of thermal imaging systems as part of the company's involvement in the UK Ministry of Defence thermal-imaging common modules programme. Both Class 1, direct viewing, and Class 2, remote viewing, systems have been developed under this programme and Thorn EMI has produced a lightweight multi-role military system using common module elements.

The basic thermal-imaging sensor comprises a cadmium mercury telluride detector array which is series/parallel scanned by a mechanical polygonal rotating mirror system. The detector is cooled by a Joule Thomson minicooler, using compressed air from a 0·6 litre bottled supply which is charged to a pressure of 300 atmospheres. One charge of the compressed air bottle provides an operating duration of not less than 2½ hours and a typical duration from a single charge is about 4½ hours. Power is supplied from a 1 A-hour battery which, at an imager power consumption of approximately 7 watts, provides a 3½ hour minimum duration.

The system operates in the 8 to 13 micron band and has a detection sensitivity of 0·5 K over an operating temperature range of minus 19 to plus 52° C. Two selectable focusing ranges are provided, one with a narrow field of view and with a focal range from 30 metres to infinity, the other focusing from ten metres to infinity with wide field. Image polarity of either white/hot or black/hot is also switch selectable. A focus athermalisation facility is provided; this control enables focus to be maintained when switching between wide and narrow field of view, at a given temperature. A temperature offset control and a window enables the operator to emphasise different temperature ranges within the thermal scene under surveillance.

When used in the direct viewing role, the imager's display is on a scanned light-emitting diode display but, as is more likely for airborne application, an indirect viewing system is also provided. This system comprises a further module which can be added to the basic imager to provide CCIR/625-line television compatibility. Single or multiple indirect television monitors may be used.

The system has been flight-tested aboard a Scout helicopter of the Royal Aircraft Establishment, Farnborough, with a view to using the system in possible remotely piloted vehicle applications.

Dimensions: (thermal imaging unit) 19 × 11 × 7½ inches (480 × 280 × 190 mm)
(indirect viewing module) 7½ × 5½ × 2 inches (190 × 140 × 60 mm)
Weight: (thermal imaging unit (including compressed-air supply bottle and battery)) 25·3 lb (11·5 kg)
(indirect viewing module) 2·86 lb (1·3 kg)

STATUS: under development.

UNITED STATES OF AMERICA

General Electric

General Electric Company, Aircraft Equipment Division, French Road, Utica, New York 13503

AN/ASQ-145 low-light level television system

The AN/ASQ-145 is a low-light level television system designed to provide fire-control facilities for the US Air Force/Lockheed AC-130 Hercules gunship aircraft, on which it is a primary sensor. The system is unusual in that it uses a dual camera installation in order to provide both wide and narrow fields of view simultaneously. General Electric was the contractor responsible for system integration, test and alignment as well as production.

STATUS: in production and service.

Fairchild

Fairchild Weston Systems Inc, 300 Robbins Lane, Syosset, New York 11791

AN/AXQ-16V cockpit television sensor

The Fairchild AN/AXQ-16V cockpit television sensor has been developed to provide a means of recording real-time gunsight or head-up display information and cockpit audio in fighters and tactical-strike aircraft. It comprises a small, all-solid-state television camera which employs a charge-coupled device (CCD) photosensor array to achieve high sensitivity and dynamic range, together with an electronic unit which may be used either as a camera mounting-platform or may be remotely sited. A militarised airborne video tape recorder is used to record imagery signals generated by the camera. The camera can also drive a monitor to provide a real-time presentation of visual data for rear cockpit display purposes. The camera's optical system employs a 31 mm, f/2·8 lens with automatic iris control, and can operate at reduced light levels.

The airborne tape recorder recommended for use with the cockpit television sensor is a TEAC V-1000AB-R, a remotely-controlled unit which uses U-Matic ¾-inch video tape cassettes for the recording medium. Recording times of either 20 or 30 minutes are possible and the tape provides dual audio tracks for an accompanying voice commentary. This recorder has been designed for extended stand-by time operation with instant record start-up time.

The system can be used to replace existing film cameras or to interface with any head-up display. The split design permits a mounting to provide minimum obscuration of the pilot's view and for remote mounting applications a variable-angle, low-profile sensor head replaces the higher profile unit used when the sensor is mounted directly onto the system's electronics unit. The cockpit television sensor has been flight-tested in a number of US Air Force Tactical Air Command aircraft and is specified for the F-14, F-15, F-16 and the A-10. It has also been proposed for remotely-piloted vehicles, automatic missile guidance, and as general-purpose flight-test instrumentation. The system is marketed in the United Kingdom by Marconi Avionics.

Dimensions: (video sensor head) .3¼ × 9/10 inches dia (81·83 × 22·35 mm dia)
(video sensor head and electronic unit combined overall dimensions) 7⅖ × 4 inches (188 × 102 mm) with height variable according to installation
(video tape recorder) 6⅕ × 13 × 9⅗ inches (157 × 330 × 244 mm)
Weight: (combined sensor head and electronic unit) 2·43 lb (1·1 kg)
(video tape recorder' 23 lb (10·45 kg)

STATUS: in production.

Flir Systems

Flir Systems Inc, 11830 S W Kerr Parkway, Lake Oswego, Oregon 97034

System 1000A thermal detection/surveillance system

Flir System's 1000A forward looking infra-red is a television-compatible thermal-imaging system designed for helicopter application. Like most other airborne thermal detection and identification systems, the 1000A is a multi-role equipment which may be used for night-time law-enforcement purposes such as covert surveillance, intruder detection and border patrol; search and rescue; industrial work such as power-cable and pipeline inspection; and for agricultural and forestry purposes including crop disease detection, wildlife surveys and fire protection.

The system's sensor comprises a single, repeating-scan mercury cadmium telluride detector which is sensitive to emissions in the 8- to 12-micron band. Detector cooling is cryogenic, using compressed argon contained in a 3-inch (76·2 mm) cylinder which provides from six to eight hours nominal continuous operation from a single charge. The sensor is housed in a weather-resistant polyester sphere which is mounted under the nose of the helicopter and which, it is claimed, produces only low aerodynamic drag coefficients at airspeeds up to 120 knots.

Infra-red radiation is admitted to the sensor sphere via a shatter-proof germanium window. The sensor's field of view is 17 degrees in elevation and 27 degrees in azimuth; the sphere however is mounted in a yoke and gimbal system which permits a field of regard of plus or minus 105 degrees in azimuth and from plus 30 to minus 90 degrees in elevation. A cockpit-mounted joystick control is used for sphere steering and a light-emitting diode display adjacent to this control indicates the angular position of the sensor in 15-degree increments. The sensor system's focal range is adjustable from 5 feet (1·5 metres) to infinity and the depth of field is 25 feet (7·6 metres) to infinity at the infinity focus.

The standard display for the 1000A system is an EIA RS-170 525-line television monitor but an optional CCIR 625-line system may also be used. Thermal polarity of either white-hot or black-hot may be selected as desired to optimise thermal background discrimination, and the images may be video-tape recorded for post-flight analysis.

Once aircraft installation has been completed, the 1000A system's sensor pod can be removed or replaced in a matter of seconds by simple engagement or release of a locking bar. This feature permits quick transfer to be made from one aircraft to another or for the system to be removed in the field for use on a tripod mounting or for attachment to a land vehicle or surface vessel.

Flir Systems Inc has been energetically promoting the application of thermal-imaging technology in civil helicopters and the company has discovered a number of unusual roles for the equipment in the course of its research. Among applications proposed by the company is the use of such systems in aerial structural surveys to determine the water content in roofing and to detect concrete delamination in road beds and bridge structures. During recent eruptions of the volcano Mt St Helens in Washington State, a Flir system installed in a JetRanger helicopter was flown into the volcano's crater and round the lava dome, the infra-red imagery revealing many hot spots on the dome where magma had pushed up from beneath the surface.

Weight: total system excluding power supply 57·5 lb (26·08 kg)

STATUS: in production and service.

System 100A infravision thermal detection/surveillance system

Flir's 100A thermal imaging system is generally similar to the company's 1000A system (see preceding entry), making use of the same technologies and some of the equipment modules. It is, however, slightly lighter and more easily tailored to specific applications. The basic system comprises three modules, a scanner, a control unit and a television monitor. The infravision control unit governs six functions; on/off, reverse/normal polarity, brightness, contrast, detail and line trim. The reduction of the number of controls, says the manufacturer, renders the system as simple to operate as a standard television monitor.

For airborne applications, a spherical pod mounted in a yoke and gimbal assembly, as in the case of the 1000A system, is also available. The triangular support plate is a standard Spectrolab 'Nightsun' model which permits bolt fixing to be made either through holes at each apex or through four centrally-located apertures. The entire sphere can be removed or replaced in the yoke by release of a retainer bolt at the yoke axis. The control unit and television monitor may be mounted at any convenient location within the aircraft. As in the case of the 1000A model, a control stick and position indicator are also available, the latter indicating sensor pointing angles in both elevation and azimuth at 15-degree increments on a light-emitting diode display. The control stick is designed to be strapped to its operator's leg in an aircraft environment. An auxiliary switch on the control stick can be connected to a high-intensity lamp which is brought into operation by pressing a switch.

The Model 100A Infravision system is built and tested to MIL-SPEC 810-C, and has a temperature operating range of between minus 15 and plus 55° C. Currently, says Flir Systems, the mechanical and electro-servo control systems are undergoing FAA approval for use on Hughes 300 and 500 and Bell 206 JetRanger helicopters.

Field of view of the 100A is the standard Flir Systems format of 27 degrees azimuth × 17 degrees in elevation. The standard lens is a poly-crystal germanium unit with a high-efficiency anti-reflective coating. This has a focal range of 1·5 metres to infinity and an

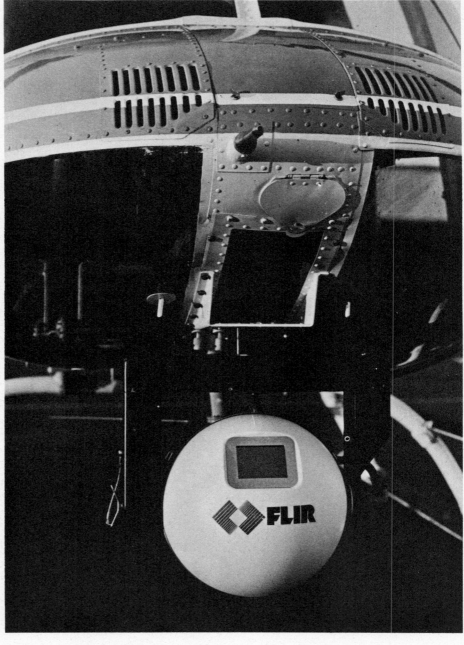

Flir Systems 1000A sensor under nose of Bell 206 JetRanger helicopter

F-number of 1. Telescopic lenses are also optionally available and include a ×2·5 unit, offering a field of view of 11·2 degrees azimuth × 6·8 degrees elevation, and a ×5 lens with azimuth and elevation field of view of 5·6 degrees and 3·4 degrees respectively.

The 100A model is also being marketed in certain parts of the world by the British-based company Lasergage, which manufactures a hand-held thermal imaging system particularly suitable for aircraft applications.

Weight: (position indicator) 1·5 lb (0·68 kg)
(control stick) 0·75 lb (0·34 kg)
(gimbal and yoke) 13·25 lb (6·02 kg)
(sphere containing scanner and argon supply) 17·75 lb (8·068 kg)

STATUS: in production and service.

Ford

Ford Aerospace and Communications Corporation, Aeronutronic Division, Ford Road, Newport Beach, California 92663

Forward looking infra-red system for F/A-18

The Ford F/A-18 flir system has been designed to provide the US Navy's McDonnell Douglas F/A-18 Hornet fighter with a day, night and adverse-weather attack capability by presenting the pilot with real-time thermal imagery in a television formatted display for the location and identification of tactical targets. The system enables selected targets to be automatically tracked and provision is incorporated for inclusion of laser target-designation.

The system's sensor and electronic section is contained in a pod mounted on the lower left side of the aircraft's fuselage at a Sparrow guided-missile station. In the cockpit, the pilot can select a displayed field of view of 3 × 3 degrees or 12 × 12 degrees and is able to control the system's stabilised line-of-sight over a field spanning plus 30 to minus 150 degrees elevation with plus or minus 540 degrees of roll freedom. Targets may be automatically tracked in either field of view throughout the field of regard.

The F/A-18 flir interfaces with the aircraft and its avionics systems through the aircraft mission computer, receiving commands directly over a MIL-STD-1553A multiplexed digital-data bus. The system provides the mission computer with accurate target line-of-sight pointing angles, angle-rates and a complete pilot-initiated and automatic periodic built-in test evaluation of readiness. The system is the first US Navy flir attack sensor designed for operation in a supersonic environment, and the pod is subject to dynamic flexure at such speeds. To compensate for this, boresight compensation is employed to ensure appropriate pointing accuracies throughout the flight envelope.

Ford says that special emphasis has been placed on reliability and maintainability of this system which is expected to exceed a 200-hour mean time between failures contractor demonstration requirement. The system comprises ten weapon replaceable assemblies which can be readily accessed and replaced without the need for calibration, alignment, special tools or handling equipment. The mean time to repair is predicted to be significantly less than the 12-minute requirement for replacement of these assemblies.

Full-scale development of the F/A-18 flir

commenced in March 1978 and progressed through completion of one development unit and six pre-production models. Initial flight testing on a Rockwell T-39D/Sabreliner, amounting to 50 hours of airborne operation, was conducted during the period November 1980 to mid-January 1981 and demonstrated satisfactory results with regard to stabilisation, automatic tracking and video quality. Further flight tests on F/A-18 aircraft commenced at the US Navy's Patuxent River Air Test Center in June 1981 and were completed in December that year, the system having successfully shown compliance with the specification. Full-scale development, including formal qualification, was being completed by mid-1982 with the full production award being due about the same time, following the placing of a long-lead production contract the previous June. Delivery of the first production equipment is scheduled for late 1983.

Dimensions: 72 × 13 inch dia (1829 × 330 mm dia)
Weight: 340 lb (154·5 kg)

STATUS: in development and production.

Mini-FLIR forward-looking infra-red system

Ford Aeronutronic Division's Mini-FLIR is a lightweight, compact, television-compatible equipment designed principally for light fixed-wing aircraft, helicopter and remotely piloted vehicle applications. There is no definitive Mini-FLIR system as such; rather, Ford offers a variety of options, based on a single type of sensor unit, which can be assembled in such a manner as to meet particular customer requirements. For example, a range of systems may be combined with the basic sensor, according to the detection ranges required. Likewise, various sensor cooling systems may be applied, again to meet a particular requirement.

The sensor unit comprises a mechanical raster scanner, detector array, and the associated electronic circuitry. The basic scanner is a cylindrical unit 4 inches (101·6 mm) in diameter and 5 inches (127 mm) in length and with a weight of 3·5 lb (1·59 kg). It operates in the 8 to 12 micron band and has a field of view of 30 × 40 degrees. Mechanical scanning of the field of view is accomplished with a 525-line raster at a field/frame rate of 60/30 per second. Serial scan techniques are employed using two multi-element arrays of mercury cadmium telluride detectors, with the two arrays time-shared and combined through an accoustic delay line. The number of detectors used depends on the overall system sensitivity required.

The detector arrays are mounted in a glass Dewar flask and cryogenic methods are used to cool them to a temperature of 77 K. Here again, depending on factors such as operating time, logistics, mounting position and cost, any of several differing cooling systems can be employed. Typical techniques include a Joule-Thomson cryostat used in either an open or closed cycle system, direct contact-cooling

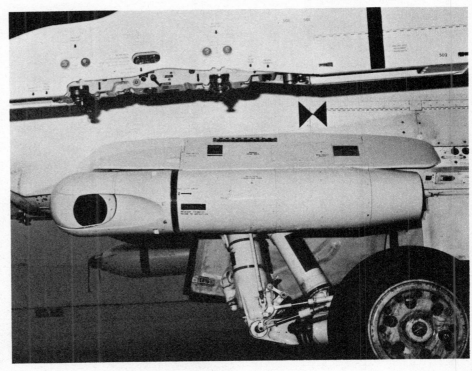

Ford's flir pod installed on US Navy/McDonnell Douglas F/A-18 fuselage by left main-wheel

with a liquid coolant supply, or the use of a closed-cycle mechanical refrigeration system. The basic viewing system elements are completed by a standard television display unit. Video tape recording facilities or a data-link can be incorporated into the system if required for any particular application. If the system is equipped with telescopic optics, then its use in a moving vehicle may bring about a need for sightline stabilisation and this can be allowed for in the system design.

A number of such systems, tailored for specific applications, have been produced by Ford. An example is that of the units manufactured for the US Forest Service which employs the Mini-Flirs as a night navigation aid for helicopters. The sensors in these systems provide the standard 30 × 40 degree field of view but are steerable in elevation from plus 10 to minus 30 degrees from the horizon. Resolution and sensitivity are sufficient for them to be employed as terrain search and navigation aids and the capability of the sensors to penetrate smoke and haze further aids visibility under poor conditions. The sensor, refrigeration, electronics and power-control modules are installed in the nose bays of the helicopters while the control panels and twin displays are cockpit mounted. All sub-systems are interchangeable in the field.

Ford has also produced specialised equipment for mini-rpv use. These systems have triple field of view optics and mechanical stabilisation to isolate the operator's sightline and sensor field of view from perturbations of the carrying vehicle. The stabilisation system

also provides means of aiming at and of tracking objects through wide azimuth and elevation pointing angles. The optical system provides a wide-angle, 65-degree field of view for navigation purposes, a 15-degree angle for detection and target acquisition, and a 4-degree narrow-angle view for target tracking and recognition. The two high-resolution, narrower fields of view are downwards-looking for use with a stabilised sight and the widest field of view is horizontally aimed through a separate aperture, as a navigation aid.

This existing design, says Ford, is also appropriate (with increased aperture for longer range) for fuselage-integrated or pod-mounted installations on high-performance manned aircraft. For the latter type of installation, aircraft modification requirements would be limited to provisions for cockpit display and controls and for possible wiring changes to the pod-mounting station. Pod-mounted precision pointing and tracking systems developed by Ford Aeronutronic are currently in use aboard a variety of helicopters and fixed-wing aircraft.

The basic Mini-FLIR scanner sensor with 0·45 inch (11·4 mm) aperture has a spatial resolution of two milliradians with a thermal resolution of 0·2 to 0·4° C. The output is a 525-line television-compatible signal suitable for use with most standard monitors.

Dimensions: (sensor scanner) 5 × 4 inch dia (127 × 101·6 mm dia)
Weight: (sensor scanner) 3·5 lb (1·59 kg)

STATUS: in production and service.

Hughes

Hughes Aircraft Company, Electro-Optical and Data Systems Group, PO Box 90515, Los Angeles, California 90009

AN/AAS-33 target recognition and attack multi-sensor (tram)

The Hughes AN/AAS-33 target recognition and attack multi-sensor (tram) detecting and ranging set is an integrated day/night weapon delivery system developed for the US Navy's A-6E Intruder aircraft, under a contract from that service's Air System's Command. The equipment comprises a forward looking infra-red sensor, a laser designator-ranger and a laser receiver, all housed in a precision-stabilised turret mounted under the nose of the aircraft.

The tram system is designed to provide sighting and guidance facilities for a wide range of laser guided and conventional weapons. When aligned and operated with the aircraft's radar system, tram also fulfils a navigation and target-location function.

The flir system is said to be sensitive enough to detect the quantities of oil in storage-tank depots on the ground, solely through temperature differentials caused by varying tank levels. In operation, the weapon-operator/navigator acquires the target by using the aircraft radar, and then engages the bore-sighted forward looking infra red. The latter is fitted with a zoom lens which brings the target into close-up view for final recognition purposes. Full travel of the optical system from

minimum to maximum field of view and vice-versa takes less than three seconds and during zoom, the scene image remains on the display so that no time is lost reacquiring a target.

Once a target is satisfactorily acquired and recognised by the forward looking infra-red element, the weapon-operator uses the laser (which is also bore-sighted) for designation, marking the target with a laser spot and so providing a radiant source for a laser-guided bomb to home on. The tram system provides either autonomous attack facilities for the aircraft on which it is fitted, or permits the aircraft to attack a target which is laser illuminated by another aircraft or by a ground laser source. Gyro-stabilisation of the tram turret ensures that both forward looking infra-red and laser sensors remain accurately aligned on the target during high-G attack manoeuvres.

The tram system is produced at the Hughes plant at El Segundo, California.

STATUS: in production and service.

AN/ASB-19(V) angle rate bombing set (arbs)

The Hughes AN/ASB-19(V) angle rate bombing set (arbs) was designed for US Marine Corps aircraft to improve the day and night bombing accuracy when operating in the close-support role using unguided weapons. Under the conditions, the system provides accurate delivery, irrespective of target velocity, ambient wind, target elevation or dive angle. It is also compatible with guided ordnance and can also be used to direct gunfire and air-to-ground rockets. Arbs was originally designed for application to the US Marine Corps' McDonnell Douglas A-4M light attack bomber, but is also compatible with the AV-8B Harrier V/STOL aircraft to be procured for the US Marine Corps.

Arbs comprises three main sub-systems: a dual-mode tracker, weapon-delivery computer, and control units. The tracker comprises laser and pilot-controlled television tracking equipment, both using a common optical system. The dual mode tracker automatically tracks targets designated by a laser either ground-based or in a partner aircraft, or targets which are television-designated by the pilot of the arbs-equipped aircraft itself. The tracking system's common optics features enable transition from laser to television tracking mode to be accomplished without losing the target.

Tracking information is passed from the dual mode tracker to the weapon delivery computer, a 16 000-word, high-speed, general-purpose digital computer in the aircraft's nose. The weapon delivery computer performs computations for weapon trajectory and fire control, position control of the dual mode tracker itself during target acquisition, digital filtering of the dual mode tracker angular rate signal outputs and automatic fire or weapon-release signals to the armament and other systems.

The weapon delivery computer receives aircraft-target line-of-sight angle and angle rate data from the tracker when that unit has achieved target lock-on. This data, combined with true airspeed and altitude information from the air data computer, and processed by the weapon delivery computer, yields the weapon delivery solution. Target position, weapon release and azimuth steering information are generated and presented to the pilot via a head-up display. The function of the arbs control units is to interface with the weapon delivery computer and to provide the pilot with control of the tracker modes, entry of display and target information, navigation and maintenance information.

When operating in the laser mode, the sensor automatically acquires the target, which is illuminated from an external laser source. The sensor, via the weapon delivery computer, presents steering signals to the pilot on the head-up display. In television mode, visible light from the common optical system is directed to form an image on the television vidicon. Tracking of small, low contrast, poorly defined moving targets is said to be possible even with changes in aspect ratio, in the presence of competing clutter or when the target is partially lost to view behind ground obstructions.

Following head-up acquisition in the television mode, the pilot is shown a ×7 magnified image of the target on the cockpit television monitor display. He may at this stage use a hand control to slew the tracker gate onto a new track point or onto a nearby alternative target, if so desired.

Weapon release in either laser or television tracking mode may be made automatically or

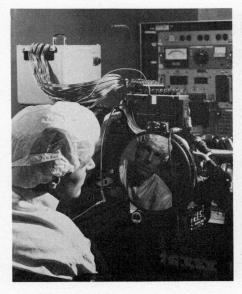
Alignment of optics on Hughes tram system

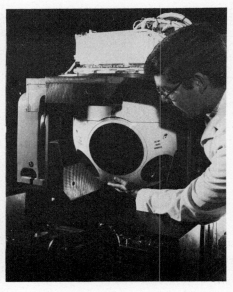
Hughes tram sensor package installed in turret

Hughes arbs installed in nose of US Marine Corps/McDonnell Douglas A-4M Skyhawk

manually, with a weapon release time and a continuously computed impact point being generated by the weapon delivery computer. A weapons-data insert panel provides entry into the weapon delivery computer of weapon characteristics and rack type for those stores being carried for a specific mission.

Arbs permits delivery of any type of weapon at any altitude and airspeed combination at any dive angle or in level flight. The system accuracy is said to be sufficient for first-pass precision delivery in close air support or interdiction missions against land or seaborne targets, but navigation steering commands are also generated to provide second passes against hardened targets. This reattack facility provides the pilot with head-up steering information for return to a designated target, the location of which is retained in the system memory until a new target is designated by the pilot.

The system's television tracking element also provides a limited air-to-air capability for daylight operations, its ×7 magnification being particularly useful for visual identification and tracking of airborne targets.

Angular coverage of arbs extends from plus or minus 35 degrees in azimuth and from plus 10 to minus 70 degrees in elevation at roll

angles of plus or minus 45 degrees. These limits are sufficiently wide to allow extensive flexibility in weapon selection and delivery profiles and permit multiple-roll manoeuvres to take place on the attack approach run without loss of target lock-on. The tracking equipment's laser is NATO-coded, but is compatible with other current laser designators such as tram or mule at ranges adequate to ensure the success of first-pass strikes.

The arbs weapon delivery computer processor is a 16 000 word capacity system which may be extended to 32 000 words capacity. At standard 16 K capacity, the system has far more capability than is required for the functions of weapon release and steering command computation and may be used for other, additional purposes. One of these is the pilot initiated self-test of the ARBS system which performs operational readiness testing and fault-isolation functions. By modification of the computer input/output scaling factor, the arbs can be configured to interface with aircraft avionics other than those of the A-4M and the AV-8B. Three basic aircraft interfaces are required: a vertical reference to provide pitch and roll data, true airspeed and a servoed optical sight or head-up display on which to present azimuth steering commands, continuously computed impact point, and bomb-release information.

Arbs systems therefore can be configured, in either internally- or pod-mounted versions, to suit the requirements of other aircraft. Several pod configurations have in fact been formulated for candidate aircraft with attachment hooks at 13·98 inches (355 mm) and 30 inches (762 mm) standard centres, compatible with standard pylon or fuselage centreline mountings. Where existing cockpit controls cannot be used, two small arbs control units may be fitted at any accessible position. The A-4M control units are configured with standard 5¾ inch (146 mm) widths adaptable to most cockpit consoles.

Exploratory development of arbs initially began in 1965 at the US Navy's Weapons Test Center, China Lake, California in 1965 although concept feasibility was not demonstrated until 1970. Hughes was awarded an engineering development contract for the construction and test of six production-standard prototypes in 1974 following a competitive flight-test programme with other systems. A full production contract was awarded in 1979 following extensive flight testing at both China Lake and the Naval Air Test Center at Patuxent River. Production commenced in late 1979 and deliveries were due to commence in 1982 for retrofit to A-4Ms and, eventually, to US Marine Corps AV-8B aircraft. The Royal Air Force has also expressed interest in the system for possible application to its own AV-8Bs, and has conducted an evaluation of the system during a series of special test flights from the US Marine Corps Air Station at Cherry Point, North Carolina.

Although the system has been developed under the auspices of the Hughes Aircraft Company's Missile Systems Group headquartered at Canoga Park in California, quantity production of arbs is undertaken at the company's plant in Tucson, Arizona.

Dimensions: 2½ ft³ (0·07 m³)
Weight: 128 lb (58·18 kg)

STATUS: in production and service.

FACTS (flir-augmented Cobra TOW sight)

FACTS, a flir-augmented weapon sighting system, was developed by the Hughes Aircraft Company's Electro-Optical and Data Systems

FACTS flir system on test bench at Hughes' factory

Setting up Hughes FACTS flir system on optical rig

Group in order to provide an enhanced night and adverse visibility capability for Cobra helicopters equipped with TOW missiles. FACTS employs US Army thermal-imaging common modules which have been manufactured by Hughes under contract from the Army's Night Vision and Electro-Optics Laboratories, Fort Belvoir, Virginia. Modules designed and produced under this programme have been applied to a number of night-vision and thermal detection and sighting systems for service with armoured fighting vehicles including the XM-1 tank and those armoured fighting vehicles equipped with the fighting vehicle system turret armed with TOW/Bushmaster weapon systems.

The FACTS system has been successfully tested aboard Bell AH-1S helicopters during trials conducted at Fort Hood, Texas and Fort Polk, Louisiana. According to Hughes, the system performed well and during live firing tests five hits were achieved in the same number of firings.

FACTS also provides the Cobra with a target sight for use with unguided rockets and during conventional gun attacks. The system has also been used during exercises to effectively monitor 'hostile' forces and to direct ground

Nose-mounted FACTS sensor in Bell AH-1S Cobra light attack helicopter

troops from the helicopter, resulting in envelopment of the 'enemy' by 'friendly' forces. Although developed primarily for the US Army, the system has also been selected for service by the US Marine Corps and armed forces of other nations.

STATUS: under development.

Martin Marietta

Martin Marietta, Orlando Aerospace, PO Box 5837, Orlando, Florida 32855

Lantirn (low-altitude navigation and targeting infrared for night) system

Lantirn is a navigation and target identification system designed to provide strike aircraft with under-the-weather, autonomous day/night navigation and precision weapon delivery against tactical ground targets. It is intended to permit single-seat aircraft to acquire, track and destroy ground targets using both guided and unguided weapons. Initially developed for the US Air Force F-16 fighter and A-10 battlefield aircraft, it will be equally applicable to other single- or two-seat aircraft operating in the air-to-ground role

The system comprises two basic elements contained in two pods suitable for underwing or under-fuselage attachment. Either or both pods can be carried, depending on the particular mission requirement. This option enhances flexibility and diminishes support demands.

The first, navigation, pod contains a wide field of view forward looking infra-red unit and a terrain-following radar installation, together with the associated power supply, pod control computer and environmental system. Flir imagery from the pod is displayed on a new,

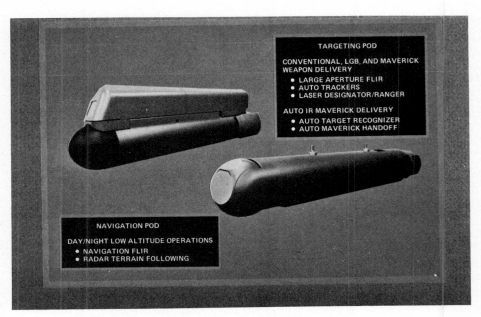

External details of the Martin Marietta Lantirn targeting and navigation pods

wide field of view holographic head-up display now being developed by Marconi Avionics for the Lantirn system. This provides the pilot with night vision for safe flight at low-level. The Texas Instruments Ku-band terrain-following radar permits operation at very low altitudes with en-route weather penetration and blind let-down capability. Terrain-following is accomplished manually by means of symbology presented to the pilot on the head-up display, neither the F-16 nor the A-10 having an afcs designed for automatic terrain-following. Automatic implementation can be accomplished, says Martin Marietta, but is not included as part of the present programme.

Lantirn's targeting pod contains a stabilisation system, wide- and narrow-field flir, laser designator/ranger, automatic target recogniser, automatic infra-red Maverick missile hand-off system, environmental control unit, and power supply equipment. The targeting pod interfaces with the aircraft controls and displays as well as the fire-control system to permit low level day/night semi-automatic target acquisition and weapon delivery of guided and unguided weapons. It may be arranged as a laser designator-only pod, for use with laser-guided munitions and conventional weapons, simply by deleting the automatic recogniser and the Maverick hand-off subsystem.

Both pods have environmental control units to ensure that their systems will function satisfactorily over a wide range of temperatures. Aircraft interfaces include a MIL-STD-1553 multiplex data bus, video channels, and power supplies. Service ground support employs typical three-level maintenance: organisational, where no special test equipment is required; intermediate, which employs automatic test equipment; and depot or base servicing.

Development of Lantirn commenced in the latter half of 1980, under a $94 million contract awarded to Martin Marietta by the US Air Force, and is expected to take three years to complete. The programme is directed by the US Air Force Aeronautical Systems Division, Deputy for Reconnaissance/Electronic Warfare Systems, at Wright-Patterson Air Force Base, Ohio. Under the development contract, six prototype units are to be designed, developed and tested and initial production commenced. Full scale production and field deployment is planned for 1985-87.

As well as Martin Marietta, Marconi Avionics and Texas Instruments, a number of other companies are also involved in the Lantirn development programme. These include Hughes Aircraft, responsible for the target recogniser and boresight correlator for Maverick hand-off; General Motors Delco Division, which supplies computers; Sundstrand Aerospace which provides the environmental control; and International Laser Systems, suppliers of the laser designator.

STATUS: in development.

TADS/PNVS target sight and night-vision sensor

Martin Marietta's tads/pnvs is designed to provide day, night and limited adverse-weather target information and navigation capability for attack helicopters. Development is being undertaken for the US Army with a view to equipping that service's Hughes AH-64 Apache advanced attack helicopter.

Tads/pnvs comprises two independently functioning systems known as the target acquisition designation sight and the pilot night vision sensor. Tads provides the co-pilot/gunner with search, detection and recognition capability by means of direct-view optics, television or forward looking infra-red sighting systems which may be used singly or in

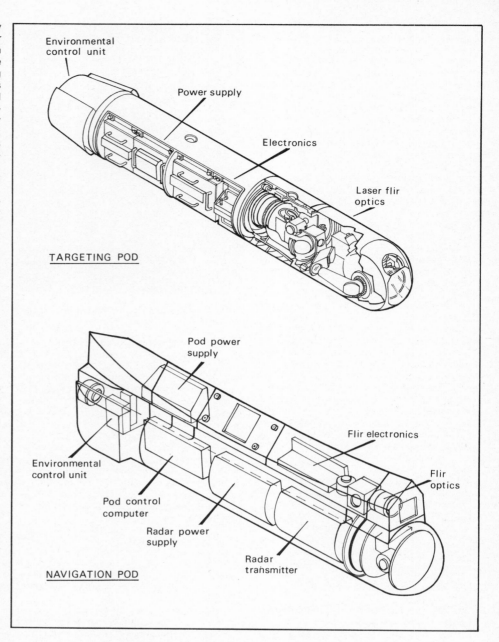

Internal details of the Martin Marietta Lantirn targeting and navigation pods

Martin Marietta tads/pnvs installation on nose of Hughes AH-64A Apache attack helicopter

combinations according to tactical, weather or visibility conditions. The pnvs system provides the pilot with flight-guidance symbology which permits nap-of-the-earth flight to,from and in the combat area at altitudes low enough to prevent or delay detection by enemy forces.

Tads consists of a rotating turret, mounted on the nose of the helicopter and containing the sensor sub-systems, an optical relay tube located at the co-pilot/gunner station, four electronic units in the avionics bay, and cockpit controls and displays. Tads turret sensors have a field of regard covering plus or minus 120 degrees in azimuth and from plus 30 to minus 60 degrees in elevation. By day, either direct-vision or television viewing may be used. The direct-vision system has a narrow field of view, 4 degrees at a ×16 magnification, and a wide field of view, 18 degrees at ×3·5 magnification. The television system provides a narrow field of view of 0·9 degree and a wide field of view of 4 degrees. For night operations the tads sensors (presumably flir imagers) have three fields of view: narrow of 3·1 degrees, medium of 10·1 degrees, and wide of 50 degrees.

Once acquired by the tads, targets can be tracked manually or automatically for autonomous attack with guns, rockets or Hellfire anti-tank missiles. A laser may also be used to designate targets for attack by other helicopters or by artillery units firing the laser-guided anti-armour Copperhead weapon.

The pnvs consists of a forward looking infrared sensor system packaged in a rotating turret mounted above the tads, an electronics unit located in the avionics bay, and the pilot's display and controls. The system covers a field of plus or minus 90 degrees in azimuth and from plus 20 to minus 45 degrees in elevation. Field of view is 30 × 40 degrees.

Tads is designed to provide a back-up pnvs capability for the pilot in the event of the latter's failure. Either the pilot or the co-pilot/gunner can view, on his own display, the video output from either tads or pnvs, raising the probability of mission success. Although designed primarily for combat helicopters flying nap-of-the-earth missions, pnvs may also be used as a single entity in tactical transport and cargo helicopters.

Since the tads/pnvs systems are designed for aircraft engaged in a high intensity combat environment, special attention has been paid to reliability, says Martin Marietta. Likewise, the ability to remove and replace units easily and rapidly on the line is also strongly emphasised. Tads/pnvs underwent final evaluation during the latter part of 1981 and a decision affecting future production plans was being made during 1982.

STATUS: in development.

Northrop

Northrop Corporation, Electro-Mechanical Division, 500 East Orangethorpe Avenue, Anaheim, California 92801

See Hawk forward looking infra-red system

See Hawk is a thermal-imaging system using tri-service common-module forward looking infra-red sensors to provide high-resolution imagery. Designed for service with both aircraft or surface vessels, the system's principal application to date has been aboard a US Coast Guard/Sikorsky HH-52A helicopter on which a prototype unit has been installed for trials purposes. Its primary role in service would be search and rescue although possible subsidiary tasks could include law-enforcement, maritime environmental control and reconnaissance, fisheries control, disaster relief, marine traffic control, and navigational assistance. The present series of trials however, is aimed at defining the Coast Guard requirement for a night-vision system to equip the service's Aérospatiale HH-65 Dolphin short-range recovery helicopters which are to be used for day and night search and rescue operations and general maritime surveillance.

As applied to the HH-52A, the See Hawk flir comprises a nose-mounted, sealed turret containing the sensors and an electronic processing and display system carried in the cabin. The internally-cooled detector array has two selectable fields of view, a wide field of 30 × 40 degrees and a narrow field of 10 × 13 degrees. The field of regard covered by the turret is plus or minus 90 degrees in azimuth and from plus 30 to minus 80 degrees in elevation. The sensor system caged in a gyrostabilised gimbal mounting is protected by a window when not actually operating. Two display systems are used in this trial installation, a 5-inch (127 mm) unit on the cockpit centre instrument panel and a 10-inch (254 mm) system in the helicopter's main cabin. The displays presently used are standard television-type cathode ray tubes and facilities for video-tape recording the field of view are also incorporated into the current system.

Control of the system can be exercised from the cockpit or cabin positions. A control stick with top-mounted thumb switches is used to control sensor scan in both azimuth and elevation, focus, scan centring, and field of view selection. An automatic scan facility which provides constant search coverage in elevation and azimuth is also controlled from switches on the control stick. The auto-search mode is enhanced by inclusion of automatic lock-on

Northrop See Hawk flir in scanning turret on US Coast Guard helicopter

which reacts to either large or small targets, as selected by the operator. Small dots appear around the display's crosswire centre-reticle when lock-on is made, and these commence flashing if break-lock occurs. Controls for the auto-search, scan rate, scan angle and continuous or step scan are above the cathode ray tube display. As an aid to target recognition, the See Hawk display system has selectable white-hot or dark-hot display polarity.

Weight: (flir turret) 70 lb (31·8 kg) (flir electronics and power supply unit) 35 lb (15·9 kg)

STATUS: in development.

Tiseo electro-optical target-identification system

Northrop's Tiseo (target identification system electro-optical) system, claimed by the company to be the first operational electro-optical system, is a television-based, passive, daytime automatic target acquisition and tracking system. Comprising a high resolution closed-circuit television sensor combined with a two-fields-of-view telescope, Tiseo is mounted on the port wing leading-edge of many US Air Force/McDonnell Douglas F-4 fighters. More than 500 Tiseo systems are known to have been delivered to the US Air Force.

STATUS: in production and service.

Tcs television camera

The Northrop Tcs (television camera system) is similar to the Tiseo (see preceding entry) but offers enhanced capabilities. It is a passive, daytime, automatic search and acquisition system which can be either manually operated or slaved to an air-interception radar. The system not only acquires targets automatically, but presents multiple fields of view and is to be fitted on US Navy/Grumman F-14 Tomcat fighters

STATUS: in production.

Vats video tracking system

Under a contract from Ford Aeronutronic Division, Northrop is working on the engineering development of a system called Vats (video augmented tracking system), designed for retrofit to the Pave Tack system used by US Air Force F-4E and F-111 fighters. Pave Tack is an air-to-ground, laser-designated weapons de-

Northrop Tiseo system mounted on US Air Force/McDonnell Douglas F-4 fighter

livery system mounted in a pod containing a forward looking infra-red system. Vats would automatically track ground targets, eliminating or reducing the weapons operator's need to follow the monitor by eye. It is a microprocessor controlled system with a self-contained power supply and a digital interface to the aircraft's flight computer system.

STATUS: under development.

Irsts infra-red search and track systems

Northrop is currently involved in the development of Irsts (infra-red search and track systems), designed for the passive long-range acquisition of targets at night and in adverse weather. Irsts equipment is known to have a multi-target capability. Other roles envisaged for these systems, about which few details have been disclosed, include weapon delivery, surveillance and fire-control.

STATUS: under development.

Texas Instruments

Texas Instruments Inc, PO Box 226015, Mailstop 3177 Dallas, Texas 75266

Texas Instruments has for many years been involved in the development of infra-red detection and recognition technology, having developed its first infra-red line-scan mapping system as early as 1958. The company also claims to have developed the first forward looking infra-red system in 1964 and by 1967 Texas-manufactured forward looking infra-red systems were serving aboard US Air Force gunship aircraft in South East Asia. In 1972, at the request of the US Department of Defense, Texas conducted studies to examine the possibilities of reducing maintenance costs associated with thermal sensor systems. The results of these studies emerged in the form of a family of common modules that can be tailor-packaged to meet particular custom requirements. A wide range of common module-based equipments is now in production.

OR-89/AA forward-looking infra-red system

The Texas Instruments OR-89/AA forward looking infra-red system sensor is a direct development of the company's earlier AN/AAD-4 and AN/AAS-28 forward looking infra-red sensors which were used extensively for the US Air Force gunship programmes. Development of the sensor, for use on the US Navy/Lockheed S-3A sub-hunter aircraft was commenced under a contract from Lockheed-California in August 1969 and the first prototype was delivered in March 1971. The system was subjected to extensive qualification testing, in addition to US Navy testing which continued over 22 months. The first production equipment was delivered to Lockheed in August 1972 and the system was introduced into the fleet early the following year. Deployment continued for at least the next five years, with well over 400 S-3A or derivative systems, together with support equipment and spares, being delivered.

Texas Instruments OR/89/AA flir system

The OR-89/AA forward looking infra-red uses mercury cadmium telluride detector arrays packaged in a gimbal system with self-contained attitude stabilisation. This provides an azimuth coverage of plus or minus 200 degrees and zero to minus 84 degrees in elevation. Output is shown on an 875-line RS-343 composite television display. The system comprises three basic modules or weapon replaceable assemblies, the infra-red viewer, power supply video converter, and infra-red control converter.

The system is controlled by a general-purpose digital computer which sends outputs

in serial word form to the forward looking infra-red system control converter, which translates them into set-control commands or analogue servo commands. Set-control commands provide such control functions as standby/off, activate, servo on/off, polarity, gain level and field of view. Servo commands control azimuth and elevation drive and brake functions. Position data is supplied by the control converter, which converts it into digital format and then transmits it to the general purpose digital computer for status purposes. On certain configurations, off-line ancillary units produced by Texas Instruments are used for control and display. These include set and slew control weapon replaceable assemblies, a position indicator, cathode ray tube displays, and a forced-air shock-mounting tray.

The OR-89/AA's primary function aboard the S-3A is that of detecting and classifying surface vessels at night, but the obvious use of the system led to the development of a family of derivative systems to fulfil a wide range of aircraft and application requirements. These derivatives have been successfully installed and operated in 12 other types of aircraft ranging in size from light twins to large transports and including helicopters and high-performance jets. Examples include the incorporation of the system in the US Navy/Lockheed P-3B Orion for anti-submarine warfare missions and for non-military use with the US Customs Service in 'Linebacker' operations and other special missions.

Weight: 266 lb (120·9 kg)

STATUS: in service.

OR-5008/AA forward looking infra-red system
The Texas Instruments OR-5008 forward looking infra-red system is a derivative of the Texas OR-89/AA equipment (see preceding entry), adapted for use on the Lockheed CP-140 Aurora asw aircraft operated by the Canadian Armed Forces. It was designed to support a variety of mission requirements in the maritime patrol field including search and rescue, shipping and fisheries surveillance, mapping, ice reconnaissance and defence surveillance.

The system is mounted in the lower part of the CP-140 aircraft's radome in a similar manner to that of the AN/AAS-36 forward looking infra-red system used in the US Navy's P-3 aircraft. The specification is similar to that of the OR-89/AA except that it uses the US Navy-type P-3 interface casting, provides for a plus five degree up-look capability, uses off-line control for the extend-retract function, and has additional composite video outputs. Additionally, says Texas, it has improved mean time between failures and to repair. Special-to-type support equipment is provided by Texas Instruments for the OR-5008/AA system.

STATUS: in service.

AN/AAQ-10 forward looking infra-red system
Another derivative of the Texas Instruments OR-89/AA system (see preceding two entries), the AN/AAQ-10 forward looking infra-red system is an adaptation of the original equipment which has been designed for use in helicopter search and rescue applications. A single-window turret has been provided to supply the increased field of coverage necessary to optimise the helicopter's capability in this role and special attention has been given to the design of the stabilisation servo elements to counter the much greater vibration levels in helicopters. Video output is shown on a RS-343A 875-line television display.

Development of the AN/AAQ-10 sensor began in late 1974 with the prototype system being delivered to the US Air Force in mid-1975. Flight test and familiarisation missions with this system continued for a 32-month period and in March 1978 a production contract was awarded for ten systems. The first production unit was delivered in December that year.

The AN/AAQ-10, as configured for use on the US Navy/Sikorsky HH-53 helicopter, consists of an infra-red receiver, power supply, electronic control amplifier, control indicator and mounting base. All controls and monitoring functions are off-line and contained in the control indicator, allowing the operator complete one-hand authority over sensor operation and servo slew commands.

Texas Instruments common-module flir system AN/AAQ-9

The system is a completely self-contained, slewable infra-red sensor requiring only external power, a gimbal position indicator and a video display to become fully operational. Accurate gimbal position indication is a critical factor in defining a target location to the helicopter's navigation computer and this may be accomplished either by three-wire synchro outputs or a buffered voltage. A Texas Instruments-manufactured position indicator, using the synchro outputs and a video display, is optionally available.

STATUS: in service.

AN/AAQ-9 infra-red detecting set
The Texas Instruments AN/AAQ-9 infra-red detecting set is a forward looking infra-red system developed for the US Air Force Pave Tack programme to provide autonomous target acquisition for delivery of standard and laser-guided (Paveway) bombs. Beside this purely target-designation capability the system offers the weapon systems operator the capability of locating, acquiring, tracking, and directing accurate weapon fire against enemy ground targets in day, night or adverse weather, and forms an integral component of the AN/AVQ-26 Pave Tack target designator system.

The system is designed primarily for use on US Air Force F-4E, RF-4C and F-111 fighters and strike aircraft. The Royal Australian Air Force has also shown interest for possible F-111 application, and a number of other international customers are understood to be considering it for use with various other airborne weapon designation and delivery systems.

Development of the AN/AAQ-9 commenced during the mid-1970s, with Texas Instruments being selected as the sole-source supplier following a competitive design and flight test programme. The system entered production in 1977 and the first production unit was flown at Eglin Air Force Base in Florida in April 1979.

Design is based on the Texas Instruments-produced US Department of Defense common modules. Features include high resolution, automatic thermal and range optical compensation and a two-field-of-view optical system. The optical system is designed for pod installation and derotation compensation is also provided with a coverage of plus or minus 190 degrees to 100 degrees per second. Video output is provided to either 525- or 875-line RS-343 composite television displays, either single-ended or differential. Selectable ×1 or ×2 optical magnification is provided.

The AN/AAQ-9 system has undergone particularly stringent reliability, qualification and flight testing with a view to achieving high reliability and, to date, a mean time between failures rate of over 730 hours has been attained. The system is of modular construction in the interests of enhanced maintainability and is supplied together with special-to-type test equipment to permit field maintenance and alignment down to intermediate servicing level. For rapid in-the-field harmonisation, the equipment is mounted on an adjustable boresight platform.

Weight: (receiver) 107 lb (48·63 kg)
(control electronics unit) 28 lb (12·72 kg)

STATUS: in production and service.

AN/AAS-36 infra-red surface vessel detection set
The Texas Instruments AN/AAS-36 infra-red detection set is a forward looking infra-red system designed for US Navy/Lockheed maritime patrol P-3C aircraft to detect surface vessels, surfaced or snorkeling submarines and drifting survivors during periods of darkness and limited visibility. The system was initially designed and developed to meet a P-3C update programme requirement, but the equipment has also been fitted to earlier P-3C and P-3B aircraft.

Production of the system commenced in 1977 following a testing, evaluation and demonstration programme which used ten pre-production systems to assure the US Navy that design specifications were either met or exceeded. The Service's Initial Operational Capability (IOC) was realised during the early part of 1979.

Based on Texas Instruments-manufactured US Department of Defense common modules, which employ mercury cadmium telluride detectors, the AN/AAS-36 is a stand-alone system requiring only aircraft power for operation. The common module infra-red receiver is mounted in an azimuth-over-elevation stabilised gimbal and provides lower hemisphere coverage of plus or minus 200 degrees in elevation and from plus 15 to minus 82 degrees in elevation. Additional weapon replaceable assemblies provide system power, servo control, forward looking infra-red system control, slew commands and a real-time video display. The display presentation is on a 875-line RS-343 composite television monitor which permits the operator to identify, as well as observe, vessels.

Features include automatic optical temperature compensation, gimbal-pointing outputs for servo platform slaving, self-contained stabilisation, and a two-field-of-view optical system. A digital computer interface is available for on-line gimbal control. The system contains built-in self test facilities which permit checkout down to weapon replaceable assembly level, and these themselves are compatible with automatic test equipment. Mean time between failures rate is 300 hours.

According to Texas Instruments, the system is being supplied to many non-US operators of the P-3 who are acquiring the system to upgrade their fleets to US Navy standards. The receiver-converter weapon replaceable assembly of AN/AAS-36 has also been fitted to Cessna Citation light twinjets, to the CH-53 helicopters and to other, unspecified aircraft.

Weight: 300 lb (136·36 kg)

STATUS: in production and service.

AN/AAS-37 infra-red detection set

The Texas Instruments AN/AAS-37 equipment is a version of the company's AN/AAS-36 detection set but is a somewhat more sophisticated system, being combined with a laser ranging facility. Developed for the US Marine Corps/Rockwell OV-10 forward air control aircraft, the AN/AAS-37 provides infra-red vision for day or night operations under degraded environmental conditions, automatic target-tracking and laser target-designation. The laser provides ranging and illumination of ground targets for laser-guided weapons such as the Paveway bomb or the Hellfire missile, either on an autonomous basis for the AN/AAS-37-equipped aircraft itself or for a co-operating aircraft armed with appropriate weapons.

Specification of the forward looking infra-red sensor and associated equipment is virtually identical to that of the AN/AAS-36 system from which its was developed. It does however possess a number of additional features, derived mainly from incorporation of the laser section. These include direct-readout laser-ranging and designation capability. The system can be used as a target sight aligned with the aircraft boresight by means of electronic adjustment, and line-of-sight depression may be set by the operator for precision air-to-ground delivery of weapons. There are interfaces with aircraft systems for a radar altimeter and remote gyroscope, and an accelerometer provides the system with rate-aided automatic target tracking capability using an adaptive gate and a centroid tracker. Offset tracking from a target or another landmark is also possible. The system's display is said to be sufficiently bright for daylight viewing.

Weight: 240 lb (109 kg)

STATUS: in production and service.

AN/AAR-42 infra-red detecting set

The Texas Instruments AN/AAR-42 is a forward looking infra-red system designed for use on the US Navy's A-7E strike aircraft. It is designed to provide a night window or bomb-sight which permits the pilot to perform single-seat night attack, close air support and reconnaissance missions by day or night and during poor weather conditions. The system is said to permit the night delivery of conventional bombs with accuracy better than or equal to that demonstrated in day bombing using more conventional sighting methods.

The AN/AAR-42 is installed in a pod, with a gimballed forward looking infra-red unit which

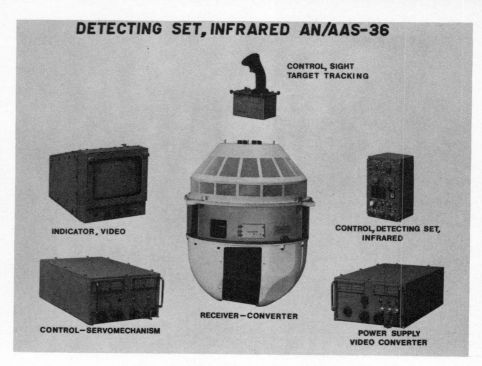

Texas Instruments AN/AAS-36 flir system

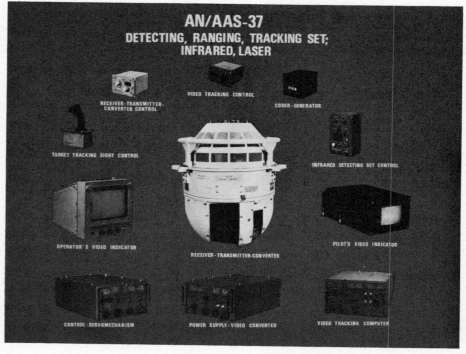

Texas instruments AN/AAS-37 flir system

Texas Instruments AN/AAR-42 flir system

provides stabilised imagery on the pilot's head-up display system (specially developed by Marconi Avionics for the A-7). The forward looking infra-red system uses components of the US Department of Defense common modules developed by Texas Instruments. It provides an azimuth coverage of plus or minus 20 degrees and an elevation coverage from plus 5 to minus 35 degrees. Both wide and narrow selectable fields of view are provided. The wide field of view, giving a ×1 magnification, is employed for pilot orientation, navigation update and target acquisition, while the narrow field of view, with a ×4 magnification, is used for target identification and weapon delivery. Features include automatic thermal focus compensation and sensor window de-icing.

Deliveries of AN/AAR-42 flir systems commenced in late 1977 following US Navy operational evaluation earlier that year. The system is currently operational on the A-7E and is said to exceed the specifications in both performance and in reliability, with a mean time between failures of 200 hours.

A version of the system for the US Air Force/Fairchild A-10 battlefield bomber has also been developed. In an evaluation programme for the US Air Force, a production AN/AAR-42 flir system was adapted for use on Fairchild's two-seat N/AW company demonstrator. The system was pod-mounted on the A-10, boresighted with the aircraft's General Electric GAU-8 30 mm gun, and interfaced through a MIL-STD-1553A digital data-bus. The wide field of view of the forward looking infra-red system was matched to the raster head-up display, providing a correct image of the scene for pilot orientation, navigation, target acquisition and weapon firing, while a narrow field of view provided target identification. A six-month evaluation programme conducted at Edwards Air Force Base, California and Eglin Air Force Base, Florida was completed in March 1980 with what Texas Instruments terms 'positive results' in performance and reliability of the forward looking infra-red sensor.

Weight: (canister assembly) 210 lb (95·45 kg)
(servo electronics) 40 lb (18·18 kg)

STATUS: in production and service.

Helicopter fire-control system
In 1975 Texas instruments initiated a privately funded demonstrator programme to develop an integrated helicopter fire-control system combining forward looking infra-red and a day television sensor with a laser-ranging and designation system. This integrated package was designed to meet the requirements of attack helicopters. Texas Instruments' objectives were to solve the problems associated with helicopter-borne, stabilised fire-control systems with technologies and capabilities existing within the company and to demonstrate publicly those capabilities with a view to securing an advantageous position in what was becoming recognised as a new and potentially large and important market.

Equipment design began in August 1975 and manufacture commenced the following month. By March the following year, integration testing was complete and flight test and demonstration flying started in April 1976, since when numerous flight demonstrations have been given. The company says that the performance goals set out in the design stages have been achieved in all areas including stabilisation, three-field-of-view forward looking infra-red sensor, laser-ranging and designation, and minimised operator control.

The forward looking infra-red and day television system's three-field-of-view coverage provides a wide angle view of 30 × 40 degrees, a mid-angle field of 6·8 × 9 degrees and a narrow pencil of 2 × 2·8 degrees. Azimuth range is plus or minus 200 degrees and the elevation range is from plus 15 to minus 85 degrees. The mercury cadmium telluride forward looking infra-red sensor array is based on US Department of Defense common modules developed by Texas Instruments and output is on an 875-line RS-343 television video display.

Further developments of the system include the incorporation of an automatic video tracker (see below). This equipment uses a digital adaptive gate centroid contrast tracker using the forward looking infra-red television output to control the gimbal system. Its major advantage is that it provides a significant decrease in operator workload while continuing to maintain a precise target track for laser designation. Variations of this tracker, using the Texas Instruments SBP-9900 micro-processor, are in production for the AN/AAS-37 flir for the Rockwell OV-10 forward air control spotter, and are under development for application to the AN/AAS-38 flir for the US Navy/McDonnell Douglas F/A-18 multi-role strike fighter.

STATUS: under development.

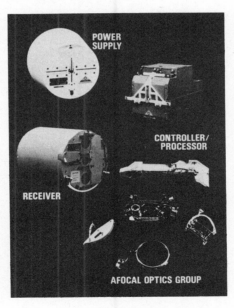

Texas Instruments AN/AAS-38 flir system

Automatic video tracker
The automatic video tracker equipment has been developed by Texas Instruments as an optional 'add-on' to forward looking infra-red systems to improve performance, especially weapon delivery accuracy. The micro-processor controlled system processes forward looking infra-red video signals and provides servo control commands, allowing hands-off operation. It also provides display symbology and offers a number of operating modes.

Typical symbology options offered by the automatic video tracker include a target cursor, automatic tracking gate and coast mode indication (all with adjustable display brilliance, and the latter also with size adjustment), narrow field of view indicators, and range read-out.

Four different tracking modes are provided. In manual mode, the operator maintains track via a slew stick control but he can select the computer-aided mode which assists him in target acquisition and maintenance of manual track by removing effects of aircraft motion. In automatic track mode, the track is accurately maintained under micro-processor control while the operator is free to deal with other tasks. The offset automatic track mode is similar to that of normal automatic track but permits the operator to hold a target at the centre of the reference crosshairs by locking the tracker onto another, more stable reference target in the system's field of view

The automatic video tracker is claimed to have sufficient tracking accuracy for laser designation and other fire-control operations, and the sensitivity is adequate for the tracking of either large or small targets with high or low contrast against their background. In the latter case, the system's use is enhanced by automatic target polarity selection. The system's so-called 'coast mode' is also a useful operational feature. If a target is screened momentarily, or if its apparent size or shape changes rapidly, then the automatic video tracker estimates its probable position to allow rapid, automatic reacquisition. Such facility is of value in high-speed manoeuvres, particularly during tactical combat at low altitudes.

Control of the automatic video tracker system is exercised by a Texas Instruments SBP-9900 micro-processor, a MIL-qualified system, which provides a high degree of flexibility in that parameters may be optimised for particular variations of the system by simply changing the software. Use of the micro-processor has also permitted significant improvements and additions to the traditional adaptive gate, centroid tracking algorithms.

STATUS: in production.

Lasers

FRANCE

Cilas

Compagnie Industrielle des Lasers, Route de Nozay, 91460 Marcoussis

TCV 115 laser rangefinder

This is a low repetition-rate, short-wavelength laser system, suitable for use in ground vehicles as well as helicopters, and using an avalanche-photodiode detector unit. The airborne version is integrated with the gyrostabilised APX M334-04 helicopter sight. In addition to the basic rangefinder, the system also has an optical assembly which matches the range-finder and sight, so ensuring that the cross hairs are projected into the eyepiece of the sight. Operator eye-protection, when the laser is in use, is ensured by an optical filter that provides 70 dB attenuation at the operating wavelength.

Dimensions: 390 × 190 × 180 mm
Weight: 17·6 lb (8 kg) with optical assembly
Operating wavelength: 1·06 micron
Measurement range: 450-19 900 m

Accuracy: ±10 m
Pulse duration: 25 × 10⁻⁹s
Energy: 100 mJ
Peak power: 4 MW
Repetition rate: 1 measurement every 2s for 3 successive shots
12 measurements/minute
Power requirements: rangefinder 19-28 V dc graticule projector 115 V 400 Hz

STATUS: in production.

Thomson-CSF

Thomson-CSF, Division Equipements Avioniques, 178 Boulevard Gabriel-Péri, 92240 Malakoff, Paris

Hades laser angle-tracker for helicopters

This is a passive laser receiver consisting of a detector module and an electronic module. The detector is associated with an optical sight and the latter supplies angular-error data to tracking servo-mechanisms. Hades can perform a wide-angle search in a designated area, to acquire illuminated targets and provide automatic target-tracking through continuous aiming of the fire-control sight at the target. When used in conjunction with a laser rangefinder, an enlargement of the firing envelope of any weapon used is claimed. Target acquisition is achievable in a ten degrees elevation by plus or minus ten degrees bearing field-of-view.

Weight: 4·4 lb (2 kg)
Power requirement: 28 V dc

STATUS: in development.

Atlis laser designator/ranger pod

This is a high-precision air-to-ground designation system which comprises both television and laser sensor apparatus mounted in a single pod. It is suitable for single-seat aircraft, and has been demonstrated on Jaguar and F-16 fighters. Development of the system was undertaken jointly by Thomson-CSF, which developed the monitoring/control system and displays, and Martin-Marietta Orlando Aerospace, which developed the sensor pod, using a laser illuminator by the French company Cilas. It is designed for low-level, high-speed attack by single-seat aircraft, operating autonomously (ie without guidance or control from ground stations or other aircraft).

Initial flight tests took place in France during late 1976 and early 1977, the equipment being installed in a single-seat Jaguar. These tests assessed the pod's ability to mark targets for laser-guided missile attack at ranges of up to ten kilometres. The Mk 1 Atlis pod used in these trials has been superseded in development by the shorter and lighter Atlis 2 pod. It is more suitable for use on Jaguar and F-16, and flight-testing was completed successfully in late 1980, using laser-guided bombs and AS30 laser missiles.

The television sensor provides a × 12 magnified picture to the pilot, and tracks targets using area-correlation techniques, performed at

Thomson-CSF Atlis pod mounted under General Dynamics F-16 fuselage

long-ranges and aided slightly as a high-contrast scene is produced by the particular electro-optic spectrum used. The laser shares a common optical port with the television sensor and can be directed either for designation or range-information. Sightline prediction capability built into the system allows dead-reckoning tracking of targets obscured temporarily by high-ground or patches of cloud.

The most common application of the system is expected to be as a designator for autonomous operation by a single-seat aircraft using laser-guided bombs or missiles, such as the Texas Instruments' Paveway or Aérospatiale AS30.

Production of the Atlis 2 pod has been authorised for French Air Force Jaguars, and the system is being marketed by Martin-Marietta as a future sensor for F-16.

Atlis 2
Pod length: 2·52 m
Max dia: 0·3 m
Weight: 160 kg
Max operating range: ≥10 km
Attachments: standard 760 mm NATO bomb rack
Angle of regard (Mk 1 system):
  roll −160 to +120°
  pitch −145 to +15°
  yaw −15 to +15°
Note: The Mk 1 system was 2·934 m long and weighed 240 kg, including a coolant-tank which needed to be refilled between missions.

STATUS: in production.

TAV-38 laser rangefinder

Ranges of up to ten kilometres can be measured to an accuracy of about five metres using this system, which has been installed in the French version of Jaguar and Mirage F1 fighter-bombers exported by France. The system comprises two units, the power supply and the laser head.

The former is largely a power-supply transformer and distribution centre, providing auxiliary supplies and a high energy direct current voltage output for the flash exciter. The laser head, developed by Cilas, houses the laser cavity, which comprises a neodymium rod (producing 1.06 micron emissions), flash exciter and triggering device, plus receiver elements such as the avalanche photodiode, digital range-counter and optics. The range counter uses a clock frequency of 29·98 MHz, corresponding to distance increments of five metres. The optics can steer the laser beam in a cone with a semi-apex angle of ten degrees, using a beam-deflection system developed by Marcoussis Laboratories CGE research centre.

Weight: (laser head) 12 kg
(power unit) 8·5 kg
Operating wavelength: 1·06 micron
Max range: in the order of 10 km

STATUS: in service.

GERMANY, FEDERAL REPUBLIC

Eltro

Eltro GmbH, Postfach 102120, 69 Heidelberg

CE 626 laser transmitter/rangefinder

Eltro has developed a laser transmitter for rangefinding in a wide variety of ships, aircraft, helicopters, tracked- and wheeled-vehicles applications.

It is a nitrogen-cooled neodymium-YAG laser (1·06 micron wavelength) which normally operates at 9 to 11 Hz pulse-repetition fre-quency, but with a 40 Hz maximum capability. Initial application is in German-operated Panavia Tornado variable-geometry bombers.

STATUS: in development.

UNITED KINGDOM

Ferranti

Ferranti Limited, Electro-optics Department, Robertson Avenue, Edinburgh, EH11 1PX

Laser ranger and marked-target seeker (LRMTS)

Development of this significant item of UK equipment began in 1968 under a government contract, and prototype units were first flown in 1974. Deliveries to the Royal Air Force, for installation in nose housings of both the Harrier V/STOL and Jaguar ground-attack aircraft and accounting for over 200 units, was completed recently.

LRMTS is a dual-purpose unit, which can be used as a self-contained laser ranger, or as a target-seeker with simultaneous rangefinding. In the latter role it can be used to detect and attack any small target characterised by ground troops using a laser-designator, and in this respect the system is well suited to battlefield close-air support.

The LRMTS is a neodymium-YAG laser mounted in a stabilised cage, which allows beam-pointing and stabilisation against aircraft movement. The seeker can detect marked targets outside the head-movement limits. It operates at a relatively high pulse-repetition frequency, thus allowing continuous up-dating of range information during ground-attack. As range is a crucially important parameter for accurate weapon-delivery, and yet virtually unobtainable on a non-laser equipped aircraft, the LRMTS is a vital additional sensor.

In a typical operation, the ground operator with pulsed-laser target-designation equipment, who is called the forward-air controller, directs the aircraft to a location within laser-detection range before switching on the ground-marker equipment. Direct radio communications are used between the forward-air controller and aircraft crew, to assure positive identification of small, and even hidden, camouflaged or moving targets.

Once the LRMTS has recognised the target from reflected laser energy, it provides steering commands to the pilot on the head-up display. Ranging data is also shown and fed directly into weapon-aiming computations for the accurate and automatic release of weapons.

LRMTS is easy to install and harmonise with other aircraft systems, and is said to be more effective than any alternative sensor during operations at the grazing angles used in low-

Ferranti LRMTS optics in nose of BAe Harrier GR3

level ground-attack. By improving weapon-delivery accuracy, the sensor assures a high probability of success in single-pass, high-speed, attacks.

Associated with the LRMTS head is an electronics unit which contains power supplies, and ranging and seeker processing. The laser needs a transparent window, and in Jaguar is mounted behind a chisel-shaped nose, with two sloping panels. For Harrier, where there is more chance of debris accumulation during VTOL operations, the optics are protected by retractable eye-lid shutters.

In addition to the systems in service with the Royal Air Force, equipment is also used by two overseas air forces, and a variant is being developed for RAF-operated Tornados interdictor/strike bombers.

LRMTS head
Dimensions: 300 × 269 × 607 mm
Weight: 47 lb (21·5 kg)
Operating wavelength: 1·06 micron
Pulse repetition frequency: 10 pulses/s
Angular coverage: (elevation) +3 to –20° (azimuth) –12 to +12°
Roll stabilisation: ±90°
Detection angle: ±18° from aircraft heading
Max range: >9 km
Electronic unit
Dimensions: 330 × 127 × 432 mm
Weight: 32 lb (14·5 kg)
Power requirements: 700 VA at 200 V 400 Hz 3-phase plus 1 A at 28 V dc

STATUS: in service.

Type 105 laser ranger

The Type 105 laser ranger has been privately developed by Ferranti as a low-cost but accurate target-ranging sensor for ground-attack aircraft. It can supplement other target-detection sensors on new aircraft, or be retrofitted to existing aircraft.

The Type 105 comprises a single fixed unit, and employs internal steering of the beam optics within a 10-degree semi-apex angle field-of-view. The beam is fully roll-stabilised. Two installations of the basic design are available, one in-line and one T-shaped, the latter being most suitable for aircraft with limited installation length available. The reduction in length is achieved by repositioning the beam-steering equipment.

Basic sensor is a convection-cooled neodymium-YAG laser capable of operating at two pulses a second, and of sustaining short bursts at ten pulses a second. Higher pulse frequencies are possible if a liquid-cooling block is added.

Performance is claimed to provide instantaneous measurements of range with a better than five metres standard deviation accuracy, at ranges equivalent to naked eye visibility, out to five kilometres. The unit has a low power consumption, is effective at shallow sightline grazing angles, and has a mean time between failures of not less than 500 hours. A Type 105 laser ranger has been flown on the Fairchild A-10 two-seat night-attack battlefield-bomber demonstrator.

Dimensions: 233 × 180 × 447 mm
Weight: 20 lb (9 kg)
Operating wavelength: 1·06 micron
Pulse repetition frequency: 2pps continuous 10pps burst
Angular coverage: within 10° semi-apex angle cone
Roll stabilisation: ±360°
Operating range: ≤5 km
Accuracy: ±5 m standard deviation
Power requirement: 2 A at 28 V dc

STATUS: in development.

Ferranti LRMTS mounted in nose of a Jaguar GR1

Type 105D laser ranger

This is a private development of the earlier Type 105 laser ranger system, and it uses a smaller transceiver than the original unit. In July 1980 the Type 105D was ordered by the Royal Danish Air Force for inclusion in the Weapon Delivery and Navigation System refit programme associated with the Service's Draken fighters.

Operating characteristics are virtually identical to the Type 105, although a mean time between failures of 1000 hours is quoted and the basic pulse-repetition frequency is now ten pulses a second.

Dimensions: 210 × 264 × 308 mm
Weight: 19·2 lb (8·7 kg)
Operating wavelength: 1·06 micron
Pulse repetition frequency: 10 pps
Angular coverage: within 10° semi-apex angle cone
Roll stabilisation: ±360°
Operating range: ≤5 km
Accuracy: ±5 metres standard deviation
Power requirements: 9 A at 28 V dc

STATUS: in production.

Ferranti Type 105D laser ranger for Royal Danish Air Force Draken

Type 106 laser seeker

This unit comprises basically the passive elements of the LRMTS used in Royal Air Force Jaguar and Harrier close-support aircraft. The system permits a crew to identify a laser-designated ground target and can present appropriate director information on the head-up display or gunsight. Its use is claimed to significantly increase the effectiveness of close-air support against difficult targets, and to reduce the briefing time required between the forward-air controller on the ground, and the pilot.

The laser receiver optics are contained in an approximately cylindrical pressurised unit which needs to have a forward field-of-view equivalent to the optic angular slewing limits. A small electronic unit is also installed containing power supplies and target-seeking electronics.

To find a marked target, the optical sensor is set at a pre-determined depression angle and scanned horizontally until the designation is detected. The laser receiver then locks onto the target and produces steering commands to the pilot, also assuring accurate target-recognition at the earliest opportunity. Blind attacks can be carried out.

Ferranti has proposed the Type 106 laser seeker with the ISIS D-209 RM gyro gunsight as part of a low-cost target acquisition system for ground-attack aircraft.

Dimensions: 275 × 247 × 312 mm
Weight: 46 lb (21 kg)
Operating wavelength: 1·06 micron
Angular coverage: (elevation) –25 to +20° (relative to roll axis)
(azimuth) –18 to +18° (relative to heading)
Roll stabilisation: ±90°

Ferranti Type 105 laser ranger

Ferranti Type 106 marked-target seeker is passive version of LRMTS

Power requirements: 300 VA at 200 V 400 Hz 3-phase, plus 3 A at 28 V dc

STATUS: in development.

Carbon-dioxide lasers

Ferranti has developed pulsed carbon-dioxide (CO_2) laser-ranging systems suitable for ground operations, which have pulse-repetition frequencies and operating ranges similar to airborne laser-ranging units. All carbon dioxide lasers operate at 10·59 microns, which is in the far infra-red range and therefore eye-safe (1·06 micron lasers, used by all current US and UK airborne equipment, can cause eye damage, although the probability of this happening, due to the narrow beam width, relatively long ranges, and intermittent short pulses used, is very small. Even so, peacetime restrictions on the use of lasers in training operations are very stringent).

Carbon-dioxide lasers, because of their longer wavelengths, can also penetrate dust, smoke and haze more effectively than existing systems. They can be configured to share the same optical port as an 8 to 13 micron flir sensor, and this could simplify optical harmonisation and operational use of a joint flir/carbon-dioxide laser system. It is expected that, as more carbon-dioxide-based ground lasers are introduced and suitable airborne munition seeker equipment is developed, then carbon-dioxide lasers will become more widely accepted in airborne applications.

STATUS: in development.

Laser pods

Ferranti, in conjunction with Portsmouth Aviation, has announced two podded-laser installations suitable for ground-attack aircraft. These could be used to retrofit existing ground-attack aircraft, or to equip fleets which are only occasionally assigned ground-attack duties, in which case a few pods shared by a larger number of aircraft is more economic financially and on aircraft internal space than having each aircraft laser-equipped.

Two pods are proposed. They are based on either the Ferranti laser ranger and marked-target seeker (LRMTS) or Type 106 laser seeker. The pod is a proven unit with known operational clearances. Within the pod would be the appropriate laser head at the front, an electronic unit at the rear, and a conditioning unit in the centre. A special-to-type electronic interface unit can also be included, if necessary. Optical alignment to within 3 mrads with on-board sensors is claimed.

Pod length: 7½ ft (2·29 m)
Max dia: 1¼ ft (0·38 m)
Weight (with LRMTS): 309 lb (140 kg)
(with Type 106): 276 lb (125 kg)
Pod clearance-symmetric installation

	Sea level	15 000 ft	20 000 ft
Max speed	M0.95	M1.10	M0.80
Vertical acceleration limits	+8·0/−4·1	—	+4·5/−2·1
Max rate of roll	150°/s	120°/s	200°/s

STATUS: in development.

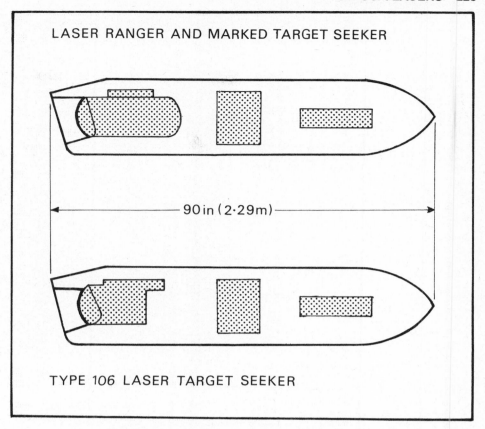

Two podded laser options based on Ferranti equipment

Lasergage

Lasergage Limited, Newtown Rd, Hove, East Sussex, BN3 7DL

LP7 laser rangefinder

The Lasergage LP7 laser rangefinder is intended primarily for artillery and mortar fire-control but, owing to its low weight and size, is also suited for use aboard helicopters for air-to-ground ranging use. The instrument is similar in size to standard 7 × 50 binoculars, and is used by the observer in similar fashion. When held to the eyes, a crosshair reticle may be laid on the target and on depression of a trigger button the range, to the nearest five metres, is displayed in the left-hand eyepiece. Due to the combination effect of the eyes, the four-digit light-emitting diode which displays the range appears to the observer to be superimposed on the target as seen through both eyepieces.

The transmitter element of the LP7 is a Q-switched neodymium-YAG laser which directs an eight-nanosecond, high-intensity, invisible infra-red pluse of 1064 microns wavelength at the target. A small portion of the light pulse is diverted to the receiver, initiating action of the range-counter timing electronics.

On striking the target, some of the transmitted light is reflected back to the instrument and is diverted to the timing circuit via the objective lens and a prismatic optical system, stopping the range counter. Range is computed from the elapsed time from transmission to reception of the light pulse. The receiver circuitry employs a silicon avalanche photodiode and operates on a clock frequency of 29·97 MHz. Maximum range of the system is 9995 metres. The telescope optical system provides a magnification of ×7 and eye protection, comprising a dichroic beam-splitter and absorbtion filter glass in the eyepiece, is incorporated. Transmitter output energy has been kept to a minimum in order to reduce the eyesafe distance to 500 metres, calculated in accordance with recommendations of STANAG 3606.

Power for the system is supplied from a 12-volt NiCd rechargeable battery which provides sufficient for more than 600 ranging shots per charge. Following each range measurement the light-emitting diode display is automatically shut down after three seconds in order to conserve battery power. The display itself has a brightness adjustment so that intensity can be varied according to the user's requirements and to reduce the possibility of detection during use in darkness. In addition to the range readout, indicator lamps also warn that more than one target is in the beam; that one or more targets have been gated out by a minimum range control facility; that laser output power is low; or that the battery requires recharge.

The LP7 can be interfaced with other sighting units to form an integral part of a comprehensive fire-control system. It has an operating temperature range from −30 to +55°C and is currently in service with the British Army and the armed forces of many other nations.

Dimensions: 200 × 200 × 90 mm
Weight: 0·9 lb (2 kg)

STATUS: in production and service.

UNITED STATES OF AMERICA

US laser research

More airborne laser weapon-aiming sensors have been developed in the USA than in any other country, about a decade and a half of experience having now been consolidated. Much of the early work was conducted under the impetus provided by US Air Force involvement in the Viet-nam War. The US Air Force Aeronautical Systems Division at Wright-Patterson Air Force Base (ASD) has managed many programmes and these have introduced laser target-designation, laser-guided weapons and other forms of high-accuracy, or blind-attack technology under project titles headed by the word Pave. The following brief survey illustrates the rapid build-up of technical capability which took place between 1967 and 1973. All subsequent developments have tended to be refinements on the principles evolved in this period.

Pave Alpha was the first US Air Force ASD study into terminal-guidance technique, and was completed in 1967. It indicated the feasibility of laser designation for certain types of target, and led to the formation of a **Pave Light** development team in the same year. This was the first US laser target-designation research programme, and a counterpart activity, also launched in 1967, was **Pave Way**, to define the operational requirements for laser-guided bombs. The latter programme gave rise to a family of munitions, produced by Texas Instruments, which is described in more detail below.

The **Pave Light** programme matured, in hardware form, as the Martin-Marietta AN/AVQ-9 laser target-designator, which itself was the forerunner of the **Pave Spike** electro-optic acquisition and laser target-designation pod, described in the Westinghouse entries below.

The most significant programme funded by ASD in 1968 was **Pave Knife**, which led eventually to the AN/AVQ-10 laser designator. Although bulky and now superseded, this device saw much active service in Viet-nam during the early-1970s, in conjunction with **Paveway** weapons. In the same year projects were initiated which set the scene for many later refinements. These included **Pave Arrow**, which defined target-seeker/tracker requirements, and **Pave Spot** which considered forward-air controller laser-ranging/designation requirements for close-air support over the battlefield.

By 1969 the technology was progressing sufficiently rapidly for a laser-seeker system to be boresighted with a US Air Force/McDonnell Douglas F-4D Phantom attack radar in a demonstration programme called **Pave Sword**. In the following year laser-designator projects aimed at providing equipment for the US Air Force/Rockwell OV-10 Bronco forward-air controller aircraft **(Pave Nail)** and AC-130 Hercules **(Pave Pronto)** were initiated.

The period between 1967 and 1970 can be regarded as having been the years of consolidation, as from 1971 onwards there was more emphasis on the development of production items. Both **Pave Penny** and **Pave Spike**, which are currently-used items, respectively described in the Martin-Marietta and Westinghouse entries below, were initiated in 1971.

A notable project was **Pave Storm**, which commenced in 1973, and was an ambitious electro-optic/laser-guided cluster bomb. It included delayed fuzing, reflecting the US requirement to attack deeply-located bunkers in Viet-nam, but this project apparently died when the Viet-nam conflict was terminated. One other major ASD system contract first let in 1973 was **Pave Tack**, which is now a very-

One of the earliest US laser projects was Pave Way, which led to development of laser-guided bombs. Five Paveway II weapons are shown under Northrop F-5

capable first-line system in operational service, and described below.

Subsequent developments have been concerned with improving the reliability and operational capability offered by the techniques pioneered in the earlier programmes. These changes have resulted in improvements such as effective single-seat operations, or smaller and lighter installations.

Apart from the use of lasers as sensing adjuncts to weapon-delivery systems, the very high energy-densities contained in the narrow beams was recognised very early on to represent a considerable destructive potential, and work is under way to turn laser devices into practicable weapons. The US Air Force has been and remains in the forefront of the Defense Department's high-energy laser effort, and a modified Boeing 707 has been suitably fitted out for operational-environment test and development under the designation Airborne Laser laboratory. Initial demonstrations are being conducted against missile and drone targets at short ranges.

Ford Aerospace

Ford Aerospace and Communications Corporation, Ford Road, Newport Beach, California 92663

AN/AVQ-26 Pave Tack laser designator/ranger

Pave Tack employs forward-looking infra-red (flir) target-acquisition sensors and laser designation/ranging. Ford Aerospace provides the pod structure, electronic control, environmental-control system and stabilisation equipment and is responsible for integration with the sensors. The forward-looking infra-red unit is manufactured by Texas Instruments, and the laser is produced by International Laser Systems. Important to overall operation is a virtual-image display sub-system developed by General Electric.

The project was initiated by a request for competitive tenders, issued by the US Air Force Aeronautical Systems Laboratories, which led to an initial $15 million award to Ford Aerospace in 1974. Flight-testing began two years later, and in 1977 evaluation was stated to be 95 per cent complete for US Air Force/McDonnell Douglas RF-4C and F-4E Phantom operations by day, night and in various weather conditions. A US Air Force/General Dynamics F-111F installation was at the same stage by August 1978 and the first Pave Tack production contract, worth $48·5 million, was let in the same year. Production deliveries began in August 1980 with Phantom initial operational clearance planned for late 1980, and F-111F in early 1981.

Total US Air Force production is expected to be 149 pods, of which 20 will be used on RF-4C Phantom, 80 on F-4E Phantom and 49 on

F-15E Strike Eagle will be able to use Ford's Pave Tack laser designator/ranger to achieve highly-accurate delivery of large conventional-weapon or laser-guided payloads

F-111Fs. The number of aircraft converted to accept the Pave Tack pod is expected to comprise 60 RF-4Cs, 180 F-4Es and 100 F-111Fs. Pods, with support equipment, have been stated to cost $153·5 million, and aircraft modifications an additional $84·3 million. The first production order in 1978 embraced 23 pod sets, a further 48 sets were ordered in February 1979 and 79 sets were contracted in February 1980. Additional production is under way to deliver $26·8 million worth of Pave Tack equipment for use on Royal Australian Air Force F-111s.

The aim of the Pave Tack system is to provide unvignetted full lower-hemisphere coverage for the sensors installed, and thus to permit the crew almost total freedom in its choice of flight path, both approaching and leaving the target area. Delivery techniques can be varied also to suit the weapon and the stand-off range desired, the crew having the choice of dive, loft, glide, toss or level-release of weapons, with virtually no manoeuvrability or flight-envelope constraints.

The system has the optical port at the aft end of the pod, retracted into its most streamlined position when not in use to minimise cruise drag. The pod is in two sections; the base-section assembly (the forward portion), and the head-section assembly (the rear portion).

The base section contains a digital aircraft interface unit, which permits direct integration with either the Phantom ARN-101 digital avionics system, or F-111 avionics. There is also a cathode ray tube interface unit. The electronics provide precise target-location data to the weapon-system computer aboard either aircraft type. Special modes, incorporated as software modifications, permit Pave Tack to be used as a terrain-following sensor. In this mode of operation the pod line-of-sight is pointed along the aircraft velocity vector. There is also a tape recorder which records the crew's video display, and which can furnish bomb-damage assessment data.

The head section contains an optical bench and turret structure. Forward looking infra-red, laser, range-receiver and stabilised-sight equipment is mounted on the optical bench, which is so linked to the turret structure and base section that 180 degrees of pitch movement is possible by using turret motion, and roll motion is provided by rotating relative to the base section.

Both the forward looking infra-red and laser are boresighted and stabilised. Flir-imagery provides a wide field-of-view display for target acquisition, but also has a narrow field-of-view, high magnification, target identification and target-tracking mode. The AN/AVQ-25 laser designator/ranger is compatible with existing 1·06 micron laser-guided weapons.

Cockpit-display equipment consists of several high-resolution cathode ray tubes clustered together, and viewed by the crew via a multi-element binocular optics system. Avionics data can be written in alpha-numeric form on them, and the system constitutes a multi-function display surface in the cockpit where the crew can view radar, forward looking infra-red, television and weapons data. Initial design was based on taking the space occupied by the AN/APQ-144 indicator/recorder system in the F-111F, and the configuration has now been adapted to replace the AN/APQ-120 radar-plot indicator used in the F-4 Phantom rear cockpit.

Pave Tack is a day or night, clear and adverse weather sensor which enhances target acquisition and permits accurate release of laser-guided or conventional weapons onto any target visible to the electro-optics system.

STATUS: in service. The system has been evaluated on the F-15E Strike Eagle.

Hughes

Hughes Aircraft Company, PO Box 90515, Los Angeles, California 90009

AN/AAS-33 target-recognition and attack multi-sensor (tram)

Installed on a large number of US Navy/Grumman A-6E Intruders, the system consists of a 51-cm diameter stabilised turret which extends below the nose radome. It contains a forward looking infra-red receiver, a laser designator/ranger, and a separate laser seeker which is used to detect and track targets designated from other sources.

Tram development led to an initial $21 million production contract for 36 sets in 1976, and this was followed by a US Navy technical and operational evaluation. The first production set was delivered in 1979, and equipment was intended to enter service later that year. Current orders for tram, all from the US Navy, are for about 70 sets at an estimated total cost of $135 million.

Operations are optimised for the maritime offensive role and the multi-sensor tram unit is complemented in operations by the A-6E's maritime radar. Target acquisition can be made by radar and the forward looking infra-red sensor used for identification and attack. The laser ranger assures high weapon-aiming accuracy on a single-pass attack. During an attack, the tram system seeks the designator signal and then uses the active laser unit to measure accurately angular position and range.

For cases where laser-guided weapons are used the tram turret can view any point in the full lower hemisphere, allowing freedom for manoeuvre to the crew after weapon release. The flir also provides a recorded view, and hence permits the attack to be video-recorded.

STATUS: in service.

Laser-augmented airborne TOW (laat) sight

The standard M65 TOW-missile sight produced by Hughes Aircraft for use on US Army/Bell AH-1S Cobra light-attack helicopters contains a thermal-imaging system. To improve its operational effectiveness a miniature laser transmitter has been developed and fitted to late versions of the unit, and this is designated the laat sight.

The laser is enclosed in a box measuring 130 × 130 × 40 mm. It produces four pulses per second for up to five seconds, then switches off for a 5-second cooling period. The unit provides ranging data to objects in the laat sight, information being fed directly to the fire-control computer for accurate target sighting, irrespective of helicopter or target motion.

A novel cooling technique is employed to reduce size and weight: the laser flash-tube is embedded in a heat-conductive and highly-reflective layer of material, and this is sufficient to avoid the need for more conventional cooling equipment which is bulky and heavy.

STATUS: in production.

International Laser Systems

International Laser Systems Inc, 3404 N Orange Blossom Trail, Orlando, Florida 32804

Lantirn laser designator/ranger

Lantirn (low-altitude navigation, targeting, infra-red, night) is a pod-mounted system specially designed for the US Air Force F-16 fighter and A-10 battlefield bomber to permit single-seat ground-attack operations with both types. Two pods will be used; a navigation unit (containing a wide field-of-view forward looking infra-red sensor and terrain-following radar) and a targeting unit (containing stabilised optics for a narrow field-of-view forward-looking infra-red sensor and laser designator/ranger). Martin-Marietta is responsible for integrating the whole sensor system and was awarded an initial development contract by the US Air Force in September 1980 worth $94 million. This is expected to lead to production and deployment of lantirn equipment by the mid-1980s.

When operational the equipment will confer on these two very different aircraft the ability to acquire, track and destroy ground targets with both guided and unguided weapons. Operations will be extended to the hours of darkness, and low-level approaches and attacks, en-route weather penetration, and blind let-downs on return to base should also be possible.

The targeting pod has, in addition to a narrow field-of-view, steerable, forward-looking infra-red sensor, a laser designator/ranger produced by International Laser Systems. Few technical details of the laser have been revealed, although for compatibility with existing, 1·06 micron, weapons, such as the Maverick air-to-ground missile it seems certain to be a neodymium-YAG laser, with a maximum range in the order of ten kilometres.

Lantirn system
Navigation pod
Wide-angle flir (Hughes/Martin-Marietta)
terrain-following radar (Texas Instruments)
Targeting pod
narrow-angle flir (Hughes/Martin-Marietta)
laser designator/ranger (International Laser Systems)
Other units
fire-control computer (Delco)
holographic wide-angle head-up display (Marconi Avionics)

STATUS: in development.

Martin-Marietta

Martin-Marietta, Orlando Aerospace, PO Box 5837, Orlando, Florida 32855

Laser-spot tracker for F-18

Developed specifically for the US Navy/McDonnell Douglas F-18 Hornet carrier fighter, the laser-spot tracker permits crews to identify laser-designated targets illuminated by either ground forces or co-operative aircraft, and to achieve more accurate tracking and weapon-release performance than a non-laser equipped aircraft. Target-position data from the laser-spot tracker is fed directly to the F-18 mission computer and used to provide weapon-aiming and ordnance-release information.

The laser-spot tracker optics are stabilised using attitude data passed to the unit from the aircraft inertial navigator. The detector is a four-quadrant photodiode type, mounted behind a hemispherical dome. Laser-pulse decoder electronics are also contained in the detector.

Also located in the same pod, for which Martin-Marietta is prime contractor, is a strike-camera. This unit is in the pod aft section and has a wide field-of-view in the lower hemisphere. It can be slaved to the laser-spot tracker or independently controlled by the mission computer. Video data from this unit permits rapid strike-damage assessment after aircraft attacks.

Units have been flown for test and evaluation purposes since 1980, and production deliveries to the US Navy and Marines are expected to commence in 1983. For strike operations with lst/scam the F-18, the laser-spot tracker/strike-camera pod will be mounted on the starboard fuselage side, and an infra-red target acquisition/tracker pod will be carried on the port fuselage side.

Lst/scam pod length: 7½ ft (2·29 m)
Body diameter: ⅔ ft (0·2 m)
Weight: 162 lb (73 kg)
Operating wavelength: 1·06 micron
Scan-patterns: pre-programmable/pilot-selectable.

STATUS: in production.

Mast-mounted laser for Hughes 500MD

A mast-mounted sight has been developed, and a flight-verification programme has been completed, on a laser/silicon-vidicon television sensor package mounted above the rotor mast of a Hughes 500MD light helicopter. The installation permits the helicopter to hover behind cover (trees, embankments, hillocks, etc) and yet to present to the crew an almost complete panoramic view of the surrounding terrain.

The laser unit is a relatively low-power designator/ranger which is suitable for guiding Copperhead and Hellfire anti-armour missiles. Demonstrations with these weapons were made at Orlando in January 1980. The system has been jointly developed by Hughes Helicopters and Martin-Marietta.

System weight: (over rotor) 50 lb (22·6 kg)
(electronics) 40 lb (18·1 kg)
(controls) 40 lb (18·1 kg)
Operating wavelength: 1·06 micron
Scan coverage: (elevation) ±15°
(azimuth) ±160°
Max operating range: (designating) 3 km
(ranging) 5 km
Total system power requirement: 600 W

STATUS: in development.

AN/AAS-35(V) Pave Penny laser-tracker

This is an advanced miniaturised day/night laser-target identification set. Used in conjunction with a laser-designation system, either ground-based or in a co-operating aircraft, targets can be recognised and identified rapidly, and accurate steering-data provided to assure quick pilot reaction and accurate delivery of weapons. The equipment was installed initially on US Air Force/Fairchild A-10 close-support aircraft and later on LTV A-7D Corsair IIs. Pave Penny is also to be used on the General Dynamics F-16, and is suitable for F-4 Phantom, F-5 and Alpha Jet.

A silicon pin diode detector head is used, with full lower-forward hemisphere coverage, plus some look-up capability. Pilot's controls permit selection of several seeker scanning patterns to increase the likelihood of early designator recognition. It can be used to improve the accuracy of conventional weapon-delivery, or to lock-up laser-guided munitions.

Pave Penny consists of a relatively small pod, which is usually fuselage-mounted to allow easy harmonisation with other on-board sensors. Pre-flight boresighting of A-10 units is claimed to allow pod attachment in a matter of minutes. An aircraft adapter module is used to integrate the sensor data with on-board processors and a pilot's control panel provides for easy use and built-in test operations.

The first operational Pave Penny/A-10 unit was delivered in March 1977 and the US Air Force plans to equip all 733 A-10 aircraft in

F/A-18 Hornet development flying has included evaluation of a Martin Marietta laser-spot tracker/strike camera pod, visible just aft of starboard air-intake lip

Martin-Marietta Pave Penny laser tracker installation on Fairchild Republic A-10 Thunderbolt uses special pylon on front starboard fuselage

service. Operations began with the 354th Tactical Fighter Wing in January 1978. Up to 380 US Air Force A-7D Corsair light bombers are also to have Pave Penny installed, and Pave Penny/F-16 development is complete.

Pod length: 2¾ ft (0·833 m)
Max diameter: ⅔ ft (0·2 m)
Weight: 32 lb (14·5 kg)

Operating wavelength: 1·06 micron
Scan coverage: (elevation) –90 to +15°
(azimuth) –90 to +90°
Selectable scan patterns: wide, narrow, depressed, offset
Output: direction cosines of line-of-sight
Power requirements: <18 A at 28 V dc

STATUS: in service.

Northrop

Northrop Corporation, Electro-Mechanical Division, 500 East Orangethorpe Avenue, Anaheim, California 92801

AN/AVQ-27 laser target-designator set (ltds)

The ltds has been developed by Northrop for installation aboard two-seat F-5B and F-5F light fighters destined for foreign military forces. Several customers have selected this system which permits target-designation during operations using laser-guided munitions.

The system is installed in the rear cockpit, and can be removed when not required. It has a retractable periscopic telescope to direct the laser optics, shared with a 16-mm mission-recorder camera. The ltds package is a relatively large piece of equipment, but it is only 100-mm wide, and therefore fits in place of equipment normally installed in the port console. A viewfinder/sight swings across into the pilot's field-of-view, so that the crewman can track targets. This can be conducted manually, using a two-axis hand controller and with rate or rate-aided tracking modes. One-hand operation is claimed. Normal operations would call for designation by one aircraft, with weapon-delivery conducted by an accompanying aircraft.

STATUS: in production.

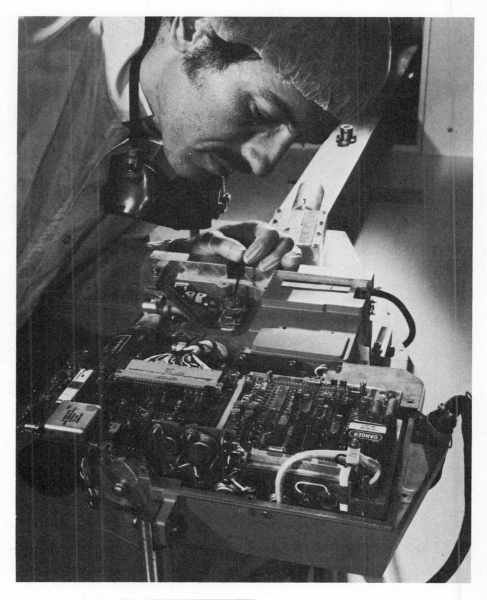

Northrop-developed ltds for F-5 fighter/strike type is assembled by Hughes technician

Perkin-Elmer

Perkin-Elmer, Electro-Optics Division, 100 Wooster Heights Road, Danbury, Connecticut 06810

Dole laser-warning receiver

Dole (detection of laser emissions) is an experimental system which aims to demonstrate the feasibility of detecting laser emissions using a dispersed set of sensors on an airframe in the same manner as an all-aspect radar-warning receiver. Development is being sponsored by the US Air Force Wright Aeronautical Laboratories, and flight evaluation conducted by the 4950th Test Wing. A combined radar/laser warning receiver set is being evaluated using a Bell/Dalmo-Victor AN/ALR-46A radar-warning receiver display to show threats determined by the radar and laser-warning system processors.

STATUS: in development.

Rockwell

Rockwell International, Autonetics Strategic Systems Division, Defence Electronics Operations, 3370 Miraloma Avenue, Anaheim, California 92803

Airborne laser-tracker (alt)

This is a laser-seeker head, with associated components, which Rockwell produces for the target acquisition designation system/pilot's night viewing system (tads/pnvs) in the US Army Hughes AH-64 armed helicopter and for the AH-1S Cobra light attack helicopter.

Production of AH-1S equipment was initiated in September 1981, and first delivery is scheduled for May 1983. Tads/pnvs production began in June 1981 and deliveries were expected to commence in October 1982.

The seeker is a wide field-of-view unit, sensitive to 1·06 micron radiation, which can detect target designations from ground troops or co-operative aircraft. Coded-pulse data is used to minimise the possibility of jamming, and a four-quadrant silicon detector head is used.

Scanning field-of-view: (elevation) –60 to +30° (azimuth) –90 to +90°
Instantaneous field-of-view: (elevation) 10° (azimuth) 20°
Optical port diameter: 5 inches (127 mm)
Focal length ratio: 0·3:1
Power requirements: 400 Hz and 28 V dc

STATUS: in production.

Texas Instruments

Texas Instruments Inc, PO Box 226015, Dallas, Texas 75266

AN/AAS-37 detecting, ranging and tracking system (drts)

This consists of an infra-red target acquisition and tracking device, to which has been added a laser designator/ranger. It can provide guidance for the operation of Paveway bombs and Hellfire anti-armour missiles. A direct readout of laser range is available to the crew and automatic centroid, adaptive-gate or offset tracking facilities are provided by the infra-red equipment.

Texas Instruments initially developed the laser designator/ranger to provide accurate range information and laser designation for Paveway guided weapons. The system comprises a transmitter-receiver, high-energy converter, receiver amplifier, and a cooling-electronics unit together with a cockpit-mounted code-generator. The units may be flexibly packaged, using standard laser sub-assemblies, so that the system can be optimised for a range of forward-looking infra-red system applications. The system is acousto-optically Q-switched.

Over 3000 hours of reliability testing were completed on the laser designator/ranger prior to the award of a production contract. Production deliveries to the US Marine Corps began in late 1979 and the system is installed in Rockwell OV-10D forward air control aircraft. According to Texas Instruments, numerous successful missions have been flown using both conventional and laser guided bombs with the laser designator/ranger providing range and target designation.

Laser designator/ranger
Azimuth coverage: ±200°
Elevation coverage: –82 to +16°
Volume: turret mounted sub-assemblies 0.32 ft³ (0.009 m³)
Weight: 25 lb (11·36 kg)

STATUS: in production.

Paveway laser-guided bombs

Texas Instruments Inc is credited with being the first to propose the concept of laser-guided weapons in 1965. After US Air Force research programmes which defined operational requirements were completed, and the need for such a weapon was determined, simultaneous weapon, sensor and illumination requirement studies were conducted. Pave Way was the first laser-guided bomb programme, and was initiated in 1967.

As now available, in either Mk II or Mk III configuration, Paveway is based on standard US Air Force 'iron' bombs. Texas Instruments supplies nose sections and new tail surfaces which convert the conventional weapons into laser-guided munitions.

The nose section is a self-contained laser-seeker, with command/autopilot electronic unit, actuator and small flightpath guidance fins. A set of retractable fins is added to the rear of the bomb in order to maintain stability with the extra nose mass. All Paveway bombs to date have been based on US Air Force Mk 80 series bombs (Mk 82, 83 and 84) and the UK Mk 13 and 540 lb bomb. Typical cost of a Paveway II laser-guided bomb (early 1980s prices) was put at $10,270 per round, compared to $2,270 for a comparable conventional weapon.

During operational use, the target to be attacked must be illuminated by a laser. This can come from an operator on the ground, the launch aircraft or another aircraft. The seeker head determines the direction to the target, and makes forward-fin actuator adjustments to steer towards it.

Stand-off delivery is possible, energy imparted to the bomb at release influencing the range and altitude combinations which can be used with particular delivery techniques; loft, glide, dive, toss or level-release. Lofted deliveries, pulling up from level at up to nine kilometres from the target, are possible. At least two kilometres stand-off is necessary at typical release speeds, which can vary between 300 and 700 knots.

Paveway bombs do not need any interface with aircraft systems. The seeker acquires the target once it is within field-of-view and guides the weapon after it has been released from the

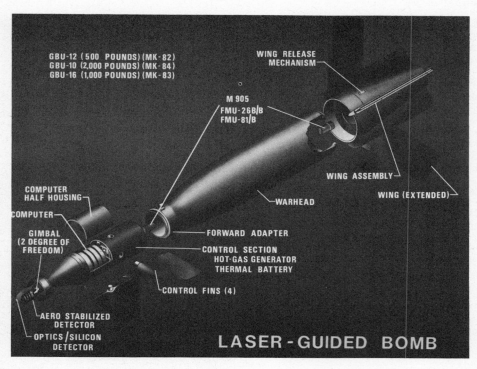

Texas Instruments Paveway II laser-guided bombs use conventional bomb as warhead with retractable rear-fins and laser-nose/canard fins added

aircraft. Up to ten years' shelf-life and reliability figures of 96 to 98 per cent are claimed. Paveway bombs were used extensively to attack small or high-value targets in Viet-nam, in the early 1970s. Pursuit and proportional seekers are available, choice depending on the launcher/designator location used in a particular mission.

Paveway I and II are certified for Mirage III, Buccaneer, F-5, F-16, Jaguar, F-4 Phantom, A-7 Corsair, A-4 Skyhawk, Harrier, F-111, A-6 Intruder and A-10 aircraft. Fit checks have been completed on Alpha Jet, Tornado, Mirage F1 and F-18 Hornet.

Users of the equipment include the United States (the US Air Force, Navy and Marine Corps), Royal Air Force, Royal Netherlands Air Force, Iranian Air Force, Greek Air Force,

Turkish Air Force, Canadian Armed Forces, Royal Saudi Arabia Air Force, Republic of Korea Air Force, Spanish Air Force, Royal Australian Air Force, Royal Thailand Air Force, and Republic of China (Taiwan) Air Force.

Compatible laser designators include Atlis II, Ferranti LRMTS, Pave Tack, lantirn targeting pod, tram (AN/AAS-33), ltds (AN/AVQ-27), drts (AN/AAS-37), Pave Spike, and Hughes Aircraft ground laser locator/designator and modular universal laser equipment (mule).

Paveway III is a low-level version of the earlier weapon, which is especially suitable for low-level stand-off delivery. It can have pursuit or proportional laser-seeker heads, inertial control or I²R guidance.

US Air Force designations for Paveway weapons are as follows;

	Weight	Basic weapon	Paveway version
Paveway I & II			
	500 lb (225 kg)	Mk 82	GBU-12/B
	1000 lb (450 kg)	Mk 83	GBU-16/B
	2000 lb (900 kg)	Mk 84	GBU-10/B
Paveway III			
	500 lb (225 kg)	Mk 82	GBU-22/B
	1000 lb (450 kg)	Mk 83	GBU-23/B
	2000 lb (900 kg)	Mk 84	GBU-24/B

STATUS: in service.

Westinghouse

Westinghouse Electric Corporation, Aerospace Division, PO Box 746, Baltimore, Maryland 21203

AN/ASQ-153 Pave Spike laser designator/ranger

The Pave Spike development programme was initiated in 1971 and delivery of 156 pod sets to the US Air Force was completed by August 1977, by which time 327 aircraft had been converted to accept the system. A further 82 sets were delivered up to September 1979, for foreign use, including a substantial number to the Royal Air Force and some to the Turkish Air Force.

Each pod has a nose section, which revolves about the pod axis to provide roll stabilisation, and a cylindrical forward portion which rotates in pitch to provide elevation stabilisation. Virtually complete lower-hemisphere coverage is thus provided in a relatively compact and light arrangement. The nose section is sealed

RAF Buccaneer equipped with Pave Spike laser-designator/ranger and Paveway II laser-guided bomb mounted respectively on inboard and outboard pylons

and pressurised with nitrogen, maintained at a constant temperature for optimum sensor performance. The centre section provides umbilical connections between the nose and rotating sections, the aircraft and the aft electronics system. In the aft section is a cold plate onto which are mounted the electronic

line-replaceable units. These comprise a low-voltage power-supply and pod control, servo drivers, laser control, laser power-supply, and interfaces.

The pod contains a television-tracking sensor and laser designator/ranger. The television sensor can be used for target acquisition and the designator permits accurate delivery of laser-guided munitions. Laser-ranging can be used to improve the delivery accuracy of conventional weapons.

Initially the unit was procured for use only on US Air Force/McDonnell Douglas F-4D and F-4E Phantoms, but it is now employed on several other types, including Buccaneer in the UK. Ferranti has held a technical support contract for systems operated by the Royal Air Force since August 1978.

A shorter model, Pave Spike-B, was offered for the General Dynamics F-16 fighter, but has not been placed in production, and Westinghouse also proposes a '24-hour' version of the system, Pave Spike-C.

The overall AN/ASQ-153 system comprises the AN/AVQ-23 pod, and several system components in the aircraft. These include a line-of-sight indicator, control panel, range indicator, modified radar-control handle and weapon-release computer. The system can be used with Paveway laser-guided bombs and several other laser-guided munitions.

Pod length: 12 ft (3·66 m)
Pod diameter: ⅚ ft (0·25 m)
Weight: 425 lb (193 kg)
Operating wavelength: 1·06 micron

STATUS: in service.

Pave Spike target-designator/ranger under port engine air intake of F-4 Phantom

Data Processing Systems

This section comprises sub-sections on computers, data-transmission and programming languages. In the main, only general-purpose airborne computers have been considered, and further specialist computers are described under their appropriate equipment entries.

Data-transmission techniques have taken on a new importance in recent times and are in the process of establishing a recognisable order. A brief introduction to the major standards now governing work in this area was seen as a valuable annex to the computer descriptions.

Finally, the programming languages used in avionics applications are in the process of being rationalised and a brief survey is given of the most common languages currently in use and planned for the future.

Computers

Two major computer instruction-set architecture standards are in the process of implementation in the military industry. These will have far-reaching influences on future computer developments in both military and civil fields. A brief description of each is given prior to descriptions of hardware in production.

MIL-STD-1750 16-bit instruction-set architecture

The US Air Force has standardised an instruction-set architecture for use in all new computer procurements. The elements of this procedure are embodied in MIL-STD-1750. This does not require a standard computer system, but by standardising the techniques used to accomplish major computational functions, advantages will result from software life-cycle-cost savings and more frequent hardware competition, which should, in turn, lead to a more rapid infusion of electronic technology into airborne computing.

MIL-STD-1750 procedures are being used to develop a common set of support software applicable to all computers, irrespective of their manufacturer or date of manufacture. This has not been possible before, and cross-compilers have become commonplace to translate between machines using customised versions of the same programming language. As experience with MIL-STD-1750-compatible machines is built up, so the risks of operating errors will reduce as more mature techniques are continually being used. This should in turn reduce the risk involved and the time needed to develop software for new applications. Furthermore, software should become a standard item, allowing computer hardware to be updated for example, and merely transferring to new equipment the software developed for a previous generation machine. This should substantially reduce cost, software development being a labour intensive and hence very expensive item.

MIL-STD-1750 developments were begun in

1976 at Wright-Patterson Air Force Base, Air Force Avionics Laboratories, using a Strawman architecture ('Strawman' was the name given to a particular stage in the development of a high-order language, see sub-section below 'Programming Languages'). The initial MIL-STD-1750 standard architecture-set was developed with industrial participation and first officially released on 21 February 1979.

Current activity includes a MIL-STD-1750 instruction architecture-set User's Group, which is responsible for developing revisions to the initial release standard.

MIL-STD-1750 is a general-purpose register architecture, which can be used for both fixed-point and floating-point operations. It is applicable to 16- and 32-bit fixed-point and 32- or 48-bit floating-point operations. A total repertoire of 100 instructions is envisaged, and this should be adequate for all foreseeable airborne computing tasks. The basic standard is satisfactory for a 65 536 word main-storage unit, but this is not a limitation. Applications with more storage requirements, up to 1 megaword limit, can be handled by expandable addressing facilities. The standard has deliberately refrained from defining programming protocol for input/output channels, thus assuring continuing use of channels and devices already in production.

US programmes which are to use MIL-STD-1750 protocol in the near future include: A-10, F-16, F-111 and F-46 (Wild Weasel) update/upgrade programmes, Lantirn (F-16 and A-10), F-5G development, KC-135 improvements, the new HX combat helicopter project, and modular automatic test equipment (mate) for the US Air Force CX advanced transport.

MIL-STD-1862 32-bit instruction-set architecture

The US Army initiated MIL-STD-1862 developments, and these are now being implemented by multi-service groups and made complementary to the 16-bit MIL-STD-1750 instruction-set architecture.

Development commenced in 1979, when the US Army commissioned Carnegie-Mellon University to develop a 32-bit instruction-set architecture called 'Nebula'. A preliminary issue of MIL-STD-1862 was made as soon as March 1980, and since then the US Army has requested proposals for computers which are compatible with this standard, but which it would not envisage to be available until about 1985, by which time it is thought that 32-bit micro-processor technology will have matured to full-scale production phase.

IBM has published the following comparisons of likely computer capability of these machines by 1985:

	1980	1985
	Nato Awacs	
	CC-2 computer	**AN/UYK-41**
Speed	2 Mips	3 Mips
Storage	2·5 megabits	2 megabits
Power	6500 W	100 W
Volume	37 ft³	0·6 ft³
	B-52 oas*	
	computer	**AN/UYK-49**
Speed	400 Kips	500 Kips
Storage	250 kilobits	1 megabit
Power	550 W	20 W
Volume	0·9 ft³	0·12 ft³
	B-52 Dbns**	**single-module**
	computer	**computer**
Speed	400 Kips	500 Kips
Storage	128 kilobits	128 kilobits
Power	450 W	5 W
Volume	0·9 ft³	0·02 ft³
		(one card)

*Offensive avionics system
**Doppler bombing/navigation system

The objectives of the MIL-STD-1862 programme are the same as those for MIL-STD-1750. In summary, the standard should result in the availability of universally-applicable software support facilities, reduced software developments risks and timescales and easy interchangability of software between computers.

FRANCE

Electronique Serge Dassault

Electronique Serge Dassault, 55 Quai Carnot, 92214 Saint Cloud

The activities of EMD (as the company was known up to early 1982) in computers and aerospace digital systems started in 1958 with the production of the **Type 4070** central computer for the Mirage IV strategic bomber.

Since 1965, when the company designed the **Sagittaire** computer (claimed to be the first European airborne computer to use integrated circuits), all French ballistic missiles have been fitted with EMD general-purpose computers, or (since 1966 when a co-operative agreement was signed) with **EMD-Sagem** computers.

In 1976, the increasing need for large computing capacity and programming flexibility led EMD, as it was then, to promote in France new component and packaging technologies, and so to create a new generation of general-purpose computers. These include the **M182** for the Mirage F1, the **2084** and **USG 284** for the Mirage 2000, and the **1084** and **CM84** for missiles.

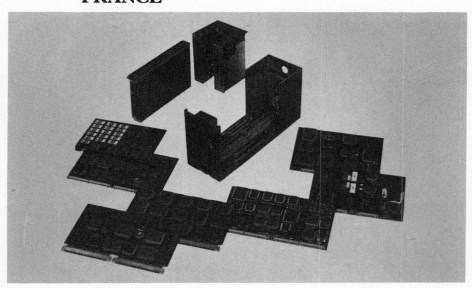

EMD M182 processor is used in export versions of Mirage F1

More than 1000 airborne computers for aircraft and missiles have been ordered or delivered.

The current range of general-purpose, digital, airborne computers possesses the following characteristics:

Operand formats: (fixed-point) 16- and 32-bit (floating-point) 32-bit
Word length: 16- or 32-bit
Cycle time: 1 μs
Typical execution speeds (16-bit):
  (add) 1·5 μs and 2·25 μs
  (multiply) 6·75 μs
  (divide) 7·5 μs
Central memory: 16 K 16-bit words, extendable to 128 K
Typical installation: ½ ATR (with 32 K core memory)
Weight: 19·8 lb (9 kg)
Software support: LTR, Fortran, Atlas

STATUS: in service.

EMD 2084 and USG 284 processors are used in Mirage 2000

Sfena

Sfena, Controls and Systems Division, Aérodrome de Villacoublay BP59, 78141 Velizy-Villacoublay, Cedex

Series 7000 general-purpose processors

Sfena Series 7000 computers are characterised by a modular and adaptable architecture, which is claimed to minimise hardware and development costs in any application, and avoid technical obsolescence by providing for continuous updating of standard modules. They use a flexible memory design, allowing a wide-range of memory types, capacity and access-time characteristics to be matched to any application. The input/output system is completely modular, providing again for a high degree of customisation, and there is a highly-developed operating system to support all applications.

The basic central-processing unit is the 16-bit CPU 7800, which is organised to operate with a memory bus and an input/output bus. A more recently developed unit is the 7068 central-processing unit, which is available for especially demanding tasks. Parallel processors are able to operate simultaneously with either central-processing unit for applications requiring the very highest performance. The interrupt system also allows processing in up to eight hierarchic levels, with an extension to 16 levels.

Normal addressing capacity is 64 K words, extendable up to 256 K words. Mass memories can be added externally, and internal memories can be semi-conductor (with or without battery back-up), programmable read-only memory or reprogrammable read-only memory.

Software support is available in LTR.

CPU Characteristics

	7800 CPU	7068 CPU
Operation:	Binary, two's-complement, fixed- and floating-point	
Word Length:	(fixed-point) 16-bit (floating-point) 32-bit	
Max address range:		
(basic)	64 K words	64 K words
(option)	265 K words	8 Megawords
Instruction set:	Up to 150	
Execution times:		
(load)	0·9 μs	—
(store)	1·2 μs	—
(fixed-point multiply)	7·5 μs	—
(floating-point multiply)	13·2 μs	—
Input/output options:	large range of possibilities, including ARINC and MIL-STD-1553	
Typical installation:		
(4 cards)	½ ATR	(2 cards) ½ ATR
Power requirements:		
	6.3 A at 5 V dc	2.5 A at 5 V dc

STATUS: in production.

Thomson-CSF

Thomson-CSF, Compagnie D'Informatique Militaire Spatiale et Aéronautique (Cimsa), 10-12 Avenue De L'Europe, 78140 Velizy

15M/05 general-purpose processor

This unit has been designed for process-control or signal-processing applications. It is less capable than the 15M/125 which is also made by Cimsa. Typical avionic applications are radar, sonar or electronic-warfare processing. M-rack and ATR-case versions of the processor are produced and can be used for ground-based or airborne applications. Software languages for which support is available are Mitra and LTR.

Computer type: binary, two's-complement, fixed- and floating-point
Word length: (fixed-point) 16-bit (floating-point) 32-bit
Max address range: 32 K words (extendable to 64 K)
Instruction set: 98
Approximate instruction run times:
  (RAM or PROM/TTL) 1·2 μs
  (RAM/HMOS) 1·2 μs
  (RAM/MOS (4 K/board)) 1·7 μs
  (RAM/MOS (32K blocks)) 1·8 μs
Micro-instruction run times: 110, 165 or 220 ns
Typical installation: ½ ATR, 1 ATR or Rack 2 M
Power requirements: 300 W at 24-48 V dc

STATUS: in production.

Thomson-CSF Cimsa 15M/05 processor

15M/125 F general-purpose processor

There are two versions of this processor. The 15M/125 is a central processing unit with main memory extendable to 512 K words, and the 15M/125F is a version with a built-in fast arithmetic operator. The unit is complementary to the 15M/05 unit and is better-suited to high-speed processing duties or tasks which require a large amount of rapidly accessible memory. Joint-service operations in weapon-command and control, communications management, navigational aids, detection systems (radar, sonar, electronic warfare) and command aids are possible.

Software support is available for Mitra and LTR programming languages.

Computer type: binary, two's-complement, fixed- or floating-point
Word length: (fixed-point) 18-bit (16-bit usable) (floating-point) double-length
Max address range: 128 K words, extendable to 512 K words core or semi-conductor
Instruction set: 128
Typical installation: ½ ATR or 1 ATR
Power requirements: 24-28 V dc or 115-200 V 400 Hz

STATUS: in production.

15M/125X general-purpose processor

This is the most recent addition to the Cimsa range of general-purpose units, and employs many internal operating procedures common to previous units, although it is a 32-bit machine with up to 1 million operations a second (1 MOPS) capability. A 1 Mega-byte memory can be addressed and versions of the processor are typically installed in a 1 ATR case. LTR real-time software support is provided, and numerous applications are expected in high-performance military equipment.

STATUS: in development.

125MS general-purpose processor

A development of the 15MS/125 range, specifically packaged for a space application (Spacelab). It uses the 16-bit word central processing unit of the 15M/125F and is installed with all peripherals in a single box.

Broadly similar to 15M/125F except:

Instruction set: 135
Addressable memory: 64 K words
Installation: 190 × 280 × 500 mm
Weight: 67 lb (30.5 kg)
Power requirement: 390 W at 24-32 V dc

STATUS: in production.

Two versions of Thomson-CSF Cimsa 15M/125F processor

Thomson-CSF Cimsa 15M/125X is used in Atlantique Nouvelle Generation tactical data-processing system

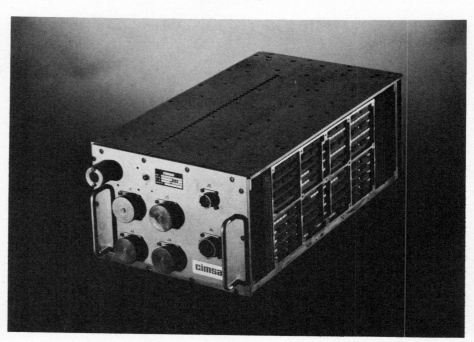

Thomson-CSF Cimsa 125MS will be used to control scientific experiments in Spacelab

GERMANY, FEDERAL REPUBLIC

Litef

Litef, Lorracherstrasse 18, 774 Freiburg

LR-1432 digital computer

This is a fast, micro-programmed, general-purpose computer which uses state-of-the-art large scale integrated circuits and is designed for the military. Fixed- and floating-point options are available. The main memory holds both program and data in EPROM/RAM-type storage with a maximum capacity of 64 K 16-bit words.

A bus interface is available to connect the computer to a digital data-bus system. This is fully compatible with dual-redundant MIL-STD-1553B, allowing operation as bus controller or remote terminal. Input/output circuits are provided for 16 various discrete signals. The whole processor, including a power-supply unit and 64 K of storage can be accommodated in a ½ ATR Short case.

The LR-1432 is in production as the main computer for the trinational Panavia Tornado (1½ ATR) and the LR-1432K is used as a navigation computer in the Franco-German Alpha Jet strike/trainer (½ ATR).

Computer type: binary, fixed- or floating-point
Word length: (fixed) 16-bit (floating-point) 32-bit
Max address range: 64 K words (extendable to 512 K)
Instruction set: 69 (including floating-point)
Store access times:

core	semi-conductor
(8/16 K module) 450 ns	(16 K RAM/PROM) 300 ns
(32 K module) 350 ns	(16 K RAM/EPROM) 450 ns

Input/output options: various analogue converters, plus digital to MIL-STD-1553, ARINC (Dits), Panavia digital links, NTDS, serial inputs/outputs (eg: CVR, TV-tab) and discretes
Typical installation: ½ ATR Short, ¾ ATR Short or 1½ ATR Short
Power requirements: 28 V dc or 115 V 400 Hz

STATUS: in service.

This version of Litef LR-1432 is main computer used in Tornado

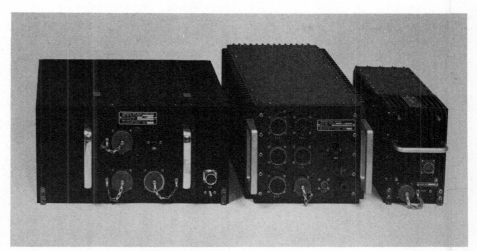

Three versions of the Litef LR-1432 processor. Left is Tornado main computer, centre is a tank computer under evaluation, and right is Alpha Jet navigation computer

ISRAEL

IAI

IAI, Elta Electronic Industries Limited, Ashdod

EL/S-8600 computer system

There are several versions of the basic EL/S-8600 main computer, all of which use similar central processing units and architectures, and are customised to various military applications. The EL/S-8610 has 15 modules (each with two printed circuit boards) in a 1 ATR box, and the EL/S-8611 has only four modules, in a customised box. In addition to their airborne applications, other computers in the range are used for land and sea operations. Typical applications are inertial navigation and weapon-delivery systems, artillery tactical fire-control system, message-switching centres and mobile communications control. Software support is available for 'C', Fortran-77, Iso-Pascal, APL and Cobol.

EL/S-8610
Computer type: binary, two's-complement, micro-instruction (15 register), fixed-point
Word length: 16-bit (32-bit instruction words)
Max address range: 64 K byte (expandable to 2 Mbyte with mapping unit)
Instruction set: 90
Micro-instruction cycle time: 225 ns

Micro-instruction depth: 1024 words (expandable to 2048 words)
Typical execution times:
(add/subtract) 0·66 μs
(load) 1·34 μs
(multiply) 6·56 μs
(divide) 9·26 μs
Input/output options: wide variety, customise to use
Installation: 394 × 257 × 194 mm
Weight: ≤47·3 lb (21·5 kg)

STATUS: in service.

ITALY

Selenia

Selenia, Special Equipment and Systems Division. Via Dei Castelli Romani 2, 00040 Pomezia, Rome

SL/AYK-203 general-purpose computer

This is a modular architecture, high-speed computer designed for real-time military applications. Processing power, memory and input/output modules can be selected in order to tailor the system to a wide-range of requirements. Three existing configurations are:

SL/AYK-203/1: ½ ATR Short, 8 K words memory, 18·7 lb (8.5 kg)
SL/AYK-203/3: ¾ ATR Short, 192 K words memory, 26·4 lb (12 kg)
SL/AYK-203/3: 1 ATR Short, 256 K words memory, 37·4 lb (17 kg)

Computer type: binary, fixed- and floating-point
Word length: (fixed-point) 8-bit (floating-point) 16-bit
Max address range: 256 K words (1 Megaword option in development)
Typical speeds:

Selenia SL/AYK-203 airborne computer

	203/1 800 KIPS	203/2 1·6 MIPS	203/3 2·4 MIPS
Typical execution times:			
(load/add)	0·6 μs	0·3 μs	0·2 μs
(multiply (8-bit))	15 μs	8 μs	5·2 μs
(multiply (16-bit))	26 μs	14 μs	8·7 μs

Input/output options: MIL-STD-1553
Tape-recorder interface
Parallel and serial channels
Status check multifunction interface
Analogue/digital multichannel interface
Frequency/digital multichannel interface
Power requirement: 115 V 400 Hz or 28 V dc

STATUS: in production.

UNITED KINGDOM

Ferranti

Ferranti Computer Systems Limited, Western Road, Bracknell, Berkshire, RG12 1RA

FM1600 general-purpose processor

The FM1600 designation covers a family of general-purpose computers, of which the FM1600D is a unit developed for airborne use. It can be employed in avionics systems that require high operating speeds, sophisticated interface structures and rapid peripheral responses. An instruction set particularly suited to highly complex real-time applications has been developed with over 330 instructions based on 60 basic forms including powerful arithmetic and logical shift and jump instructions. The power offered by the instruction set facilities is enhanced by a three-address architecture and hardware implementation.

Input/output peripheral control facilities ensure rapid interrupt response and efficient peripheral handling. Peripherals are connected via high-speed serial, star-connected interface channels, and control is exercised by way of a dedicated controller which can monitor up to 16 channels. Priority-scanning to assure short response time for high-priority periphal equipment is available.

Memory is organised in blocks of different sizes, speeds and technology, covering RAM, ROM, and core storage. Each block can have up to four interface ports, of which one is allocated to the processor/memory highway.

Extensive software is available to support Coral 66 and Fortran IV operations. The sole airborne application to date is as the processor to the EMI Searchwater radar in RAF Nimrod MR2 sub-hunters.

Computer type: binary, two's-complement, fixed- and floating-point
Word length: (fixed-point) 24-bit (floating-point) 48-bit
Typical speed: 350 KIPS
Max address range: ≤ 32 K words core (expandable)

Input/output options: up to 16 B-serial input/output channels
Installation: 393 × 191 × 318 mm
Weight: 44-55 lb (20-25 kg) depending on options
Power requirements: 360 VA at 115 V 400 Hz

STATUS: in service.

Argus M700 general-purpose processor

Argus M700 is the latest member of the Ferranti family of avionics computers. It is a military system that embodies all the advantages to be gained from a fully modular building-block approach to design. Argus system architec-

Ferranti FM1600D processor as used in EMI Searchwater radar

Two versions of Ferranti Argus airborne processor. Background cards comprise Argus M700/20, and hand-held in foreground is Argus M700/40

tures can be tailored to specific requirements, using a wide range of standard computer modules. Thus, equipment may be developed in which the different sub-systems have identical, fully-interchangeable circuit cards, the same programming language and the same software development system.

Processing power of existing systems is provided by the Argus M700/20, a central processor that combines a powerful architecture with an advanced implementation to provide high-performance with reduced space and power demands. M700/20 features 16-bit word length, multiple operating modes, a sophisticated interrupt structure and hardware-implemented double-length arithmetic.

Argus M700/40 is a recent update, and will be the first processor in the Argus range to be implemented entirely in large-scale integrated form to significantly enhance the power and versatility of Argus M700-based systems. The Argus M700/40 military micro-processor will be fully compatible with all other Ferranti Argus M700 series products, and can therefore replace other M700 processors in existing systems as well as forming the basis of new configurations.

Argus M700 computers may be connected to other elements of an overall system via a MIL-STD-1553B digital data-bus, while internal communication uses Eurobus 18/A.

All memory technologies can be incorporated in Argus M700-based computers, including RAM, PROM and core store. A maximum of 256 K words can be connected to each Eurobus and a further 64 K words of private memory can be provided for each processor. Private memories, used to hold the main part of operational programs, are accessed via a special high-speed interface.

Software support is available for Coral 66 applications and is associated with Mascot modular software construction facilities. Argus M700 has been developed in close co-operation with UK Ministry of Defence research establishments and is a preferred computer for UK future defence programmes.

Computer type: binary, fixed- and floating-point
Word length: (fixed-point) 16-bit (floating-point) 32-bit
Typical speed: (M700/20) 250 KIPS (M700/40) 1.6 MIPS
Instruction set: over 120
Input/output options: wide range of customised modules
Typical installation: 380 × 190 × 193 mm
Weight: 28·6 lb (13 kg)
Power requirements: 200 VA at 115 V 400 Hz

STATUS: in production.

F-100L micro-processor

This is a 16-bit micro-processor system, comprising a micro-computer set with memory interface, timer and interrupt controller, multiply/divide unit and clock-generation chips. It can accommodate a 32 K word address range and has comprehensive software support facilities including cross-product and resident development software, Coral 66, sub-routine library and hardware-test programs. Using these, single-unit or multi-processor systems can be developed.

The F-100L is part of an overall system concept. Its structure has been chosen to provide the facilities necessary to enable fast, real-time systems, as well as less complex applications, to be developed in the simplest possible manner. To meet this aim a comprehensive instruction set is available, plus program interrupt and reset/start facilities. One of

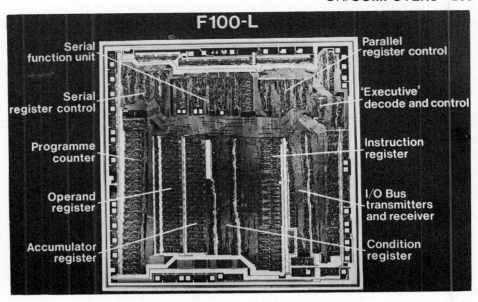

F100-L

Serial function unit · Serial register control · Programme counter · Operand register · Accumulator register · Parallel register control · 'Executive' decode and control · Instruction register · I/O Bus transmitters and receiver · Condition register

Close-up of F-100L processor chip

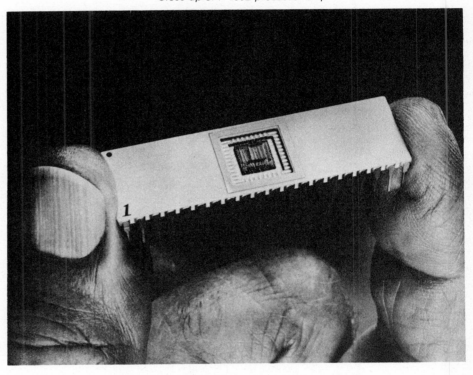

Ferranti F-100L military micro-processor in 40-pin pack

several addressing modes can be used and fast instruction times are possible.

F-100L development was sponsored by the UK Ministry of Defence, and the system meets many military requirements. Applications are under way for fire-control equipment in land/sea operations and for the digital control of turbine engines in aircraft.

Operation: 16-bit, binary
Max address range: 32 K RAM plus 1·2 megabyte mass storage
Instruction set: 153
Clock rate: 13·5 MHz
Memory access time: 1·2 μs/word
Execution time
   (multiply) 9-13·5 μs
   (divide) 12-16·2 μs
Radiation hardness: immune to latch-up, high resistance to neutron and cumulative ionisation damage.

STATUS: in production.

F-100L processor hybrid FBH 5092

The Ferranti F-100L micro-processor, plus clock-generator, multiply/divide unit and two bus-buffer chips, are available as a single hybrid package, providing about 50 per cent space-saving compared with the volume required in mounting the same components on a printed-circuit board. System reliability is increased both by the reduction in the overall number of soldered joints, and by the testing of the chip set prior to integration. The hybrid pack is also claimed to allow more heat dissipation with significantly lower average temperature and reduction of high-spot temperatures compared with printed-circuit board-mounted assembly.

Dimensions: 81·3 × 22·9 × 4·6 mm
Operating temperature: (commercial) 0 to 70°C
(military) –55 to +125°C

STATUS: in production.

UNITED STATES OF AMERICA

Control Data

Control Data Corporation, Aerospace Division, 3101 East 80th Street, Minneapolis, Minnesota 55440

AN/AYK-14 CDC 480 general-purpose processor

A member of the CDC 480 computer family, the AN/AYK-14, has been selected by the US Navy as its standard airborne computer.

The designation CDC 480 describes a family of modular, micro-programmable computers which represents low life-cycle cost, standard computing elements for a broad range of military systems and environments. Functional partitioning for the computer results in a set of standard modules which are configured and packaged to the user's specification and requirements. Basic shop-replaceable assembly types provide a general complement of computer module building blocks.

The CDC 480 general-purpose capability can be used with 16- or 32-bit instruction sets, since the modules can be configured as a 32-bit computer (horizontal expansion). The emulation task can also be divided between two processors (vertical expansion) to increase processing throughput. Configurations range from 16-bit single card IOP, to 32-bit high-speed processor with addressability to 512 K words and extensive input/output capability. Memory modules are available in configurations of 16 K or 32 K words. Memory types (core or semi-conductor) are also interchangeable.

The CDC 480 is adaptable to a variety of enclosures for airborne, shipboard and ground-based applications as a general-purpose processor, emulator, controller, dedicated processor or algorithm unit. Representative applications include weapon delivery/fire control, guidance, communications, navigation,

F-18 Hornet avionics system incorporates two CDC 480 (AN/AYK-14) mission computers

display sub-system control, radar or sonar processing system control, electronic countermeasures and electronic surveillance measures management, and digital flight control. It has been selected for the US Navy/McDonnell Douglas F/A-18 Hornet fighter (which has two CDC-series mission computers), Sikorsky SH-60B Seahawk Lamps Mk III helicopter, and eight other airborne applications.

Computer type: binary, fixed- or floating-point
Word length: 16- or 32-bit
Typical speed: 300-800 KOPS
Max address range: 512 K words

Typical execution times (high-speed floating-point):
  (add/subtract) 2·2 μs
  (multiply) 3·7 μs
Input/output options: MIL-STD-1553A (two buses)
32 i/o discretes
8 interrupts
NTDS fast, slow, ANEW and serial
Typical installation: 194 × 257 × 356 mm
Weight: 34 lb (15·5 kg)
Power requirements: 350 W typical

STATUS: in service.

Delco Electronics

Delco Electronics Division, General Motors Corporation, Santa Barbara Operations, 6767 Hollister Avenue, Goleta, California 93017

Delco manufactures several related computers under the generic name of Magic III, including the Delco M362F, M362FL, M362S and M372. A new range of units with the title Magic IV is being introduced. Those units with significant airborne applications are described here.

M362F general-purpose processor

This is a general-purpose unit which uses parallel, binary, floating-point, two's-complement, 16-bit processing. A typical instruction mix yields an operating speed of about 340 KOPS.

The M362F can be tailored to particular operations by specifying from a wide choice of memory types and standard input/output circuit modules. It is mechanised on two operating modules. The instruction repertoire can be varied by adding micro-program memory, or changing the existing micro-program memory, which consists of ten integrated circuits. The M362F will operate with core and/or semi-conductor memory. The processor provides 72 basic machine instructions, including nine special-purpose types that are mechanised using the basic input/output instructions. Micro-instructions are held in a 512-word control memory, but this can be expanded to 2048 words if necessary, providing for such operations as byte (8-bit); control jumps, skips and transfers; register-variable shifts, register/register floating-point arithmetic; logical immediate and register/-register; and macro-instructions. The latter

provides square root and trigonometric functions much faster than by using sub-routine executions.

Development of M362F software can be accomplished on support equipment in the form of mini-computer-directed systems, or on IBM 360/370 or similar facilities. Mini-computer equipment enables software development on a stand-alone basis in the laboratory.

The M362F is used as the General Dynamics F-16 fire-control computer and has a comprehensive set of Jovial or assembly-coded software facilities.

Computer type: parallel, binary, fixed- and floating-point, two's-complement
Word-length: 16-bit
Typical speed: 340 KOPS
Max address range: 65 536 (64 K) memory locations
Instruction set: 72 (expandable)
Execution times (fixed point):

	Semi-conductor memory	Core memory
(add/subtract/load)	1·46 μs	2·4 μs
(multiply (32-bit))	4·6 μs	4·6 μs
(multiply (64-bit))	20·5 μs	22·7 μs
(divide)	8·2 μs	8·7 μs

Input/output options: analogue/digital/analogue converter
analogue input/output multiplexer
discrete input/output (28 V)
MIL-STD-1553 bus
multi-purpose serial-digital processor
Typical installation: processor + 16 K core memory + 4 K ROM in ½ ATR case
Weight: 14·2 lb (6·5 kg)
Power requirements: 28 V dc

STATUS: in service.

M362S general-purpose processor

Related to the M362F, this is a 32-bit, high-speed, general-purpose processor. It uses micro-programmed, parallel, binary, fixed- and floating-point, two's-complement operations, and has a typical operating speed of 750 000 operations a second (750 KOPS).

The processor provides 96 basic machine instructions, micro-programmed in a 512-word control memory, with expansion to 1024 words possible. A random-access memory system is available and comprises semi-conductor cmos memory modules, memory controller and an error-detection and -correction unit. Up to 65 536 (64 K) memory locations (16-bit) can be accommodated.

Packaging is either for forced-air cooling, as in aircraft bays, or a radiant cooling frame for space applications. The system has been selected for Nasa's Inertial Upper Stage programme. A comprehensive set of Jovial and assembly-coded software facilities are available.

Computer type: binary, fixed- and floating-point, two's-complement
Word length: 32-bit
Typical speed: 750 KOPS
Max address range: 65 536 (64 K) memory locations
Instruction set: 96 (expandable)

Execution time (fixed-point)	Semi-conductor memory
(add/subtract/load)	0·88 μs
(multiply (32-bit product))	2·72 μs
(multiply (32-bit product))	4·29 μs
(divide)	5·66 μs

Input/output options: analogue/digital/analogue converter
analogue input/output multiplexer
discrete input/output (5 or 28 V dc)
MIL-STD-1553 bus
multi-purpose serial-digital processors
Typical installation: 365 × 365 × 152 mm
Weight: 54 lb (24·5 kg)
Power requirements: 220 W at 28 V dc

STATUS: in production.

M372 general-purpose processor

This is a development of the M362F used as the fire-control computer of the General Dynamics F-16 fighter. It has been designed for applications requiring extended performance in memory, input/output, throughput and computational speed. Operating speed is typically between 570 and 720 KOPS, and non-volatile, electrically-alterable memory of 32 K, 64 K, 128 K or 256 K words capacity can be accommodated and addressed. In addition to analogue and discrete input/output channels, up to three MIL-STD-1553 dual-redundant digital data-bus channels can be accommodated, and the computer executes the new US Air Force MIL-STD-1750 standard instruction-set architecture. Initial applications will be supported by Jovial software, and installation dimensions are expected to be between 1/2 and 1 ATR, depending on the facilities incorporated. Potential application is the F-16 updated and with Lantirn-equipment capability.

STATUS: in development.

Magic IV general-purpose processor

The hardware used in this new processor range is unrelated to earlier Delco processors. Magic IV is a high performance, all large-scale integrated micro-computer system, which promises to provide lower cost, reduced power, weight and size, greater modularity and increased reliability advantages over existing machines. Typical operating speed is 250 000 operations a second (250 KOPS). There are three basic machines, the M4116, M4124 and M4132, respectively available with 16-, 24- and 32-bit word architectures. They are regarded as being particularly suited for remote-terminal or sensor-orientated processor applications.

Almost exclusive use of large-scale integrated circuits has resulted in a great reduction in the number of components compared to a

Delco M372 general-purpose processor

conventional processor. The large-scale integrated units are:

	M4116 (16-bit)	M4124 (24-bit)	M4132 (32-bit)
Cpu-control unit	1	1	1
Cpu-arithmetic unit	4	6	8
Input/output control unit	2	3	4
Memory controller	2	3	4
TOTAL	**9**	**13**	**17**

Additional large-scale integrated circuits may be needed to meet programmable communication-interface and digital-input requirements. Mean time between failure estimates ranges from 27 000 hours for a simplex configuration to 150 000 hours for a dual-redundant system. NMOS large-scale integrated circuits are used with typical component densities of 1000 gates and 5000 transistors per 250-mil chip, with pair-gate delays below 5 nanoseconds. Delco claims better nuclear-radiation tolerance than contemorary dynamic NMOS circuits due to the static-logic design, substrate bias design and exclusive use of nor logic. Digital inputs compatible with ARINC 561, 575 and 583 can be provided.

Computer type: binary, parallel, fixed-point, two's-complement
Word length: 16-, 24- or 32-bit
Typical speed: 250 KOPS
Max address range: 32 768 (32 K) words
Instruction set: 89 (expandable)
Execution time (125°C with semi-conductor memory)
  (add/store/jump) 2·5 μs
  (store (double-length)) 3·5 μs
  (add (double-length)) 5·5 μs
  (multiply) 11 μs
Typical installation: 167 × 69 × 127 mm
Weight: 3 lb (1·4 kg)
Power requirements: 25·3 W
STATUS: in production.

Data Transmission

As digital processors have replaced analogue equipment there has been a greater proportion of digital data-transmission to and from equipment. Whereas individual links (cables) are necessary to carry separate data in an analogue system, the same information can be sent as a sequence of data-words using digital data-transmission. This has led to the definition of internationally-recognised digital data-transmission standards to ensure that individually manufactured systems have compatible interfaces.

A résumé of the most significant digital data-transmission standards is given below.

ARINC 429

This is a single-source, multi-sink uni-directional, data-transmission standard. Digital avionics equipment suppliers are the almost exclusive users in the commercial airliner field, but where equipment commonality runs across into military applications, the standard can also affect that category of aircraft also. It is sometimes referred to as digital information transport standard (dits), and ARINC 429 has superseded ARINC 575, which was an earlier development proposal.

An ARINC 429 data-bus consists of a single, twisted, shielded-cable pair, with the shield grounded at ends and all breaks. One end of the data-bus is at the transmitting line-replaceable unit, and there may be up to 20 receiver ends, each configured as a 'stub' on the basic data-bus.

Information can be carried in one of two data-rate bands. The high-speed data-rate band is 100 K bits a second (plus or minus 1 per cent) and the low-speed data-rate band is 12 to 14 K bits a second (plus or minus 1 per cent on selected value).

Modulation is RZ-polar which means that three states can exist: hi, null or lo. In any bit interval, a hi-state returning to a null-state represents a logic '1' and a lo-state returning to a null-state represents a logic '0'. Hi or lo states can be between plus or minus 13 volts to plus or minus 5 volts, and null can be between plus or minus 2.5 volts.

All information is sent as 32-bit words, with a parity-bit included (odd parity required). The identity of each word is transmitted first and a gap between words equivalent to at least four-bit time is necessary.

In any 32-bit word the bits 1 to 8 are a label, and bits 9 and 10 are a source/destination identifier. Bit 32 is the partity digit, and a sign/status matrix precedes this, being either two or three bits (bit numbers 31 and 30, or 31 to 29) depending whether data is ISO-alphabet, BCD or binary. Bits 11 to 28 or 29 (18 or 19 bits in total) are available to carry actual data. Special word formats can be provided, the most commonly used being latitude/longitude data. The ARINC standard defines such details together with the mandatory labels assigned to all data.

Each ARINC 429 transmitting unit on an aircraft has a data-bus associated with it. On the new-generation commercial transport aircraft there may be in the order of 100 ARINC 429 data-buses.

ARINC 453

This standard relates to a very high-speed data-bus used specifically to supply data from a weather-radar receiver unit to its associated display. Pulse characteristics are similar to ARINC 429, but users are not constrained to a 32-bit word format.

MIL-STD-1553A

This is a military, multi-source, multi-sink, bi-directional, very high-speed data-bus specification. Since formal definition in 1975 it has been widely used, and examples of its application can be found on the Space Shuttle, Rockwell B-1 bomber, General Dynamics F-16 and McDonnell Douglas F/A-18 fighters, and updated B-52 bombers.

A MIL-STD-1553A data-bus consists of a single twisted cable pair, and although not required by the specification it will usually have at least one layer of shielding, grounded at all ends and breaks. The data-bus is terminated by resistors, and along its length equipment is connected by transformer-coupled stubs. Up

to 32 stubs can be accommodated, and the total data-bus length must not exceed 100 metres. In practice fewer stubs and much shorter bus installations are used.

Information is carried at 1 MHz (ie 1-million bits a second) data-rate, using Manchester bi-phase pulse-coded data. Transmitted pulses are between 18 and 27 volts peak/peak, and the detectable peak/peak voltage at any stub is in the range 6 to 9 volts.

All information is sent as 16-bit words, with a parity bit (odd parity required) added and preceded by two synchronisation pulses which occupy a time equivalent to three bits; up to 50 K words a second can therefore be carried.

As any unit can transmit to any other unit, the potential for clashed signals is too high to allow uncontrolled use of the data-bus. A bus controller function is accordingly allocated to one stub, and this may be a dedicated unit, or a software function hosted in a powerful processor. The remaining stubs are referred to as remote terminal positions. The majority of traffic on a MIL-STD-1553 data-bus is carried by three types of data transfer. These are remote terminal to remote terminal, remote terminal to bus controller, and bus controller to remote terminal.

Different data-transfer types are recognisable by the structure of messages, which can comprise three types of words. These are command, status and data words. Each type of word is characterised by its structure:

Command Word Can only be generated by the bus controller. Synchronisation pulses are identical to status word. The 16-bit data section is divided into four groupings, thus:

(a) a single bit, set to zero or one, designates whether the terminal addressed will transmit or receive.(Depending on how this bit is set, the whole word is called a transmit command or receive command word.)

(b) five bits identify the remote terminal being addressed (address is in the range zero to 31 inclusive).

(c) five bits identify the sub-address associated with the remote terminal (again, the address range is zero to 31).

(d) five bits convey information on the number

Boeing 767 is first commercial transport aircraft to use large numbers of ARINC 429 digital data-buses

of words to be transmitted or received. This imposes a 31-word limit on any data-transfer.

Status Word Can only be generated by remote terminals. Synchronisation pulses are identical to command word. The 16-bit data section includes five bits which identify the address of the originating remote terminal. The remainder of the bits is dedicated to various functions which illustrate (to the bus controller) various operating data; these are chiefly concerned with efficiency of operations.

Data Word Can be sent in any message. Synchronisation pulses are opposite to command/status words. Sixteen-bits can be used to convey any selection of data, eg BCD, binary, ISO-alphabet and so on.

The operating procedure for each data-transfer type is as follows:

(a) Remote terminal to remote terminal
Bus controller sends receive command word to appropriate remote terminal (RTr)
Bus controller sends transmit command word to appropriate remote terminal (RTt)
RTt sends a status word, followed by the commanded number of data words.
RTr sends a status word.

(b) Remote terminal to bus controller (RT to BC)
Bus controller sends transmit command word to appropriate remote terminal (RTt)
RTt sends a status word, followed by the commanded number of data words.

(c) Bus controller to remote terminal (BC to RT)
Bus controller sends receive command word to appropriate remote terminal (RTr)
Bus controller sends appropriate number of data words
RTr sends a status word.

MIL-STD-1553 requires a previous knowledge of the remote terminals in use, how much information will pass to/from each remote terminal and remote terminal or bus controller

combination, and how frequently each transfer will take place. Based on this information the bus controller has to be pre-programmed to provide the desired mix of data-transfer types, and at the appropriate data-rates. Typical operations are based on a maximum data-rate interwoven with lower data-rates of exactly half, quarter-value etc; eg 50 Hz, 25 Hz, 12·5 Hz, 6·25 Hz.

Each unit on the data-bus has to know its remote terminal address. This can be generated automatically by socket pins as the unit is installed, or programmed in software.

Only a few MIL-STD-1553 data-buses are used in a single aircraft. Usually one or two will suffice to integrate a distributed-computing system, and there may be extra data-buses dedicated to special functions, eg stores management, flight controls and so on.

MIL-STD-1553B
All the procedures and capabilities of MIL-STD-1553A are retained in this subsequent standard, first issued in 1978, but two additional capabilities were added:

Dynamic bus control The bus controller function can be passed to different remote terminals. This can complicate some aspects of the software needed, but is designed to provide graceful degradation of data-bus operation in the event of the primary bus controller unit, or the host computer, failing.

Broadcast mode It is often found that some information is being transmitted several times by one remote terminal to several other remote terminals. Using broadcast mode, the bus controller tells all the terminals to prepare to receive, and one transmission therefore reaches all destinations. This may be useful only in the case of non-critical data, as there is no remote terminal status word response to assure the bus controller that the data has been received.

Ginabus
An acronym from Gestion (des) Informations Numeriques Aéroportées (Management of Digital Avionics Data), Ginabus is a military data-bus standard which has been developed by French industry, and is used extensively in the new Dassault Mirage 2000 and 4000 fighters. Other French Air Force, Navy and Army applications are anticipated.

The system operates at 1 MHz (ie 1 million bits a second), and has two twisted cable-pairs, each shielded with two mesh screens, either side of a mu-metal wrap. One cable-pair conveys data while the other transmits protocol messages, similar to the command and status words used in MIL-STD-1553. Bus ends are terminated with 75-Ohm loads, and nominal pulse characteristic is plus or minus 6 volts with a 1-pulse starting positive, and a 0-pulse starting negative. A broadcast mode is available and French industry has developed sets of Ginabus visi components.

Dual-redundant data-bus installations
Almost all military aircraft which use, or are intending to use, the MIL-STD-1553A/B information-transmission system have specified dual-redundant data-bus installations.

In this configuration two identical data-bus systems are installed between dual-configuration remote terminals and bus controller units. If any single remote terminal, bus controller, or data-bus component fails, communication is assured via one route between any two points in the aircraft. More complex failure-survival capability can be incorporated by appropriate use of bus controller intelligence. The aim of a dual-redundant data-bus installation is to provide a high degree of data availability.

Programming languages

General-purpose processors have to be programmed to do specific tasks, and whereas until about a decade ago, all airborne processing was so specialised that these programming instructions were written in 'assembler' or 'machine' code (which are machine-dependent techniques, and very time-consuming), most processors today can be programmed using a high-order language.

The high-order programming language used to generate most software in the engineering industry, and familiar to many computer users, is **Fortran**. This is a recognised language in the USA (Fortran 78 is defined in MIL-STD-1753), but its aeronautical use has generally been restricted to non-airborne, software-development tasks.

Programming with high-order language has many advantages over lower-level languages.
(a) The program structure is clearly visible, both to the software designer and subsequent users, so modifications to it can be accommodated more easily.
(b) Fewer statements need to be written to conduct most tasks, so increasing programming productivity.
(c) Programming staff is easier to train and can become familiar with a high-order language much more rapidly than with any lower-level programming method.
(d) The program is, or should no longer be, machine-dependent. It can be loaded into any machine which has the appropriate compiler facilities, and therefore software development and testing can be conducted before airborne hardware is available, providing the host machine has similar operating capabilities.

Notwithstanding the advantages quoted above and apparently contradictory to (b) at first sight, a high-order language program, after compilation, takes up considerably more machine storage than a program written in a lower-level language; neither is it as efficient, in that it operates less rapidly. However, these disadvantages have been more than offset by the enormous increase in storage and operating speed made available in airborne processors in the last decade or so. High-order language software applications have therefore been very closely associated with technical improvements in processor hardware.

A great number of processor/software combinations are offered by the manufacturers of digital equipment used in commercial aircraft, largely because these are sold as discrete systems with relatively well-tried interfaces, or

customised analogue discrete interfaces to other systems. Many companies, for reasons of commercial security, and because programs are relatively small, find it advantageous still to use machine-dependent software techniques. This situation is unlikely to last indefinitely and commercial software rationalisation could follow the standardisation example being set by the military fraternity, where currently there are several languages (typically a nationally-preferred high-order language in each major nation), but an internationally-recognisable high-order language is being developed. A review of high-order languages in use in several nations currently, and the possibility of a new internationally-recognised high-level language are presented below.

France
French military suppliers are encouraged to prepare software in **LTR**. This is a high-order language recognised by the French Army, Air Force and Navy, and a root compiler is maintained by the French military authorities. It is available at no charge to suppliers who are required to furnish LTR software to French military projects. No notable users of this language exist outside France, but it will remain the preferred high-order language for future French military software.

Germany, Federal Republic
Suppliers of programmable military items to West Germany are encouraged to use **Pearl**, a high-order language developed and used exclusively in West Germany. It employs a slightly different approach to most other high-order languages, the specification having two sections. One is the problem division, which defines all functions, algorithms, etc, and is designed to be compatible with all types of processors. The system division describes the communication with hardware and program interfaces, so it is machine-dependent.

Although support for Pearl is strong in West Germany, it is expected to be superseded by any future internationally-recognised high-order language.

United Kingdom
The preferred British high-order language is **Coral 66** (British Standard 5905). Although well documented and supported by the UK Ministry of Defence for Army, Air Force and Navy use, only a small proportion of Coral experience has been accumulated on airborne hardware. Despite this, it is probably still the most-well

established European high-order language.

The Royal Signals Research Establishment (formerly the Royal Radar Establishment) at Malvern, with Ministry of Defence support, has developed a support facility. This is called Mascot (modular approach to software construction, operation and test), and by pools/channels concepts it aids visualisation of programming requirements. It can also be used to make automatic checks, assuring the consistency and completeness of programs. The portability of programs between processors is also considerably simplified.

Coral/Mascot is still the preferred British high-order language and support technique, but there is a very firm commitment to the development of a new, internationally-recognised language.

United States
The bulk of all types of military digital processor experience has been with US projects, and the US Department of Defense already operates a list of approved languages and control agents. These are:

Language	Control Agent
Fortran	Assistant Secretary of Defense (Comptroller)
Cobol	National Bureau of Standards
Tacpol	US Army
CMS-2	US Navy
SPL/1	US Navy
Jovial J3	US Air Force
Jovial J73	US Air Force

This list is very shortly to be revised, deleting Tacpol and Jovial J3, and will add the new US Department of Defense language, Ada, applicable to all services. (See below.)

The languages used most frequently for airborne applications have been CMS-2 and Jovial J3/J73. CMS-2 is used in current US Navy computers, eg AN/AYK-14 in the F/A-18 fighter and Lamps III helicopter, and has, generally, a wide range of support facilities. Jovial probably represents the most well-used airborne high-order language, and Jovial J73 is defined in MIL-STD-1589B. It is a language especially suited to real-time processing applications and was developed largely from Algol. Even so, the US Air Force is committed, via the US Department of Defense, to the development of a new, internationally-recognised high-order language.

General Dynamics F-16 uses a MIL-STD-1553A data-bus to link major avionics sub-systems. Bus controller, in this installation, is in the Delco M362F fire-control computer. This demonstrator aircraft is also carrying Paveway laser-guided bombs, and a laser designator

Future internationally-recognised language

Since January 1975, when the US Department of Defense formed a high-order language working group, with representatives from each US military agency and various Department of Defense agencies, the possibility of a US-supported high-order language, available for world-wide use, has been great.

Development continued between 1975 and 1978 via stages referred to as 'Strawman', 'Woodenman', 'Tinman', 'Ironman', and 'Steelman'. After evaluation of languages available, world-wide, a predominantly French-used high-order language, developed by CII-Honeywell-Bull, was selected as most able to meet the 'Steelman' requirements. This is expected to form the basis of new internationally-recognised high-order language, and will be referred to as **Ada***. It is defined in MIL-STD-1815.

After inital development, the first reference document *Reference Manual for Ada Programming Language* was published by the US Department of Defense in July 1980. This was due to be superseded by a definitive ASCII-standard manual in late 1982, and ISO-Standardisation documents are expected in 1984. So far, all countries (except France) have indicated a willingness to adopt Ada as the future high-order language for airborne software.

Software development using Ada is likely to begin in the USA in 1985, and in Europe after 1986. Considerable effort is being devoted by the US Department of Defense to the development of an Ada programming support environment. This is not dissimilar to the Mascot method use to support Coral programming in the UK. The Ada programming support environment, it is claimed, will bring large life-cycle cost savings in future software, and will offer users the ability to produce software which is portable to new host hardware and operating systems.

The US Army has let a contract covering the design of an Ada compiler and environment for DEC machines, and operating documents were expected during 1982. A similar contract, let by the US Air Force, relating to IBM/370 hardware, should be operational in 1983. Ground-development of airborne real-time processing software will be initiated following completion of these programmes.

In the long term, US Department of Defense intends that Ada shall be the sole US military high-order language for real-time processor operations.

*Ada is a Trademark of the US Department of Defense.

A number of mid-1960s/early 1970s military projects such as the F-4 Phantom preceded large-scale use of digital software, but are early candidates for application of a new programming language if updated digital avionics are introduced

Communications

Radio Communications
FRANCE

Sintra-Alcatel

Sintra-Alcatel, 9 rue HG Fontaine, 92600 Asnières

Saram 7-82 uhf radio

The Saram 7-82 is a uhf radio communications set for high performance aircraft which covers 7000 channels in the band 225 to 400 MHz. It provides in the am mode A2, radio-telephony and radio-telegraphy A3, A9 NRZ and diphase, adf homing and beacon control transmission. In the fm mode it provides F9 NRZ and diphase, F1, L4 and L11. The system incorporates a guard receiver which can monitor any fixed frequency selected between 238 and 248 MHz.

A remotely-controlled set, the 7-82 is of solid-state, modular construction and employs digital synthesis frequency-generation techniques. It is a member of a family of airborne military radio systems produced by Sintra-Alcatel, modules of which are interchangeable within the family. For example, the synthesiser section of the 7-82 also generates vhf frequencies and may be used directly in either uhf or vhf systems.

The 7-82 is said to be particularly suited to co-located operation. It produces a low level of wide-band noise in transmission mode and, in receive mode, has no channel inhibition, irrespective of whatever frequency combination may be selected among the receivers.

In-flight, self-test facilities for both the transmitter and receiver sections are incorporated.

STATUS: in production and service.

TRT

Télécommunications Radioélectriques et Téléphoniques (TRT), Defence and Avionics Commercial Division, 88 rue Brillat-Savarin, 75640 Paris Cedex 13

A leading French producer of airborne electronic systems, TRT is an associate of the international Philips Group and manufactures large quantities of avionic equipment for both military and civil markets. Products include airborne tactical radars, ground proximity warning systems, radar altimeters and air traffic control transponders.

ERA 7000 vhf/uhf radio

TRT's ERA 7000 equipment is a combined vhf/uhf transmitter-receiver which covers the 118 to 143·975 MHz vhf band and the 225 to 399·975 MHz uhf band. It is designed for use by high performance military aircraft in air-to-air and air-to-ground tactical communications. In the vhf band it provides 1040 channels and in uhf, 7000 channels. An independent guard receiver is incorporated within the main equipment and continuously monitors a single channel in the 228 to 248 MHz uhf band. Pre-selection is possible for up to 20 channels from the combined vhf/uhf frequency range. Power output is 5 watts.

The ERA 7000 provides speech and data facilities on vhf and uhf together with a homing facility on uhf only.

The system is remotely controlled from one of three types of controller: a main control unit with both channel pre-selection and individual frequency selector, a secondary controller with preset channel selection only, and a manual controller with no pre-selection facility.

An optional broadband amplifier, the ERA 7400, is also available and may be used to uprate both vhf and uhf transmitted power output to 25 watts.

The ERA 7000 system is all-solid-state and is of modular construction. It employs digital synthesis frequency generation for uhf and vhf bands with the exception of the uhf guard receiver in which a single crystal frequency reference is used. Integrated built-in test equipment permits in-flight confidence checking, pre-flight test and fault-finding down to module level.

Dimensions: ¼ ATR Long
Weight: 8 kg

STATUS: in production and service.

ERA 7200 uhf radio

This equipment is derived directly from the ERA 7000 system and is one of a series of radio systems which make up what is referred to by TRT as the 7000/8000 family.

The principal difference between the ERA 7000 and ERA 7200 equipment is that the latter contains no vhf element, being designed for aircraft which have no communications requirement in this band. In all respects, the specification is that of the uhf section of the ERA 7000 system. It is not offered with the ERA 7400 power amplifier although there should be no reason why this optional equipment should not be used with the ERA 7200. It is reasonable to assume that this amplifier, or a smaller, lighter variant, will be made available by TRT.

TRT ERA 7000 vhf/uhf transmitter-receiver

TRT ERA 7200 transmitter-receiver with control unit

Dimensions: (transmitter-receiver) ¼ ATR Medium
(main controller) 146 × 762 × 156 mm
(secondary controller) 146 × 444 × 156 mm
Weight: (transmitter-receiver) 6·8 kg
(main controller) 1·5 kg

STATUS: in production and service.

ERA 8250 uhf radio

Yet another member of the 7000/8000 family, the ERA 8250 is a uhf radio that conforms closely to the general technical specification of the ERA 7200, with the important exception that transmitter output power is rated at 25 watts.

The equipment may however be operated at lower power settings if required, namely at reduced power, 5-watt carrier, and 'mini-power', 3 mW.

ERA 9000 vhf/uhf radio

The ERA 9000 is a versatile system with a wide frequency coverage of both the vhf and uhf bands and a comprehensive range of facilities. Described by TRT as a 'modular growth multi-mode transceiver', the ERA 9000 uses inter-changeable 'slice' modules which enable the user to tailor the system to his needs.

Frequency coverage and operating mode options comprise: 26 to 88 MHz vhf fm; 100 to 400 MHz vhf/uhf fm and am; 225 to 400 MHz uhf fm and am; 26 to 400 MHz all-bands fm and am; and 400 to 410 MHz for search and rescue reception. A single channel guard receiver monitors a frequency within the band 238 to 248 MHz. Channels are spaced at 25 and 50 kHz intervals, and the transmission power output is 15 watts in am mode or 25 watts in fm.

The range of facilities includes clear and secure speech; automatic direction finder; radio telephone terminal; electronic counter-countermeasures; data/frequency shift keying; NATO links; loss of signal and satellite communications; search and rescue receive; radio relay; and sonobuoy control. It is probable that other modules for further facilities will be introduced since growth space has been left available.

The system is remotely controlled and has storage capacity for up to 20 pre-selected channels, although individual frequency selection is retained. At least two types of controller can be supplied, with or without pre-selection capability.

A microprocessor is used for mode and frequency selection and management, continuous self-test, and automatic fault identification by software. It is also used to aid the manual test procedure. The entire system is solid-state and TRT says that no hybrid or special components are employed; all are standard available types.

The system meets a number of military standard specifications and can operate over a temperature range from –40° C to +80° C at pressure altitudes up to 70 000 feet.

Dimensions: ½ ATR Short or Long, according to version
Weight: 8–14 kg according to version

STATUS: in production and service.

Power amplifier for TRT ERA 7400 vhf/uhf system

TRT ERA 9000 multi-mode transmitter-receiver

GERMANY, FEDERAL REPUBLIC

Rohde and Schwarz

Rohde and Schwarz GmbH, Muhldorfstrasse 15, D-8000 Munich 80

Rohde and Schwarz is a leading West German research development and manufacturing company in the field of radio communications and electronic measuring equipment. As well as providing equipment for general surface and airborne communications and air-traffic control, the company produces radio monitoring equipment and is noted for precision direction-finding equipment. It is a major supplier to the West German armed forces and has provided equipment for Europe's tri-national Tornado bomber, the Franco-German Alpha Jet strike/trainer, Phantom F-4, and the BO-105 VBH and PAH helicopters.

Vhf/uhf radio-equipment family

Acting on a common requirement for all branches of the West German armed forces, Rohde and Schwarz developed a family of vhf/uhf equipment for airborne and surface land mobile and shipboard communications purposes. It is based on a range of functional modules which may be combined on a 'building-block' basis to form a system for any particular application. Advantages are gained not only in terms of operational flexibility and logistics but also in the ability to interchange modules for servicing and repair. All members of this family of equipment employ high-density integration techniques and, with one exception, use digital synthesis for frequency generation.

XT 3000 vhf/uhf radio

The XT 3000 is a combined uhf/vhf air-to-air and air-to-ground transmitter-receiver covering the frequency ranges 100 to 162 MHz and 225 to 400 MHz, in both fm and am modes, for voice and data. Frequency increments are spaced at 25 kHz intervals but the channel spacing is switchable by increments of 25, 50 or 100 kHz as required. Transmitter output is 10 watts. Up to 28 operational channels together with the international distress frequencies at 121·5 and 243 MHz, may be pre-selected at remote control units. The system incorporates built-in test equipment and inputs for special-to-type and automatic test equipment.

The XT3000 comprises a single vhf/uhf

transmitter-receiver, vhf and uhf amplifiers, two control units for remote operation, and a channel/frequency indicator. It meets MIL-E-5400 and MIL-STDs-810B,461/462/463/781B and VG 95211.

Dimensions: ½ ATR Short plus two ¼ ATR Short units
Weight: 30 kg

STATUS: in production and service.

XT 3011 uhf radio

The XT 3011 represents what may be regarded as the uhf section of the foregoing XT 3000 system, with which it is almost identical. Exceptions are the lack of a vhf transmitter-receiver and appropriate amplifier module, and the inclusion of an integrated but independent guard receiver.

Frequency range is from 225 to 400 MHz with a frequency spacing of 25 kHz. Channel spacing is normally 50 kHz but this may be modified to give 25 kHz increments. Modulation mode is am only.

The system is operable from either one or two remote-control positions and there are three different types of controller available. Provision is also made for a remote frequency/channel indicator.

Dimensions: (transmitter-receiver) ½ ATR Short controllers in accordance with MIL-STD-25212
Weight: 7·5 kg

STATUS: in production and service.

XT 3013 vhf radio

This system is the vhf counterpart of the XT 3011 uhf equipment, the modules for uhf operation having been replaced by vhf equivalent units.

The XT 3013 covers the vhf band from 100 to 156 MHz at 25 kHz frequency setting and a channel spacing also of 25 kHz or, optionally, at 50 kHz increments.

Dimensions: (transmitter-receiver) ½ ATR Short controllers to MIL-STD-25212
Weight: 7·5 kg

STATUS: in production and service.

XT 3010 uhf radio

This system is basically a cockpit-mounted variant of the XT 3011 equipment, in which a standard control panel is fitted to the main electronic module stack for direct user control. Provision is also made for remote operation from another flight crew position and a remote frequency/channel indicator may also be used with this system.

Dimensions: in accordance with MIL-STD-25212
Weight: 6·2 kg

XT 3012 vhf radio

As the XT 3010 uhf equipment relates to the XT 3011, so does its vhf equivalent, the XT 3012, relate to the XT 3013. Again, it is a cockpit-mounted vhf equivalent of the XT 3010 system and conforms to the broad vhf specification of this entire range of equipment.

Dimensions: in accordance with MIL-STD-25212
Weight: 6·2 kg

STATUS: in production and service.

XT 2000 uhf radio

The XT 2000 system is a cockpit-mounted uhf am emergency transmitter-receiver operating on the international distress frequency of 243 MHz. It is the only member of the Rohde and Schwarz vhf/uhf family which is non-synthesised in terms of frequency generation. Although two other frequencies in the range 242 to 244 MHz may be used, they must be selected by replacement of crystals. The distress channel may be externally switched, with

priority, irrespective of which channel is selected.

Power output of the XT 2000 transmitter is 3 watts. The unit conforms with the standards specifications of the other equipment in the Rohde and Schwarz vhf/uhf family.

Dimensions: in accordance with MIL-STD-25212
Weight: 2·3 kg

STATUS: in production and service.

Member equipments of the Rohde and Schwarz vhf/uhf family; left, uhf power amplifier; centre, transmitter/receiver; right, vhf power amplifier. Two control units and a power supply are also shown

Rohde and Schwarz XT 3011 uhf transceiver with control units

The Rohde and Schwarz XT 2000 uhf emergency transmitter-receiver with uhf/vhf family controllers

Rohde and Schwarz XT 3012 vhf radio

XK 401 hf ssb radio

The XK 401 has been developed jointly by Rohde and Schwarz with Siemens AG for the tri-national Panavia Tornado bomber, with the aim of providing reliable air-to-air and air-to-ground communication over long ranges.

The system covers the hf band from 2 to 29·999 MHz in 100 Hz increments, providing more than 280 000 channels, of which any 11 are pre-selectable. All channels are individually selectable by means of decade switches. Operational modes are usb, A3J (duplex), and continuous wave. Transmitter power output is 400 watts peak power for modulated transmission and 100 watts in carrier wave mode transmission.

The XK 401 comprises three basic units: a transmitter-receiver XK 041, a remote controller GB 041 and a power amplifier VK 241. Additionally, two optional units, an antenna tuner FK 241 and a control frequency selector, are available and are recommended for optimum operation. The system is of modular construction and solid-state components are used.

The transmitter-receiver section uses digital synthesis frequency-generation techniques and the synthesiser itself is said to possess outstandingly good noise characteristics. Two intermediate frequencies, 72·03 MHz and 30 kHz, are used for both transmission and reception paths. The receiver section has automatic squelch which operates if the hf ssb level exceeds an adjustable threshold. In the transmitter section, the lower sideband is suppressed by mechanical filters. Built-in test equipment is incorporated.

The remote controller, which may be up to 50 metres from the other major units, contains all necessary system controls as well as storage facilities for the 11 pre-selectable channels.

Two identical amplifier modules, with a common output matching circuit, form the power amplifier section, this duplication being applied in the interests of reliability. In normal operation both modules are in use; and in the event of one module failing, the only consequence is reduction of output power rather than total power output breakdown. Thermal and mismatch overload protection circuits, which include open and short circuit protection, are provided. Heat dissipated by the amplifier section is extracted by an external ventilator.

Use of the optional antenna tuner and the control frequency selector considerably enhances system performance. The former unit matches the base impedance of the antenna to transmitter-receiver output impedance and during reception periods acts as a pre-selector. The control frequency selector is used to control the digital tuning information. Average tuning time for the system is less than half a second but use of the control frequency selector unit enables tuning information to be stored for all pre-selected channels and the tuning time is thus further reduced. No power is radiated during tuning, resulting in radio silence being maintained during such operation.

Dimensions: (transmitter-receiver) 124 × 194 × 319 mm
(power amplifier) 257 × 194 × 319 mm
(controller) 146 × 86 × 165 mm
Weight: (transmitter-receiver) 11·2 kg
(power amplifier) 17·2 kg
(controller) 1·8 kg

610 series vhf/uhf radios

The Rohde and Schwarz 610 family of vhf/uhf radio communication systems comes in two basic versions, a single panel-mounted unit or a remotely located transmitter-receiver with a panel-mounted control unit. The controllers for

Control unit for the Rohde and Schwarz XK 401 hf system

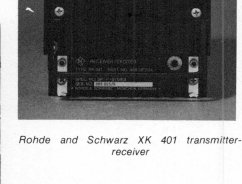

Rohde and Schwarz XK 401 transmitter-receiver

Rohde and Schwarz XK 401 ssb transceiver

Rohde and Schwarz uhf antenna switch unit

Rohde and Schwarz 610 vhf/uhf radio

the vhf and uhf variants are identical, both electrically and mechanically. They permit parallel operation of both a vhf and a uhf transmitter-receiver from a single controller and/or the operation of a single transmitter-receiver from two control units. All systems in the 610 series provide not only radio-telephonic communication but also incorporate a 16 k/bit baseband data transmission and adf facilities.

Technical specifications of the vhf and uhf variants are virtually identical. Each type has a transmitter power output of 105 watts peak power and carrier power of 10 watts at normal supply voltage of 28 volts, or 1 watt carrier wave with a reduced emergency power supply of 16 volts. Frequency range of the vhf systems is from 100 to 155·975 MHz and that of the uhf equipments from 225 to 399·975 MHz. Guard receivers cover the emergency channels of 121·5 and 242 MHz respectively. Frequency-setting increments are 25 kHz in each case and the channel spacing is also 25 kHz, although in the case of uhf systems the spacing is optionally adaptable to 50 kHz. Up to 30 channels, plus the guard channel, may be pre-selected and a remote frequency/channel indicator is an optional accessory.

All 610 series transmitter-receivers are compatible with one another, and the various modules have exactly-designed interfaces to permit easy replacements to be made without

the requirement for adjustment. The systems are especially suitable for retrofitting and Rohde and Schwarz have produced replacement kits, with tailored adaptor trays, for such aircraft as the McDonnell Douglas F-4 and RF-4E and the Lockheed F-104G Starfighter. A range of special-to-type test equipment is available with first-line test sets which can isolate faults down to module level, and full-scale automatic test equipments for base repair facilities.

Dimensions: (XU-610 vhf panel-mounted) 5⁷/₁₀ × 4⁹/₁₀ × 6¹/₂ inches (146 × 124 × 165 mm)
(XU-611 vhf remote controlled) 5 × 4⁹/₁₀ × 6¹/₂ inches (127 × 124 × 165 mm)
(XD-610 uhf panel-mounted) 5⁷/₁₀ × 4⁹/₁₀ × 6¹/₂ inches (146 × 124 × 165 mm)
(XD-611 uhf remote controlled) 5 × 4⁹/₁₀ × 6¹/₂ inches (127 × 124 × 165 mm)
(GB-600 remote control unit) 5⁷/₁₀ × 3 × 4³/₁₀ inches (146 × 76 × 110 mm)
Weight: (XU-610 vhf panel-mounted) 9·13 lb (4·15 kg)
(XU-611 vhf remote controlled) 8·14 lb (3·7 kg)
(XD-610 uhf panel-mounted) 9·13 lb (4·15 kg)
(XD-611 uhf remote controlled) 8·14 lb (3·7 kg)
(GB-600 remote control unit) 4·18 lb (1·9 kg)

STATUS: in service.

UNITED KINGDOM

Chelton

Chelton (Electrostatics) Ltd, Marlow, Buckinghamshire SL7 1LR

Chelton, established in 1947, specialises in the design, development and manufacture of vhf and uhf guard receivers, homing systems, antennas and a wide range of avionic auxiliary equipment and components. The company is a supplier to more than 150 armed forces and major airlines in over 60 countries. Customers include the British and United States Governments. The company is particularly recognised for work on static dischargers and special-to-type antenna units. Aircraft for which Chelton has supplied equipment include Sea Harrier, Concorde, Boeing 737, British Aerospace 125, General Dynamics F-16 and the Boeing Vertol Chinook and Westland Lynx helicopters.

System 7 homing system

Chelton's System 7 is a 'building block' homing system which provides broad-band and emergency guard channel homing facilities. In its basic form, it comprises a homing indicator unit, an antenna feed unit, an antenna system and an aircraft receiver.

This system provides broad-band homing over the frequency spectrum of the aircraft receiver. If an independent emergency guard channel homing system is required, the Chelton Series 7-28 two-channel receiver is incorporated.

The System 7 can be interfaced with both the main aircraft receiver and the 7-28 unit, providing both broadband and guard channel homing. System 7 interfaces with all am receivers such as ARC-116, ARC-159, ARC-164 and PTR-1751. It will also interface with fm equipment possessing an am facility.

The Series 7-28 emergency guard receiver is available in three variant forms: the 7-28-1 operating at 121·5 MHz and test frequency, the 7-28-3 operating at 243 MHz and test frequency, and the 7-28-7 operating at 156·8 MHz and test frequency.

Two differing types of 'steer left/right' indicator unit are available, one of which enables the homing signal to be heard on the aircraft's audio system, together with four types of indicator feed unit intended for conducting the signal directly to existing aircraft navigation indicators or flight-directors. Irrespective of the system configuration used, the System 7 automatically reverts to communications mode if the transmit key is depressed.

Weight: <2·2 lb (1 kg)

STATUS: in production and service.

Chelton System 7 homing system, with range of aerials for other equipment

Marconi Avionics

Marconi Avionics Limited, Airport Works, Rochester, Kent ME1 2XX

ARC-340 (AD 190) vhf fm radio

Marconi's ARC-340 radio, marketed by the company under the designation AD 190, is a tactical vhf fm equipment designed specifically for air-to-ground communication between military aircraft and ground forces. It is in service with the British Armed Forces, in conjunction with Clansman series vhf equipment, and is in operation in a wide range of fixed-wing and helicopter aircraft in many parts of the world.

The system provides clear and secure speech communication, data transmission, automatic re-broadcast for extending the range of tactical communication, and homing facilities. The homing mode may be used simultaneously with a communications channel without mutual interference. The range capability of the homing facility is equivalent to communications range.

The AD 190 covers the vhf band from 30 to 75·975 MHz at selectable channel spacing of either 25 or 50 kHz increments. Tuning is silent and instantaneous and transmitter output power is selectable at either 1 or 20 watts.

Multi-station, co-located operation is pos-

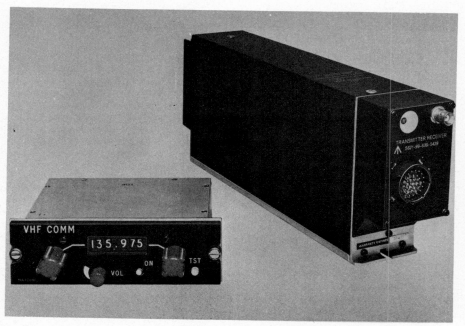

Marconi Avionics AD 120 vhf radio

sible with a maximum of three systems in the same aircraft. This is subject to the proviso that antennas must be sited at least 3 feet apart and frequency separation of 3·5 per cent for three systems or 3 per cent for two systems must be maintained. The AD 190 incorporates diagnostic built-in test equipment.

Dimensions: (transmitter-receiver) 15 × 7³/₅ × 4⁹/₁₀ inches (385 × 197 × 125 mm) (controller) 5⁷/₁₀ × 2³/₅ × 3³/₁₀ inches (146 × 67 × 85 mm)
Weight: (transmitter-receiver) 20 lb (9·2 kg) (controller) 2 lb (0·9 kg)

STATUS: in production and service.

AD 120 vhf am radio

Originally designed for civil aviation applications by the King Radio Corporation in the USA, Marconi has requalified this vhf am system for military roles and, manufacturing under licence from King, markets the system under the designation AD 120. It serves aboard most United Kingdom military aircraft, both fixed-wing and helicopter, and has also been supplied to a number of overseas customers.

Covering the frequency band 118 to 135·975 MHz, the AD 120 provides 720 channels at a channel spacing of 25 kHz. Services provided are double sideband am voice communication and selcal. Power output can be varied between 10 and 20 watts. The system is an all-solid-state equipment of modular construction and is designed for easy installation and maintenance. It is also exceptionally simple to operate. The system's standard remote controller possesses only five controls: an on-off switch, volume control, a test button and two rotary switches for frequency selection. Tuning is instantaneous. Automatic squelch and gain control eliminate the need for manual adjustment. Optionally, a keyboard-type tuning controller may be substituted for the standard equipment.

The self-test facility may be used during operation as a crew member's confidence check.

Dimensions: (transmitter-receiver) 2¹/₃ × 4·95 × 12 inches (60 × 127 × 315 mm) (controller) 5⁷/₁₀ × 1⁵/₆ × 3¹/₂ inches (146 × 47 × 90 mm)
Weight: (transmitter-receiver) 5·06 lb (2·3 kg) (controller) 1·32 lb (0·6 kg)

STATUS: in production and service.

AD 1550 communications control system

The AD 1550 communications control system is a single-box unit which governs an aircraft's entire intercommunication system and all incoming and outgoing audio services. This function permits all crew control station boxes to become 'passive' in nature and also enables an aircraft manufacturer to tailor the station-boxes to match the design of the crew stations, while the electronic system concerned may be accommodated in the avionics bay.

The system's main feature is the incorporation of special techniques to improve speech intelligibility against the background noise invariably encountered on aircraft flight-decks. Although the system was designed initially for civil aircraft, these speech-intelligibility techniques have a likely read-across to military aircraft, in which the background noise environment can be appreciably more severe.

Dimensions: ³/₈ ATR

STATUS: the AD 1550 system was introduced by Marconi in mid-1981 and has so far been chosen for the British Aerospace BAe 146 airliner and the BAe 125-700B business-jet.

AMRICS communications control system

Although few details have been released regarding the electronic equipment status of the Royal Air Force's Nimrod Mark 3 AEW early-warning reconnaissance aircraft, it is

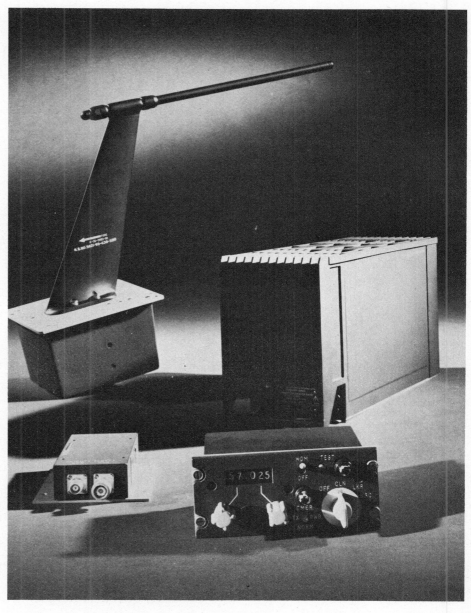

Marconi Avionics ARC 340 (AD 190) vhf radio system with antenna and switching unit

Marconi's AD 1550 clear speech central audio unit and headphones

known that Marconi Avionics has made the major contribution to the aircraft's communication system, as well as developing and supplying its radar sensors.

Nimrod AEWs have an extensive and sophisticated communications suite. To deal with the operational management of this equipment, Marconi has developed AMRICS (Automatic Management of Radios and Intercom System). A Nimrod crew comprises two categories of personnel, the flight crew, responsible for the conduct of the aircraft, and the electronic systems operators who are responsible for managing sensor equipment, tactical co-ordination and communications.

Marconi AMRICS tactical control panel

Both require access to a variety of radios with multi-mode capabilities, and AMRICS serves the needs of both categories of crew-member.

To use the communications systems, operators simply select a channel number, high or low power, clear or secure communication. A microprocessor-based logic unit then selects and tests a radio of the correct frequency band for the channel selected, tunes the radio, ties in any special equipment requested, coupling it to an appropriate antenna, and finally feeds back the status to the operator's panel.

The system reduces the operator's work load and makes the best use of the available radio and antenna systems. AMRICS incorporates an internal intercommunication system which provides individual or conference nets for the crew.

The aircraft are thought to carry uhf and vhf fm and am radios, hf ssb radios, data-link hf and uhf systems together with associated modems and data-control units, secure speech facilities and digital data-bus systems. To maintain security where encryption facilities are fitted, isolation is required between the secure and clear circuits within the aircraft and Marconi is known to have developed suitable equipment for this purpose. It is likely that the company has supplied such systems for the Nimrod AEW 3.

STATUS: in production.

ASD 40 data-link

The Marconi ASD 40 is an inexpensive, lightweight two-tone data-link for maritime helicopters operating in the air/surface-vessel (asv) role, probably for passing over-the-horizon target-data between helicopters and ships or other surface craft.

The ASD 40 operates on either uhf or hf at data-rates of up to 1·6 bits per second. It consists of two hard-mounted units, a data-control system and an associated modem which are both convection cooled and draw a power of 50 watts each.

Dimensions: (data-control unit) ½ ATR Short (modem) ½ ATR Short

STATUS: under development.

AD 130 sonar homing and direction finding receiver

The AD 130 sonar homing and direction finding receiver is designed for both fixed-wing aircraft and helicopters. It provides sonobuoy-homing and direction-finding facilities on 99 channels over the frequency range 136 to 173·5 MHz with a channel spacing of 375 kHz.

The system comprises a receiver with computer interface and a remote control unit together with associated antenna switching controls. Additional items can include a direction-finding antenna and a left/right and fore/aft antenna for 'sonobuoy on top' indication, plus cross-pointer and radio magnetic indicators.

Either computer or manual channel-selection may be employed. In computer mode, selection is accomplished by decoding of the serial bit stream and in manual, by means of pairs of thumbwheel switches. A light-emitting diode readout indicates the channel selected. Dual receivers, with independent channel selection, may be operated from a single controller. The homing system provides left/right and fore/aft deflections on a meter and the direction finding section detects and processes am signals from a direction finding antenna for an azimuth display. Built-in self-test circuitry continuously monitors all modules, with interruptive test-switching for homing and direction finding facilities, a failure mode indication for both homing and direction finding, and slew switching for checking the direction finding operation.

The system is said to be compatible with all known and planned NATO and allied services sonobuoys.

Radio, control and integration units produced by Marconi for the British Aerospace Nimrod AEW3 early warning aircraft

Nimrod systems test-rig by Marconi

Dimensions: (receiver) $7^7/_{10} \times 4^9/_{10} \times 7^1/_2$ inches (196 × 124 × 324 mm)
(control) $5^9/_{10} \times 3^9/_{10} \times 7^9/_{10}$ inches (152 × 101 × 202 mm)
(switched antenna, main) $1^9/_{10} \times 3^1/_{10} \times 7^1/_2$ inches (50 × 80 × 193 mm)
(antenna switch fore and aft) $1^9/_{10} \times 3^1/_{10} \times 7^1/_2$ inches (50 × 80 × 193 mm)
Weight: (receiver) 17·46 lb (7·94 kg)
(control) 5·06 lb (2·3 kg)
(switched antenna, main) 1·76 lb (0·8 kg)
(antenna switch fore and aft) 1·76 lb (0·8 kg)

STATUS: in production and service.

AD980 central suppression unit

The AD980 central suppression unit is designed for co-ordination of all transmitter and receiver equipment in military aircraft in order to prevent mutual interference between high-power systems such as radar, iff and TACAN and communications systems. It can deal with up to 30 input and 30 output channels and can combine up to eight outputs into a single one.

Each transmitter when in operation outputs a 'suppression' pulse and can accept a blanking pulse from other systems. The central suppression unit receives and transmits these pulses and controls the blanking pulses with a hard-wired logic matrix. The matrix is programmed for particular aircraft configurations and may be changed to meet new requirements such as equipment retrofits.

Dimensions: $1^7/_{10} \times 11^7/_{10} \times 2^4/_5$ inches (43 × 288 × 72 mm)
Weight: 3·3 lb (1·5 kg)

STATUS: in production and service.

Plessey Avionics

Plessey Avionics and Communications, Vicarage Lane, Ilford, Essex IG1 4AQ

PTR 1751 uhf am radio

The PTR 1751 is a lightweight vhf am radio designed for all types of military fixed-wing aircraft and helicopters. It provides 7000 channels in the frequency band 225 to 399·975 MHz at increments of 25 kHz and a channel spacing of 50 kHz, and is available with either 10- or 20-watt transmitter-outputs.

Both versions comprise a single transmitter-receiver unit with either a manual controller or, optionally, a manual/preset controller. Each controller provides full selection of the range of 7000 channels together with control of the built-in test functions; the manual/preset unit additionally permits pre-selection of up to 17 channels, including guard frequency, through incorporation of a non-volatile memory store.

Options include continuous am monitoring on 243 MHz (the international distress frequency) via a separate guard receiver module which plugs directly into the main transmitter-receiver chassis, an external homing unit and a wide-band secure-speech facility which requires no additional interface equipment.

The PTR 1751 conforms generally to DEF-STAN-07-55 and operates satisfactorily over a temperature range from –25 to +70° C. It is of all-solid-state modular construction with high reliability as a principal design aim, and Plessey claims a current mean time between failures of 800 hours.

Dimensions: (transmitter-receiver 10 W version) 1/2 ATR Short × 61/4 inches (160 mm) high (20 W version) 1/2 ATR Medium × 61/4 inches (160 mm) high
(manual controller) 53/5 × 14/5 × 41/5 inches (146 × 48 × 108 mm)
(preset controller) 53/5 × 37/10 × 41/5 inches (146 × 95 × 108 mm)
Weight: (transmitter-receiver 10 W version) 12 lb (5 kg)
(20 W version) 14·74 lb (6·7 kg)
(preset controller) 3·08 lb (1·4 kg)
(manual controller) 1·5 lb (0·7 kg)
(guard receiver module) 0·66 lb (0·3 kg)

STATUS: in production and service.

PTR 1741 vhf am radio

The PTR 1741 equipment may be regarded as the vhf counterpart of the Plessey PTR 1751 system, both units being members of the same family of Plessey military airborne radios. Covering the frequency range from 100 to 155·975 MHz, the PTR 1741 provides 2240 channels at 25 kHz separation. If desired however, channel spacing can be set at steps of 50 kHz to ease interface with older equipment which does not have the same frequency stability as new-generation systems. As in the case of the PTR 1751, the PTR 1741 is available with a choice of manual or manual/preset control unit; however only a single power output level (10 watts) is available.

The PTR 1741 shares a number of common features with the PTR 1751 in regard to build standard, operating range, options etc, although no secure speech facility is available for the vhf system. Both are suitable for dual uhf/vhf installation, their electrical interfaces and mechanical dimensions being identical. A single controller may be used to operate the two transceivers in such a dual installation.

Dimensions: (transmitter-receiver) 1/2 ATR Short × 61/5 inches (160 mm)
(manual controller) 57/10 × 15/6 × 41/5 inches (146 × 48 × 108 mm)
(preset controller) 57/10 × 37/10 × 41/5 inches (146 × 95 × 108 mm)
Weight: (transmitter-receiver) 11 lb (5 kg)
(manual controller) 1·54 lb (0·7 kg)
(preset controller) 3·08 lb (1·4 kg)
(guard receiver module) 0·66 lb (0·3 kg)

STATUS: in production and service.

Plessey PTR 1751 20-watt (left) and 10-watt uhf radios with alternative controllers

Plessey PTR 1721 combined vhf/uhf system

PTR 1721 uhf/vhf radio

The PTR 1721, a combined uhf/vhf radio covering the uhf band 225 to 400 MHz and the vhf band 100 to 156 MHz, is designed for all types of military aircraft. Plessey says that the ability to communicate on vhf in addition to uhf offers increased flexibility to air forces which, on occasion, may need to operate from civil airfields equipped with vhf facilities only. It is noteworthy that the PTR 1721 has been chosen for British and Italian Tornado multi-role bombers.

The PTR 1721 offers up to 9240 channels, 2240 on vhf and 7000 on uhf. Channel spacing in the vhf band is at 25 kHz and standard spacing in uhf is at 50 kHz although an optional interval of 25 kHz spacing in uhf is available. Frequencies are selectable directly from the remote-control unit which also provides pre-selection of up to 17 channels with the addition of the uhf and vhf guards at the international distress frequencies of 243 and 121·5 MHz respectively. Frequency synthesis techniques ensure good frequency stability and channel selection characteristics.

The system is entirely solid-state in construction and, wherever possible, employs conventional technology in the interests of reliability enhancement. The receiver is varactor-tuned and the frequency synthesiser is compared against a single reference oscillator.

A sealed case houses the transmitter-receiver unit and access to the modules is gained by removal of the sides of the case, to which the modules themselves are attached. Heat is conducted from the modules via the chassis, which acts as a heat sink, to the sides of the case and hence to external air. Forced cooling is provided by the mounting tray installation which also acts as a vibration insulator. Forced-air cooling may be dispensed with in less demanding aircraft environments.

Switching for dual control operation is external to the system and in two-seat aircraft, identical control units may be installed at each crew position. Options include an antenna lobe switch for azimuth homing requirements and a uhf antenna switch for automatic direction finder operation.

The environmental temperature range extends from –40 to +70° C ambient and the normal operational altitude is to a maximum of 50 000 feet, although operation is possible for short periods up to altitudes of 70 000 feet.

Dimensions: (transmitter-receiver) 42/5 × 73/5 × 131/5 inches (125 × 194 × 339 mm)
(controller) 37/10 × 57/10 × 69/10 inches (94 × 145 × 175 mm)
Weight: (transmitter-receiver) 24·75 lb (11·25 kg)
(controller) 3·96 lb (1·8 kg)

STATUS: in production.

UNITED STATES OF AMERICA

Brelonix

Brelonix Corporation, 106 North 36th Street, Seattle, Washington 98103

Brelonix Corporation manufactures radio communications equipment and ancillaries for aircraft, marine and land applications. It specialises in hf ssb equipment and has supplied systems to the Royal Norwegian Air Force, the South African Air Force, and to many scheduled or charter airlines in Canada and Alaska.

SAM 100 hf ssb radio

The Brelonix SAM 100 basic equipment provides ten-channel hf communication within the frequency band 2 to 14 MHz. Operating modes are ssb-A3J suppressed carrier and am-A3H with 6 dB carrier signal and in each case output power is 100 peak power. Two channels are available for semi-duplex operation. The system is remotely controlled and has voice detecting squelch control. Frequency stability remains within plus or minus 20 Hz over the temperature range –30° C to +60° C.

Three differing antenna couplers are available, the BD model which is manually operated and which uses power peaking for fine tuning, and the BR and WBR models which are both remotely operated. The latter is recommended by Brelonix for wider frequency response. Options include a 5-watt audio insulation amplifier and a choice of fixed, trapped dipole or whip antennas.

Dimensions: (transmitter-receiver) 6¼ × 7½ × 17 inches (159 × 191 × 432 mm) (controller) 2½ × 6¼ × 3½ inches (64 × 159 × 89 mm)
Weight: (transmitter-receiver) 12·5 lb (5·7 kg) (controller) 1 lb (0·45 kg)

STATUS: in production and service.

SAM 70 hf ssb radio

This is a manually operated hf unit which offers broadly the same facilities as the SAM 100 system but with only five channels, two of which may be semi-duplex, and a lower power output. Modes are the same as those for the SAM 100 equipment but power output has been increased to 40 watts peak power for both ssb and am modes, with a 20-watt carrier signal in the latter case.

As with the SAM 100 system, three antenna couplers are available for the SAM 70 but in this instance are designed for five channel operation. The same range of antenna options is also available.

A new version of the SAM 70, the SAM 70R, was introduced by Brelonix in May 1981. This is a remotely controlled version with a miniature controller which provides the same operational facilities as the direct manual control model.

Dimensions: (transmitter-receiver) 4¼ × 6¼ × 11 inches (108 × 159 × 279 mm) (controller) 1 × 6¼ × 2½ inches (25 × 159 × 64 mm)
Weight: (transmitter-receiver) 5 lb (2·3 kg) (controller) 0·5 lb (0·2 kg)

STATUS: in production and service.

Collins

Collins Avionics Divisions, Rockwell International, 400 Collins Road NE, Cedar Rapids, Iowa 52406

628T-1 hf ssb radio

The Rockwell-Collins 628T-1 is an hf ssb transceiver for long-range transport aircraft operating on overwater routes or other areas over which reliable extended-range communications are required. It provides upper sideband or am voice or data communications on any of 24 200 channels, at 1 kHz increments, in the 2·8 to 26·999 MHz band. Tuning is automatically controlled through a remote control unit. Nominal transmission power is 200 watts peak power in ssb or 100 watts average in compatible am.

The 628T-1 is an all-solid-state system which uses digital synthesis techniques for frequency generation. High stability is maintained through temperature compensation of the frequency standard.

Mechanical design is aimed at maximising maintainability. The transmitter/receiver is housed in a case with hinged tray and fold-out doors for easy accessibility. Plug-board circuitry is used extensively and bench-testing is simplified through the provision of a built-in test connector at the rear of the casing. Special provision is made for the replacement of the power amplifier transistors without the necessity of removing the entire power amplifier board.

Transmitter cooling is achieved through a heat sink and filtered, forced-airflow, while the receiver section relies on conventional convection cooling and is not dependent on a cool air supply.

An am selcal facility is provided by means of a special audio output through which selcal signals are monitored irrespective of the selected operating mode. Options include automatic antenna tuning couplers, to permit antenna performance optimisation over the frequency spectrum covered by the 628T-1, and a 999W-1/A1 adaptor unit which permits interchangeability with the Rockwell-Collins 618T-2/5 transceivers without disturbance to aircraft wiring, racks, connectors, antenna couplers or frequency selector.

Dimensions: ¾ ATR Short
Weight: 26 lb (12 kg)

STATUS: in production and service.

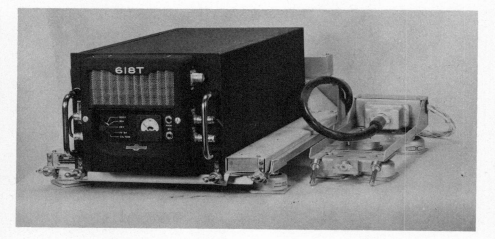

Rockwell-Collins 618T equipment in mounting tray

Rockwell-Collins 618M-1 vhf transceiver

Rockwell-Collins 628T-1 hf transmitter-receiver

628T-2 hf radio

The Rockwell-Collins 628T-2 is based on the 628T-1 and to some extent as well on the earlier 618T-2/5 equipment which is still to be found on many transport aircraft. It uses much of the operational and design experience derived from these systems. The 628T-2 however was designed to offer certain advantages over the earlier equipment, the principal of which are higher power output, extended coverage (and hence a greater number of channels) of the hf band, and more options in terms of operating modes.

Transmitter output power is 400 watts peak power and full coverage of the hf band, from 2 to 30 MHz, permits use of up to 280 000 channels at 100 Hz increments or 28 000 channels at a separation of 1000 Hz.

Operational modes include usb, lsb, am, carrier wave and data. Full 400 watts peak output is available in the sideband modes with 125 watts average in compatible am and 125 nominal in continuous wave mode.

In mechanical and electronic design, the 628T-2 is very similar to the 628T-1 although Collins has incorporated a number of improvements with regard to sensitivity, crossmodulation elimination, intermediate frequency translation, and heat dissipation in the power amplifier stage. All are aimed at extension of performance and enhancement of reliability.

Dimensions: 6 MCU in accordance with ARINC 600
Weight: 28 lb (12·72 kg)

STATUS: in production and service.

628T-3 and 628T-3/A hf radio

Like the 628T-2, the 628T-3 and 628T-3/A are based on the 628T-1 design and represent the latest standard in this series. The principal difference between the 628T-3 and the 628T-3/A variant is the narrower bandwidth intermediate frequency filtration of the latter, resulting in heightened selectivity over part of the operating spectrum and slight differences in audio response.

As in the case of the 628T-2, each system provides full coverage of the 2 to 30 MHz hf band and offers 280 000 channels at 100 Hz increments. There is however no option for a smaller number of channels at 1000 Hz separation, and transmitter output is 200 watts peak power.

Operating modes are voice, and voice and data in both upper and lower sidebands, with compatible am and carrier wave in usb only. Selcal facilities are as those of the 628T-1.

Dimensions: ¾ ATR Short
Weight: 25 lb (11·36 kg)

STATUS: in production and service.

HF-200 hf radio

First introduced in 1978, the Collins HF-200 radio is designed primarily for the light aircraft segment of the general-aviation market.

It is an all-solid-state radio of 100 watts peak power transmitter output and provides 20 pre-programmed channels plus two manually tunable channels, all in the 2 to 22·999 MHz hf band. Frequency increment between any two adjacent channels is 100 Hz. Synthesis techniques are employed for frequency generation. Channel pre-programming is undertaken by Collins' dealers to user requirements by programming of the synthesiser card, while manual tuning may be carried out by approved individuals and no special equipment is required for the latter operation.

Normal operating mode is usb but of the 20 available channels, 12 are capable of half-duplex operation, which provides radio-telephone patch-through facility. Operation in am mode is also incorporated, in which case output is 25 watts average.

The system comprises a transmitter-receiver, power amplifier, antenna coupler unit, and a controller, of which three optional types are available to cater for differing panel-mounting requirements.

Dimensions: (transmitter-receiver) 5·04 × 4 × 11¾ inches (128 × 102 × 298 mm)
(power amplifier) 5·04 × 5 × 11⁷/₁₀ inches (128 × 127 × 297 mm)
(antenna coupler) 5·06 × 7·54 × 10·96 inches (129 × 195 × 278 mm)
(controller) 1³/₅ × 6³/₅ × 4²/₅ inches (412 × 161 × 114 mm)
Weight: (transmitter-receiver) 6·5 lb (2·94 kg)
(power amplifier) 7·25 lb (3·3 kg)
(antenna coupler) 10 lb (4·55 kg)
(controller) 1 lb (0·45 kg)

STATUS: in production and service.

Rockwell-Collins 628T-3 hf transceiver in mounting tray

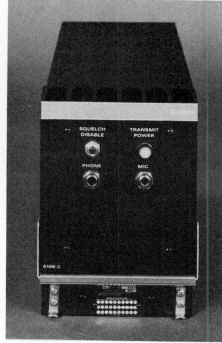

Rockwell-Collins 618M-3 hf radio

Complete Rockwell-Collins HF-220 hf installation

HF-220 hf radio

The Collins HF-220, which was introduced in April 1980, is an hf set designed for general aviation fixed-wing aircraft and helicopters. It covers the hf spectrum from 2 to 22·999 MHz, providing 210 000 channels at 100 Hz frequency separation. Up to 16 channels may be preset with the aid of a special programmer and all preset channels may be further programmed for half-duplex operation for radio telephone patch-through on either the US A3A mode or in the International A3J format.

Normal operating mode is usb but am operation is also possible. Transmitter outputs are 100 watts peak power and 25 watts average respectively.

The HF-200 is a fully solid-state, synthesised equipment comprising a transmitter-receiver, power amplifier, antenna coupler and control unit. Emphasis is on simplicity, with simplified controls and a light-emitting diode readout of channel or frequency selected. Antenna tuning is fully automatic and the antenna coupler is compatible with antenna lengths of 10 to 30 feet.

Dimensions: (transmitter-receiver) 5·04 × 4 × 11¾ inches (128 × 102 × 298 mm)
(power amplifier) 5·04 × 5 × 11⁷/₁₀ inches (128 × 127 × 297 mm)
(antenna coupler) 5·06 × 7·54 × 10·96 inches (129 × 192 × 278 mm)
(controller) 2⁵/₈ × 5³/₄ × 4³/₁₀ inches (67 × 146 × 109 mm)
Weight: (transmitter-receiver) 6·5 lb (2·94 kg)
(power amplifier) 7 lb (3·17 kg)
(antenna coupler) 10 lb (4·5 kg)
(controller) 1·5 lb (0·68 kg)

STATUS: in production and service.

VHF-250 vhf radio

The Collins VHF-250 is a 720-channel transmitter-receiver which covers the vhf band from 118 to 135·975 MHz in increments of 25 kHz. It is a panel-mounted, compact, lightweight system of 10-watt nominal transmitter output and is designed for easy integration into a stack of companion navaid/communication equipment.

The system is contained in a single unit with an integral control panel. Frequencies are generated by a single-crystal synthesiser and integrated circuitry is extensively employed throughout. Low power consumption (0·5 A in receive mode) with consequently reduced heat dissipation is a notable characteristic.

Features include automatic gain control, audio levelling and modulation compression to ensure good intelligibility over a varying range of speech characteristics and mouth-to-microphone distances.

Display of the selected frequency is on a standard mechanical digital presentation, but the VHF-250 may be upgraded to accept the display and storage facilities of the VHF-251 variant with no changes to circuitry or the mounting tray assembly.

Dimensions: 3·12 × 2³/₅ × 12·45 inches (73 × 66 × 32 mm)
Weight: 3·3 lb (1·48 kg)

STATUS: in production and service

VHF-251 vhf radio

This equipment is a variant of the VHF-250 but possesses additional store and recall facilities which permit pre-planning of frequency changes when high workload conditions, eg traversing a terminal area, are anticipated. This feature is valuable when rapid frequency changes for handover to approach or departure control are necessitated in regions of dense traffic. It is also possible to store an in-use frequency for immediate recall when attempting to contact a new centre.

In appearance, the most distinguishing feature between the VHF-250 and VHF-251 is the latter's variable-intensity led frequency display which automatically adjusts its brightness according to cockpit ambient lighting conditions. In all other respects the system is virtually identical to the VHF-250.

Dimensions: 3·12 × 2³/₅ × 12²/₅ inches (79 × 66 × 316 mm)
Weight: 3·4 lb (1·54 kg)

STATUS: in production and service.

VHF-253 vhf radio

The VHF-23, which was introduced in late 1981, is a vhf transmitter-receiver covering the band 118 to 135·975 MHz at selectable 25 or 50 kHz increments. It is designed for general aviation fixed-wing aircraft and helicopters.

The system is microprocessor-controlled and is claimed by Collins to be the first panel-mounted radio to offer direct and remote-control access to six vhf frequencies while the active and a preset frequency are simultaneously and continuously displayed. Values of the active and preset frequencies are presented on a liquid-crystal display which is back-lit for night-time operation.

As frequencies are selected for storage, on pilot demand, they appear in the preset (right-hand) window of the display and can be stored in any of four locations by pressing the appropriate storage-location button. Changes between active and preset frequencies are accomplished by depressing a transfer button at which the frequencies briefly alternate on the display until an aural tone confirms completion of the transfer. Recall into the preset window is made by depressing the appropriately numbered button.

Alternatively, recall can be accomplished from a remote switch mounted on the pilot's control yoke or, in the case of a helicopter, the cyclic-pitch control. This permits the pilot to run through the stored frequencies and transfer them to the active frequency position.

Direct tuning of the desired active frequency is also possible, if preferred. A non-volatile memory which retains all six frequencies (active, preset and four stored) guards against temporary power-loss or momentary power-supply interruption.

Rockwell-Collins VHF-253 equipment in dual configuration aboard a Beechcraft turboprop twin

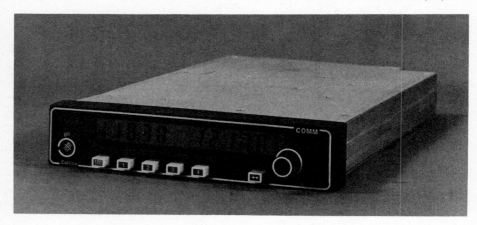

Rockwell-Collins VHF-253 system

Large scale integration circuitry is used extensively in the VHF-253 and is expected to result in markedly improved reliablity. The use of lcd as opposed to led presentation of frequency selection is also claimed to result in lower power consumption, reduced heat generation and further reliability enhancement. Other system features include audio-gain control, 'stuck-microphone' protection and adjustable microphone gain.

An alternative version, designated VHF-253S, is available for operations in locations where ground transmitters may exhibit frequency instability. This unit can accommodate variations of up to 13 kHz from the nominal transmitted frequency.

STATUS: in production and service.

AN/ARC-186(V)/VHF-186 vhf am/fm radio

The Collins AN/ARC-186/VHF-186 is a tactical vhf am/fm radio communications system designed for all types of military aircraft. It has been selected by the US Air Force as the standard equipment for all US Air Force aircraft requiring vhf am/fm capability and the first aircraft to be so equipped include the F-15 and F-16 air-superiority fighters, the A-10 battlefield bomber, and the C-130 tactical transport.

The basic ARC-186(V) is a solid-state equipment of modular construction which provides 4080 channels at 25 kHz spacing. The 2320 am channels are contained in the 30 to 88 MHz band and 1760 fm channels within the range 108 to 152 MHz. A secure speech facility can be used in both am and fm modes, and the equipment is compatible with either 16 or 18 kilobit secure systems in diphase and base-band operation.

Up to 20 channels may be programmed for pre-selection on the ground or in the air. Pre-selection is accomplished through incorporation of a non-volatile metal-nitride-oxide semiconductor (MNOS) memory which continues to retain data in the event of a loss of power supply. Two dedicated channel-selector switch positions cover the fm and am guard channel frequencies of 40·5 and 121·5 MHz respectively.

Either panel mounting, with direct control through an integral controller, or remote mounting, with an identical control-panel presentation, is possible. A half-size remote controller which contains the same control functions is also available and a typical configuration in a two-seat aircraft would comprise a full-panel mount in the pilot's cockpit with a half-size controller at the co-pilot's position. In these dual-control configurations, a manual take-control switch provides full communications control for either crew member.

Configuration changes requiring conversion from panel- to remote-mount control are accomplished by removal of the panel-type controller and its replacement by a plug-in serial control receiver module. A typical conversion is said to require less than five minutes working time. Frequency displays on both types of controller are immune to fade-out during periods of low voltage.

Circuitry of the ARC-186(V) is of modular design. Seven module cards are held in place by the body chassis or card cage and are electrically interconnected by a planar card in which all hard wiring has been virtually eliminated. All radio frequency lines in the interconnecting planar card have been buried to minimise electro-magnetic interference. Individual module cards are readily removable and may be replaced in the field to reduce fault-finding and repair time.

Current options for the ARC-186(V) include am fm homing facility, but growth capability has been designed into the equipment from the outset and possible future developments could include selcal, burst data, target hand-off, and sincgars. One present simple modification, carried out by replacement of the decoder module in the remote transceiver, permits the radio to be directly connected to a MIL-STD-1553 digital data bus and it is claimed that the system will be equally compatible with suites of future-generation equipment. An additional possibility is the uprating of transmitter output power.

A currently available variant is the VHF-186, which differs from the basic equipment only in that it extends the am frequency coverage up to 155·975 MHz.

A principal design objective for the ARC-186 series equipment was that it should be capable of easy retrofit in existing installations. Since the system is considerably smaller than the equipment it is designed to replace, this is accomplished by use of plug-in adaptor trays which permit rapid replacement without disturbance to existing aircraft wiring harnesses.

Dimensions: (remote-mounted transmitter-receiver) 5 × 6½ × 4⁴/₅ inches (127 × 165 × 123 mm) (half-size remote control) 5¾ × 3¾ × 2¼ inches (146 × 95 × 57 mm)
(panel-mounted transmitter-receiver 5¾ × 6½ × 4⁴/₅ inches (127 × 165 × 123 mm)
Weight: (transmitter-receiver, panel- or remotely-mounted) 6·5 lb (2·95 kg)
(remote control) 1·75 lb (0·79 kg)
(fm homing module) 1 lb (0·45 kg)

STATUS: in production and service.

VHF-20 series vhf radio

Designed primarily for general-aviation aircraft of all types, the Collins VHF-20 transmitter-receiver is a remotely-controlled, rack-mounted equipment of 20-watt transmitter output. It is available in two versions, the A equipment covering the vhf band from 118 to 135·975 MHz and the B variant from 118 to 151·975 MHz. Channel spacing may be at either 25 or 50 kHz increments, implementable on installation.

The VHF-20 uses digital synthesis frequency-generation techniques and is of all-solid-state construction. It provides automatic carrier-plus-noise squelch and automatic gain control and is designed to drive cabin audio systems of all types. Principal attractions are low weight, compactness and particularly, the low power consumption of 6·5 A during transmission. Consequently it requires no forced air supply and electronic section cooling is carried out by a combination of heat sink and convective air flow.

Either hard- or soft-mounting may be used and all connections are made through a single connector on the rear of the casing.

Dimensions: 3½ × 3½ × 12½ inches (89 × 89 × 317 mm)
Weight: 5·3 lb (2·4 kg)

STATUS: in production and service.

Rockwell-Collins ARC-186 military vhf system in both direct and remote control configurations

618M-3 and 618M-3A vhf radio

The 618M-3 radio and its variant the 618M-3A are air-transport category vhf transceivers of 25-watt nominal transmitter output. They were developed by Collins as retrofit replacements for the earlier 618M-1 radio and for similar Arinc 546 and 566 systems as a response to the introduction of 25 kHz spacing between channels for vhf air traffic control communication.

Coverage of the vhf band by the 618M-3 is from 118 to 135·975 MHz, while the 618M-3A version coverage extends from 116 to 151·975 MHz. The former provides 720 channels, the latter 1440 channels, in each case in 25 kHz incremental steps. Besides basic voice communications, each system also posesses data and selcal facilities.

Frequency generation is supplied from a digitally-controlled frequency standard. Low component density and use of solid-state circuitry is a feature and heat-sinking with cooling vanes assists in maintaining low transmitter temperatures. All of these factors result in enhanced reliability and the calculated mean time between failures is greater than 4000.

Speech compression is used to ensure good intelligibility irrespective of the user's voice characteristics or microphone technique and overmodulation is avoided by input signal amplitude limiting. Carrier-to-noise, carrier-override squelch control and automatic gain control are also incorporated.

With a cooling air supply, the continuous operating temperature range is from –54° C to +55° C; without cooling air, the continuous upper maximum limit is +30° C. On a one minute to two minute transmit-to-receive ratio duty cycle, operation can continue to a maximum temperature of +55° C. Maximum operational pressure altitude is 55 000 feet.

Dimensions: ½ ATR Short
Weight: 10 lb (4·53 kg)

STATUS: in production and service.

346B-3 audio control centre

The Rockwell-Collins 346B-3 audio control centre is designed to provide selection and control functions for aircraft audio equipment. The system, which is rack-mounted and remotely controlled, permits remote selection, isolation and amplification of 12 receiver audio

inputs plus microphone and interphone inputs and has provision for two further inputs. Sidetone inputs for three transmitters are included together with provision for three more. Inputs may be combined in different combinations to meet various installation requirements, eg remote-receiver selection, isolation, cockpit-speaker amplification, passenger-address amplification and a five-station interphone.

Volumes of speaker and interphone channels are separately controlled and compression amplifiers are employed to maintain desired voice outputs irrespective of input level. Outputs are 10 watts nominal into speaker loads and 100 mW into 600 ohm headset loads. Public address functions may be used for cabin, flight-deck, wheelwell, or other external speakers. The interphone for pilot, co-pilot and up to three other stations can be implemented by normal keying or by 'hot' voice-activated microphones. The 346B-3 equipment is of all-solid-state construction.

Dimensions: ¾ ATR Short
Weight: 3·5 lb (1·59 kg)

STATUS: in production and service.

387C-4 audio control unit

This system is designed for a wide range of general-aviation aircraft, from light twins to high altitude jets in which it provides flexible audio and transmitter-receiver control. With the exception of the adf, all receiver, interphone and passenger address signals are processed by dual-channel limiter amplifiers to a standard 100 mW, 600 ohm level. When the desired speaker and interphone audio level is selected, the system maintains the level. Adf inputs continue to provide the normal build and fade type of signal. A built-in range-voice-both switch is incorporated for the adf filter. Marker-beacon 'hi-lo' sensitivity and six-position microphone selection switches are included.

The 387C-4, which is a self-contained unit, may be panel-, eyebrow-, or console- mounted without modification.

Dimensions: 5¾ × 1⁷/₈ × 4⁷/₁₀ inches (146 × 48 × 119 mm)
Weight: 1·6 lb (0·73 kg)

STATUS: in production and service.

618M-1 vhf transmitter-receiver

The Collins 618M-1 is a vhf transmitter-receiver covering the band 118 to 135·95 MHz in its basic version. A number of variants are available and coverage of the vhf band from 116 to 149·975 MHz is accomplished in steps of either 25 or 50 kHz.

The system is designed for air transport, military or general aviation applications. It is remotely controlled, normally from a Collins 313N-3D dual controller which is also used to govern dme/vor navigation receivers simultaneously with 618M-1 communications equipment. Transmitter power is 25 watts.

With the exception of the driver and the power amplifier sections, which retain conventional valve electronics, the system is otherwise solid-state in construction. Features include dynamic squelch, which automatically and continuously adjusts squelch to the prevailing background noise level and audio-limiting which maintains constant modulation regardless of the speech characteristics of the operator or the type of microphone used. Automatic gain control is also incorporated. Front-panel test points are included, together with a test meter which together permit many functional checks to be completed without the need to remove the equipment from the aircraft.

Dimensions: 1/2 ATR Short
Weight: 19 lb (8·62 kg)

STATUS: in service.

618M-2B/D vhf transmitter-receivers

The Collins 618M-2B/D systems are designed for air transport applications and are available in a number of versions which cover the bands 118 to 135·975 MHz with 720 channels in the 618M-2B version, and 116 to 151·975 MHz in the 618M-2D version in which 1440 channels are provided. The former has a transmitter power output of 25 watts minimum and the latter 20 watts minimum. Each variant may have either 25 or 50 kHz channel separation.

The systems are of fully solid-state construction and use digital synthesis techniques for frequency generation. Features include a dynamic squelch control using carrier/noise ratio and carrier squelch and compression circuitry which maintain intelligible speech levels irrespective of signal changes or speech characteristics. It is remotely controlled and has an optional power amplifier if required. A connector for special-to-type automatic test equipment is provided in the front panel of the transmitter-receiver unit to simplify test procedures without removing the equipment from the aircraft.

Dimensions: 1/2 ATR Short
Weight: 17·3 lb (7·84 kg)

STATUS: in service.

618T hf ssb transmitter-receiver

The Collins 618T and its variants form a family of radio systems appropriate to a wide range of airborne applications, both civil and military. Fundamentally, they are all of similar design and construction; some units however are specifically tailored for new installations, some for retrofit and others for specialised applications. Priority is given below to a description of the basic version and details of the variant forms follow.

The 618T is a high frequency single sideband transmitter-receiver which provides voice, compatible am, cw or data communications in the 2 to 29·999 MHz band. It is automatically tunable to any one of 28 000 channels in this range in increments of 1 kHz. Nominal transmitter output power is 400 watts pep in single sideband or 125 watts in compatible am. The system is remotely controlled.

A notable aspect of this particular radio is that although designed for airborne use, it has been produced in large quantities for transportable, mobile, shipborne and semi-fixed stations in many parts of the world.

Collins 618M-2B/D vhf transmitter-receiver complete system

Accurate frequency control and stability is attained by the use of phase-locking circuits in which, by use of phase-comparison techniques, all injections to transmitter and receiver sections are related to a single accurate frequency standard which is maintained on-frequency by temperature compensation.

The equipment is cooled by a filtered air supply which delivers a metered quantity of air to all parts requiring it from a blower installed in the front panel. However, an exhaust port is provided for use with central cooling systems, in accordance with ARINC 404. Electronic construction is of the plug-in module type, and transistor circuitry is used wherever possible.

Features include selcal and data transmission facilities. Use of selcal on am is made possible by way of a special audio output which enables the signals to be monitored irrespective of the mode-selection switch setting. With regard to data transmission, the 618T is suitable for frequency-shift keying or other signalling modes at rates of up to 100 words per minute. A connection is provided to an external frequency standard if so required. Selection of the data mode on the control unit automatically adjusts the receive sensitivity for maximum gain and causes the system to operate in upper side band mode.

A choice of optional accessory antenna tuning units is available to ensure maximum performance over the frequency spectrum employed, and a Collins 49T adaptor is also available to facilitate retrofit installation aboard aircraft wired for the earlier Collins 618S high frequency system.

The different variants of the basic 618T equipment are as follows:

618T-1 high frequency radio

This model includes a power-supply module for 400 volts, 1500 Hz primary source and is used only for certain retrofit installations. More usually, power is furnished by a Collins 516H-1 power-supply unit which provides the 115 volt, 400 Hz and the direct current of 27·5 volts required to operate the equipment. This two-package system may also be used with the 49T-4 retrofit adaptor to effect direct retrofitting into a 614C-2 harness without alterations to existing cable installations.

618T-2 high frequency radio

Intended primarily for new installations, the 618T-2 has a high-voltage power-supply module using three-phase, 400 Hz power, together with a direct current of 27·5 volts. No external power source is required.

618T-3 high frequency radio

This also is a self-contained system, intended for new installations although retrofitting is possible in some cases through employment of the 49T-3 adaptor. It is equipped with a 27·5 volt electronic inverter-type high-voltage power supply module and requires a direct current of 27·5 volts, 35 A and 115 volts, 1-phase, 400 Hz, 1 A supply.

HF-101 high frequency radio

This comprises a 618T-1 transmitter-receiver and a 714E-2 control unit (see subsequent entries). Operation is from a direct current of 27·5 volts, 35 A and 115 volts, 1-phase, 400 Hz, 2 A power supply.

HF-102 high frequency radio

The HF-102 configuration uses a 618T-2 transmitter-receiver and a 714E-2 controller with a self-contained power supply for operation from a direct current of 27·5 volts, 4 A and 208 volts, 3-phase, 400 Hz, 1000 watts source.

HF-103 high frequency radio

This system consists of a 618T-3 transmitter-receiver and a 714E-2 controller with a self-contained power supply for operation from a direct current of 27·5 volts, 35 A and 115 volts, 1-phase, 400 Hz, 1 A source.

AN/ARC-94 high frequency radio

Consisting of a 618T-2 transmitter-receiver and a 714E-2 controller with a self-contained power supply, this configuration operates from a direct current of 27·5 volts, 4A and 208 volts, 3-phase, 400 Hz, 1000 watts source.

AN/ARC-102 high frequency radio

618T-3 transmitter-receiver and 714E-2 controller with a self-contained power supply for operation from a direct current of 27·5 volts, 35 A and 115 volts, 1-phase, 400 Hz, 1 A source.

AN/ARC-105 high frequency radio

This is a pressurised communication system designed for tactical jet aircraft. It has performance characteristics comparable with the 618T system. It operates from 115 volts, 400 Hz, 3-phase, 4-wire, 1000 watts power source.

Dimensions: (basic 618T) 10 1/4 × 7 4/5 × 22 1/4 inches (260 × 200 × 565 mm)
(618T-2) 1 ATR
Weight: (618T-1) 50 lb (22·65 kg)
(618T-2) 52 lb (23·56 kg)
(618T-3) 51 lb (23·1 kg)

STATUS: in service.

714E high frequency control units

The 714E control units used in conjunction with the 618T series transmitter-receivers provide remote selection of the available 28 000 channels. Frequencies are indicated in a direct-reading digital display and can be selected in 1 kHz increments throughout the 2 to 29·999 MHz range. Frequency selection is accomplished by rotating four knobs until the desired frequency appears in the window. A function selector and radio frequency sensitivity adjustment are included.

The 714E-2 control unit may be used either in new installations or as a replacement for the 614C-2 control unit in retrofit applications.

The 714E-3 is used with equipment operated in the cw or data mode. Two lighting options are available, the 714E-3 uses a black, edge-lighted panel with red lamps and the 714E-3B has a grey panel with blue lamps.

Dimensions: 5³/₄ × 2¹/₂ × 6¹/₄ inches (146 × 65 × 159 mm)
Weight: 1 lb (0·45 kg)

STATUS: in service.

516H-1 high frequency radio power supply

The 516H-1 is a solid-state power supply designed to replace the original 416W-1 in Collins 618S-1 retrofit applications. It is on the same shockmount used for the former installation. A self-contained blower unit is incorporated to ensure adequate cooling air. Output is 400 volts, 1500 Hz for the 618T-1 transmitter-receiver, which needs a high voltage for the power amplifier.

STATUS: in service.

490T-1 high frequency antenna coupler

The 490T-1 is a general purpose high frequency automatic antenna coupling unit for 25 feet (7·62 metres) or longer whip and wire antennas in the 2 to 30 MHz frequency range. Shorter antennas may be accommodated by the use of suitable loading coils. It features a short tuning cycle of five seconds maximum and three seconds average tuning time. Such rapid tuning capability reduces the overall rechannel time and keeps transmission to a minimum in the interests of radio silence.

The system contains an antenna transfer relay and an antenna grounding relay which can be used as needed in dual installations. The 490T-1 will operate with the 437R-1 helical monopole antenna and optional applications include exchange with either the 180L-3 or 180L-3A antenna coupler.

The unit's rapid tuning time enhances reliability since the operating elements are only energised for brief periods. The servo system is controlled by a demand surveillance technique which causes the coupler to retune if the antenna impedance changes appreciably but does not require the servo system to remain in constant operation.

The 490T-1 comprises four radio frequency assemblies, three modules, a chassis, front panel and dust cover. Solid-state logic circuits, capable of fast decisions with high speed switched and variable elements, are used to ensure reliable high speed tuning. All assemblies are removable for maintenance and repair purposes.

Dimensions: 10¹/₁₀ × 7⁵/₈ × 12¹/₂ inches (257 × 193 × 319 mm)
Weight: 19·7 lb (8·93 kg)

STATUS: in service.

AT-101 high frequency antenna tuning system

The AT-101 antenna tuning system is for use in aircraft employing a tailcap antenna. It consists of a 452A-1 pressurised lightning arrester and relay assembly a 180R-4 pressurised antenna coupler assembly and a 309A-1 control unit. The lightning arrester and relay assembly serves as a mounting for the antenna coupler assembly. Provision is made for a second coupler for installations in which two transmitter-receiver units are required, permitting operation of the two receivers simultaneously on a common tailcap although allowing only one transmitter to be operated at any one time. Two optional type 156G-1 receiver modules plug into the 309A-1, permitting additional receivers to be used for monitoring purposes.

The antenna coupler assembly contains servo-controlled loading and phasing elements for resonating the antenna and matching its impedance at various operating frequencies. Maximum tuning time required is ten seconds. Protective circuits are incorporated to safeguard the equipment against loss of pressure or an excessive rise in temperature. There are no conventional glass envelope valves, transistors or diodes within the lightning arrester assembly.

Dimensions: (lightning arrester and relay unit) 7¹/₂ × 10⁵/₈ × 16²/₅ inches (190 × 270 × 417 mm) (antenna coupler) 7²/₅ × 5¹/₂ × 11¹/₈ inches (189 × 141 × 282 mm)
(coupler control unit) ³/₈ ATR Short
Weight: (lightning arrester and relay unit' 10 lb (4·54 kg)
(antenna coupler) 13 lb (6·15 kg)
(coupler control unit) 12·75 lb (5·78 kg)

STATUS: in service.

180R-6 high frequency antenna coupler

Used in conjunction with the 309A-2D antenna coupler control, the 180R-6 will automatically resonate 45 to 100 feet (13·71 to 30·48 m) antennas over the 2 to 30 MHz frequency range. The addition of optional plug-in 156G-1 modules in the coupler permits the use of up to three additional receivers for monitoring other frequencies.

Dimensions: 7 × 9¹/₂ × 17¹/₄ inches (177 × 241 × 437 mm)
(coupler controller) 3⁵/₈ × 7⁵/₈ × 14¹/₂ inches (93 × 193 × 368 mm)
Weight: 21·51 lb (9·75 kg)
(coupler controller) 12·25 lb (5·56 kg)

STATUS: in service.

180R-12 high frequency antenna coupler

The 180R-12, together with the 309A-9 coupler control unit, automatically matches a probe antenna in the 2 to 30 MHz frequency range. The system is automatically tuned in a maximum of 16 seconds, but the typical tuning time is five seconds. The 180R-12 antenna coupler was designed for the Boeing 727 commercial transport but can be retrofitted into the earlier Boeing 707, using the 309A-9A unit.

The servo loop is activated only during tuning or when the voltage standing-wave ratio exceeds preset limits. This contributes to increased component life, but all components are tested to provide a calculated mean time between failures of 2000 hours.

High-voltage protection is provided by a ball gap which fires at a voltage lower than that which causes either external or internal arcing. This activates a circuit which cuts off transmitter power within 50 milliseconds. If the protective circuit functions due to gap-firing, the transmitter can be channelled to a new frequency and, if the excessive voltage does not exist at that frequency, the coupler will tune correctly. A sensor is incorporated to cut radio frequency power if internal air temperature rises to exceed 100° C.

Dimensions: 8¹/₄ × 7¹/₂ × 18³/₄ inches (211 × 190 × 476 mm)
Weight: 21 lb (9·53 kg)

STATUS: in service.

51X-2 vhf receiver

The Collins 51X-2 is a vhf navigation/communications receiver designed for air transport, general aviation and military aircraft applications. It covers the frequency range 108 to 151·95 MHz at increments of 50 kHz. In the communications mode, the equipment provides full coverage of the band stated unless coverage of only the 108 to 135·95 MHz band is required, in which case only a partial crystal complement is supplied.

In navigation modes, localiser coverage is over the band 108·1 to 111·9 MHz at odd-tenth MHz intervals and tvor coverage is in the same band at even-tenth MHz intervals. Vor reception is at 50 kHz steps over the band 108 to 117·95 MHz. In all navigation modes, an output is available for either Collins 344A-1, B-1 or D-2 vor/ils instruments. Voice communication reception continues during navigation mode operation.

The system is remotely controlled by a choice of three types of controller, the 614U-1, 614U-3 or 614U-7. The 614U-1 provides remote frequency selection of both the 51X-2 and its associated 17L-7 vhf transmitter in single- and double-channel simplex and double-channel duplex communications system. The controller covers the range 118 to 151·95 MHz in 50 kHz steps. The 614U-3 and the 614U-7 controllers channel the 51X-2 through both the aircraft navigation and communication frequencies. In addition, both units provide for automatic selection of glideslope frequencies whenever an ils channel is selected. The 614U-7, in addition to the above, also contains circuits for automatic selection of dmet frequencies whenever a vor channel is selected. Both controls channel the 51X-2 in 50 kHz steps over the entire vhf frequency range from 108 to 151·95 MHz.

The 51X-2 is of modular construction and is based on a bridge-type chassis, stressed at the sides. The three modules (radio frequency amplifier, variable intermediate frequency and frequency selector and fixed intermediate frequency and audio, and power supply) plug into the chassis with their interconnecting wiring locating between the two removable decks. Hold-down screws keep the modules in place. The electronic sections use conventional glass-envelope valves.

Dimensions: ³/₈ ATR Short
Weight: 10·5 lb (4·77 kg)

STATUS: in service.

17L-7 vhf communications transmitter

This transmitter, a companion unit to the 51X-2 receiver (see previous entry) is a 680-channel unit covering the vhf band from 118 to 151·95 MHz with 50 kHz spacing. It uses a total of 38 crystals for frequency generation and has an integral power supply.

Dimensions: ³/₈ ATR Short
Weight: 14 lb (6·36 kg)

STATUS: in service.

344B-1 instrumentation unit

This unit, used with the 51X-2, furnishes conversion and instrumentation of tvor, vor and localiser signals. A combination 51X-2 and 344B-1 is available as a single package. Type number is 51R-4.

Dimensions: ³/₈ ATR Short (51R-4 is ¹/₂ ATR Standard)
Weight: 12·5 lb (5·68 kg)

STATUS: in service.

137X-1 vhf communications/navigation antenna

The 137X is the standard antenna unit for the Collins 51X-2 radio receiver and provides essentially the same electrical characteristics as the 37J-3 navigation and 37R-1 communications antennas. Standing wave ratio (SWR) at the communications terminal is less than 2:1 from 118 to 136 MHz and at the navigation terminal is less than 5:1 from 108 to 122 MHz. Both the navigation and communications sections have omni-directional azimuthal patterns, the communication terminal being vertically polarised and the navigation terminal being horizontally polarised.

The antenna is designed to withstand the aerodynamic forces encountered at Mach 0·9 at sea level at attitudes of 5 degrees pitch and 5 degrees yaw. Drag is approximately 2·66 lb (1·2 kg) at 250 mph (402·33 kph) at sea level with no pitch or yaw and 15·3 lb (6·95 kg) at 600 mph (965·6 kph). The unit is plastic filled and has a glassfibre covering.

Dimensions: 24$\frac{1}{8}$ × 25$\frac{7}{8}$ × 10$\frac{7}{8}$ inches (614 × 657 × 275 mm)
Weight: 4·7 lb (2·13 kg)

STATUS: in service.

37R-2 vhf communications antenna

This unit is designed for use on aircraft cruising up to 600 mph (965·6 kph) and is compatible with the Collins 51X-2 receiver. It is a vertically polarised antenna and provides a standing wave ratio of 2:1 or less over the range 116 to 152 MHz. It is used for both receiving and transmitting and can handle a maximum input power of 125 watts. Drag is approximately 0·518 lb (0·23 kg) at 250 mph (402·33 kph), sea level, zero angle of attack. Under the same conditions, but at 400 mph (643·7 kph), drag rises to 1·32 lb (0·6 kg). Printed circuitry and foamed-in-place plastic are used in construction of this antenna.

Dimensions: 12$\frac{1}{4}$ × 3$\frac{3}{4}$ × 11$\frac{1}{2}$ inches (312 × 94 × 292 mm)
Weight: 21 lb (0·9 kg)

STATUS: in service.

390E shockmount

The 390E-2 shockmount provides dual mounting for the 51X-2 receiver and its companion 17L-7 transmitter while the 390E-1 shockmount does the same thing for the 51X-2 and the associated 344B-1 instrumentation unit. Load isolators for the 390E type shockmounts have been selected for correct smoothing of g forces. An extractor mechanism is available to prevent damage during insertion or removal of equipment into these mountings.

Weight: 2·4 lb (1·09 kg)

STATUS: in service.

394H-4 shockmount

Again for use with the Collins 51X-2 receiver, this shockmount is designed to accommodate single $\frac{3}{8}$ ATR cases and to prevent 'bottoming' under high g conditions. A floating-type rear connector and coded guide pins are incorporated to safeguard against damage during removal or replacement of units.

Weight: 1·4 lb (0·63 kg)

STATUS: in service.

Genave

Genave Inc, 4141 Kingman Drive, Indianapolis, Indiana 46226

Genave manufactures a wide range of mobile, hand-held and aircraft transmitter-receiver radio-communications equipment. The aviation products are aimed at the general aviation and light aircraft markets.

AirCom vhf radio

The Aircom is a hand-held, battery-powered vhf am transmitter-receiver covering the frequency range 118 to 128 MHz, or optionally to 139·975 MHz, in which range it provides four channels. Channel separation is 25 kHz and the frequency spread is 10 MHz. Transmitter output power is 3 watts peak power. Manual squelch adjustment is provided.

Power is supplied from eight nicad batteries which may be recharged by a Genave charging unit. Electronic circuitry is designed for simplicity and consists mainly of a two-sided printed circuit board layout. The AirCom is a low-cost radio particularly suited for general aviation and related environments and would provide a useful emergency or standby communications set for those aircraft which are not fitted with dual communications equipment.

Dimensions: 1$\frac{4}{5}$ × 2$\frac{9}{10}$ × 9$\frac{1}{10}$ inches (46 × 29 × 80 mm)
Weight: 0·99 lb (0·45 kg)

STATUS: in production and service.

Alpha 6 vhf radio

The Alpha 6 vhf am transmitter-receiver provides six channels in the 118 to 136 MHz band with a channel spacing of 25 kHz at a frequency spread of 18 MHz. This hand-held equipment is battery powered, of similar construction to the Genave AirCom system (see previous entry), and aimed at the same market segment. Power output however is 2 watts carrier and ten nicad batteries are used as the power source.

A variant, the Alpha 6U, covers the uhf band from 350 to 400 MHz and is also available for export markets, though not for use within the USA.

Dimensions: 7$\frac{9}{10}$ × 2$\frac{3}{5}$ × 1$\frac{3}{10}$ inches (203 × 66 × 12 mm)
Weight: 2·24 lb (1·02 kg)

STATUS: in production and service.

Genave Alpha 6 hand-held radio

Genave AirCom hand-held radio

Alpha 12 vhf radio

The Genave Alpha 12 is a panel-mounted vhf am transmitter-receiver for light aircraft. It has a low power consumption which renders it especially suitable for aircraft with a limited electrical generation capacity, such as gliders, certain agricultural aircraft, or homebuilts. It provides 12 channels in the band 118 to 135·975 MHz at a channel spacing of 25 kHz. Transmitter power output is a nominal 4 watts carrier, with 3·3 watts minimum.

Features include a mosfet, track-tuned front-end and crystal intermediate-frequency filtering. A light-emitting diode is incorporated to act as a 'transmit' indicator.

Dimensions: 2½ × 6½ × 10 inches (63 × 165 × 254 mm)
Weight: 4 lb (1·81 kg)

STATUS: in production and service.

Alpha 100 vhf radio

The Genave Alpha 100 is a panel-mounted vhf am transmitter-receiver providing 100 channels in the band 118 to 127·9 MHz at a channel spacing of 100 kHz. Transmitter power output is 8 watts peak power, 2 to 3 watts carrier. Construction is fully solid-state and the receiver section is of the double conversion, super-heterodyne type and is crystal controlled. Facilities include a manually adjustable squelch disable and automatic gain control. Frequencies are selected by means of a dual knob selector with digital readout and a light-emitting diode is employed as a 'transmit' indicator.

The system has a low power requirement, in common with other Genave equipment, making it suitable for aircraft with little or no electrical generation capacity.

Dimensions: 6½ × 2½ × 9 inches (165 × 63 × 228 mm)
Weight: 4 lb (1·82 kg)

STATUS: in production and service.

Alpha 720 vhf radio

The Genave Alpha 720 is a panel-mounted vhf am transmitter-receiver providing 720 channels in the 118 to 135·975 MHz band at a channel separation of 25 kHz. Transmitter output power is 4 watts nominal. It is designed for the general aviation and light aircraft market. The system is a single-crystal unit using digital phase-locked synthesis techniques for frequency generation. Construction is fully solid-state, with extensive employment of integrated circuitry. Features include a transformerless series modulator in the transmitter section, a single conversion receiver and field-effect transistor front-end and mixer circuitry. Facilities include automatic squelch disable and active impulse noise limitation to reduce external interference effects. Channel selection is performed by use of a dual control frequency-selector knob on the equipment's front casing, and the selection is confirmed by a dimmable incandescent readout display.

Like many Genave products, the Alpha 720 is a low power consumption system suitable for aircraft with limited electrical power. A variant, the Man-Pack, designed for portable use, is produced for use in gliders, homebuilts and agricultural aircraft with no electrical systems.

Dimensions: 2½ × 6½ × 10 inches (63 × 165 × 254 mm)
Weight: 4 lb (1·81 kg)

STATUS: in production and service.

GA/1000 vhf communications/ navigation system

The Genave GA/1000 is a vhf am communications/navigation system for light aircraft. It can be panel-mounted as a single unit or, alternatively, be installed so that the vor/loc indicator section can be retained as a panel instrument with the control head mounted

Genave Alpha 12 aircraft hand-held transmitter-receiver with microphone

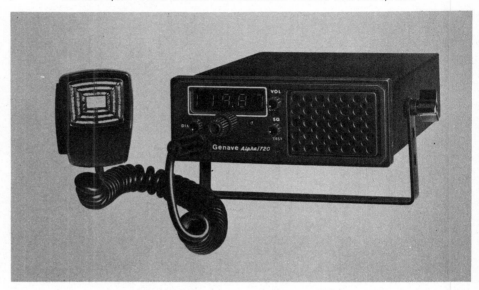

Genave Alpha 720 synthesised transmitter-receiver

Genave Alpha 720 transmitter-receiver with microphone

elsewhere in the cabin.

The system's communications section covers 720 channels in the vhf band 118 to 135·975 MHz at a 25 kHz channel separation. The independent navigation receiver covers the band 108 to 117·95 MHz, with 50 kHz separation and covers 200 navigation channels (160 vor and 40 localisers). The use of separate receivers for communications and navigation functions permits radio operation without disrupting reception of navigation signals. Both sections employ hot filament digital readout displays for conformation of the frequency selection. This is carried out in each case through use of dual frequency selector knobs on the front case of the control unit.

Transmitter output power is a nominal 4 watts carrier signal and a radio frequency actuated light-emitting diode is incorporated as a transmit indicator. Communications receiver gain is automatically controlled and automatic squelch with manual disable is also provided. Audio output may be either through a 4-ohm speaker for which a 3-watt output is available, or into 600-ohm earphones from a 100- mW output. The same values are applicable to the audio outputs from the navigation receiver section. In vor mode, transmissions from the

system's transmitter section cause no visible deflection of the course-deviation indicator needle. Both vor and localiser have ARINC standard autopilot outputs.

The GA/1000 uses solid-state integrated circuitry and a single crystal digital synthesiser is employed for frequency generation. Receiver demodulation circuitry is of the single con-

version type. A range of antenna systems and associated antenna coupler units is available to match differing aircraft installation requirements and a speaker muting relay is also obtainable.

STATUS: in production and service.

Genave GA/1000 communications/navigation radio panel

Genave Alpha 120/Alpha 1200 remote transmitter-receiver

Genave Alpha 120/Alpha 1200 control head

Kearfott

Kearfott Division, The Singer Company, 1150 McBride Avenue, Little Falls, New Jersey 07424

Joint tactical information distribution system (JTIDS)

JTIDS is a United States joint-service jam-resistant, secure communications system. Using time-division multiple access (tdma) technology, it provides a flexible, multiple-user tactical information exchange service between both surface and airborne military units and their command. It consists of digital data and voice links operating in the 960 to 1215 MHz band and employs spread-spectrum techniques for increased resistance to jamming.

Each member of a JTIDS network is allocated time slots to accommodate his input messages into the system. Such inputs would automatically include identification and navigation data in addition to specialised tactical information such as target acquisition details. All users can continuously monitor and sample the data-base for such information as they require.

Participants routinely inject information into the net through their regular broadcast slots without necessarily knowing the identity or location of the ultimate recipients who, likewise, do not necessarily know the whereabouts or identity of the provider of the data. Typical JTIDS messages from an airborne data source would cover identity, location, altitude, speed and heading; then, tactical information regarding the mission such as target acquisition, weapon or stores availability, fuel remaining and equipment status would follow.

JTIDS communication security is maintained by coding and frequency hopping techniques. Each transmission consists of a pulse stream, the leading pulses of which are synchronisation pulses for locking transmissions to the hopped frequency sequence. Identification pulses and the message itself then follow.

An advanced version selected for use by the US Navy uses distributed tdma which transmits the pulses in a pseudo-random fashion over a longer time basis, further increasing the inherent security of the system.

JTIDS is one of the most ambitious communications projects, civil or military, ever undertaken. It permits the interchange of essential tactical information between aircraft, surface vessels, mobile or fixed-base land stations and, most importantly, the command, on a scale not previously envisaged. The technical problems which have had to be resolved in order to implement the system have required an exceptional research and development effort on the part of the manufacturers

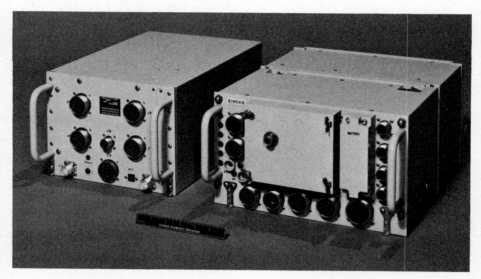

A JTIDS Class 2 airborne transmitter/receiver terminal manufactured by Singer's Kearfott Division in association with Rockwell-Collins

producing the equipment for the system.

In regard to the airborne elements, the principal contractors are Tadcom (a joint management company formed between Hughes and ITT acting under contract from Boeing, the prime contractor to the Electronic Systems Division of the US Air Force's Systems Command) and The Singer Company's Kearfott Division, prime contractor and systems integrator, working with Rockwell Collins which provides engineering support and development as an alternative production source.

Responsibilities for the surface elements are divided among the companies concerned in various ways, according to differing terminal classifications and individual service requirements. A principal contractor with regard to ground-based air-defence radars within NATO, is IBM.

Initial emphasis in development of JTIDS hardware was on the Class 1 terminal system, which is intended for land-based and ship-borne application as well as airborne use. Its most notable application to date is aboard the earlier US Air Force/Boeing E-3A Sentry Awacs early-warning aircraft.

Developed by Hughes Aircraft, the Class 1 terminals underwent considerable development refinement during the earliest history of the programme. Work on the initial Class 1 terminal, designated AN/ARC 181, commenced under a contract placed in 1974 and first deliveries of the resultant hardware took place in January 1977. While this proved

successful in over 7500 hours of operation, which included flight tests aboard an E-3A, work was already proceeding on the improved B-waveform model, airborne development of which was completed by May 1978. Tests included live jamming and were completed ahead of schedule with all requirements and specifications being successfully satisfied, an impressive performance in view of the considerable technical challenge presented by the task.

The ARC-181 terminal comprises three electronic units, together occupying approximately five ATR rack spaces, together with a controller. Hughes, however, was already working on a privately-funded JTIDS Class 1 terminal project, in parallel with the main programme, with the aim of size and weight reduction. This resulted in the Hughes Improved Terminal (HIT) with a reduction of some 40 per cent in volume, a corresponding weight decrease and no compromise in terms of performance, while maintaining full TDMA compatibility. These HIT terminals are now current standard equipment for US Air Force and NATO E-3A Awacs aircraft.

A further development of Class 1 terminals has been undertaken by Hughes, again on a privately-funded basis. The resulting fourth-generation terminals have demonstrated more size and weight reduction, mainly through the employment of advanced lsi techniques combined with modular construction, and are understood to approximate to a single ATR-sized case. Hughes claims that the latest terminals have four times the data capacity of earlier units and twice the original message access rate.

Known as Advanced Time Division Multiple Access (ATDMA) terminals, these devices maintain a local data-base on targets within the user's vicinity, allow participation in different multiple networks on a time-shared basis and have an integrated tacan navigation function. Hughes also claims a higher processing efficiency for these units, which are compatible with existing equipment on an aircraft via either a platform-unique or exchangeable interface module.

Airborne Class 2 terminals for the US Air Force and Army are being developed by the team of Singer-Kearfott and Rockwell-Collins under a $50 million contract placed in June 1981. Under this contract, 20 terminals are due to be delivered for flight test aboard US Air Force F-15 and F-16 air-superiority fighters in mid-1983. A proportion will subsequently be passed to the US Army for further tests as part of the hybrid Position Locating and Reporting System (PLRS)/JTIDS system, known as PJH. Development of both PLRS and PJH is a Hughes Aircraft responsibility.

Under the JTIDS Class 2 contract, Collins will supply transmitter-receiver sub-systems to Kearfott, which is developing processor and input-output units. Kearfott is responsible for systems integration as well as overall pro-

gramme management and has received some $30 million worth of the total contract value.

The Kearfott Class 2 terminal has applications in both surface and airborne roles. It comprises a two-box package occupying about 1·6 cubic feet (0·045 cubic metres) and weighing approximately 95 lb (43 kg). The Collins-built transmitter-receiver section contains all tdma radio frequency functions. An independent receiver-processor provides a tacan function which operates simultaneously with JTIDS communications. The Kearfott-constructed equipment is divisible into two line-replaceable units, the data processor and input/output adaptors. The former carries out signal and digital message processing; the latter provides a unique interface with the host platform.

A software-controlled relative navigation function is incorporated and automatically provides position information with respect to other co-operating tactical entities, during both active and passive operation.

The key output of the system is the display of information received from the JTIDS network, and the US Air Force is evaluating a number of possibilities. One of these is a multi-function display which will present JTIDS information as selected by the pilot. The important factor remains the ability of JTIDS to permit target assignment by a command and control authority, by self-assignment, or by co-ordination between members of a flight. One particular

area of evaluation, according to Kearfott, is the intra-cockpit correlation of radar tracks and JTIDS data display together with inter-cockpit exchange of radar and JTIDS data between members of a flight.

While the Rockwell-Collins/Singer-Kearfott team proceed with development of US Air Force/Army Class 2 terminals, the US Navy has selected the Hughes/ITT Tadcom consortium for development of several classes of terminal under an $87 million contract awarded in February 1982 by Naval Electronic Systems Command. Eventual value of US Navy JTIDS terminal procurement is expected to exceed $1000 million.

The development programme is for about 50 terminals of Class 1, Class 1A and Class 2 categories for test and evaluation purposes. In the US Navy case, the Class 1 terminals are for surface vessels. The Class 1A units are for service aboard US Navy/Grumman E-2C Hawkeye early-warning aircraft and the ligherweight Class 2 terminals for the Navy's Grumman F-14 Tomcat and McDonnell Douglas F-18 Hornet fighters.

Terminals to be developed under the Tadcom contract will be of the Phase II distributed tdma type which, while permitting the use of higher data throughputs and multiple net communications, remain fully compatible with Phase 1 conventional JTIDS nets.

STATUS: in development.

King Radio

King Radio Corporation, 400 North Rogers Road, Olathe, Kansas 66062

King is one of the principal US manufacturers of airborne radio equipment and produces large quantities, mainly for the general aviation market. As well as communication systems, King develops and produces a wide range of navigation aids including airborne dme, adf, atc transponders and radar altimeters. Its current civil communications equipments cover hf and vhf equipments.

KHF-950 hf radio

Introduced in February 1981 King's first hf system, the KHF-950, originally covered the hf band from 2 to 26·999 MHz, but has since been given extended frequency coverage up to 29·999 MHz and can consequently offer a full choice of 280 000 frequencies at 100 Hz spacing. The system operates in usb, lsb and am modes. Transmitter output in each ssb mode has been uprated to 150 watts pep and to 35 watts average for am transmission over the full frequency range.

The KHF-950 employs synthesised frequency-generation techniques and employs microprocessor control for easy in-flight operation. It possesses a non-volatile memory which allows pre-selection storage for up to 99 channels and their appropriate modes but, in addition, allows the user to manually tune to any other frequency within the covered bandwidth without disturbance to the stored presets. The initial version of the KH-950 had storage capacity for 24 channels.

Operation may be either simplex, for normal air traffic or similar communication, or optionally, in duplex which permits patch-through into public utility telephone circuits. Provision is also made for selcal and facility and dedicated such circuits enable continuous selcal monitoring to be maintained without the necessity of having am mode selected.

A feature of the KHF-950 is its automatic antenna tuning capability, an operation carried out simply by keying the microphone. King claims that the system has minimised the normal hf requirement for long wire antennas and that the system will operate satisfactorily on a 10-foot antenna. It will also tune to fixed-rod aerials and towel-rail antenna used on helicopters. King is also developing equipment

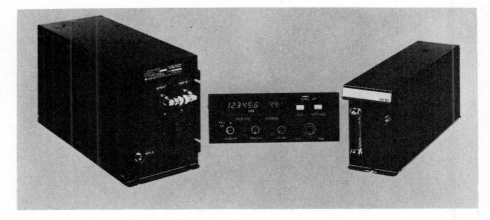

A King KHF-950 hf radio system pictured with a standard control unit

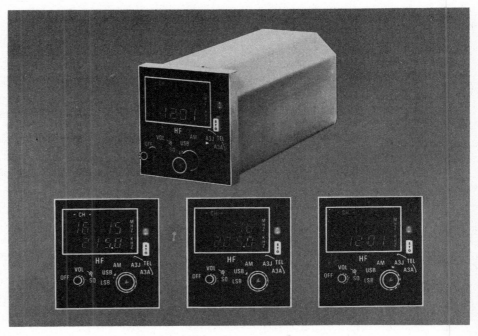

Optional controller for the King KHF-950

to facilitate operation from shunt and notch-type antenna systems.

The KHF-950 system comprises the KCU-951 remote controller, the KAC-952 power amplifier/antenna coupler, and the KTR-953 transmitter-receiver. The remote controller (or frequency selector) is designed for panel mounting and presents channel, frequency and mode-selection data on a self-dimming gas discharge numeric display.

For duplex operation, the KCU-951 controller is replaced by a KFS-954 unit which is also panel-mounted. This unit contains storage for all 176 ITU maritime radio-telephone channel selection plus an additional pre-selected simplex air traffic or conventional airborne communication channels. By pre-programming the ITU channels, it is possible for the operator to call any radio-telephone station without having to manually select the separate transmit and receive channels required. He merely selects the radio-telephone mode and the required channel. The KHF-950 also interfaces with teletype and facsimile systems and a recently-introduced dual installation equipment allows dual-frequency reception from a single antenna.

Weight: 19·59 lb (8·93 kg)

STATUS: in production and service.

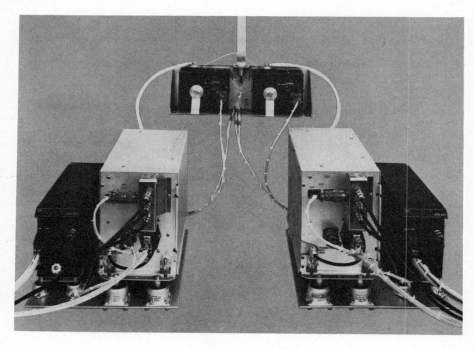

King KHF-950 radio shown in dual configuration

KTR 908 vhf radio

King's KTR 908 equipment is a general aviation vhf transmitter-receiver handling 720 channels in the 118 to 135·975 MHz band. Channel spacing is normally 25 kHz but 50 kHz spacing may also be used. Also, optionally, a variant is available with extended coverage in the vhf band up to 151·975 MHz. Power output over the 118 to 135·975 MHz range is nominally 20 watts.

The system comprises a panel-mounted KFS 598 controller and a KTR 908 transmitter-receiver which may be remotely located. The controller possesses a non-volatile memory which permits storage of one standby frequency while an active frequency is in use. The two frequencies may be interchanged between active and standby status by use of a 'flip-flop' transfer button and remain stored when power is removed. Tuning is controlled by a single-chip microprocessor and is fully automatic. Display of the operational and standby frequencies is made through a self-dimming, gas-discharge numeric readout display.

The transmitter-receiver is likewise fully solid-state and is of compartmentalised construction, the different units being separated by function and the compartment casings providing radio frequencies screening between circuits. A single crystal digital synthesiser is used for frequency generation. Automatic gain con-

King KHF-950 radio: left to right, KAC-952 power amplifier, KTR-953 transmitter-receiver, and KCU951 cdu

trol and squelch are incorporated with manual squelch disable and override.

Dimensions: (transmitter-receiver) 11¾ × 1¾ × 5 inches (299 × 45 × 127 mm)
(controller) 5½ × 2⅕ × 2⅓ inches (140 × 56 × 60 mm)

Weight: (transmitter-receiver) 3·5 lb (1·59 kg)
(rack and connectors) 0·3 lb (0·66 kg)
(controller) 0·5 lb (0·227 kg)

STATUS: in production and service.

Magnavox

The Magnavox Government and Industrial Electronics Company, 1313 Production Road, Fort Wayne, Indiana 46808

Military radio manufacture within Magnavox, a subsidiary of North American Philips, is vested in Magnavox Electronics Systems Company at Fort Wayne. Since 1957 the company has developed a wide range of military communication systems and has produced them in large quantities, mainly for the United States armed forces.

ARC-164 uhf radio

In strict terms, the ARC-164 is the basic member of a family of radio-communication equipment and sub-variants, each designed for particular applications, yet all with a high degree of commonality.

This basic ARC-164 covers the uhf band, providing 7000 channels over the range 225 to 400 MHz in 25 kHz increments. Any 20 channels may be pre-selected. Standard power

output is 10 watts, although this may be readily uprated to 30 watts.

A fully solid-state system, the ARC-164 is distinguished by its unique 'slice' module construction, in which a series of modules, connected by a flexible harness, are simply bolted together to form the desired electronic configuration. A typical, simple system would comprise transmitter, receiver, guard-receiver, and synthesiser. The control unit may either form part of this consolidated package or be remotely located. The modular approach adopted implies 'growth' capability, extra modules being added as required. A range of optional facilities, such as data transmission and secure speech is available by the addition of the appropriate slices.

A number of directly-connected or remote-control units are produced for the ARC-164. These include a simple frequency selection controller; a 32-channel preset control with led readout of the selected channel; a 20-channel preset unit with provision for two-cockpit take-control; and a microprocessor controller with

Magnavox ARC-164 uhf radio

400 preset channels (for uhf, vhf, am or fm), lcd channel and frequency readout, and the capability of controlling up to four systems simultaneously. Additional remote frequency/channel indicators are available. Magnavox produces a variety of mounting trays to suit

differing aircraft installations for new types of aircraft and for the updating of older aircraft equipment.

Although designed for airborne application, the ARC-164 is also the basis for a number of surface radio-communication systems, including base stations, vehicular equipment, shipborne systems and satellite terminals.

Perhaps the most notable feature of the ARC-164 is its high mean time between failures. In a 100 000 hour life-cycle cost verification programme, conducted by the US Air Force on T-37, T-38, F-100 and C-130 aircraft, the ARC-164 has demonstrated a mean time between failures of 2000 hours.

Dimensions: (transmitter-receiver, 10 W version) ½ ATR × 7 inches (178 mm), (30 W version) ½ ATR × 14¾ inches (374 mm) (controller) ½ ATR × 3⅓ inches (83 mm)
Weights: (transmitter-receiver, 10 W version) 8·1 lb (3·7 kg), (30 W version) 15 lb (6·8 kg) (controller) 4·3 lb (2 kg)

STATUS: the system is fitted to a wide range of US Air Force aircraft, including the General Dynamic F-16 fighter, and that service has placed orders for over 18 000 sets. Current production total is in excess of 20 000 sets.

The ARC-164 is marketed in the United Kingdom and certain other non-US territories by Philips Electronics MEL Division, which is supplying it for the Royal Navy's Sea King helicopters and Sea Harrier V/STOL combat aircraft and for the Hawk, Jaguar and other Royal Air Force aircraft. Further Hawk aircraft produced by British Aerospace for overseas customers, notably those delivered to Kenya, are also fitted with the ARC-164, as are Strikemaster strike/trainers which have been refurbished by that manufacturer. The latter includes Strikemasters operated by the Royal Saudi Arabian Air Force.

Magnavox ARC-164 installation (by pilot right hand) in British Aerospace Hawk strike trainer

ARC-195 vhf radio

This radio may be considered as a variant of the uhf ARC-164, with which it has a component commonality of 93 per cent. It also is available in 10- and 30-watt output versions. The major differences between the two systems consist of some component value changes and the substitution, in the ARC-195, of a synthesiser with a frequency standard appropriate to the vhf section of the radio frequency spectrum.

The ARC-195 covers the vhf band from 116 to 156 MHz, providing 1750 channels at a frequency separation of 25 kHz. In other respects it is almost identical to its uhf counterpart. Control units, again, are almost identical and certain units from the ARC-164 range of controllers may be used for combined uhf/vhf operation.

STATUS: in production and service.

Radio Systems Technology

Radio Systems Technology Inc, 10985 Grass Valley Avenue, Grass Valley, California 95945

Radio Systems Technology (RST) is a small organisation, founded during the early 1970s to produce low cost avionic equipment for the private light aircraft market. Most of RST's systems are provided in self-build kit form but, if required, the company provides fully-assembled and tested equipment complete for installation.

RST-542 vhf radio

The RST-542 radio is designed for air-to-ground communications in all categories of light aircraft, including gliders. It provides crystal-controlled, six-channel communication in the vhf band from 118 to 136 MHz, using two crystals per channel. Transmitted power output is 6 watts peak power.

Comprising a single-box panel-mounted unit, the solid-state RST-542 features extensive use of integrated circuit technology. It incorporates automatic gain control and adjustable squelch facilities and provides a receiver audio output power of 4 watts minimum to drive either a 40-ohm cabin speaker or a 600-ohm headset. A principal feature, rendering it particularly suitable for gliders, is low power consumption, 6 mA in receive and 0·6 A in transmit mode.

Dimensions: 9 × 3 × 3 inches (229 × 76 × 76 mm)
Weight: 1·5 lb (0·68 kg)

STATUS: in production and service.

RST-571/572 vhf navcom radios

The RST-571 and -572 are light aircraft vhf navcom radio systems which vary from each other only in the number of channels provided by their communications sections. Each covers the band 118 to 135·95 MHz; the RST-571 provides 360 channels at 50 kHz increments, and the RST-572 provides 720 channels at

Radio Systems Technology RST-542 six-channel transmitter-receiver

increments of 25 kHz. The common navigation receiver section covers 200 channels in the band 108 to 117·95 MHz, in increments of 50 kHz. In transmit mode, the transmitter-receiver section has a transmitted power output of 10 watts peak power or 2·5 watts in continuous wave.

The RST-571 and -572 are panel-mounted units using thumbwheel switches for selection and display of the channel and for a chosen vor radial. Both the navigation and communication sections of an equipment may be used simultaneously. To prevent the display of spurious signals, the navigation section is automatically

Presentation of Radio Systems Technology RST-571 360 channel nav/comm system

disabled when the transmitter-receiver section is in transmit mode.

Construction is solid-state and digital synthesis techniques are used for frequency generation in both the communication and navigation-receiver sections. Employment of digital techniques permits the phase-locking of the vor/loc circuits to the navigation receiver signal, with consequently enhanced accuracy. Automatic gain control and adjustable squelch facilities are incorporated in the communications section.

RF Products

RF Products Inc, Davis and Copewood Streets, Camden, New Jersey 08103.

RF Products (formerly TRW RF Filter Products) specialises in the development and production of tunable bandpass filters and multi-couplers. These are produced mainly for the military market, in which the proliferation of communications systems aboard aircraft often creates problems caused by the need to site radios and antennas in close proximity. Major programmes to which RF has contributed include the US Air Force/NATO/Boeing E-3A Sentry (AWACS), US Air Force/Boeing E-4B Advanced Airborne Command Post, US Army Guardrail and EH-1X Quick-Fix and the RAF Nimrod AEW aircraft.

Automatically tuned uhf multi-couplers

RF's airborne antenna multi-couplers use a building-block, modular approach design in which plug-in tunable filters are used interchangeably as single-channel devices or in two-, three-, four- or five-port multi-couplers, covering the uhf band from 225 to 400 MHz. The radio frequency power capability of these systems is 120 watts peak or 50 watts carrier wave. Good selectivity characteristics are claimed for these units, with the following performance:

	399 MHz	300 MHz	225 MHz
3 dB	± 2·9	± 2·7	± 2·5
45 dB	± 12·5	± 9·5	± 8·0
80 dB	± 38·0	± 33·0	± 30·0

For 50 dB attenuation, channel separations are 15 MHz at 399 MHz, 12 MHz at 300 MHz and 10 MHz at 225 MHz. The tuning times are a maximum of ten seconds but six seconds is said to be typical.

RF's tunable bandpass filters use digital techniques for remote tuning. This is accomplished by converting digital information from the radio equipment's digital synthesiser, or another source, to rotational command information. The only tuning element in the filter is an air-variable capacitor which can be stepped to a position corresponding to a pre-calibrated frequency by a digitally-controlled stepping motor. Digital logic circuitry permits fast and accurate tuning by precise positioning of the filter tuning shaft to within plus or minus 0·04 degree. Angular shaft position information is provided to the digital control circuitry by a position-sensing circuit comprising a light-emitting diode and a photocell.

On receipt of a 'tune-initiate' command, the stepping motor rotates the filter tuning shaft to the upper end of the tuning range which is used as a reference. The control circuit then counts a number of pre-determined steps corresponding to the frequency code input, stopping the motor at the desired shaft position for the frequency desired. On completion, a 'tune-cycle complete' signal is generated. An optional self-initiated tuning module can be made available.

Dimensions: (five-port multi-coupler) 14 × 12 × 15 inches (350 × 300 × 380 mm) (four-port multi-coupler) 14 × 12 × 15 inches (350 × 300 × 380 mm) (three-port multi-coupler) 9 × 14 × 15 inches (230 × 350 × 380 mm) (two-port multi-coupler) 7 × 10 × 15 inches (180 × 250 × 380 mm) (single filter) 7 × 7 × 11 inches (180 × 180 × 280 mm)
Weights: (five-port multi-coupler) 62 lb (28 kg) (four-port multi-coupler) 57 lb (26 kg) (three-port multi-coupler) 42 lb (19 kg) (two-port multi-coupler) 28 lb (13 kg) (single filter) 15 lb (7 kg)

STATUS: in production and service.

Type 6 tunable uhf filter

The RF uhf Type 6 tunable filter covers the band 225 to 399·9 MHz and is provided with a 243 MHz guard channel by-pass filter to ensure reception of emergency transmissions regardless of the programmed operating frequency. Its radio frequency power handling capability is 100 watts peak or continuous wave.

The unit comprises a pressurised four-pole cavity band-pass filter which is compatible with modern uhf transmitter-receivers both dimensionally and electrically. It is thus particularly suitable for retrofit applications in which a compact transmitter-receiver/filter packaging is required. The system has a mean time between failures of 5000 hours.

Dimensions: 6½ × 3½ × 11 inches (165 × 89 × 279 mm)
Weight: 4·5 lb (2 kg)

STATUS: awaiting certification.

Dimensions: 7 × 3½ × 14⅘ inches (177 × 88 × 375 mm)
Weight: 13 lb (5·9 kg)

STATUS: in production and service.

Type 7 two-, three-, and four-port uhf/vhf multi-couplers

RF Type 7 multi-couplers are available in versions covering the military vhf band from 110 to 160 MHz and the uhf band from 225 to 400 MHz. They are designed to ensure reliable operation of multiple transmitter and receiver installations in severe co-location environment situations and may be used interchangeably as single-channel, two-port, three-port or four-port units. This is accomplished by use of modular building blocks.

The systems are remotely tuned by use of digital control circuitry combined with stepper motor-driven filter tuning shafts (see entry for RF automatically tuned uhf multi-couplers). Power handling capability is 150 watts maximum and the average tuning time is five seconds with a maximum of ten seconds. Adjacent channel operation is 3 MHz minimum. Two-port multi-couplers from the Type 7 series are used aboard the US Air Force/Boeing E-3A Sentry AWACS aircraft.

Dimensions: 14 × 15 × 26 inches (355 × 381 × 660 mm)
Weight: (three-port) 105 lb (47·72 kg) (four-port) 140 lb (63·63 kg)

STATUS: in production and service.

Type 7 digitally-tuned uhf filter

RF's Type 7 digitally-tuned uhf filter covers the frequency range 225 to 399·99 MHz. It is designed to permit multiple transmitter-receiver operation in severe co-location situations. The system is available in two versions, a 0·7 per cent bandwidth version with a radio frequency input power limit of 150 watts continuous wave, or a two per cent bandwidth unit with 200 watts peak power. The filter is of the four-pole cavity type and is capacitively loaded and tuned. Tuning is digital parallel in 100 kHz steps.

Dimensions: 21½ × 6¾ × 6½ inches (546 × 171 × 165 mm)
Weight: 28 lb (12·72 kg)

STATUS: in production and service.

Terra

Terra Corporation, 3520 Pan American Freeway North East, Albuquerque, New Mexico 87107.

TPX-720 vhf radio

Terra's TPX-720 transmitter-receiver is a hand-held unit with built-in antenna and speaker. Covering the vhf band from 118 to 135·975 MHz, the TPX-720 provides am communications on 720 channels in that band. It is designed principally for light aircraft with limited or no electrical or radio facilities and as a back-up system for aircraft with radio. In addition to the 720-channel two-way communications facility, the receiver section also covers 200 vor channels. Channel separation is 50 kHz over the communications range, and in the navigation range, a 25 kHz separation applies.

Two transmitter power-output levels are available. In the high range, output is 8 watts peak power with a carrier signal of 2·5 watts. Alternatively, the low range may be used with the lower 2-watt peak power output but with an increased carrier signal power of 5 watts. The unit is powered from ten rechargeable nicad batteries.

All controls are in the head of the case and a miniature digital read-out thumbwheel switch is used for channel selection. Digital synthesis techniques are employed for frequency-generation and electronic construction is of cmos integrated circuitry. The receiver is of double conversion super-heterodyne design and uses crystal monolithic filtering to achieve high selectivity. A wide range of optional accessory equipment is available. It includes adaptors for operating directly from an aircraft electrical power supply, headsets with or without microphone attachment, separate microphone, charging units, and a push-to-talk switch for use on sticks or control wheels.

Dimensions: 3³⁄₁₀ × 1⁹⁄₁₀ × 9½ inches (83 × 48 × 241 mm) (antenna) 9¼ inches (235 mm)
Weight: 2·1 lb (0·954 kg)

STATUS: in production and service.

Tracor

Tracor Aerospace, 6500 Tracor Lane, Austin,
Texas 78721

Radio management system

Tracor's radio-management system (RMS)
provides centralised cockpit or flight-deck
control of radio communications and navigation aids in a single unit and is suitable for
aircraft of all types, general aviation, commercial transport or military. The RMS permits a
high degree of avionic system integration with
consequent reduction in the panel space
occupied by conventional controllers, and a
decrease in pilot workload.

The system can control all or part of a range
of Arinc 700 or 500 (2/5 tuned) series radios
including two vhf transmitter-receivers, two
dme interrogators, two adf receivers, two atc
transponders, two hf transmitter-receivers and
two auxiliary vhf VOR/ILS equipment. It
comprises two units, the radio management
display (RMD) and an interface adaptor unit
(IAU). The RMD, which is the centralised
cockpit controller, interfaces directly with Arinc
700 equipment and with the IAU which, in turn,
provides the interface with Arinc 500 series
radios. Each independent RMS contains all
frequency and mode information and in a dual
system configuration all systems can be tuned
by either RMS, thus providing full system
redundancy. Each system element is isolated to
localise the effects of a unit failure.

The RMS front panel consists of a four-
window vacuum fluorescent display which,
claims Tracor, is the first of its type to be
applied to avionics equipment, and a numeric
keyboard. Equipment to be tuned is selected by
depressing selector keys adjacent to each
window display and the frequency or atc code
then entered via the keyboard. Pre-selection of
frequencies can be made for vhf communications and for VOR/ILS. The system also
permits frequency selection of VOR, ILS and
dme from an external remote source such as a
flight-management computer. A transfer func-

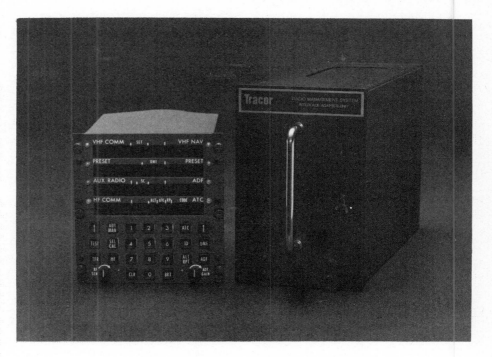

Tracor radio management system and interface unit

tion permits control of a complete 'on-side' or
'off-side' bank of radios from the RMS which
displays the frequencies and modes in use in
that particular bank.

Other facilities include a selcal *signal-
received* alert, and means of controlling the
identification mode of the aircraft's atc transponder.

The RMS uses a digital data-bus for transmitting frequency and status data to the IAU
and Arinc 700 system radios. The system is
controlled by a large scale integrated circuitry
microprocessor which contains a non-volatile
memory to safeguard against loss of stored

information in the event of interruption to the
power supply. The system incorporates built-in
hardware and software self-test facilities. The
system's low power consumption permits it to
be safely battery operated under emergency
conditions.

Dimensions: (RMD) 5¾ × 6 × 9 inches (146 × 152
× 229 mm)
(IAU) 4⅞ × 7⅔ × 12¾ inches (124 × 194 ×
324 mm)
Weights: (RMD) 6 lb (3 kg)
(IAU) 9·2 lb (4·2 kg)

STATUS: in production and service.

Wulfsberg Electronics

Wulfsberg Electronics Inc, 11300 West 89th
Street, Overland Park, Kansas 66214

Wulfsberg Electronics manufactures vhf and
uhf radio communication systems for the civil
aviation market. As well as producing equipment for the air transport sector, Wulfsberg is
active in the manufacture of equipment for the
general aviation market, particularly for
heavier, longer-range corporate aircraft.

WT-200 vhf radio

The Wulfsberg WT-200 is a general aviation air-
to-ground communications transmitter-receiver suitable for fixed-wing aircraft and
helicopters. In essence, the system is a general
aviation variant of the Wulfsberg WT-2000
equipment, an Arinc standard radio designed
primarily for the airliner market segment. The
WT-200 however possesses no data input
facility (although Wulfsberg claims that it will in
fact operate with ACARS, the Arinc data
network) and, while it does not meet full Arinc
specification, it is in fact virtually identical to the
Arinc standard WT-2000 model.

Systems such as the WT-200 appeal to the
more sophisticated end of the general aviation
market and it is noteworthy that Wulfsberg is
marketing the system as a triple-equipment
package, with a suitable controller (the TR 342),
following current air-transport operation practice. One such aircraft to offer such a package,
known as Flitecomm 3, is the new Learjet.

The WT-200 is a vhf equipment providing 720
am voice channels in the frequency range 118
to 135·975 MHz at 25 kHz frequency spacing.
Transmitter output is 20 watts. It possesses
automatic gain control and manual carrier

Wulfsberg's RT-7200 vhf radio with C722 controller

squelch adjustment. The system is remotely
controlled and has manual individual frequency selection. It is of all-solid-state, modular construction and uses synthesised frequency generation techniques.

Dimensions: 2⅞ × 5 × 13¼ inches (73 × 127 ×
337 mm)
Weight: 5·6 lb (2·54 kg)

STATUS: in production and service.

WT-2000 vhf radio

This is the Arinc-standard variant of the WT-200 system. Specification of the WT-2000 is identical to that of the WT-200 except that it has marginally higher transmit and receive current values, possesses data input and output facilities to meet Arinc 566A and is housed in an Arinc standard box.

The WT-2000 has been selected by Boeing as standard equipment for the 727, 737 and 747 aircraft.

Dimensions: ½ ATR Short
Weight: 8·9 lb (4·04 kg)

STATUS: in production and service.

TR-342 switching system

Although not strictly speaking a communications systems in its own right, this unit merits inclusion in the context of being an optional addition to a trio of WT-200 systems operating as a triple-comm installation.

The TR-342 simplifies the installation of the three radio units without the need for changes to the two-comm microphone control, audio select panel, control heads and antenna. Also available is the Wulfsberg MT-342-2 mounting rack which permits the three WT-200 units to be fitted as a composite stack.

STATUS: in production and service.

Wulfsberg RT-30 vhf transmitter/receiver

RT-30 vhf fm radio

Wulfsberg's RT-30 transmitter-receiver covers the lower vhf band providing fm air-to-ground communication over a frequency range 29·7 to 49·99 MHz, giving 2030 channels at 20 kHz spacing. The system is however capable of tuning to a narrower incremental stepping of 10 kHz.

The RT-30 incorporates a guard receiver with a similar specification to the receiver section of the main equipment or, optionally, a plug-in, single-channel, crystal-controlled guard receiver assembly.

Nominal transmitter power output is 10 watts and the system can operate in simplex mode for normal air traffic control communications, or, alternatively, in semi-duplex for radio-telephone connection communication. Duplex operation is entered by use of a push-to-talk button on the controller. Tuning of channels, with transmit at one extreme of the band and receive at the other, is fully automatic.

Controlled by a remote C-200 unit, the system offers the user a choice of 32 pre-selected programmable channels within the covered bandwidth. It has automatic signal-to-noise squelch with manual override. Construction is all-solid-state and frequencies are generated by synthesis techniques.

The RT-30 is suitable for large or small fixed-wing aircraft or helicopters, especially corporate aircraft in which the duplex operational mode is of special use.

Dimensions: 4³⁄₈ × 10½ × 5 inches (111 × 267 × 127 mm)
Weight: 8·3 lb (3·76 kg)

STATUS: in production and service.

RT-7200 vhf fm radio

The Wulfsberg RT-7200 transmitter-receiver is an air-to-ground communications equipment suitable for a wide range of fixed-wing aircraft or helicopters. It provides fm operation over a choice of 7200 channels within the band 138 to 174 MHz at either 25 or 30 kHz increments but can be tuned to the much lower incremental value of 5 kHz. Like other equipment in the Wulfsberg range, the RT-7200 provides a semi-duplex facility with automatic push-to-talk tuning.

In simplex operation, frequency selection may be made on an individual dial-up basis using conventional thumbwheel controls; a digital readout on the C-722 remote controller confirms the selection. Alternatively, up to 15 channels are available on a programmable pre-selection basis. Power output is operatable selectable at either 10 or 1 watt.

The system has a built-in guard receiver which conforms to the same general specification as the main equipment's receiver section but, as an optional alternative, a two-channel, crystal-controlled, plug-in guard receiver assembly is also offered. There is no frequency separation restriction applicable if the optional guard receiver is fitted.

Automatic signal-to-noise squelch, with manual override, is provided and a separate input is included for encoder, voice-scrambler, data or other systems. Construction is all-solid-state and digital synthesis is employed for frequency generation.

Dimensions: 5 × 12³⁄₅ × 5¹⁄₅ inches (127 × 320 × 132 mm)
Weight: 9·3 lb (4·21 kg)

STATUS: in production and service.

RT-9600 vhf fm radio

This equipment is virtually identical to the RT-7200 system (see preceding entry) in its general technical specification. The major difference is in its frequency coverage and channel capacity; ie it covers the frequency range 150 to 174 MHz but provides a total of 9600 channels in this band. Normal frequency spacing is, as in the case of the RT-7200, 25 or 50 kHz but the RT9600 is in fact capable of tuning to a finer 2·5 kHz incremental spacing, unlike the RT-7200 which offers a minimum frequency space increment of 5 kHz.

Furthermore, the RT-9600 is offered with a

choice of three remote controllers, the C-920, the C-961 or the C-962 units. Other differences between the two equipments are of an extremely minor technical nature.

Dimensions: 5 × 12³⁄₅ × 5²⁄₅ inches (127 × 320 × 132 mm)
Weight: 9·3 lb (4·21 kg)

STATUS: in production and service.

Flitefone 40 communications radio system

Wulfsberg's Flitefone 40 is a civil air-to-ground communications system designed to provide two-way, multiple channel contact between an aircraft and co-operating surface mobile stations. By its nature, the system is particularly suitable for helicopters in use by law-enforcement agencies, forestry commission authorities, fire-fighting departments and other such public bodies.

The Flitefone 40 comprises a number of transmitter-receiver units covering the vhf and uhf bands, combined with a series of compatible controllers which together may be assembled to provide a system tailored to requirements. All transmitter-receivers operate in fm mode and each is limited to a maximum power output of 7 watts, in compliance with United States legal requirements. All are operable over a temperature range from –40° C to +50° C and are equally suitable for use as ground-base stations in addition to their design role as airborne equipment. Each can operate to a pressure altitude of 45 000 feet. The transmitter-receiver units comprise:

RT-19 A vhf equipment which provides up to six channels in the 30 to 50 MHz band. It is capable of either simplex or duplex operation with unrestricted frequency spacing between channels.

RT-15 Another vhf radio, similar to the RT-19. This unit however is a high-band system in that it covers the 143 to 174 MHz part of the vhf spectrum. Again, it provides simplex or duplex operational modes and has unrestricted channel frequency separation.

RT-16 A uhf system offering up to 12 channels of push-to-talk duplex operation in the 450 to 512 MHz band. Three or five MHz duplexing is available.

RT-17 A uhf set providing up to six channels in the same 450 to 512 MHz band, with maximum frequency spreads of 2 MHz on receive and 7 MHz on transmit.

All units in the series are of modular all-solid-state construction and feature plug-in exchange of modules for easy maintenance. They contain protection devices to safeguard against open- or short-circuits between transmitter and antenna. The receiver sections provide carrier-to-noise squelch facility.

A range of remote-mounted control units is available, enabling the user to choose a

Components of a Wulfsberg Flitefone 40 system

controller which matches his particular requirement. These comprise:

The C-119, a simple two-channel toggle-switch controller with a ganged switch-deck to permit antenna or ctcss tone switching; the C-117, a rotary switch unit for six- or twelve-channel selection, also with a ganged switch-deck; the C-120, another six- or twelve-channel rotary switch control but with independent tone selection; the C115/116 controller, a six- or twelve-channel unit with annunciating push-button selection. An optional fitting is the SP-40 speaker-amplifier which, powered from the transmitter-receiver in use, provides an independent 15 watt audio output.

Dimensions: (all transmitter-receivers of Flitefone 40 system) 3½ × 5 × 12 inches (89 × 127 × 305 mm)

(C115/6 controller) 5¾ × 1⅛ × 4½ inches (146 × 29 × 114 mm)

(C117 controller) 3 × 1¾ × 4⅝ inches (76·2 × 44·4 × 117·4 mm)

(C120 controller) 3⅞ × 1¾ × 4⅝ inches (98 × 44·4 × 117·4 mm)

Weight: (all transmitter-receivers of Flitefone 40 system) 7 lb (15·4 kg) maximum

(all controllers of Flitefone 40 system) 0·75 lb (0·34 kg)

STATUS: in production and service.

Flitefone III and IV uhf fm radio telephones

Wulfsberg Flitefone III and Flitefone IV are full-duplex uhf fm airborne radio-telephone systems which operate in conjunction with ground stations, providing almost total coverage of the United States and part of the more heavily populated areas of Canada. The systems, which provide air-to-ground and ground-to-air communication via the public telephone networks, are installed, according to Wulfsberg, in the large majority of corporate aircraft built in the USA and Canada. Also, many are fitted to European-manufactured corporate aircraft destined for US users. The equipment is also used in other countries.

The latest system is the Flitefone IV, which uses many of the same major components of the earlier Flitefone III system. These include

Wulfsberg Flitefone IV system with radio-telephone handset

Wulfsberg C200 'flex-comm' controller

an RT-18A transmitter-receiver of 10 watts output, a MT-28A rigid mounting and a AT-461 blade antenna. The RT-18A uses printed circuit board technology of a MIL-standard epoxy-impregnated glass-fibre composition. It has been tested for service aboard aircraft rated to altitudes of up to 51 000 feet. The unit permits operation from multiple cockpit or passenger cabin controls.

Cockpit control is exercised from a C-118 remote controller which provides pushbutton channel selection, intercom and transmit control and hand-set volume variation.

Operation from a passenger cabin position is accomplished by use of a WH-4 controller. This provides pushbutton channel selection, with a digital led confirmation readout, and incorporates electronic-tone output to a speaker to indicate upcall or intercom signalling. This may be channelled through either a public-address cabin speaker system or a dedicated speaker unit. The WH-4 controller, although designed primarily for Flitefone uhf radio-telephone application, is also available in an hf version. When employed in conjunction with a full-duplex Flitefone installation, it is used as a normal telephone hand-set but when operated on hf, a press-to-talk switch is necessary.

C-200 remote controller

This cockpit or flight-deck mounted remote control unit is designed for use with a number of Wulfsberg vhf and uhf fm airborne communication systems. It provides programmable pre-selection of up to 32 simplex or duplex channels and stores transmit and receive frequencies and transmit and receive ctcss tones. Additional switches are provided for control of external functions, ie main or guard receiver selection, squelch override etc.

The C-200 can operate with two transmitter-receivers in parallel. Channels may be randomly assigned to each with the control unit selecting the correct frequencies regardless, either vhf or uhf or both. Two of the channel selection rotary switch positions are dedicated to guard frequencies.

Replacement of a transmitter-receiver by another covering different bands is possible providing that the controller is appropriately re-programmed. Initial and subsequent programming is undertaken with a PR-200 programming unit.

Dimensions: 5¾ × 2¼ × 4⅓ inches (146 × 57 × 109 mm)

Weight: 1·25 lb (0·568 kg)

STATUS: in production and service.

Internal Communications

UNITED STATES OF AMERICA

Collins

Collins Avionics Divisions, Rockwell International, 400 Collins Road NE, Cedar Rapids, Iowa 52406

346D-1B passenger address amplifier

The Collins 346D-1B passenger address amplifier is a silicon fully-solid-state unit with an output of 60 watts. Five inputs are provided, each with its own tone level and the system has continuously variable tone control. Inputs are arranged in order of priority so that the no 1 input (the flight crew's) may override control of the other four. Likewise, no 2/2A input has priority over no 3 and so forth. Dual chimes are provided to signal a forthcoming announcement or to draw attention to visual notices, for example *fasten seat belt* or *no smoking* signs. A compression circuit is employed to provide maximum uniformity of voice output irrespective of input levels, and with appropriate connections, gain is automatically increased by 6 dB when engines are running. An auxiliary amplifier provides 5 watts output at attendants' stations.

Self-test is provided for checking the outputs, and a single technician can check system parameters by means of a test tone signal. A function switch and meter are on the front panel for this purpose.

Dimensions: ¾ ATR Short
Weight: 9·5 lb (4·31 kg)

STATUS: in service.

346D-2/2B passenger address amplifier

The Rockwell-Collins 346-2 passenger address amplifier is a single-box system designed to replace earlier Collins products such as the 346D-1B equipment. Use of jfet transistors in place of photo-resistors and relays, combined with integrated circuitry instead of discrete transistorised circuits, has led to substantial weight reduction and a doubled mean time between failures.

The 346D-2 provides 120 watts rms output (30 watts continuous) into a 41·5 ohm load of 60 watts rms (15 watts continuous) into a standard 83 ohm load. Five inputs are provided with priorities arranged so that a flight crew member (no 1) input takes priority over attendants' no 2/2A inputs.

The 346D-2B variant can accept taped announcements or taped music through inputs nos 3 and 4 respectively. Pilot and attendant priority, in that order of precedence, can override these two taped inputs. Programmable dual chime circuits permit high, low or combined chime tones to be used for such functions as passenger call, attendant call or *seat belt/no smoking* sign attention attraction.

Gain is automatically increased by 6 dB when engines are running and a further 3 dB increase is available for use during a cabin decompression emergency. The attendant station is provided with a 20-watt output from an auxiliary amplifier.

Self-test facilities are incorporated for both ground and in-flight testing. A function switch and a light-emitting diode display are provided on the front panel to facilitate tests.

Rockwell-Collins states that one unit will suffice for narrow-bodied aircraft. Wide-bodied aircraft may be wired to accept one or two systems as required.

Dimensions: ¾ ATR Short
Weight: (346D-2) 6·.1 lb (2·75 kg)
(346D-2B) 6·.6 lb (2·97 kg)

STATUS: in production and service.

Hughes

Hughes Aircraft Company, Microelectronic Systems Division, 2601 Campus Drive, Irvine, California 92715

AN/AIC-27A(V) intercommunication set

This equipment provides internal communication between aircraft crew-members and direct access for external communication. It comprises a centralised audio-switching system controlled by multiplexed data from six crew-station units, but may be expanded to accommodate up to 20 crew positions.

Audio switching and mixing is carried out by a central control unit which interfaces with up to 30 radio communication channels. Switching within the central control unit permits each crew member to simultaneously monitor channels as required. Channel selection and volume level is controlled at each crew station unit and signalled to the central control unit in multiplexed format. The system also provides an intercom network for ground crew maintenance staff via four maintenance station units.

Each crew station unit permits transmit and receive access to all radio and intercom channels and enables the crew member to select the channels for monitoring, select the channel for transmission, to individually adjust volume level of each channel, and continuously transmit without use of the push-to-talk switch by use of a 'hot microphone' function.

Maintenance station units provide an interphone channel to remote aircraft locations for maintenance purposes or as a remote crew interphone station.

The central control unit is an audio switching and control unit which accepts multiplexed data from all crew stations simultaneously, identifying channels selected for transmission or monitoring and volume control.

Other facilities provided by the system are simultaneous clear or secure speech, alarm and call audio alert to all station units, a composite audio input to a flight-data or crash-recorder and intercom with a tanker aircraft via a boom intercom line.

All crew station units are connected to the central control unit by three twisted shielded cable pairs. The central control unit accepts data from all crew stations, supplying each with a synchronising signal and the stations respond via the same cables with the control data in time-divisioned multiplexed format. An interlock in the central control unit prevents secure speech from being overheard at crew stations using an insecure channel. All electronic units are constructed in modular form in which hybrid circuitry is extensively employed.

Dimensions: (CCU) 19½ × 4⁹/10 × 7³/5 inches (498 × 124 × 194 mm)
(CSU) 7¾ × 5¾ × 4½ inches (197 × 146 × 114 mm)
(MSU) 4³/10 × 3⁴/5 × 4³/5 inches (111 × 98 × 118 mm)
Weight: (CCU) 14·3 lb (6·5 kg)
(CSU) 2·5 lb (1·13 kg)
(MSU) 1 lb (0·45 kg)

STATUS: in production and service.

Model 1103000-110 (HAC-110) advanced passenger entertainment and service multiplexing system

The Hughes HAC-110 passenger entertainment/passenger service system is derived from the earlier HAC-100 system which has flown more than a million hours on Boeing 747 and Lockheed 1-1011 wide-body airliners. The system provides 14 entertainment channels of stereo or monaural audio with four pre-emptable motion-picture channels and one public-address input for each zone of the aircraft. Single wire distribution is used throughout the aircraft. The service portion of the system provides reading-light and attendant call-light control and a chime alert for the attendants. There are independent sub-systems in each cabin compartment.

A built-in test function allows convenient testing of each of the system's components directly from the attendant's panel. This enables fault finding to be efficiently carried out down to line replacement unit level.

The system consists of a main multiplexer, a zone multiplexer in each cabin compartment, and a seat electronics box for each seat group throughout the aircraft. The main multiplexer receives up to 14 audio input signals from a tape-reproducer which may be pre-programmed into monaural or stereo pairs. Inputs are converted into a pulse-code modulated digital bit stream for transmission along a single miniature coaxial cable. At each movie zone or compartment, the data enters a zone multiplexer which provides four additional input channels for the local movie audio, independent of the tape-deck inputs. These also are converted into digital data for insertion into the digital data stream. Passenger-address override is available at each zone multiplexer for attendant announcements. Data from each zone multiplexer is directed via coaxial cable to the seat columns where at each group of one to three seats the data is received by a seat electronics box. This processes the digital data to recreate the individual analogue input signals, amplifies them and presents them to a passenger control unit and thence to the passenger. The passenger control unit permits each listener to select from up to 12 programmes and provides individual volume control of audio delivered to the passenger headset.

The passenger service system consists of a series of independent sub-systems in each cabin compartment. Each sub-system consists of a column timer/decoder driving up to three columns of seat encoders and overhead decoders. The column timer/decoder generates timing signals for transmission to the seat columns via the same cable harnesses used for the passenger entertainment system. These signals interrogate the encoder portion of the first seat group to determine the condition of the various passenger commands. This status is then transmitted back to the column timer/decoder which decodes the information to initiate the outputs of chime and master call lights, then re-encodes the information for transmission to the passenger service unit decoder. The latter then carries out the instructions relayed to it (for example *turn on reading light 1*). Sequentially, the column timer/decoder interrogates the next seat group and proceeds down the columns. When self-test procedures are initiated, the reading and call lamps themselves serve as indicators for fault isolation purposes.

Hughes has enjoyed considerable success in the passenger entertainment and service systems market and over the past decade has supplied equipment to 55 major airlines. Customers include all four Western wide-body manufacturers: Lockheed, Boeing, McDonnell

Douglas and Airbus Industrie. Much experience has been gained therefore on large aircraft, particularly with the HAC-100 system from which the HAC-110 has been derived. The later system is claimed to offer significant improvements over the HAC-100, with lower power requirements and weight, higher reliability, easier maintenance, and with an expanded audio capacity. Typical weight-savings quoted are: Boeing 747 105 lb (47·72 kg), Lockheed L-1011-500 65 lb (29·54 kg), Airbus Industrie A300 56 lb (25·45 kg).

The system is solid-state in construction with mos-lsi technology being extensively employed in all modules. Special attention has been paid to heat dissipation and to thermal stresses on components and this will result in considerably improved reliability. As a result, calculated mean times between failures range from 25 000 hours for the main multiplexer to 330 000 hours for a three-seat passenger decoder unit.

Weight: (main multiplexer) 2·28 lb (1·03 kg)
(zone multiplexer) 1·65 lb (0·75 kg)
(column timer/decoder) 1·62 lb (0·74 kg)
(seat electronics box) 0·68 lb (0·3 kg)
(passenger service unit decoder three seat) 0·63 lb (0·28 kg)
(four seat) 0·68 lb (0·3 kg)

STATUS: in production and service.

Model 1022000-110 (HAC-110) advanced passenger entertainment and services multiplexing system

This system was designed for the McDonnell-Douglas DC-10 wide-body transport. In virtually all respects, the Model 1022000-110 is very similar to the 1103000 (HAC-110) system (see previous entry) with the exceptions that the main multiplexer accepts up to 16 audio inputs and that certain units are referred to by different nomenclature for example zone multiplexers are referred to as submultiplexers and seat electronic boxes become demultiplexers. Operational principles are almost identical to those of the 1103000-110 system.

Weight: (main multiplexer) 2⁹/₁₆ lb (1·16 kg)
(submultiplexer) 1½ lb (0·68 kg)
(section timer decoder) 8⅛ lb (3·68 kg)
(seat demultiplexer/encoder three seat) ⅝ lb (0·29 kg)
(two seat) ⅝ lb (0·29 kg)
(overhead decoder) 11/₁₆ lb (0·31 kg)

STATUS: in production and service.

AIC-28 audio distribution system

The AIC-28 ADS audio distribution system provides COMSEC-compatible internal communications among aircraft crew-members and direct access communications between them and personnel in support of aircraft, mission and maintenance activities. The audio distribution system (ads) consists of subscriber control panels located at the various crew-stations, central switching unit(s) to route audio traffic and a programming display and test panel to control and monitor ads usage.

Ads traffic is divided into clear, classified and secure communications. Secure channels may be encrypted and transmitted via aircraft radios, while classified channels are used solely for internal aircraft communication. Clear voice traffic, both internal and external, is unclassified. Interlock circuitry prevents any subscriber from transmitting simultaneously on a clear and a secure or classified channel. The interlock also inhibits all direct access transmissions from a subscriber station initiating a pa announcement.

Any subscriber control panel or station can signal any other subscriber station for a private conversation via a four-channel selective intercom. Other subscribers can be added to form progressive conferences.

Access to radios by subscribers is accomplished remotely from the programming, display and test panel. Subscribers are provided with panel controls for selection of monitor and transmit functions, and controls for adjusting the receive volume level for up to four radios at a time. Flight-crew and other designated stations may initiate public-address announcements which are broadcast over all headsets and pa loudspeakers.

The ads retains all selective interphone and radio access operations in the event of an aircraft power failure. Built-in self-test facilities permit rapid pre-flight verification of full system operation and also provide identification of a failed unit.

Each flight deck audio panel (fdap) has talk/listen access to three programmed radio nets, one dedicated radio, one programmed mission intercom net and the selective intercom. The fdap has access to public address and, provisionally, Autovon trunks. A fdap also has dedicated access to other radios such as Tacan, vor, adf and vhf guard.

The programming display and test panel enables a control operator to selectively programme the direct access radio and intercom nets to the various subscribers. It also displays the current programme status by visual display of the channels assigned to each subscriber. The programming display and test panel permits self-checkout of the system and isolates faults down to replaceable unit level. In conjunction with the central switching unit the programming display and test panel provides all of the necessary electronic facilities to accommodate 14 separate subscriber stations within the system. The addition of a second central switching unit provides a service for up to 28 subscriber stations.

The central switching unit interfaces directly with the system's subscriber stations and serves as an audio distribution matrix. It consists of an array of audio multiplexer devices with appropriate station and channel input/output audio buffers and station data buffers. All audio switching and mixing is performed within the central switching unit so that, irrespective of the total number of channels a subscriber chooses to monitor simultaneously, composite earphone audio is conveyed from the central switching unit to the subscriber station via a single twisted cable pair. Conversely, when a subscriber transmits, his microphone audio is conveyed on a single twisted-pair to the central switching unit which distributes it to other subscriber stations and to radios on the channels on which he is transmitting. If two central switching units are employed, they operate synchronously in a master-slave relationship.

Four separate types of subscriber station panels can operate with the system: a mission audio panel (map), a special audio panel (sap), a mission maintenance audio panel (mmap), and an air-vehicle maintenance audio panel (avmap).

The map provides access to three mission intercom nets and four direct-access radio nets. Each map is equipped with controls and displays to select and simultaneously talk on any programmed channel and to monitor any combination of receive channels. The panels provide direct access to guard radios, public address and, provisionally, to Autovon trunks.

The sap is an abbreviated version of the map and is used for special subscriber stations as required to accommodate mission personnel. It provides direct access to one radio net for external communication. Mission intercom nets, a guard radio, selective intercom and a maintenance net may also be accessed through the special audio panel.

The mmap provides a separate party-line intercom for mission maintenance purposes, directly accessible from all subscribers and from flight-deck and air-vehicle maintenance stations via a selector switch.

The avmap also provides a separate party-line intercom for aircraft maintenance purposes, accessible from all subscribers via a selector switch.

All flight-deck and mission audio-panels can initiate pa announcements which are broadcast over all headsets and loudspeakers in the aircraft. When in passenger address mode, all direct-access radio transmission from the originating subscriber station is inhibited, and when originating from the flight deck the speaker nearest the originating station is muted. All-weather five-inch (12·7 cm) and eight-inch (20·32 cm) speakers are provided. The passenger address amplifier is packaged for standard air transport racking. The package contains controls and an output power-meter for testing and adjusting the 120-watt main amplifier, a 16-watt auxiliary amplifier and the speakers. Provision is made for the incorporation of recorders to log all two-way external communications.

The system has built in self-test capability at both system level and at subscriber station level. Replacement of primary units is possible during mission operations.

System features include access of subscriber stations to audio nets programmable from a single location; automatic assignment of nets on the selective intercom, together with logic to decode call digits and determine which nets are available to the respective subscribers; capability of subscribers to transmit on more than one communication net; direct access to 16 main radios and three guard radios with additional access from flight-crew stations (fdaps) to seven navigation monitors; and capability for distributing audible alarm signals to subscribers in response to sensors and switches external to the ads.

The AIC-28 audio distribution system is constructed from advanced technology micro-circuit systems such as mos-lsi. It is capable of further growth to incorporate interface with additional radio equipment and subscriber stations. An advanced automatic-test facility is available to facilitate ground maintenance and repair.

STATUS: in production and service.

Model 1150 advanced cabin interphone system (acis)

The Hughes Model 1150 advanced cabin interphone system (acis) is designed primarily for Boeing 747 wide-body transport aircraft. It provides up to 20 handset stations, one pilot station and 19 attendant stations which are normally located at strategic positions such as each door, the upper deck and in the galleys.

The system is composed of five basic types of unit: a flight-crew's handset, an attendant's handset, a central switching unit, a flight-crew's control unit and a chime light sensor.

The flight-crew's handset, used in conjunction with the pilot's control unit, provides immediate access to the flight interphone, the passenger address system, and any or all of the attendant's stations. The flight-crew's control unit is mounted in an overhead flight-deck panel and contains the dialling keyboard and push-button controls for other functions. The system is programmed to give the flight-crew immediate priority for any station, whether busy or otherwise. Attendant's handsets contain the dialling keyboard, reset and push-to-talk switches. The keyboard is used to dial any other station. The number of simultaneous calls which may be made is limited only by the number of stations installed. Operation is similar to public utility telephone systems with dial, ringing and engaged tones. A chime sounds at the called station and a call light is illuminated to indicate that the station is being called.

Every attendant's station can initiate an 'attendant's all call' which places all attendants' stations on a common party line, and an 'all call' which signals all stations, including the crew's. For direct immediate calls to the pilot, any station can place a 'pilot's alert' call giving direct communication to the pilot's station on the flight deck. In addition, each station can have access to the passenger-address system either in first class section only, coach sections only or ail together. If the pilot is using the passenger address, an attendant's station cannot interrupt but the pilot has priority to interrupt the passenger address when it is in use by an attendant. A pilot's conversation to an attendant using the passenger-address is not however broadcast over the pa system.

Howler-alert is provided to signal that a handset is 'off hook' to nearby personnel but any number of 'off hook' sets will not disrupt normal operation of the system.

The central switching unit is in the aircraft's forward electronics bay and chime light sensor units are situated near the console at each attendant's station.

The central switching unit is a fully-solid-state system under microprocessor control. To meet reliability standards, Hughes has elected to use MIL qualified components wherever possible throughout the system. Other telephone functions apart from those already included may be programmed in by modification of the software, so the system may be said to have an operational growth potential. Built-in self test facilities for the system are incorporated into the central switching unit.

Weight: (CSU) 8 lb (3·63 kg)
(pilot's control unit) 1·25 lb (0·59 kg)
(CLSU) 0·25 lb (0·13 kg)
(all handsets) 0·5 lb (0·36 kg)

STATUS: in production and service.

Inflight Services

Inflight Services Inc, 485 Madison Avenue, New York, New York 10022

V Star 3 in-flight entertainment system

The Inflight V Star 3 in-flight entertainment system has been developed jointly by Inflight and Gavi, both of New York, and the Belgian company Barco Electronics. In this collaborative programme, Inflight was responsible for design, management and production control, installation design and the provision of repair, maintenance and software services. Barco supplied expertise in the professional television and video display field, while Gavi provided audio-visual and optical technology, together with systems manufacturing capability.

The V Star 3 is a tri-standard video tape projector system compatible with NTSC, PAL and SECAM formats. Inter-format signal-switching is accomplished automatically by the incorporation of sensing and display devices. A three-lens refractive optical system is employed and the unit provides a light output of 600 lumens peak brightness. Three standards of screen are available to meet varying passenger-cabin requirements. A 2·5 gain screen provides a picture brilliance of 105 foot-lamberts (359·73 cd/m²), a 5 gain screen provides 210 foot-lamberts (719·46 cd/m²), and the 10 gain screen gives 420 foot-lamberts (1438·92 cd/m²) brilliance. The resultant picture brightness over a throw distance of 40 to 80 inches (101 to 203 cm), claims Inflight, is double that of other systems and makes viewing possible in high ambient light conditions. Actual screen size at 60 inches (152 cm) throw is 30 × 40 inches (76 × 101 cm). The screen has a flat profile and the picture is viewable over a wide angular range with no edge fall-off.

Additional facilities available from the V Star 3 include dual-language speech playback and passenger-address feed input into two independent areas. The system is remotely controlled and all audio and video functions can be governed from the control switching unit. Projector electronics comprise eight snap-in modules and Inflight claims that any module in the system can be replaced in under 5 minutes.

Dimensions: 18 × 28 × 5⁹/₁₀–9³/₅ inches taper (457 × 711 × 150–244 mm taper)
Weight: 48 lb (21·8 kg)

STATUS: in production and service.

Telex

Telex Corporation (Communications Division), 9600 Aldrich Avenue South, Minneapolis, Minnesota 55420

Telex Corporation manufactures a diverse range of electronic, electro-mechanical and acoustic products, largely for the computer peripherals industry. Its aviation products include pilot communication headsets, microphones and in-flight entertainment systems for commercial carriers manufactured and marketed through the Minneapolis-based Communications Division. Telex supplies a number of leading international airlines with such equipment.

Model Tape-II reproducer

The Model Tape-II reproducer is a 16-channel, continuous-play magnetic-tape music centre for in-flight passenger entertainment and boarding music on commercial aircraft. The system is compatible with wide-body multiplex distribution systems as well as conventionally-wired power-amplifier equipment. It is therefore particularly suitable for airlines with a mixed fleet of wide- and narrow-body aircraft which can standardise the system, using the same programme tapes on each, as well as deriving the normal logistic advantages to be gained from equipment standardisation.

The system provides intermixed stereo/monaural endless-loop programmes of up to 85 minutes each before repetition. Tapes are inserted or removed through a front panel door. Industrial fromet tape is used for both programming and duplicating and a speed of 3³/₄ inches (95·35 mm) per second is used to reduce hiss, distortion and background noise. A slow-speed, servo-controlled, single-capstan drive is employed in the interests of reliability, power efficiency and immunity from power-supply variation. The equipment is of silicon solid-state construction and has built-in self-test circuits.

Rack connections and pin functions of the Tape-II are compatible with existing 16-channel

Telex Tape-II system with covers removed

entertainment systems. An independent, balanced output is provided for the aircraft's boarding music power-amplification system in addition to the normal 16 programme outputs. Volume and channel-selection are controlled from the attendant's control panel without affecting the normal 16 programme channel outputs. Cable harnesses consist of twisted pair wiring which does not require any shielding when running next to main multiplexer or multichannel power amplifiers.

Dimensions: 1 ATR
Weight: (with four 85 minute cartridges) 28·7 lb (13 kg)
(without cartridges) 24 lb (10·9 kg)

STATUS: in production and service.

MCA-13 multi-channel amplifier

The Telex MCA-13 unit is a modular audio-distribution amplifier for wired in-flight entertainment systems. The equipment is wired for 13 audio channels with individual power amplifier plug-in modules. It is available in several channel configurations, and permits growth through the addition of plug-in amplifier modules. The full complement of 13 channels can be used for any desired combination of monaural and stereo programmes for up to 12 channels plus a monaural selection of boarding music, recorded announcements or motion picture sound track. Solid-state switching is provided to override music programmes for public-address announcements on all channels simultaneously.

The solid-state amplifier provides 25 watts rms output per channel which is sufficient to drive up to 250 transducers and meets a variety of passenger seating configurations. Circuitry is designed to prevent changes in audio volume as passengers switch in and out of programmes.

Dimensions: ½ ATR Long
Weight: 16 lb (7·2 kg) max. with 13 channels

STATUS: in production and service.

Trans Com

Trans Com (a unit of Sundstrand Corporation), Costa Mesa, California

Trans Com, part of the Sundstrand Corporation, is a leading supplier of in-flight passenger entertainment systems and claims to have more systems in daily operation in more aircraft than any other supplier. The company offers a range of video projection systems, Super 8 mm motion picture projection systems and audio reproducers. Services include entertainment programming, dubbing services and high-speed duplicating laboratory facilities.

Mark IV Super 8 mm projection system

The Trans Com Mark IV projection system is a well tried equipment with more than ten years' airline service accumulated. Using a patented film cartridge, the system is capable of projecting any film from two minutes to two hours duration. Operation of the system is automatic. A pushbutton starts the projector, which stops automatically at the end of the film and no rewinding is required owing to the continuous-loop design of the cartridge. No special skills are required to change the cartridge and any attendant can complete a full programme change in less than two minutes. A simplified control system, which consists of start, stop and focus pushbuttons, is remotely located for convenience of operation.

The projector chassis is constructed from die-cast aluminium and zinc and fits into the standard overhead baggage storage bin of wide-body aircraft. It uses a 600/700-watt xenon light source. The snap-in continuous loop film cartridge eliminates the requirement for film threading. Eastman Kodak mylar-based 'Estar' film stock is used for increased durability and Sundstrand claims that the Mark IV's sprocketless film-drive system results in a further extension of film life and reliability. A standard 20 mm optical sound track is printed alongside the film and dual sound tracks are available to provide two-language capability.

STATUS: in production and service.

Mark V video projection system

This system has been developed jointly between Trans Com and the Sony Corporation of Japan. It comprises a projector, a video tape-reproducer, a control/distribution unit, and a screen available with either flat or concave surface.

The tape-reproducer provides 180 minutes of programme material on standard video cassettes with two magnetic sound tracks providing either stereo sound or two languages. Rewind time is less than three minutes. Video formats available include NTSC (Standard), PAL and SECAM. A 130-watt projection system which operates over a variable throw length of between 58 and 87 inches (1470 and 2210 mm) provides a screen brilliance of 120 lumens. The projection system's cathode ray tube has a guaranteed life of 3000 hours. The control/distribution unit controls an entire aircraft entertainment system by means of independent signals to each projector.

Two additional options are available, an antenna and tuner for reception of broadcast television programmes and a Sony colour-camera for mounting on the flight-deck. The camera is a solid-state, two-chip unit employing charge-coupled devices as image sensors and weighs less than 3 lb. It is claimed to provide high-quality pictures under the wide range of ambient lighting conditions which exist in an aircraft environment.

Dimensions: (projector) 22 × 9³/₅ × 27³/₅ inches (56 × 24 × 70 mm)
(screen) 28 × 37 inches (71 × 94 mm)
Weight: (projector) 45 lb (20 kg)

STATUS: in production.

Sundstrand Mark IV Super 8 mm projector and 2 h cartridge

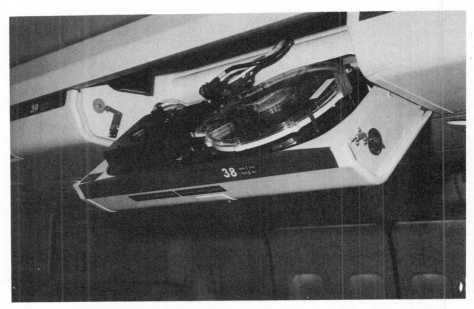

Sundstrand Mark IV Super 8 mm projection system installed in roof-rack

Sundstrand Mark V video projection system's Sony camera and video cassette

200 series audio reproducer

Trans Com's 200 series reproducer is a stereo and/or monaural music and announcement system with capacity for 16 monaural channels of pre-recorded music to an aircraft's multiplex passenger entertainment system or to an external amplifier. The 16 channels can be paired to provide up to eight stereo channels. The reproducer accommodates four magazines of pre-recorded tape and programme

Sundstrand Mark V video projection system installation in airliner roof panel

changes are made by changing these magazines. When the system is used in the announcement and/or boarding music configuration, an attendant control selects the desired programme.

Dimensions: ½ ATR Short
Weight: 12-18 lb (5·45-8·18 kg) depending on magazine and amplifier complement

STATUS: in production and service.

600 series audio reproducer

The Trans Com 600 series audio reproducer is designed for service aboard wide-body transport aircraft. It features a single reel-to-reel disposable type cassette with 12 channels of pre-recorded music programmes with a maximum capacity of 24 hours of monaural entertainment (up to two hours of music per channel) or 12 hours of stereo through the aircraft's multiplex system. Up to four of the twelve channels can be remotely selected to provide boarding music through the aircraft's public address system.

The 600 series reproducer uses one inch (25·4 mm) magnetic master tape housed in the reel-to-reel cassette. Recording format is 24 tracks interleaved to provide 12 programme channels of two hours capacity per channel. No rewinding is required. The tape is played in both directions with one hour duration each way. The playback head has 12 tracks and is shifted to play back the 12 tracks in one direction and the 12 interleaved tracks in the opposite direction. The beginning and end of the tape is sensed by a light-emitting diode and photocell to automatically reverse the tape direction.

A 165 degree tape wrap-on dual capstan provides the tape drive without any need for the use of pinch rollers, the elimination of which reduces skew and consequent flutter. Capstans are driven by a single, endless

Sundstrand 200 series audio reproducer, with twin magazines partially exposed

Kapton belt. The bi-directional play system has also eliminated the requirement for high-speed rewind mechanisms. A programme pause facility causes the recorder to stop during public address announcements, ensuring that programme continuity is not lost during use of the public address system.

Dimensions: ½ ATR Short
Weight: <15 lb (6·81 kg) including cassette

STATUS: in production and service.

CAM 202 combined announce/music reproducer

The CAM 202 combined announce/music reproducer is designed for aircraft public-address systems and meets requirements for automatic announcements and music reproduction. The system comprises three elements, an audio reproducer, an attendant control unit and music tape-magazines. It provides arrival and departure messages, mandatory announcements such as service or emergency information, and music programmes. Selection of the desired announcement or programme is

Sundstrand 600 series tape reproducer, with side-cover hinged down

Sundstrand CAM 202 combined announcement/music reproducer

controlled from the attendant's control panel.

Pre-recorded announcements are contained on 24 channels, one of which has priority over all other system functions. This channel is normally programmed with emergency announcements.

STATUS: in production and service.

Head/Handsets and Microphones

UNITED KINGDOM

Clement Clarke

Clement Clarke International Limited, Airmed House, Edinburgh Way, Harlow, Essex CM20 2ED

Airlite 62 headset

The Airmed Airlite 62 is a general-purpose headset designed for civil and military airborne and ground applications. A wide-ranging choice of microphones is available for this model, including a 300-ohm electro-magnetic differential, a 50-ohm electro-magnetic differential, a 100-ohm carbon differential, a 300-ohm moving-coil differential, a 'Dyn-a-Mike' 100-ohm moving-coil differential with amplifier, a carbon pressure, a 100-ohm carbon differential and a 100-ohm moving-coil differential with amplifier. These microphones fit into a clip at the end of the microphone boom and are readily interchanged.

For airborne use, earphone receivers with an individual impedance of 300 ohms are supplied. The earphones are pressure compensated and are fully tropicalised. Earpads are of foam plastic, for good noise attenuation, and are covered with washable linen covers for hygiene. A plastic covered, sponge-foam headpad is used and this is also detachable for cleaning purposes.

The microphone boom is fully adjustable and a one-piece locking screw is fitted to tensioning on fore, aft and lateral movement.

A single earpiece version, marketed under the name 'Unilite', is also available.

STATUS: in production and service.

Airlite 71 headset

The Airmed Airlite 71 headset offers all of the features of the Airlite 62 model, including the wide choice of microphones, switches etc (see preceding entry). It differs from the Airlite 62 in having a double headband and high attenuation earphones. The microphone boom is mounted on a universal joint to permit precise positioning.

STATUS: in production and service.

Airgard headset

The Airmed Airgard headset was introduced to complement Clement Clarke's Airmed Airlite range of headsets, and uses common components wherever possible. The Airgard has good noise-attenuation characteristics and is well suited to use in helicopters and other aircraft with noisy environments. The earphone housings are moulded in high impact material and give an average attenuation of 35 dB over the frequency range 500 Hz to 5 kHz. High-sensitivity rocking armature earphone inserts, of 300 ohms impedance each, are fitted as standard. Any of the Airmed range of clip-in microphones of the carbon or dynamic type can be fitted, but for particularly noisy environments, throat microphones are recommended. Either carbon or electro-magnetic types of 300 ohms impedance can be supplied. A wide range of press-to-talk switches is available.

The Airgard's headband comprises two

Clement Clarke Airmed Airlite 71 headset with boom microphone

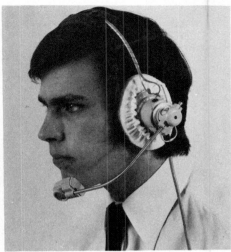

Clement Clarke Airmed Airlite 62 headset in place, with boom microphone

Clement Clarke Airmed Airgard headset with boom microphone

Clement Clarke Airmed Airgard headset with throat microphone

heavy gauge spring-steel wires held together by a plate which supports a detachable headpad. Fine adjustment of the earphone position is achieved by rotating knurled collet clamps at the ends of the headband wire. Where separate amplification is required for use with any of the dynamic microphones available, a transistor amplifier is supplied with a plug and socket, available in a number of configurations, for fitting to the main lead. Alternatively, the same amplifier can be supplied for connection to the transmitter.

Normally, a standard lead length of 80 inches (2 m) is supplied. This incorporates a terylene strain relief cord and a NATO jack-plug

STATUS: in production and service.

Airmed Dyn-a-Mike microphone

The Airmed Dyn-a-Mike is a moving-coil differential microphone designed as a clip-in replacement for carbon microphones in Airmed headsets. It features a miniature, thin-film amplifier, uhf filtering, mu-metal screening for use in an electro-magnetic sensitive environment and a bridge-type circuit which permits use of a direct current supply of either polarity.

A low-cost version with no uhf filtering is also available.

Dimensions: 1¾ × 1¼ inches (46 × 32 mm)
Weight: ¾ oz (24 g)

STATUS: in production and service.

Racal

Racal Acoustics Limited, Beresford Avenue, Wembley, Middlesex HA0 1RU

Minilite headset

This lightweight headset is available in a number of configurations and specifications. It may be worn in either single or dual earpiece form, with the microphone either left- or right-handed according to choice. Furthermore, if the user is a spectacle wearer, it is possible in

certain cases to dispense with the headband and to clip the Minilite to the side frame of the spectacles. The microphone is acoustically coupled through a flexible tube to which a sound collector, incorporating a sibilant filter, is attached. The receiver earphone is mounted on a universal joint and is fully adjustable in all directions to obtain optimum fit and comfort. Normally, the Minilite is worn with the earpiece not in direct contact with the ear, but acts rather like a miniature speaker operating in a 'free

field'. In noisy environments, the earphone can be adjusted to fit into the bowl of the ear to improve noise exclusion.

A choice of two types of earphone is available, both with a frequency range of 100 Hz to 3·5 kHz. The standard-sensitivity earphone has an impedance of 300 ohms and a maximum input of 50 mW, and the high-sensitivity unit an impedance of 600 ohms and a maximum input of 25 mW.

A magnetic microphone is supplied as

standard. This is a 300-ohm impedance model with a specially-shaped response covering the range 300 Hz to 4 kHz. An amplifier is optionally available to raise the microphone output to correspond with that of carbon microphone levels. The amplifier has a preset gain-control which provides an adjustment range of 30 dB. Another optional amplifier, designed specifically for low voltage levels, is also available. A range of microphone switches, some built into

in-line amplifiers, can be supplied.

For use in areas which are subject to magnetic interference, the Minilite may be fitted with mu-metal magnetic screens. These are particularly recommended to reduce interference with sensitive navigation equipment.

Dimensions: (amplifier) 2⁷/₁₀ × ⁴/₅ × ⁴/₅ inches (68 × 20 × 20 mm)

STATUS: in production and service.

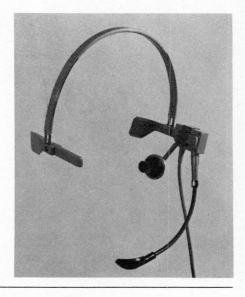

Racal Minilite single-side headset

UNITED STATES OF AMERICA

Plantronics
Plantronics, 345 Encinal Street, Santa Cruz, California 95060

MS50-14 headset
The MS50-14 is an unamplified headset designed for service in commercial transport aircraft. Receiver impedance is 600 ohms and microphone impedance is 3000 ohms. A five-foot (1·524 m) cord is terminated with a PJ068 jack connector. The headset is used in conjunction with a T50-1 amplifier.

STATUS: in production and service.

MS50-35 headset
The MS50-35 is an unamplified headset designed for use in commercial transport aircraft. Receiver impedance is 600 ohms and microphone impedance is 150 ohms. A seven-foot (2·13 m) cord is terminated with a PJ608 and a PJ055B jack connector. The microphone is of the dynamic type.

STATUS: in production and service.

MS50-75 headset
The MS50-75 is an unamplified headset designed for service in commercial transport aircraft. Receiver impedance is 300 ohms and the microphone impedance is 5-ohms. A three-foot (0·91 m) cord is terminated with a U93 A/U jack connector. The microphone is of the dynamic type.

STATUS: in production and service.

MS40/T30-1 microphone
This is a carbon microphone designed for service commercial transport aircraft. It has a five-foot (1·52 m) cord terminated with a PJ068 jack connector.

STATUS: in production and service.

MS50/T30-1 headset
The MS50/T30-1 is an amplified headset designed for use in commercial transport aircraft. Receiver impedance is 600 ohms and microphone impedance is 150 ohms. Both receiver and microphone outputs are non-adjustable, the microphone being factory-set to 120 mV. A five foot (1·52 m) cord is terminated with PJ068 jack connector. The microphone is of the carbon type.

STATUS: in production and service.

MS50/T30-2 headset
The MS50/T30-2 is an amplified headset for service in commercial transport aircraft. Receiver impedance is 600 ohms and microphone impedance is 150 ohms. Both receiver and microphone outputs are non-adjustable, the microphone output being factory-set to 120 mV. A five-foot (1·52 m) cord is terminated with both PJ068 and PJ055B jack connectors. The microphone is of the carbon type for dual-plug applications.

STATUS: in production and service.

HS0177-1 headset
The HS0177-1 is an amplified headset designed for service in commercial transport aircraft. Receiver impedance is 600 ohms and microphone impedance is 150 ohms. Both receiver and microphone outputs are non-adjustable, the microphone output being factory-set to 120 mV. A five-foot (1·52 m) cord is terminated with a PJ068 jack connector. The microphone is of the carbon type for single-plug applications.

STATUS: in production and service.

HS0117-2 headset
The HS0177-2 is an amplified headset for use in commercial transport aircraft. Receiver impedance is 600 ohms and microphone impedance is 150 ohms. Both receiver and microphone outputs are non-adjustable, the microphone output being factory-set to 120 mV. A five-foot (1.52 m) cord is terminated with both PJ068 and PJ055B jack connectors. The microphone is of the carbon type for dual-plug applications.

STATUS: in production and service.

MS57/T30-1 headset
The MS57/T30-1 is an amplified headset with a dual receiver configuration designed for service in commercial transport aircraft. Receiver impedance is 600 ohms and microphone impedance is 150 ohms. Both receiver and microphone outputs are non-adjustable, the microphone output being factory-set to 120 mV. A five-foot (1·52 m) cord is terminated with a PJ068 jack connector. The microphone is of the carbon type and incorporates a Plantronic BNS-1 background-noise suppressor which fits over the end of the voice tube, providing an additional 10 dB signal/noise

ratio. This headset is of value in high-noise cabin or flight-deck environments.

STATUS: in production and service.

MS50/T30-3 headset
The MS50/T30-3 is a general aviation headset in which the microphone is factory-set to a medium gain position. Receiver impedance is 600 ohms and microphone impedance is 50 ohms. Receiver gain is adjustable. A five-foot (1·52 m) cord is terminated with both PJ068 and PJ055B jack connectors. The microphone is of the carbon type.

STATUS: in production and service.

MS50/T30-21 headset
This is a general aviation headset conforming to the same specification as the MS50/T30-3 model (see previous entry) with the exception that the microphone is factory-set to a high gain position.

STATUS: in production and service.

MS50/T30-12 headset
This is a general aviation headset conforming to the same specification as the MS50/T30-3 model (see above) except that the cord is terminated with a PJ068 jack connector. An additional cord is terminated with the PJ055B type connector.

STATUS: in production and service.

StarSet HS0177-3 headset
The StarSet HS0177-3 is a general aviation headset with 600 ohm receiver impedance and 150 ohm microphone impedance. Receiver output level is adjustable from a two-position High/low switch which provides 10 dB attenuation. The microphone gain is also adjustable for all standard radio inputs from 45 to 185 mV. A five-foot (1·52 m) cord is used and terminates with both PJ068 and PJ055B jack connectors.

Weight: ½ oz (14 grams)

STATUS: in production and service.

Telex

Telex Corporation (Communications Division), 9600 Aldrich Avenue South, Minneapolis, Minnesota 55420

PEM-78 earset

The PEM-78 earset is designed for multi-crew aircraft in which a master volume control provides the same level for all crew members. It has an in-line volume control providing a 10 dB range for individual volume adjustment and uses a single magnetic receiver.

STATUS: in production and service.

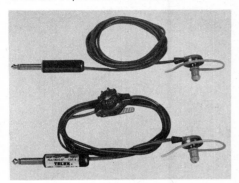

Telex PEM-77 (upper) and PEM-78 (lower) pilot earsets

HTW-2A Twinset headphone

Claimed by Telex to be the most widely used pilot's headphone, the Twinset uses dual magnetic receivers which rest on the temples, directing sounds into the ear via tubular sound arms. This arrangement is said to block background noise and to improve intelligibility of weak signals. The adjustable sound arms are mounted on a ball-and-socket joint which permits either arm to be turned away without removal of the headset.

Weight: 0·25 lb (0·113 kg)

STATUS: in production and service.

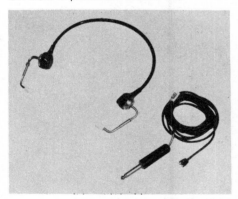

Telex Twinset extra lightweight headphone

Model A-610 headphone

The model A-610 headphone is a low-cost set suitable for private flying and general aviation. It uses dual magnetic receivers, is lightweight and of robust design to withstand hard wear.

STATUS: in production and service.

Model A-1210 headphone

This is a general purpose aviation headset designed to provide close-ear coupling to reduce ambient noise interference. Dual dynamic receivers are employed.

STATUS: in production and service.

Model TH-900 headphone

The TH-900 headphone is designed for maximum hearing protection in high noise-level environments and is therefore suitable for helicopters and open-cockpit aircraft. Good

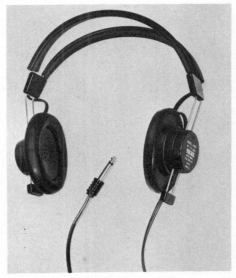

Telex A-610-1 lightweight magnetic headphone

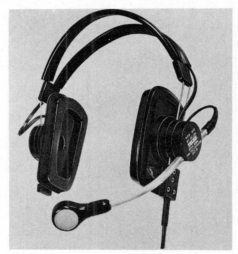

Telex MRB-600 headset

noise attenuation characteristics result from a design which features a deep ear cavity surrounded by large, padded ear cushions. The same basic unit, with receivers deleted, is used as a standard ear defender. Dual dynamic receivers are employed in this headphone.

STATUS: in production and service.

Model TAH29 headphone

This headphone, which has been designed specifically for air-transport applications, features audiometric-type dual receivers for good sound clarity and signal intelligibility. The entire receiver element is a snap-in fit into the headband and may easily be removed and replaced for field servicing.

STATUS: in production and service.

Model MRB-600 economy headset

This is a low-cost headset designed for the general aviation sector. It employs dual magnetic receivers and a noise-cancelling carbon boom microphone. A push-to-talk switch is optionally available for use in aircraft not so equipped.

STATUS: in production and service.

Model MRB-2400 headset

The MRB-2400 headset is designed for aircraft with moderately high noise levels. It uses circum-aural foam-filled ear cushions and has a noise-cancelling, amplified, dynamic boom microphone. Dual dynamic receivers are employed. A push-to-talk switch is optionally available for use in aircraft not so equipped.

STATUS: in production and service.

Telex model TAH29 headphone

Telex DBM-1000 and EBM-1400 'HearDefender'

Model DBM-1000 headset

This unit has been designed to provide high-quality reception in exceptionally noisy cockpit or cabin environments and is particularly suitable for use in helicopters. A noise-cancelling, amplified dynamic microphone on an adjustable boom, to match the ideal mouth-microphone relationship, is employed. The headset is finished in light grey to minimise sunlight heat absorbtion. A push-to-talk switch is optionally available for use in aircraft not so equipped. A choice of connectors is available; the DBM-1000 is normally supplied with a PJ-055 headphone jack and a PJ-068 microphone plug but either the U-75/U or U-93A/U plugs may be fitted.

STATUS: in production and service.

Model EBM-1400 headset

This model meets the same specification as the DBM-1000 (see preceding entry) except that the carbon microphone of the DBM-1000 is replaced with an electret-type microphone in the EBM-1400. Performance of electret microphones is considerably superior to that of carbon microphones, particularly with regard to noise-cancelling. The microphone in the EBM-1400 is said to eliminate radio frequency feedback and 400 Hz hum. It is compatible with carbon microphone circuitry. All plug options available on the DBM-1000 are also available on the EBM-1400.

STATUS: in production and service.

Model HS-500 handset

The Telex HS-500 handset is designed for interphone and paging use aboard large aircraft. It employs an amplified electret microphone and an audiometric-type dynamic receiver to ensure high-fidelity response with low distortion. The handset is said to possess superior noise-cancelling characteristics which permit it to be used close to public-address speakers without sacrifice to tonal quality or volume level. It is also claimed to be resistant to radio frequency interference and to the hum induced by 400 Hz electrical systems. The HS-500 has a bar-type push-to-talk switch built into the handset frame.

STATUS: in production and service.

Model 5 × 5 Mark IIA headset

This equipment is one of Telex's lightweight headset range which are marketed under the generic trade mark '5 × 5'. The Mark IIA model uses a single-side magnetic receiver with a soft ear-tip which is adjustable for use on either left or right side, as desired. Microphone is of the distortion-free, magnetic type and is mounted on an adjustable boom which permits it to be located at the correct mouth-to-microphone distance. The headset can be worn by attaching it to spectacle frames, in which case the headband may be removed.

STATUS: in production and service.

Model 5 × 5 Pro II headset

The Pro II headset is another of the Telex 5 × 5 range with a similar specification to the Mark II model (see preceding entry) but with an electret microphone instead of the carbon type. Incorporation of the electret microphone results in improved speech intelligibility.

STATUS: in production and service.

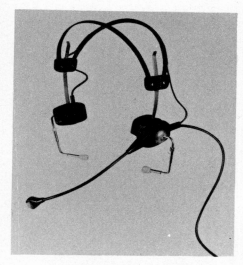

Telex 5 × 5 Pro 1R headset

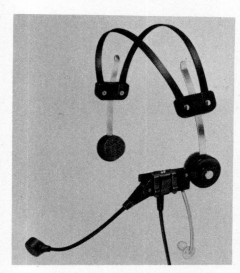

Telex 5 × 5 Mark IIA headset

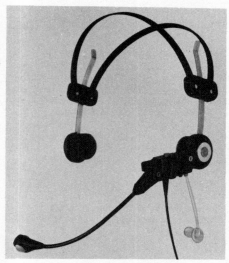

Telex 5 × 5 Pro II headset

Model 5 × 5 Pro 1R

The Pro 1R model is another headset in the Telex 5 × 5 range (see preceding two entries). Again, it is similar in specification to other 5 × 5 models and incorporates an electret microphone as does the Pro II equipment. The Pro 1R however is fully reversible so that the microphone may be worn on either the left or right side. It is also equipped with dual magnetic receivers which uses the Telex Twinset adjustable ear-tubes. These may be worn singly or simultaneously according to preference to meet varying cockpit noise levels.

STATUS: in production and service.

Model 66C microphone

The Telex 66C is a carbon noise-cancelling microphone. It has a self-contained push-to-talk switch and a microphone hanger placed in the front for quick orientation, and is designed for long service with hard usage.

STATUS: in production and service.

Model 66T microphone

The model 66T is a dynamic microphone which provides superior noise-cancellation and voice-response. It contains a solid-state built-in amplifier which can be adjusted to provide the correct transmitter modulation.

STATUS: in production and service.

Model 100TRA microphone

The model 100TRA is a dynamic microphone with an end-mounted element to eliminate 'front or back' orientation. It contains a built-in, two-stage, solid-state amplifier which may be adjusted to obtain optimum power transmission and has good noise cancellation characteristics. This model is designed to blend with styled aircraft interiors, having a beige and black finish with wood-grain trim.

STATUS: in production and service.

Telex 66C hand-held microphone

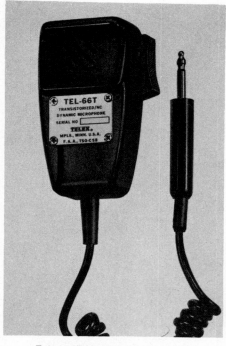

Telex 66T hand-held microphone

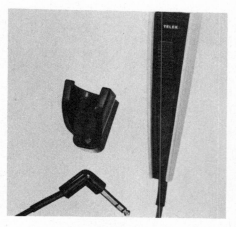

Telex 100T hand-held microphone

Model 500T microphone

The model 500T represents the most advanced of the Telex hand-held microphone range. It uses an electret element to provide first-class noise cancellation, has high resistance to electrical and rf interference, and contains an adjustable, solid-state amplifier. It is a styled model, with a black finish with woodgrain trim.

STATUS: in production and service.

Flexible-boom microphone

This solid-state, amplified, electret-element microphone is mounted at the end of a flexible boom, the base of which is at the most convenient cockpit location. It provides the advantages of a headset-mounted boom microphone, in that it can be positioned in close proximity to the user's lips, while dispensing with the need for a headset band. Hands-free

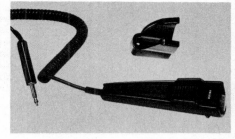

Telex 500T electret hand-held microphone

operation is permitted via the use of a controls-mounted press-to-talk switch. The microphone also has good noise-cancellation characteristics.

STATUS: in production and service.

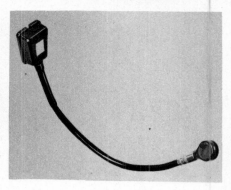

Telex flexible boom microphone

Stores Management

GERMANY, FEDERAL REPUBLIC

Teldix

Teldix GmbH, Postfach 10 56 08, 6900 Heidelberg 1

Missile control system for Tornado/Kormoran

The Teldix missile control system has been designed to control the air-to-ship Kormoran missile fitted to German versions of the Panavia Tornado aircraft. The missile control system serves as a computing and interface system between the aircraft's main digital computer with its associated peripheral systems including stores management, inertial navigation, maintenance panel, warning system and others, and the analogue input/output circuits of the missile.

The missile control system comprises three units; the missile control panel, missile control unit and launcher decoder unit. The missile control panel contains a number of controls and displays which enable the operator to monitor the status of the missiles and select those to be fired and the firing mode. The missile control unit contains the interface circuits that convert and process externally-fed data, feeding it to the missiles via the launcher decoder unit.

Teldix missile control unit

Teldix missile control launcher-decoder

Teldix missile control panel

Interface between the missile control unit and individual missiles is provided by the launcher decoder unit and one of these is therefore included in each Kormoran launcher. The launcher decoder units contain digital/analogue and analogue/digital conversion of data to and from the missiles and their power supply also.

STATUS: in production.

UNITED KINGDOM

Computing Devices

Computing Devices Company Limited, Castleham Road, St Leonards-on-Sea, East Sussex

SMS 2000 series stores management systems

The Computing Devices SMS 2000 stores management system controls a complete aircraft inventory of releaseable stores and armament including bombs, guided weapons, guns and rockets and drop-tanks. It comprises cockpit control panels, which also indicate the status of the aircraft's armament and a digital stores management computer.

The system permits in-flight pre-selection of a weapon delivery package to suit a particular target or conditions. Control facilities are also provided for selection or release modes (single or multiple stores), ground spacing optimisation in relation to aircraft speed and fuzing. Duplex selective and emergency jettison for the entire stores load is also permitted.

Hang-ups due to battle damage or for any other reason are displayed on the control panel and in such cases the system controls release of the remainder of the stores within centre of gravity and lateral balance limits. An optional serial data interface, conforming to MIL-STD-1553, permits the system to operate in conjunction with a navigation/attack computer. The level of integrity of the system ensures that no single failure results in an inadvertent release of stores.

The SMS 2000 is of modular construction and contains built-in self-test circuitry. For off-aircraft servicing and repair, an automatic test equipment is available. This initially, enables a fault to be isolated to sub-module or circuit-card level and subsequent determination of the precise component fault when the suspect module has been removed from the system.

The system is designed for a wide range of light tactical attack aircraft, and the SMS 2001/A variant has been selected as standard equipment for export models of the British Aerospace Hawk strike/ trainer.

Dimensions: (control panel) 5½ × 7¹/₁₀ × 6³/₁₀ right 3½ inches left (140 × 180 × 160 right/ 90 mm left)
(stores management computer unit) 8²/₃ × 5¹/₁₀ × 12¹/₅ inches (220 × 130 × 310 mm)

Computing Devices SMS 2000 stores management system with BAe Hawk strike-trainer

Weights: (control panel) 6·16 lb (2·8 kg)
(stores management computer unit) 15·18 lb (6·9 kg)

STATUS: in production and service.

SMS 3000 series stores management systems

SMS 3000 stores management systems are based largely on the experience gained with the company's SMS 2000 series equipments but offer greater flexibility of use in heavier aircraft. They are intended for use with both current and projected weapon systems, are micro-processor-controlled and have an expanded memory with which to cater for future requirements.

The SMS 3000 series has a dual mode of operation which has twofold intent: that of maximising safety by reducing the risk of accidental stores release and of improving mission reliability. This involves the employment of a dual redundant channel system which operates in serial form to release stores in the safety mode. When in combat, either one of two parallel systems can be used to effect release. The systems provide all of the facilities of the 2000 series equipment, but have greater inbuilt flexibility. In display terms, for example, the control panels inform the aircrew of hazards such as missile misfire, unsafe height, unsafe spacing etc. Expandability, from an equipment viewpoint, has been taken into account by providing a range of modular stores interface units to match the increased processor capacity.

STATUS: in development.

Marconi

Marconi Avionics Limited, Airport Works, Rochester, Kent ME1 2XX

Stores management system for Tornado

The weapons management system developed by Marconi Avionics for the tri-national Tornado multi-role bomber is capable of handling at least 26 different types of store with control of weapon arming or fuzing and firing or release functions.

The system has a total of 27 outputs and undertakes the timed and sequenced release of these stores, and external fuel tanks or equipment pods which may have to be jettisoned in the event of emergency.

Among the principal design aims of the Tornado stores management system were a significant reduction in crew workload (a common desirable feature of stores management philosophy; the reduction of inter-connecting cable between line-replaceable units to the minimum (there are ten line-replaceable units in the total system); high reliability in terms of operation and safety rates. The specification also called for a degree of electro-magnetic compatibility 'hardness' (resistance to electrical interference) never previously considered for a weapon system.

To achieve these targets, Marconi has developed a totally dual-channel, digital architecture employing mini-computers and digital data transmission of extremely high integrity. Wherever possible, cmos circuitry has been used for low power consumption and resistance to interference. Both channels are fully synchronised and perform in a consolidated fashion except in the case of a channel unserviceability, when the serviceable channel reverts to an independent mode.

The level of redundancy in the system is illustrated by the fact that the jettison control system is not only quadruplexed but can operate autonomously independent of the channel integration hardware. This is further underlined by the fact that each crew member has his own system control panel, thereby increasing the degree of flexibility and commensurate with a low crew workload.

STATUS: in production.

Plessey

Plessey Electronics Systems, Vicarage Lane, Ilford, Essex IG1 4AQ

Weapon control system for Harrier

Until the mid-1960s weapon arming and release systems mainly comprised manually-actuated switches controlling relay networks which completed the stores arming and release/firing sequence. The advent of new combat aircraft capable of accommodating a huge variety of stores mounted on a variety of underwing and/or fuselage stations, raised a requirement for stores management methods by which to meet the control problems caused by asymmetric release of weapons, drop-tanks, or other disposable items. These difficulties were compounded by the diminishing size of airframes while the stores themselves remained largely unchanged, thus increasing their influence.

In the United Kingdom initial studies into the problem were undertaken by the Royal Aircraft Establishment, Farnborough which later co-opted Plessey Avionics and Communications to carry out development of an electronic weapon control system. The eventual outcome of the programme was the system for the RAF/Hawker Siddeley Harrier V/STOL fighter, for which contracts probably in excess of 200 units were placed. The system was also to have been supplied to the US Marine Corps for the AV-8A version of the Harrier.

With the Harrier's weapon load distributed between wing and fuselage pylons, it is of importance that lateral balance is maintained as weapons are released, particularly during low-level strike attacks. The Plessey weapon control system computes and executes a release sequence which maintains aircraft balance under all conditions.

When stores are loaded onto the aircraft, thumbwheel switches are manually set on the weapon control system's control panel to provide the pilot with an indication of the external stores status. During the loading operation the groundcrew also move the role-selection switches on the panel to an appropriate setting for the type of stores carried. Two modes of operation, manual or automatic, are possible. In manual mode, the pilot during run-up to the target chooses, via pylon selector switches, which weapons are to be released. If two stores are carried on one pylon, then use of a single/double selector determines whether one or both stores will be dropped. Selection of fuzing is the next operation and then, if the selection of pylon, store and fuzing is valid for the weapon state and maintenance of lateral balance, a store indicator light will illuminate to confirm that the selection is correct. Actual release is made by depression of a late-arming switch situated remotely from the weapon control system panel as a safeguard against premature release. The weapon control system prevents store release if the store indicator lamp remains unlit or if the stability of the aircraft will be jeopardised. Release intervals of 20-, 40-, 80-, 160-, and 320- metres a second are

Plessey weapon control system installed in BAe Harrier V/STOL fighter

selectable from a rotary switch on the weapon control system panel.

For automatic release, the auto/manual switch interfaces the weapon control system with the aircraft's nav/attack computer providing it with data on stores status, types selected and spacing required. Release interval then comes under computer control. In either mode, the pilot may use both individual or 'clear aircraft' overriding jettison control switches to jettison stores separately or as a single salvo in an emergency.

During rocket or guided weapon release or during gun firing, the weapon control system ensures that power is supplied to the engine igniter circuits to safeguard against flame-out caused by aerodynamic flow disturbance or gas ingestion.

Built-in test circuits for functional and power supply checking are also controlled from the weapon control system panel. All switches can be operated with gloved hands and vital controls are guarded against inadvertent operation.

The system is digital-based and makes extensive use of silicon integrated circuitry. Logic circuits are of diode-transistor type and the number of relays has been reduced to the minimum required to handle the heavy switching-currents which occur at the time of weapon release.

In two-seat Harriers a monitor unit is fitted in

the rear cockpit. It duplicates the front cockpit jettison facilities and enables the control panel to be monitored during operation.

The system's capacity permits mixed loads to be carried and controlled on each pylon. The range of stores includes single or twin bombs, flares, one or two rocket launchers, two gun pods, external fuel tanks or guided weapons. Bomb fuzing signals can be transmitted to the five pylons.

Dimensions: (wcs unit) $4^{1}/_{2} \times 8^{1}/_{4} \times 12^{5}/_{8}$ inches ($114 \times 210 \times 321$ mm)
(monitor) $4^{1}/_{2} \times 8^{1}/_{4} \times 6^{3}/_{10}$ inches ($114 \times 210 \times 160$ mm)
Weights: (wcs unit) 10·25 lb (4·65 kg)
(monitor) 5·25 lb (2·38 kg)

STATUS: in production and service.

Weapon control system for Jaguar

The Plessey weapon control system for Jaguar is a development of that provided for Harrier (see previous entry) and employs the same digital, solid-state electronics technology. In the single-seat Jaguar the system is divided into two major elements, the main unit housing the control logic and the cockpit-mounted control unit which contains the indicator lamps, the pilot-operated armament switches and the individual jettison switches. In the two-seat variant, a monitor unit is fitted in the rear cockpit for training purposes.

As well as providing all of the functions of the Harrier system, the Jaguar unit has greater reliability as the result of wider use of solid-state logic. It provides fully duplicated release and jettison of individual stores, dispenses with

the need for autoselector switches on tandem store carriers, and provides a more advanced method of rocket firing control. It may be operated as an autonomous system or used in conjunction with the Jaguar nav/attack computer. Orders for Jaguar weapon control systems probably exceed 250 units.

Dimensions: (control unit) $6 \times 5^{3}/_{4} \times 6^{1}/_{6}$ inches ($152 \times 146 \times 157$ mm)
(logic unit) $6^{1}/_{8} \times 13^{3}/_{4} \times 16^{3}/_{4}$ inches ($155 \times 349 \times 425$ mm)
(monitor) $3^{3}/_{8} \times 5^{3}/_{4} \times 5^{1}/_{8}$ inches ($85 \times 146 \times 130$ mm)
Weights: (control unit) 3·875 lb (1·76 kg)
(logic unit) 32 lb (14·54 kg)
(monitor) 1·25 lb (0·56 kg)

STATUS: in production and service.

UNITED STATES OF AMERICA

Hughes Aircraft Company
Radar Systems Group, El Segundo, California 90245

AN/AQX-14 weapon control data link

The Hughes AN/AQX-14 is a two-way communication data-link to control the GBU-15 guided bomb. It provides a video and command link between the command aircraft and the weapon, enabling the systems operator to remain in the control loop while the weapon is being directed to its target. In effect, the data-link permits a command authority similar to a 'fly-by-wire' situation, whereby the operator can transmit guidance instruction all the way from launch to impact. Alternatively, he may select any one of a number of autonomous weapon-control modes, including an override mode which permits target up-dating or re-designation as required.

The extended weapon-control capability conferred by the data-link contributes to weapon-system performance in terms of stand-off range and operational utility. Stand-off range may be increased by launching the weapon outside the range that it can recognise a target or lock onto it. Target acquisition is deferred until the weapon, rather than the command aircraft, is closer to the target. Tactically, the aircraft can leave the immediate target zone immediately after launch.

The AN/AQX-14 system comprises three major elements: a data-link pod mounted on the command aircraft, a data-link control panel, used in conjunction with an existing display within the aircraft, and a weapon data-link module, mounted on the rear of the weapon itself.

The pod is an aerodynamically-shaped container mounted on a standard stores-carriage strong-point on the fuselage centreline or on an underwing station, according to aircraft type. It contains four line-replaceable units comprising an electronics section incorporating all radio frequency generating and receiving equipment, a de-multiplexer to decode all aircraft command and pod control signals, an encoder, and antenna controls; a phase-scanned array for weapon tracking in normal operation; a forward horn antenna which provides additional coverage; and a mission tape-recorder which maintains a permanent record of weapon video data. The pod is suitable for high performance aircraft, is certificated for operation at speeds in excess of Mach 1 and is also compatible with high/low-altitude operations. There is said to be no compromise of aircraft performance attributable to carriage of the pod.

Used in conjunction with an existing display system, the aircraft control-panel acts as the interface between the weapon system operator and the weapon guidance system. The panel accepts signal inputs from the aircraft as well as from its own controls, and formats these into discrete commands as required by the data link. Although the panels are tailored to the individual requirements of the aircraft type and intended customer usage, each unit accepts the standard configurations of the GBU-15 data-link and the pod.

Attached to the after end of the GBU-15 weapon is the ultimate component in the data link chain, the weapon data-link module. This simultaneously transmits video from the weapon's seeker-head and processes incoming command signals from the aircraft to the weapon. Heading changes during the wea-

pon's flight are effected through discrete command signals. Dual analogue voltage channels enable the operator to slew the weapon in pitch and yaw during approach to the target.

Digital techniques are employed in the AN/AQX-15 system and the transmitter is of all-solid-state construction. The system's electronically phase-scanned antenna array provides the data-link with high rate tactical manoeuvring capability. A comprehensive range of test equipment is provided, including a flight checkout unit for testing aircraft cables from the pod connection point, an aircraft simulator unit which permits functional checks of the control panel, and a weapon simulator unit for test of the aircraft-mounted pod and isolating faults down to line-replaceable unit level. Used together, these two latter units permit full system functional checkout.

Two primary launch modes are envisaged for operation of the GBU-15 weapon/AN/AQX-14 control combination: low-altitude penetration and high-altitude stand-off. Hughes claims that use of the data-link has demonstrated improvement in weapon delivery accuracy over non-link weaponry in various profiles from airborne platforms such as the US Air Force F-4 and F-111 fighter-bombers and B-52 strategic bomber. As well as being operationally tested with these, the system is also said to be compatible with the F-15, F-16, F/A-18 and the A-7 fighter-bombers. Potential weapon applications include Harpoon, Maverick, MRASM (medium-range air-to-surface missile) and cruise missiles.

STATUS: in production and service.

Marconi

Marconi Avionics Limited, Airport Works, Rochester, Kent ME1 2XX

Stores management system for Tornado

The weapons management system developed by Marconi Avionics for the tri-national Tornado multi-role bomber is capable of handling at least 26 different types of store with control of weapon arming or fuzing and firing or release functions.

The system has a total of 27 outputs and undertakes the timed and sequenced release of these stores, and external fuel tanks or equipment pods which may have to be jettisoned in the event of emergency.

Among the principal design aims of the Tornado stores management system were a significant reduction in crew workload (a common desirable feature of stores management philosophy; the reduction of inter-connecting cable between line-replaceable units to the minimum (there are ten line-replaceable units in the total system); high reliability in terms of operation and safety rates. The specification also called for a degree of electro-magnetic compatibility 'hardness' (resistance to electrical interference) never previously considered for a weapon system.

To achieve these targets, Marconi has developed a totally dual-channel, digital architecture employing mini-computers and digital data transmission of extremely high integrity. Wherever possible, cmos circuitry has been used for low power consumption and resistance to interference. Both channels are fully synchronised and perform in a consolidated fashion except in the case of a channel unserviceability, when the serviceable channel reverts to an independent mode.

The level of redundancy in the system is illustrated by the fact that the jettison control system is not only quadruplexed but can operate autonomously independent of the channel integration hardware. This is further underlined by the fact that each crew member has his own system control panel, thereby increasing the degree of flexibility and commensurate with a low crew workload.

STATUS: in production.

Plessey

Plessey Electronics Systems, Vicarage Lane, Ilford, Essex IG1 4AQ

Weapon control system for Harrier

Until the mid-1960s weapon arming and release systems mainly comprised manually-actuated switches controlling relay networks which completed the stores arming and release/firing sequence. The advent of new combat aircraft capable of accommodating a huge variety of stores mounted on a variety of underwing and/or fuselage stations, raised a requirement for stores management methods by which to meet the control problems caused by asymmetric release of weapons, drop-tanks, or other disposable items. These difficulties were compounded by the diminishing size of airframes while the stores themselves remained largely unchanged, thus increasing their influence.

In the United Kingdom initial studies into the problem were undertaken by the Royal Aircraft Establishment, Farnborough which later co-opted Plessey Avionics and Communications to carry out development of an electronic weapon control system. The eventual outcome of the programme was the system for the RAF/Hawker Siddeley Harrier V/STOL fighter, for which contracts probably in excess of 200 units were placed. The system was also to have been supplied to the US Marine Corps for the AV-8A version of the Harrier.

With the Harrier's weapon load distributed between wing and fuselage pylons, it is of importance that lateral balance is maintained as weapons are released, particularly during low-level strike attacks. The Plessey weapon control system computes and executes a release sequence which maintains aircraft balance under all conditions.

When stores are loaded onto the aircraft, thumbwheel switches are manually set on the weapon control system's control panel to provide the pilot with an indication of the external stores status. During the loading operation the groundcrew also move the role-selection switches on the panel to an appropriate setting for the type of stores carried. Two modes of operation, manual or automatic, are possible. In manual mode, the pilot during run-up to the target chooses, via pylon selector switches, which weapons are to be released. If two stores are carried on one pylon, then use of a single/double selector determines whether one or both stores will be dropped. Selection of fuzing is the next operation and then, if the selection of pylon, store and fuzing is valid for the weapon state and maintenance of lateral balance, a store indicator light will illuminate to confirm that the selection is correct. Actual release is made by depression of a late-arming switch situated remotely from the weapon control system panel as a safeguard against premature release. The weapon control system prevents store release if the store indicator lamp remains unlit or if the stability of the aircraft will be jeopardised. Release intervals of 20-, 40-, 80-, 160-, and 320- metres a second are

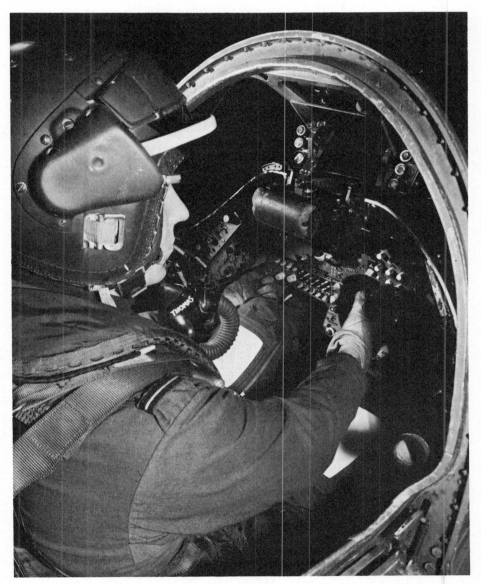

Plessey weapon control system installed in BAe Harrier V/STOL fighter

selectable from a rotary switch on the weapon control system panel.

For automatic release, the auto/manual switch interfaces the weapon control system with the aircraft's nav/attack computer providing it with data on stores status, types selected and spacing required. Release interval then comes under computer control. In either mode, the pilot may use both individual or 'clear aircraft' overriding jettison control switches to jettison stores separately or as a single salvo in an emergency.

During rocket or guided weapon release or during gun firing, the weapon control system ensures that power is supplied to the engine igniter circuits to safeguard against flame-out caused by aerodynamic flow disturbance or gas ingestion.

Built-in test circuits for functional and power supply checking are also controlled from the weapon control system panel. All switches can be operated with gloved hands and vital controls are guarded against inadvertent operation.

The system is digital-based and makes extensive use of silicon integrated circuitry. Logic circuits are of diode-transistor type and the number of relays has been reduced to the minimum required to handle the heavy switching-currents which occur at the time of weapon release.

In two-seat Harriers a monitor unit is fitted in

the rear cockpit. It duplicates the front cockpit jettison facilities and enables the control panel to be monitored during operation.

The system's capacity permits mixed loads to be carried and controlled on each pylon. The range of stores includes single or twin bombs, flares, one or two rocket launchers, two gun pods, external fuel tanks or guided weapons. Bomb fuzing signals can be transmitted to the five pylons.

Dimensions: (wcs unit) $4\frac{1}{2} \times 8\frac{1}{4} \times 12\frac{5}{8}$ inches (114 × 210 × 321 mm)
(monitor) $4\frac{1}{2} \times 8\frac{1}{4} \times 6\frac{3}{10}$ inches (114 × 210 × 160 mm)
Weights: (wcs unit) 10·25 lb (4·65 kg)
(monitor) 5·25 lb (2·38 kg)

STATUS: in production and service.

Weapon control system for Jaguar

The Plessey weapon control system for Jaguar is a development of that provided for Harrier (see previous entry) and employs the same digital, solid-state electronics technology. In the single-seat Jaguar the system is divided into two major elements, the main unit housing the control logic and the cockpit-mounted control unit which contains the indicator lamps, the pilot-operated armament switches and the individual jettison switches. In the two-seat variant, a monitor unit is fitted in the rear cockpit for training purposes.

As well as providing all of the functions of the Harrier system, the Jaguar unit has greater reliability as the result of wider use of solid-state logic. It provides fully duplicated release and jettison of individual stores, dispenses with the need for autoselector switches on tandem store carriers, and provides a more advanced method of rocket firing control. It may be operated as an autonomous system or used in conjunction with the Jaguar nav/attack computer. Orders for Jaguar weapon control systems probably exceed 250 units.

Dimensions: (control unit) $6 \times 5\frac{3}{4} \times 6\frac{1}{6}$ inches (152 × 146 × 157 mm)
(logic unit) $6\frac{1}{8} \times 13\frac{3}{4} \times 16\frac{3}{4}$ inches (155 × 349 × 425 mm)
(monitor) $3\frac{3}{8} \times 5\frac{3}{4} \times 5\frac{1}{8}$ inches (85 × 146 × 130 mm)
Weights: (control unit) 3·875 lb (1·76 kg)
(logic unit) 32 lb (14·54 kg)
(monitor) 1·25 lb (0·56 kg)

STATUS: in production and service.

UNITED STATES OF AMERICA

Hughes Aircraft Company

Radar Systems Group, El Segundo, California 90245

AN/AQX-14 weapon control data link

The Hughes AN/AQX-14 is a two-way communication data-link to control the GBU-15 guided bomb. It provides a video and command link between the command aircraft and the weapon, enabling the systems operator to remain in the control loop while the weapon is being directed to its target. In effect, the data-link permits a command authority similar to a 'fly-by-wire' situation, whereby the operator can transmit guidance instruction all the way from launch to impact. Alternatively, he may select any one of a number of autonomous weapon-control modes, including an override mode which permits target up-dating or re-designation as required.

The extended weapon-control capability conferred by the data-link contributes to weapon-system performance in terms of stand-off range and operational utility. Stand-off range may be increased by launching the weapon outside the range that it can recognise a target or lock onto it. Target acquisition is deferred until the weapon, rather than the command aircraft, is closer to the target. Tactically, the aircraft can leave the immediate target zone immediately after launch.

The AN/AQX-14 system comprises three major elements: a data-link pod mounted on the command aircraft, a data-link control panel, used in conjunction with an existing display within the aircraft, and a weapon data-link module, mounted on the rear of the weapon itself.

The pod is an aerodynamically-shaped container mounted on a standard stores-carriage strong-point on the fuselage centreline or on an underwing station, according to aircraft type. It contains four line-replaceable units comprising an electronics section incorporating all radio frequency generating and receiving equipment, a de-multiplexer to decode all aircraft command and pod control signals, an encoder, and antenna controls; a phase-scanned array for weapon tracking in normal operation; a forward horn antenna which provides additional coverage; and a mission tape-recorder which maintains a permanent record of weapon video data. The pod is suitable for high performance aircraft, is certificated for operation at speeds in excess of Mach 1 and is also compatible with high/low-altitude operations. There is said to be no compromise of aircraft performance attributable to carriage of the pod.

Used in conjunction with an existing display system, the aircraft control-panel acts as the interface between the weapon system operator and the weapon guidance system. The panel accepts signal inputs from the aircraft as well as from its own controls, and formats these into discrete commands as required by the data link. Although the panels are tailored to the individual requirements of the aircraft type and intended customer usage, each unit accepts the standard configurations of the GBU-15 data-link and the pod.

Attached to the after end of the GBU-15 weapon is the ultimate component in the data link chain, the weapon data-link module. This simultaneously transmits video from the weapon's seeker-head and processes incoming command signals from the aircraft to the weapon. Heading changes during the wea-pon's flight are effected through discrete command signals. Dual analogue voltage channels enable the operator to slew the weapon in pitch and yaw during approach to the target.

Digital techniques are employed in the AN/AQX-15 system and the transmitter is of all-solid-state construction. The system's electronically phase-scanned antenna array provides the data-link with high rate tactical manoeuvring capability. A comprehensive range of test equipment is provided, including a flight checkout unit for testing aircraft cables from the pod connection point, an aircraft simulator unit which permits functional checks of the control panel, and a weapon simulator unit for test of the aircraft-mounted pod and isolating faults down to line-replaceable unit level. Used together, these two latter units permit full system functional checkout.

Two primary launch modes are envisaged for operation of the GBU-15 weapon/AN/AQX-14 control combination: low-altitude penetration and high-altitude stand-off. Hughes claims that use of the data-link has demonstrated improvement in weapon delivery accuracy over non-link weaponry in various profiles from airborne platforms such as the US Air Force F-4 and F-111 fighter-bombers and B-52 strategic bomber. As well as being operationally tested with these, the system is also said to be compatible with the F-15, F-16, F/A-18 and the A-7 fighter-bombers. Potential weapon applications include Harpoon, Maverick, MRASM (medium-range air-to-surface missile) and cruise missiles.

STATUS: in production and service.

Flight Instruments and Equipment

Air Data Systems and Instruments

FRANCE

Jaeger

Jaeger, Aircraft Division, 2 rue Baudin, 92303 Levallois Perret

Electric vertical speed indicator

Designed for the new commercial transports using ARINC 700 characteristics, this vertical speed indicator is one of Jaeger's fully digital instruments based on micro-processors and CMOS technology. It can operate in either inertial or barometric mode and in the case of failure of the heading information, switches automatically to the digital air data computer. The barometric mode may then be selected by an external control and the system operated from two separate ARINC 429 data-buses.

Vertical speed is displayed by a white pointer moving over a non-linear circular dial, and in case of failures a large OFF flag appears. When operating in barometric mode a flag marked IVS comes into view.

All processing circuits are fully monitored and the modular design of the electronic boards makes both access and maintenance easy. The modules are all interchangeable without need for adjustment and 70 per cent of the parts are common to those of the Jaeger altimeter and rmi/adf/vor.

Dimensions: 3 ATI
Power supply: 115 V 400 Hz

STATUS: in production.

Type 65205 digital altimeter

This electronic altimeter is one of Jaeger's new line of fully digital flight-deck instruments intended for new commercial aircraft using ARINC 700 characteristics. The design reflects current technology, with micro-processors and large-scale integrated circuits for maximum reliability and ease of maintenance.

The altimeter can operate from two separate ARINC 429 data-buses and provides electrical barometric setting information by means of one or two ARINC 706 resolver outputs as well as discrete validity and baro-setting mode output signals. The instrument has a push-pull setting knob for rapid switching to the standard baro-setting level.

Altitude is indicated by a pointer making one revolution every 1000 feet (300 metres) and a four-drum mechanical counter. An internal altitude reference 'bug' is manually controlled by an external knob while a bezel mounted amber light gives an ALERT indication.

The altimeter is fully modular in construction, all modules being directly replaceable and requiring no adjustment on exchange. An optional built-in test program can reduce trouble-shooting time and the need for external equipment.

Jaeger instantaneous vertical speed indicator

Dimensions: 3 ATI, ARINC 408 – 215 mm
Power supply: 15 W 114 V at 400 Hz
Baro-setting range: 745-1050 mb 22-31 Hg
Accuracy: ± 5 ft

STATUS: in production.

Sfena

Sfena, (Société Française d'Equipements pour la Navigation Aérienne) Aérodrome de Villa-coublay, BP 59, 78140 Velizy Villacoublay, Cedex

VIP vertical speed and thrust indicator

The Sfena 'Variometre à indication de poussée' (VIP) is an instrument which uses the total-energy concept to provide the pilot with an indication of aircraft instantaneous energy balance by simultaneously displaying vertical and total speed.

This makes it possible to compare available thrust and drag at any moment, thereby increasing safety during take-off and in the event of an engine failure. It also facilitates engine control in all phases of flight, particularly during flight below minimum drag speed. The compact 2-inch diameter instrument is economical on space and may be fitted in place of a conventional vertical speed indicator.

Dimensions: $2^2/5 \times 2^2/5 \times 6^1/10$ inches
Weight: 2 lb
Power: 26 V 400 Hz

STATUS: in production.

Sfena VIP Vertical speed and thrust indicator

UNITED KINGDOM

Marconi Avionics

Marconi Avionics Limited, Airport Works, Rochester, Kent ME1 2XX

Digital air data computers

Marconi Avionics produces a series of digital air data computers of modular design based on micro-processor technology. Plug-in modules permit any output interface combination for compatibility with other equipment, while the chassis design simplifies servicing and repairs. Maintainability is aided by built-in test and the

absence of lifed components. Consequently rectification generally consists of diagnosis and module replacement without need of adjustments.

ADC 81-01-08/32 analogue air data computer

This twin-capsule analogue air data computer meets full UK military standards and is currently supplied for all versions of the Sepecat Jaguar Anglo-French light bomber.

Dimensions: 1/2 ATR
Weight: 18 lb
Power: 50 W

STATUS: in production.

DADC 50-048-01 digital air data computer

This digital air data computer has been specially configured to fit the PRC T7 fighter of the Peoples' Republic of China. It interfaces with a head-up display/weapon-aiming system and uses vibrating cylinder transducers, a 16-bit micro-processor and has both digital and analogue outputs of air data parameters. Both manually initiated and automatic built-in test equipment are incorporated.

Apart from the motherboard there are only four modules. The chassis is of a non-standard direct-mounting design.

Dimensions: 8½ × 7 × 7½ inches
Weight: 8 lb
Power: 35 W

STATUS: in development.

DADC 50-050-01 digital air data computer

This digital air data computer was introduced as a unit to interface with a head-up display/ weapon aiming system for the IAI Dagger which is an Israeli-built version of the Mirage III for Argentina and is designed to full US military standards. The computer employs twin force-balance transducers, a 16-bit micro-processor and provides both digital and analogue outputs of air data parameters. Both manually initiated and automatic built-in test equipment are provided. In this application the chassis is of a non-standard direct mounting design.

Dimension: 8½ × 4 × 7½ inches
Weight: 8 lb
Power: 35 W

STATUS: in development.

SADC 50-053-01 air data computer

This is a duplex secondary air data computer configured with twin sensing, computing and output channels. The forward section contains four transducers (two pitot, two static), pneumatic connectors, manual test buttons and indicators. A time totaliser is also included. The upper section contains the duplicated electronic modules while the lower section houses the twin power supplies and output sockets.

Dimensions: ¼ ATR
Weight: 9·5 lb
Power: 35 W

STATUS: in development.

SADC 50-027-08 air data computer

The 50-027-08 triplex transducer unit is a secondary air data computer which has been specially designed for all versions of the Panavia Tornado. It is a three-channel device which uses six transducers (three pitot, three static) to produce both analogue and discrete outputs for the aircraft systems. All computing is analogue, outputs are both triplex and duplex and the system incorporates automatic built-in test equipment. It meets full US military specifications.

Dimensions: ⅜ ATR
Weight: 14 lb
Power: 50 W

STATUS: in production.

IS 03-004 digital air data computer

This is a specially developed helicopter air data system for the updated US Army/Bell AH-1S Cobra light attack helicopter. It uses a swivelling pitot-static probe enabling speed to be measured in all axes, from zero to maximum in any direction. The sensor information is processed by a 16-bit micro-computer and then transmitted to an analogue-presentation airspeed indicator and other systems in serial digital format. Manually initiated and automatic built-in test equipment is incorporated.

The outputs of the system are longitudinal ias and tas, lateral ias and tas, vertical tas,

Marconi Avionics digital air data computer for PRC T7 fighter

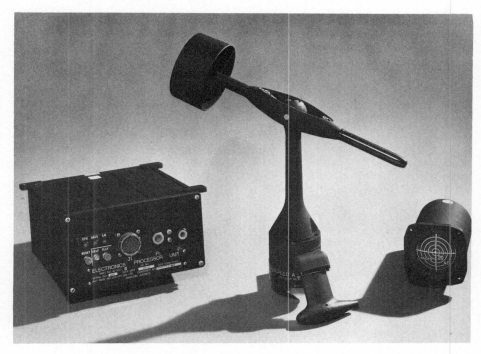

Marconi Avionics digital air data system for Bell AH-1S helicopter

downwash air speed tas, pressure altitude, static air temperature, and system-status discretes.

Air speed and direction sensor
Dimensions: 3⅘ × 12½ × 9⁷⁄₁₀ inches
Weight: 2·4 lb
Electronics processor unit
Dimensions: 4⅕ × 8⅕ × 7½ inches
Weight: 7·4 lb
Low-airspeed indicator
Dimensions: 3¼ × 6 × 3¼ inches
Weight: 11·4 lb
Power: 34 W at 28 V dc and 140 VA at 115 V for de-icing

Standard central air data computers scadc

At the end of 1981 Marconi Avionics was contracted to develop a new range of standard central air data computers to upgrade US Air

Force and Navy aircraft. The intention of the programme is to update to a common, advanced standard the great variety of air data systems of different generations currently installed on variants of ten types of aircraft.

The use of a standard range of standard central air data computer units, having on average more than 80 per cent commonality of sub-assemblies, will lower the life-cycle costs associated with air data equipment. The design also incorporates the latest standards of computer software and data-transmission now being specified for US military aircraft.

An 18-month development programme is envisaged for the design and proving of the four basic configurations, including various mounting trays for adapting them to the 27 aircraft variants. Eventually 56 prototypes will be produced for qualification and flight proving.

STATUS: in development.

UNITED STATES OF AMERICA

Astronautics

Astronautics Corporation of America, PO Box 523, Milwaukee, Wisconsin 53201

Digital air data computer

This air data computer meets full US military standards and has been produced for an undisclosed foreign military customer. It is micro-processor-based using precision solid-state vibrating-quartz pressure-transducers, with analogue potentiometers and synchro outputs as well as dual-redundant serial digital data-buses.

Built-in test equipment allows a high degree of self-diagnosis and the computer has considerable growth potential. It has been designed also as a standard air data computer for retrofit and new-aircraft programmes.

STATUS: Currently in production and operational.

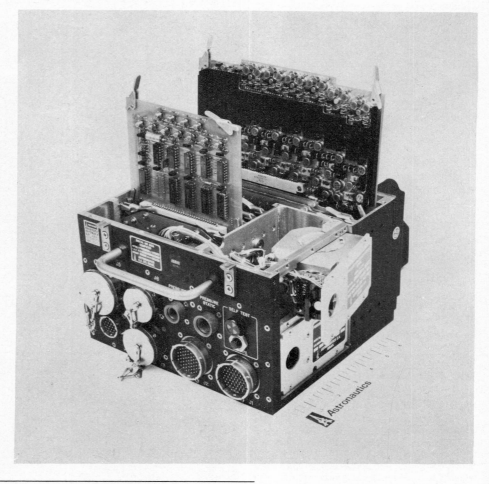

Astronautics compact digital air data computer

Collins

Collins Avionics Divisions, Rockwell International, Cedar Rapids, Iowa 52406

ADS-80 digital air data system

Collins has developed the ADS-80 air data system as part of the Pro Line range of avionics equipment. The system processes data derived from pitot and static pressures and interfaces with the aircraft flight-control and flight-director system.

To adapt the system to a variety of aircraft a single plug-in module is mounted at the rear of the computer. Operating parameters are coded in the module according to aircraft type.

Dimensions: 1/2 ATR
Weight: 6·6 lb
Power: 23 W

STATUS: in production.

ASI-800 omni-directional air speed indicator

The ASI-800 is a new omni-directional air speed indicator developed for helicopters or V/STOL aircraft. It is a 3-inch panel-mounted indicator which displays density altitude, outside air temperature and never-exceed speed (V_{NE}) using a liquid-crystal digital display. An analogue tape/needle display shows lateral and fore-and-aft air speed, the intersection of horizontal and vertical bars representing the air speed vector.

Intended for installation in conjunction with

Rockwell Collins ASI-800 air speed indicator

Pacer Systems Loras (low-range air speed system) indicator displays either omni-directional information from that system or conventional pitot-static air data. The instrument can be used to complement or replace a conventional pneumatic indicator.

STATUS: in development.

Garrett

Garrett AiResearch Manufacturing Company of California, 2525W 190th Street, Torrance, California 90509

AiResearch has developed a series of air data computers for a variety of supersonic and subsonic aircraft, drones and spacecraft. They receive pitot and static pressures and air temperature and provide output information for use in navigation, fire-control, powerplant, cockpit display and other aircraft sub-systems, and flight instruments.

Digital air data computer for Rockwell B-1 bomber

This computer was developed for the US Air Force/Rockwell B-1 supersonic bomber and features a digital mos/lsi processor and digital pressure transducers. It has an altitude reporting facility and the built-in self-test system provides continuous failure monitoring.

Dimensions: $6^1/4 \times 7^2/3 \times 19^2/3$ inches
Weight: 27·7 lb
Inputs: 19
Outputs: 92

STATUS: Developed for the B-1 supersonic bomber programme which was cancelled in July 1977 and reinstated in late 1981.

Digital air data computer for Saab JA37 fighter

This is another all-digital computer with similar features to those of the B-1 air data computer and has bi-directional digital communication with the supplied sub-systems. It contains both static and dynamic self-test facilities.

Dimensions: $6^1/5 \times 8^2/5 \times 17^3/10$ inches
Weight: 25·1 lb
Inputs: 16
Outputs 37

STATUS: in service with the Saab JA37 interceptor.

Air data computer for Grumman F-14 fighter

Although similar in basic design to other members of the AiResearch adc family, the system for the F-14 US Navy/Grumman carrier-based air-superiority fighter has additional functions. These include control of the variable-geometry wing sweep angle, provision of a flight-parameter and status display, and the generation of certain commands for the engine thrust autopilot.

Dimensions: $7^3/4 \times 5^7/10 \times 20$ inches
Weight: 27·5 lb
Inputs: 70
Outputs: 60

STATUS: in production and service.

Air data computer for Fairchild A-10A/B tank-destroyer

Designed to full military environment and MIL-STD 1553 digital data-bus standards, this air data computer features digital pressure transducer and a ferroresonant power supply. Modular programming is provided and the memory has a 6144-word read only memory and a 256-word random access memory.

Dimensions: 1/2 ATR
Weight: 13·9 lb
Power: 36 W
STATUS: in production for US Air Force/Fairchild A-10 close support aircraft.

True air speed computer

This device provides outputs of true air speed for RNav, ins and display applications. It features a vector pressure-ratio transducer which facilitates computation of true air speed using only one device. This servoed force-balance unit maintains a high level of performance under all aircraft operating conditions.

An accuracy of plus or minus 4 knots is obtained over a true air speed range of 95 to 500 knots. A static air temperature output is available as a customer option.

Dimensions: 3/8 ATR
Weight: 9·2 lb
Power: 30 VA at 115 V 400 Hz

STATUS: in production.

Honeywell

Honeywell Aerospace and Defense Group, Honeywell Plaza, Minneapolis, Minnesota 55408

Air data computers

Honeywell supplied the first digital air data computer for commercial air transports with a unit for the McDonnell Douglas DC-10 widebody. The analogue HG180 equipment still remains in production however, and is fitted in the Boeing 707, 727 and 737 transports. Now two improved air data computers have been developed based on the latest electronic systems and techniques.

HG280D80 air data computer

Intended for the DC-10 and DC-9 Super 80 transports, this is a micro-processor-based ARINC 576 type digital system that can replace the earlier HG280D5 computer. The transducers use proven D5 sensors, with high accuracy and a fast warm-up time. Reliability is claimed to be excellent, with a mean time between failures prediction of 11 000 hours.

The central processor memory and input/output functions have been reduced from the 360 integrated circuits on nine boards of the earlier D5 to just 22 circuits on a single card. Built-in test includes both continuous monitoring and manually activated self-test. The D80's non-volatile memory automatically stores information for five flights, and may be retrieved by selecting a single switch on the front panel. If a failure has occurred either of two warning lights will indicate the problem.

The D80 can also accommodate limited power interruptions. A capacitively maintained supply voltage and a cmos memory can retain critical data during power losses of several milliseconds. When power resumes the device automatically restores all outputs based on this stored data.

HG280D5
Dimensions: 1/2 ATR
Weight: 18·2 lb
Power: 101 W
HG280D80
Dimensions: 1/2 ATR
Weight: 13 lb
Power: 30 W

STATUS: both types are in production and in service.

HG480B air data computer

This digital computer is now replacing the analogue HG180 on new-production Boeing 727 and 737 transports. Heart of the system is a solid-state pressure transducer containing a diaphragm made from single-crystal silicon with diffused piezoresistive elements.

A single-board micro-computer controls the functions, and built-in test (which includes both continuous monitoring and manually activated self-tests) reduces maintenance tasks. Ground built-in-test has been expanded and includes q-pot stimulation and transmission of fixed values on all signal output lines.

The HG480 is available in four versions. The basic configuration allows for differences in aircraft wiring but is functionally identical for what are designated the B1, B2 and B3 versions. Designed to operate with conventional analogue equipment, the system also provides outputs for digital instruments and digital autopilot. The B4 version is designed primarily to interface with an all-digital flight control system, but can still operate with synchro-driven instruments.

Dimensions: 1/2 ATR
Weight: 16·8 lb
Power: 44 W

STATUS: in production and in service.

HG580 air data computer

This advanced device was originally designed as a competitive bid for the Boeing 757 new-generation transport. Although unsuccessful in this application, development is continuing for future aircraft use.

STATUS: in development.

Kollsman

Kollsman Instrument Company, Daniel Webster Highway South Merrimack, New Hampshire 03054

Digital air data computer

This is a modular device easily adaptable to other standards or requirements. A vibrating-quartz pressure-transducer, temperature-compensated by a platinum resistance element measures static and total pressures with external sensors providing basic data for air temperature and angle of attack computation.

These analogue inputs are processed by a 16-bit micro-processor at a rate of between 2 and 16 times a second according to priority. The resulting data is then transmitted in digital form to user systems such as flight instruments, autothrottle or autopilot.

The Kollsman air data computer meets the requirements of ARINC 706-2 and is suitable for all types of aircraft including helicopters.

Dimensions: 1/2 ATR
Weight: 11·2 lb
Power: 50 W

STATUS: in production.

Pacer

Pacer Systems Inc, 1755 South Jefferson Davis Highway, Arlington, Virginia 22202

Loras low-range air data converter

Loras is an omni-directional air speed system developed for helicopter applications. It comprises a mast-mounted sensor connected via electrical cable to an air data computer which provides the data-processing.

The air-data computer outputs air speed along the flight path from zero to above 200 knots regardless of direction. In addition it provides the components of air speed along the forward, rearward and sideways directions. Density altitude, temperature, pressure and other interface signals can be provided as customer options.

The original Loras equipment first flew in the 1960s in the Bell X-22 ducted-propeller tilt-engine V/STOL research aircraft. Development of the current equipment began in 1970 and evolved as the Loras Model 1000 in 1974 and now a more sophisticated version, the Model 2000, is being introduced on the US Army/Hughes AH-64A Apache anti-tank helicopter.

Model 1000
Dimensions: $5^1/_4 \times 4^3/_5 \times 6^1/_{10}$ inches
Weight: 2·88 lb

STATUS: Model 1000 in service and Model 2000 in development for Hughes AH-64 helicopter.

Sperry

Sperry Flight Systems, Avionics Division, PO Box 29000, Phoenix, Arizona 85038

ADZ air data system

Sperry produces a wide range of aircraft flight instruments and supplies complete packages of air data equipment for aircraft including the air data computer itself and a full set of associated instruments.

The ADZ air data system uses an AZ 241 or 242 computer depending on whether the aircraft is a turboprop or jet. Both systems use the Sperry patented vibrating diaphragm pressure sensor to provide outputs for altitude, air speed, vertical speed, true air speed, true air temperature and total air temperature information.

If required, a single computer can handle a dual flight director installation, consequently, no matter which director is driving the autopilot the full complement of modes are available.

AZ-241 (turboprop)
Dimensions: $7^3/_5 \times 3^1/_{10} \times 14^1/_5$ inches
Weight: 8·6 lb
AZ-242 (jet)
Dimensions: $7^3/_5 \times 4^9/_{10} \times 14^1/_5$ inches
Weight: 11·5 lb

STATUS: in production for a wide range of executive turboprop and jet aircraft, one of the latest being the Canadair Challenger.

Attitude Director Indicators

FRANCE

Sfena

Sfena (Société Française d'Equipements pour la Navigation Aérienne), BP 59, Aérodrome de Villacoublay, 78140 Velizy-Villacoublay, Cedex

Attitude director indicator

Sfena produces a wide range of attitude director indicators for both civil and military use. Current military programmes include Mirage 2000, ANG (Atlantic Nouvelle Génération) Transall, Super Etendard, Nord 262, and Aérospatiale SA 241 and 342 helicopters. Civil standard ADIs are fitted in Air France Boeing 707s, Concorde, Airbus Industries A300/A310, and Fokker F27.

The current range provides a wide choice of instruments, and features flight-director command bars, failure-flag indicators and warning lights for decision-height alerting. Some instruments include a basic yaw indicator for additional assistance in asymmetric flight.

Sfim

Sfim, 13 avenue M. Ramolfo-Garnier, 91301 Massy

All-attitude flight control indicators

Sfim has developed a range of spherical indicators on which all-attitude information is displayed to give the pilot heading, roll and pitch information on a single dial without freedom limits around the three axes.

Some versions are fitted with command bars to display signal information from navigation aids or landing systems. Four types of instruments are currently available and equip Mirage F1 and 2000 fighters, Jaguar and Super Etendard light bombers, French Navy Lynx helicopters, and Galeb, Jurom and Pucara light fighters and strike aircraft.

A special helicopter version is designated 11-2 and features a manual pitch setting mode of plus or minus 10 degrees. The various features of the series are labelled in the following table:

TYPE	810	811	816
Roll	Yes	Yes	Yes
Pitch	Yes	Yes	Yes
Heading	Yes	Yes	Yes
Ils/vor/Tacan	Yes	No	Option
To-from	Yes	No	*
Beacon	Yes	No	Option
Side-slip	Yes	Yes	Yes
Failure detection	Yes	Yes	Yes
Warning flag	Yes	Yes	Yes

* Option —
separate unit.

Type 810 and 811
Dimensions: $3^4/5 \times 3^4/5 \times 8$ inches
Power: 26 or 115 V 400 Hz plus 28 V dc
Type 816
Dimensions: $3^1/5 \times 3^1/5 \times 8^1/5$ inches
Power: 26 or 115 V 400 Hz plus 28 V dc

STATUS: in production.

UNITED KINGDOM

Ferranti

Ferranti Instrumentation Limited, Aircraft Equipment, Lily Hill House, Lily Hill Road, Bracknell, Berks RG12 2SJ

FH30/32 series attitudes indicators

Ferranti Instruments manufactures a range of artificial horizons for use either as main attitude indicators or as standby instruments. In the latter case some of the instruments are driven directly from the aircraft's direct current supply by means of an integral static inverter. Other types feature pitch and roll signal pick-offs.

The FH30/32 series is becoming a standard fit on aircraft of the Royal Air Force. They are fitted in the Jet Provost, Hawk and Tornado F2. In 1978 an order was received to retrofit the RAF C-130 Hercules fleet and they were specified for the Boeing Chinook HC 1 assault helicopter.

The instrument has sold abroad to Finland and Australia in the Valmet L-70 and AESL CT4 trainers.

The smaller FH20D instrument is fitted in RAF/McDonnell Douglas Phantoms and all Tornado GR1 strike aircraft.

STATUS: in production.

Newmark

Louis Newmark Limited, Aircraft and Instrument Division, 80 Gloucester Road, Croydon, Surrey CR9 2LD

Attitude indicators

Newmark has developed 3-, 4-, and 5-inch versions of an attitude indicator of which the 5-inch version, the Type 5755, has been built in the greatest quantity. It has been fitted in Westland Sea King Mk 2s of the Royal Navy and a derivative, the 9222, which features cross command bars, has been adopted for the Sea King Mk V.

The attitude sphere is free to rotate through 360 degrees in pitch and roll and is coloured so that the upper hemisphere is grey to represent the sky and the lower hemisphere is black. A power-failure flag is fitted and a slip indicator is mounted at the lower edge of the instrument. A pitch and roll trim knob is fitted to facilitate trimming of the sphere. An elapsed-time indicator is fitted at the rear of the instrument enabling an accurate check to be kept on the unit's total running time.

Dimensions: $5^1/4 \times 5 \times 6^3/4$ inches
Weight: 4·75 lb
Power: 115 V 400 Hz

Newmark Type 5755 attitude indicator

UNITED STATES OF AMERICA

Sperry

Sperry Flight Systems, Avionics Division, PO Box 29000, Phoenix, Arizona 85038

Attitude director indicators

Sperry has been involved in flight director displays since the very first Sperry Zero Reader cross-pointer command instrument. Now the company offers a family of instruments from which customers may choose installations spanning the range from light aircraft to commercial transports. Several types of attitude director indicators allow 'customising' or even variation of presentation from pilot to co-pilot instrument layout.

HZ-6F attitude director indicator

This instrument provides attitude reference data, flight director, and radio displacement data for control of the aircraft in pitch and roll during all phases of flight. Computed attitude commands are displayed on the horizontal and vertical flight director bars and are slaved to the flight director controller. When the controller is switched off the command bars are biased from view.

All the other displays provide constant cues unless built-in fault circuits detect a malfunction. In most failure cases pointers are biased from view and an appropriate warning flag appears.

The HZ-6F is electronically self-contained with integral power supply, amplifiers, servos and monitors, and a self-test feature ensures pre-flight display integrity. Loss of either power or an attitude signal will cause the sphere to move to an obviously erroneous position.

Dimensions: 5 × 5¼ × 8½ inches
Weight: 9·5 lb
Power: 11·1 W at 26 V 400 Hz

STATUS: in production.

AD-300B attitude director indicator

Although broadly similar to the HZ-6F, the AD-300B does not display radio altitude.

Dimensions: 5 ATI
Weight: 11 lb
Power: 15 W at 26 V 400 Hz

STATUS: in production.

AD-300C attitude director indicator

While still based on the earlier models the AD-300C displays attitude, flight director commands, instrument landing system deviation and speed commands on a 5-inch indicator.

Dimensions: 5 × 5 × 9²/₅ inches
Weight: 11 lb
Power: 15 W at 26 V 400 Hz

AD-350 attitude director indicator

The AD-350 offers a display of attitude, flight director commands, instrument landing system deviation, speed command, rate of turn and radio altitude in two versions. Differing slightly in symbology, the AD-350-903 has two decision-height warning lights while the AD-350-904 has just one. The instruments are otherwise identical.

AD-350-903/904
Dimensions: 5 × 5 × 9 inches
Weight: 10·5 lb
Power: 11 W at 26 V 400 Hz

STATUS: in production.

AD-650-600 attitude director indicator

This family of attitude director indicators represents a completely new design incorporating micro-computer and micro-servo technology to replace earlier conventional mechanisms, torque motors and gears. They are lighter, run cooler and are more reliable. Both single-cue or cross-pointer displays are available in these 5-inch instruments. The AD-650 features digital readouts of decision height and radio altitude, these facilities being absent on the AD-600.

The AD-600 series also provides a symbolic rising runway combined with an expanded-scale localiser which depicts the aircraft's position relative to the runway during the last 200 feet of descent. The glideslope pointer may be on the right, as in the AD-650A or on the left as in the AD-650B. The AD-650H is a version optimised for helicopter operations.

Dimensions: 5 × 5 × 8¾ inches
Weight: 7 lb

STATUS: in production.

AD-550/500 attitude director indicator

The 500 series instruments are 4-inch versions of the 600 series. They have all the features and technology of the larger indicators and are compatible with Sperry flight director systems and radio altimeters.

Dimensions: 4 × 4 × 9 inches
Weight: 4·7 lb

STATUS: in production.

AD-800 attitude director indicator

This attitude director indicator is an advanced technology, digitally interfaced unit offering the full range of flight data and cues. Flight director, speed command, glideslope, expanded localiser and radio altitude are presented on high-torque meter movements which are now software monitored for improved reliability.

A state-of-the-art micro-computer handles all data and can update the display at least 30 times a second to ensure smooth operation of all servo- and meter- driven elements.

Dimensions: 5 ATI
Weight: 10·2 lb
Power: 17·4 W at 115 V 400 Hz or 26 V 400 Hz

STATUS: in production.

GH-14 attitude/attitude director indicator

The GH-14 can offer a range of presentations from basic attitude indicator to full flight director indicator. It features the well established Sperry spherical attitude presentation with full 360 degrees of freedom in roll and plus or minus 80 degrees in pitch. The pitch display is expanded to provide 1.4 degrees of sphere movement for every degree of aircraft pitch

movement.

The basic GH-14-100 at one end of the range is a simple 4-inch gyro horizon which can be compared with the GH-14-50 at the other end. The latter has rising runway presentation, expanded localiser, pitch-attitude trim, cross-pointer flight-director bars, flight-director flag and pitch and roll outputs. Each configuration is available for either zero or 8 degrees panel tilt.

Internal failure-monitoring is standard and when in operation any failure is indicated by means of a gyro-warning flag.

It is suitable for a wide range of aircraft, either as a primary instrument or standby and is also suitable for helicopters.

Dimensions: 4 × 4 × 9 inches
Weight: 4·5 lb.
Power: 26 V 400 Hz

STATUS: in production.

HZ-444 attitude director indicator

The HZ-444 attitude director indicator combines pitch and roll attitude sphere display, cross-pointer flight-director bars, expanded glideslope deviation indication, expanded localiser indication and radio altitude runway bar. Failures are indicated by red warning flags.

The indicator features low-power dc torque servos and self-contained electronics, and is clamp-mounted for ease of removal and installation. The sphere has full freedom in roll and plus or minus 85 degrees in pitch. An inclinometer, test switch and pitch-attitude knob are fitted to the lower area of the bezel. The instrument features in the Sperry STARS flight-director system and has become well established in general aviation ranging from light twins to the larger corporate jet.

Dimensions: 4 × 4 × 7½ inches
Weight: 4·8 lb
Power: 115 V 400 Hz

STATUS: in production.

HZ-454 attitude director indicator

This is also a part of the Sperry STARS flight director system and is directly interchangeable with the HZ-444. It functions in exactly the same way except that a single-cue command bar provides flight direction information, and to maintain a desired course the pilot flies to centre the aircraft symbol in the 'inset' cue bar. A pitch synchronisation button is fitted that aligns the flight director command to the

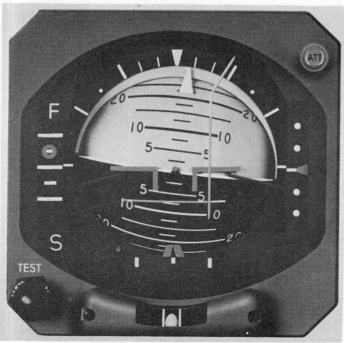

Sperry AD-300B attitude director indicator

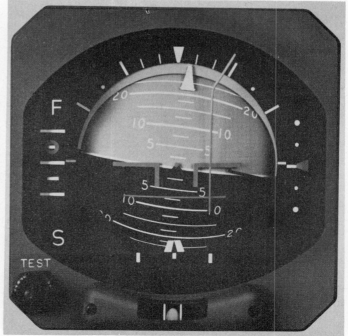

Sperry AD-300C attitude director indicator

existing aircraft attitude. This enables the pilot to maintain a constant pitch attitude during climb or descent. Pitch attitude may be selected either by a bezel-mounted push-button or by a yoke-mounted switch.

Dimensions: 4 × 4 × 8 inches
Weight: 6·2 lb
Power: 115 V 400 Hz

STATUS: in production.

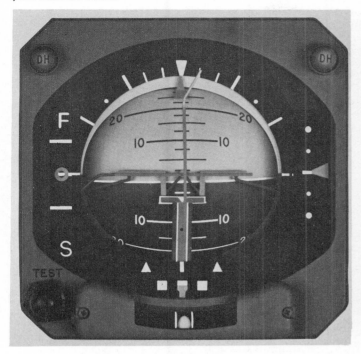

Sperry AD-350 attitude director indicator

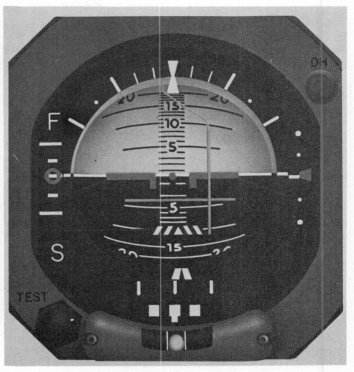

Sperry HZ-6F attitude director indicator

Horizontal Situation Indicators

JAPAN

Tokyo Aircraft Instrument Limited

Tokyo Aircraft Instrument Limited, 35-1, Izumi-Honcho 1-chome, Komae-shi, Tokyo 201.

IDS-8 horizontal situation indicator

This is a conventional course indicator with heading and course setting knobs mounted at the bottom of the bezel. Course and 'miles' indication are provided by a veeder counter and signals are provided for autopilot guidance.

Instrument failure is indicated by red flag monitors. The instrument is currently fitted on the P-2J and PS-1 anti-submarine and maritime surveillance aircraft.

STATUS: in production.

UNITED STATES OF AMERICA

Astronautics

Astronautics Corporation of America, PO Box 523, Milwaukee, Wisconsin 53201

ACA 126370 horizontal situation indicator for Lamps helicopter

This is a 5-inch instrument featuring standard ARINC 407 inputs and outputs with digital interface and extensive built-in test equipment. It was designed for the US Navy/Sikorsky Seahawk SH-60B Lamps (light airborne multipurpose system) programme and is currently in production and operational. A mean time before failure figure of 3000 hours is claimed.

STATUS: in production.

ACA 113515 horizontal situation indicator for helicopters

This has proved to be a popular instrument for helicopters and has course-bar indicator, to/from flag, glideslope pointer, two bearing pointers and course set knob with associated course selection window.

STATUS: in production and operational on Bell 212, 412, 214, Sikorsky S-76, S-61, Aérospatiale Super Puma and Agusta AB212, AB412 helicopters.

Four-inch horizontal situation indicator

British Aerospace is the customer for this instrument, which is fitted to export versions of the BAe Hawk light strike/trainer. Major features include a course-bar indicator, to/from indicator, glideslope pointer, two bearing pointers, course-set knob, course-selection window, and digital readout of range which is compatible with ARINC 568 digital input. The instrument meets full MIL-Spec standards

STATUS: in production and operational.

Astronautics 4-inch horizontal situation indicator for BAe Hawk

Astronautics ACA 126370 horizontal situation indicator

Astronautics ACA 113515 horizontal situation indicator

Astronautics ACA 123790 horizontal situation indicator

ACA 123790 horizontal situation indicator

This is a 4-inch horizontal situation indicator to military specifications and fitted in the Bell AH-1S Cobra light attack helicopter. It has similar features to other members of this company's family of instruments, including a range readout compatible with ARINC 582.

STATUS: in production and operational.

ACA 130500 horizontal situation indicator

Similar in presentation to other company horizontal situation indicators, this 3-inch diameter instrument has been supplied to McDonnell Douglas for the US Marine Corps AV-8B V/STOL light attack aircraft.

STATUS: in production and operational.

ACA 126460 horizontal situation indicator

Another 3-inch horizontal situation indicator differing only in minor detail from the ACA 130500 above, this instrument has been supplied for the Northrop F-5 and General Dynamics F-16 fighters.

STATUS: in production and operational.

Astronautics ACA 126460 horizontal situation indicator

Sperry

Sperry Flight Systems, PO Box 21111, Phoenix, Arizona 85036

Horizontal situation indicators

Sperry Flight Systems produces a wide range of horizontal situation indicators to suit all types of operation and aircraft. These instruments may be part of integrated flight systems where they are compatible with and complementary to other instruments.

RD-350J horizontal situation indicator

This horizontal situation indicator provides heading, two dme distances, radio navigation information via a displacement bar, indications of selected course and heading, and to/from indications for vor operations. Flags give indication of failures.

A Nav Mode annunciator takes the form of a rotary display in the centre of the compass card, controlled by vor/loc valid, loc tuned, to/from and back course selected signals. The annunciator displays a symbol to indicate selected modes and can also indicate when radio data is invalid.

Dimensions: 5 × 5 × 8⁹/₁₀ inches
Weight: 9·2 lb
Power: 6·4 W at 26 V 400 Hz

STATUS: in production.

Sperry RD-350J horizontal situation indicator

RD-444/445 horizontal situation indicator

This instrument provides a display of navigational information on a single 4-inch ARINC-size instrument. Dual heading-error synchros, dual course-error synchros and a course resolver are provided as well as a control datum synchro for comparator purposes. All electronics are self-contained.

Dimensions: 4 × 4 × 7 inches
Weight: 4·5 lb
Power: 115 V 400 Hz

STATUS: in production

RD-550 horizontal situation indicator

This 4-inch indicator has digital incandescent vor and dme displays. The glideslope pointer and course and distance displays can be on either the left- or right-hand side of the instrument face, depending on the model selected. The rmi pointer is coloured for improved clarity and the instrument can be supplied with (RD-550B) or without (RD-550) integral heading selector.

Dimensions: 4 × 4 × 8 inches
Weight: 6·7 lb
Power: 115 V 400 Hz

RD-600A/650A horizontal situation indicator

The 5-inch RD-600A has conventional veeder-counter readout for course and distance whereas the 4-by-5-inch RD-650A has luminous digital presentations. Both may be supplied with or without heading reference and course control knobs. A choice of remote controls for these horizontal situation indicators is available. They can be separate for each instrument in a dual flight-deck installation, or can have a single control that synchronises pilot and co-pilot heading reference while still providing independent course control.

Dimensions: RD-600A 5 × 5 × 8 inches
RD-650A 5 × 4 × 8 inches
Power: 115 V 400 Hz

STATUS: in production.

RD-700 horizontal situation indicator

The RD-700 series has been developed for airline applications in new aircraft or retrofit installations. High torque, low power-flag and shutter movements eliminate sticking displays and low-power devices coupled with open card construction result in low heat dissipation and power demands. All indications are conventional in presentation and numeric indicators feature standard veeder readouts. The following list summarises the presentation and displays of each instrument:

RD-700 horizontal situation indicator

Displays all standard navigation radio inputs, compass system and ARINC 561 ins data including digital readout of drift angle and ground speed.

Dimensions: 5 ATI
Weight: 8·5 lb
Power: 115 V 400 Hz or 26 V 400 Hz

RD-700A horizontal situation indicator

All standard navigation radio and compass data is presented together with dual digital dme readout.

Dimensions: 5 ATI
Weight: 8·5 lb
Power: 115 V 400 Hz or 26 V 400 Hz

RD-700C horizontal situation indicator

In addition to radio and compass navigation data this instrument includes ARINC 561 ins data and digital readout of time to waypoint, distance to waypoint and ground speed.

Dimensions: 5 ATI
Weight: 8·5 lb
Power: 115 V 400 Hz or 26 V 400 Hz

RD-700D horizontal situation indicator

Although very similar to the RD-700C this presentation does not include the time to waypoint counter.

Dimensions: 5 ATI
Weight: 8·5 lb
Power: 115 V 400 Hz or 26 V 400 Hz

RD-700F horizontal situation indicator

Displays all standard data including to/from, drift angle and digital readout of ground speed and distance to waypoint.

Dimensions: 5 × 5 × 8½ inches
Weight: 8·5 lb
Power: 115 V 400 Hz or 26 V 400 Hz

RD-700G horizontal situation indicator

In addition to all standard navigation radio, compass and ARINC 561 ins inputs, this instrument presents digital readout of drift angle, distance to waypoint and ground speed, though in this case the drift and ground speed presentation is at the lower area of the indicator rather than the more usual upper region of the instrument.

Dimensions: 5 ATI
Weight: 8·5 lb
Power: 115 V 400 Hz or 26 V 400 Hz

RD-700M horizontal situation indicator

In addition to standard information displays the RD-700M uses new, 11-position low-power mag-wheels for dual dme displays. A new thermal design includes a more efficient heat-sink mounting for high-power components and heat-sensitive capacitors. Electronic and mechanical sections are segregated for maintenance access, while open-board packaging in the electronic section facilitates trouble-shooting.

Dimensions: 5 × 5 × 8³/₅ inches
Weight: 115 V 400 Hz or 26 V 400 Hz

Sperry RD-700M horizontal situation indicator

Sperry RD-850 horizontal situation indicator

RD-800 horizontal situation indicator

This instrument features three digitally-driven servoed displays in conjunction with two four-digit gas-tube displays showing time and distance to waypoints. Improved versatility in navigational data processing is provided by micro-processor control.

Dimensions: 5 ATI
Weight: 8·8 lb
Power: 115 V 400 Hz or 26 V 400 Hz

RD-800J horizontal situation indicator

In this application the readout of true air and ground speed is provided by conventional counter displays and for ease of intepretation the command bars are colour-identified.

Dimensions: 5 ATI
Weight: 8·8 lb
Power: 115 V 400 Hz or 26 V 400 Hz

RD-850 horizontal situation indicator

The RD-850 horizontal situation indicator features the most recent applications of instrument technology including micro-processor control. Coloured display elements are included together with distance-to-go and ground speed counters. Automatic direction-finder annunciators are fitted in the lower instrument area.

Dimensions: 5 ATI
Weight: 10·2 lb
Power: 115 V 400 Hz or 26 V 400 Hz

Flight Performance and Management Systems

FRANCE

Sfena

Sfena (Société Française d'Equipements pour la Navigation Aérienne) BP 59, Aérodome de Villacoublay, 78141 Vélizy Villacoublay, Cedex

Performance management system

The Sfena performance management system is being proposed by Airbus Industrie for installation in the A300 B2/B4 wide-body transport, the intention being to carry out an evaluation on an Air Inter A300 B2 for a period of approximately six months, starting early in 1983.

The performance management system features a fully coupled autopilot and autothrottle system and a limited horizontal navigation capability which provides an accurate computation of the 'top-of-descent' point.

At all times the system displays the optimum

vertical profile based on data inserted prior to departure and modified by experience of actual ambient conditions encountered in flight.

The system function is really an additional cruise mode of the A300 autopilot which is also manufactured by Sfena. It is claimed that the facility minimises changes in operational procedures and simplifies crew training. It can also facilitate the normal use of the autopilot with standard monitoring scans and safety checks.

The crew has the choice of three operational modes: climb, cruise and descent. Several sub-modes are available in each of the above such as 'economy', which optimises the whole flight in accordance with a cost index selected by the crew in conjunction with the airline's flight-operations department.

The performance management system inter-

faces with all standard aircraft sensors, accepting inputs from the ins, vor, dme and air-data computer.

The complete system consists of a performance management computer, which is ¾ ATR Long, housing two micro-processors used for data acquisition, optimisation algorithms and command signal computation, and a micro-processor-controlled control and display unit which is used to enter and display performance data and select performance management modes and sub-modes. The control and display unit consists of a single, monocolour cathode ray tube consisting of 6 lines of 18 characters each.

STATUS: in development.

UNITED KINGDOM

Smiths Industries

Smiths Industries, Bishops Cleeve, Cheltenham, Gloucester GL52 4SF

Flight management computer system

The Smiths flight management computer system has been designed for full integration with other digital equipment and represents an advanced implementation of ARINC 702. Initially, a collaborative organisation was formed with Racal-Decca Navigator Company Limited to promote the system under the title of Flight Navigation Limited. Racal-Decca supplied some of the software, but by early 1982 the agreement between the two companies was under review.

The flight management computer system performs the functions of flight-planning, performance management, navigation, guidance and display processing. It is the 'nerve centre' of the flight-deck, enabling the crew to manage the aircraft for optimum performance throughout the entire flight from take-off to final approach, and makes possible significant advances in flight-deck automation and operational efficiency.

This second-generation system closes the flight control loop with coupled pitch, attitude and thrust management to ensure optimum flight profiles, and integrates navigation inputs to facilitate economical point-to-point flight. Smiths claims that a unique feature of this flight management system is its great capacity for future growth, based on advanced digital technology and a modular approach to system architecture using distributed micro-processors.

Massive storage capacity is provided by an all-solid-state bubble memory for the data bank, a cmos random access memory for work-space memory, and a UV-EPROM for the operational program. This provides a total processing capability in excess of 1 MOPS (1 million operations per second) and a memory provision of 768 K words of data-bank store, 200 K words of program store and 100 K words of working store. In each case, this allows for at

Smiths Industries flight management computer system

least 100 per cent growth after satisfying existing requirements.

Such great capacity is of considerable importance as later generation aircraft are developed with all-digital systems and increasingly complex functions. Smiths Industries predicts even greater capability as components are improved and denser memory devices, with capacities of up to 4 million bits and faster micro-processors become available.

Potential growth features could include 4-D navigation, Vlf-Omega, mls, Gps, minimum safe altitude monitoring and flight management data storage.

The crew communicates with the system by means of a control and display unit, a keyboard providing for the manual insertion of operational modes and data. A cathode ray tube display provides a read-out of selected parameters as well as verification of the entered

data. The control and display unit uses advanced digital technology, based on a 8-bit micro-processor and ARINC 429 data-highways for intercommunication.

Control and display unit design reflects an overriding objective of simple, consistent operating procedures with minimal need of button-pushing. This interactive unit leads the crew member through the procedures in a natural sequence, thereby encouraging use of the full range of facilities in a straightforward and unambiguous manner.

This flight management system has been ordered by Air France, Kuwait Airways, Saudia and British Caledonian Airways for their Airbus Industrie A310s. Future possible applications include the A300-600, the F 29 and future all-digital aircraft.

STATUS: in production.

UNITED STATES OF AMERICA

Arma

Arma Division of Ambac Industries, Roosevelt Field, Garden City, New York 11530.

Flight management systems

As part of the Hamilton Standard organisation, Arma produced the first flight management system to be certificated for airline service and deliveries began in 1977. Designed around the Lockheed L-1011 TriStar wide-bodied airliner, it is now in service with British Airways, Delta, Gulf Air, Pan Am and Saudia Airways and under the designation FMS 500 is standard equipment on all L-1011-500 aircraft.

Arma produces a family of equipment based on modular techniques that allow upgrading of a basic performance system to the full flight management system standard. In the case of the FMS 500 the navigation element was developed from the earlier Mona (modular navigation) system.

The flight management system provides navigation control by interfacing with the aircraft's navigation computer and position-sensing equipment, and can handle ins and dme information for optimum updating. En-route waypoints are stored in the flight management system computer via a control and display unit and the system automatically steers the aircraft from one waypoint to the next by providing the appropriate commands to the navigation computer and thence to the auto-pilot. This method achieves completely auto-mated 'three-dimensional' flight from shortly after take-off through to the destination ter-minal area.

The Arma systems interface with all major components of the aircraft including engines, navigation sensors, autopilot, autothrottle, air-data computer, and designated flight instru-ments.

Dimensions: (Computer unit) 1 ATR Long (Cdu) 4½ × 5¾ × 9 inches
Weight: (Computer unit) 45 lb (Cdu) 8·5 lb

Power: (Computer unit) 400 W
(Cdu) 45 W

STATUS: in production. Derivatives of the system are being developed for other aircraft both commercial and military.

PMS 500 performance management system

The PMS 500 has been designed for the Lockheed L-1011 TriStar wide-body transport providing outputs to drive the autothrottle system and pitch computers to maintain the best vertical profile. In this system there are no navigation inputs. The PMS 500 automatically controls engine exhaust pressure ratio, altitude and airspeed via the autothrottle and pitch interfaces to match commanded values. The equipment guides the aircraft in the vertical plane from take-off, in the climb, during cruise and descent.

Dimensions: (Computer unit) 1 ATR Long (Cdu) 4½ × 5¾ × 9 inches
Weight: (Computer unit) 40 lb (Cdu) 8·5 lb
Power: (Computer unit) 300 W
(Cdu) 45 W

STATUS: in production. Deliveries began in 1980.

Arma PMS 500 performance management system for L-1011

Pilot's performance system

The basic element of the 500 Series family is the pilot's performance system which comprises a control and display unit and avionics bay mounted computer unit. As in other systems initial pre-flight data is entered via the control and display unit. After selection of the appropriate flight mode, eg climb, cruise or descent, the computer calculates the altitude, airspeed and engine thrust to provide optimum fuel savings based on variables such as ambient temperature, aircraft gross weight, wind speed and direction and remaining distance to go. These performance parameters are then displayed continuously on the control and display unit and the aircraft is flown to match them.

Dimensions: (Computer unit) 1 ATR Long (Cdu) 4½ × 5¾ × 9 inches
Weight: (Computer unit) 37 lb (Cdu) 8·5 lb
Power: (Computer unit) 275 W
(Cdu) 45 W

STATUS: currently being developed for the Boeing 747 wide-body airliner though the key market segment is considered to be all wide-bodies and helicopters.

Collins

Collins Avionics Divisions, Rockwell Inter-national, 400 Collins Road NE, Cedar Rapids, Iowa 52406

FMS-90 flight management system

The Collins FMS-90 flight management system is an integrated navigation system which combines world-wide, point-to-point, terminal area and approach navigation and keyboard radio tuning. The system was introduced in late 1980 and first customer deliveries were sche-duled to start in mid-1981. It is designed mainly for business jet aircraft.

The FMS-90 allows the flight crew to navigate on either omega/vlf or vor/dme over point-to-point great circle courses and to compute independent positions from either source simultaneously for comparison purposes. Omega/vlf navigation information can also be instantaneously updated with vor/dme infor-mation. The omega/vlf section of the system, says Collins, has five times the capacity of conventional systems and can receive 80 independent data inputs from all eight omega and all seven vlf stations.

The FMS-90 has the latitude and longitude of over 5000 waypoints stored in its memory and in order to navigate in vor/dme mode the system computes an offset from any of these entries. The FMS-90 computer establishes latitude and longitude for the departure and destination points, drawing a great circle route between them. The pilot tunes selected vor stations along the route by means of an alpha identifier. Entries of offsets from these stations, as required by conventional R/NAV navigation methods, are not required.

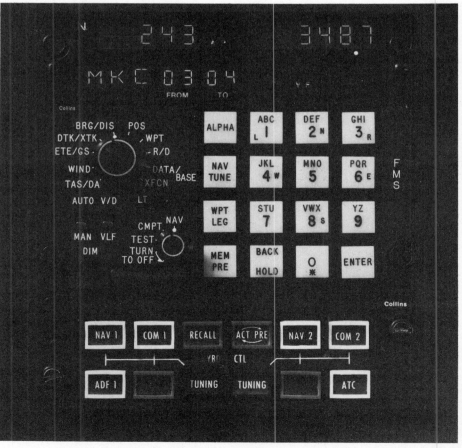

Control and display presentation for Collins FMS-90 flight management system in vlf/omega mode

In the terminal area, the FMS-90 operates in the vor/dme mode, providing precision point-to-point navigation to R/NAV standard.

According to Collins, the advanced omega/vlf receiver significantly reduces dead reckoning navigation errors. If the system should revert to dead reckoning mode, then it is capable of self-re-laning even if a cross-track or along-track error of over 200 nautical miles has accumulated.

In addition to its operation as an integrated navigational system, the FMS-90 also acts as a single control point for all flight plan parameters and radio tuning. Through the system, control can be exercised over the omega/vlf, three communication radios, dual navigation receivers, dual dmes, dual adfs and dual transponders. The system is also designed to interface with a complete installation of Collins equipment including nav/communication, navigation, weather radar and flight control systems.

All data displayed on the FMS-90's control display unit can also be displayed on the multifunction colour display cathode ray tube of a Collins WXR-300 weather radar. This unit can display 12 lines of information containing 20 characters on each. The format allows concentrated blocks of data, such as portions of flight plans, progress reports or self-test routines, to be displayed. Display on other manufacturers' weather radar equipment is possible with the aid of an appropriate interface.

The FMS-90 data base of more than 5000 waypoints, says Collins, includes over 2500 airports and all vors, vor/dmes and vortacs in the world, together with their associated latitudes and longitudes. The details are recalled from the memory by entering the appropriate three-letter identifier on the control display keyboard. Vor stations are tuned by entering the identifier code on the keyboard where the stored frequency is used to tune the radio.

The data base can be updated every four weeks through use of a hand-held loader and an update tape supplied by Collins. Updating is effected by plugging the tape loader to the FMS-90 unit and without removing any part of the system.

The FMS-90 can be integrated with the Collins FCS-80 flight control system and the ADS-80 air data system. The latter contains a new instrument, the VNI-80 vertical navigation system which, when used in conjunction with the FMS-90, provides three-dimensional navigation. Distance information from a waypoint can be supplied to the VNI-80 by the FMS-90, and a vertical intercept point for a descent or climb can be established by the FMS-90. The VNI-80 can also offset a vertical intercept point from a FMS-90 waypoint. In turn, the VNI-80 calculates the vertical speed required to reach a waypoint and displays any deviation from the vertical flight path.

This integrated use of the two systems assists the flight to remain at its selected altitude until the optimum moment for descent, which can then be made at the latest possible time in the interests of fuel conservation. On departure, the VNI-80 facilitates the task of complying with altitude restrictions when crossing intersections on instruction from air traffic control.

First application of both the FMS-90 and VNI-80 systems is aboard an Israeli Aircraft Industries Westwind 2 aircraft.

STATUS: in production and service.

Delco

Delco Electronics Division of General Motors Corporation, Milwaukee, Wisconsin 53201

Civil programmes performance management system

The Delco performance management system was initially developed for use in the Boeing 747 wide-body transport, firstly as a retrofit system and now an option in new 747 deliveries.

It is a basic two-box system, with the addition of an engine interface unit and a switching unit bringing the number of units to four. The first device processes engine discrete signals such as fuel flow into a format usable by the computer. The switching unit handles all switching of autopilot, autothrottle and instrument signals from standard aircraft configuration to performance management control.

The performance management system achieves closed loop control of the aircraft by providing commands to the autopilot for vertical steering, and either directly or via the ins to the autopilot for lateral steering, and the autothrottle for engine thrust control.

Delco has also adapted the performance management system for all models of the McDonnell Douglas DC-9 though in this case the engine interface unit is not required, and in the case of the DC-9-80 the switching unit is not necessary either. The system has also been adopted for the McDonnell Douglas DC-10 wide-body airliner and the Boeing 727 tri-jet.

Dimensions: (Computer unit) ½ ATR Long (Cdu) 5¾ × 4½ inches high (Engine interface unit) ¼ ATR Long (Switching unit) ½ ATR Long
Weight: (Computer unit) 29 lb (Cdu) 6 lb (Engine interface unit) 4 lb (Switching unit) 22 lb
Power: (Computer unit) 200 W (Cdu) 38 W (Engine interface unit) 4·2 W (Switching unit) 110 W

STATUS: in production.

Flight management system

The computer of the performance management system fitted to the McDonnell Douglas DC-9 transport (see entry above) can be upgraded to flight management standard, and in this form it becomes the basis of Delco's flight management system. This FMS has been developed for Pan Am Boeing 747 and United Airlines DC-10 wide-body transports, though there are some detail changes between the two installations. The B-747 thrust control uses the existing actuators but has updated electronic control. The system is based on information updated by dme. The DC-10 uses the existing autothrottle and is ins-dependent in oceanic operation only, using vor/dme in the domestic operation for navigation inputs.

The FMS in both cases provides flight-path optimisation in speed, thrust, altitude, climb, cruise and descent, together with associated advisory data. Initial flight-test programs featured computer-stored tables of aircraft flight-manual data and the flight-paths were developed within the constraints of existing air traffic control and airline operational procedures. The programs were later developed to include general flight-path optimisation algorithms and models of aircraft lift, drag and thrust.

Dimensions: (Computer) ½ ATR Long (Cdu) 5¾ inches wide × 4½ inches high
Weight: (Computer) 30 lb (Cdu) 6 lb
Power: (Computer) 150 W (Cdu) 38 W

STATUS: in production.

Delco performance management system for DC-9-80

Military programmes
Fuel savings advisory system

Analysis of commercial flight data derived from the performance and flight management programmes demonstrated significant fuel savings compared with previous operational methods. It seemed also that potential savings would be even greater in the military application and that other benefits could accrue. Military flight decks tend to be extremely crowded and the small size of the control and display units would not be demanding in terms of space; they could in fact replace the ins control and display units in aircraft such as the US Air Force/Lockheed C-141 and C-5A heavy transport aircraft.

Inbuilt growth capability and software flexibility to meet changing situations and the technical standard of the aircraft are additional attractions in the military role, offering greater precision and adaptability in cargo drops, refuelling missions or in the loiter role.

This theory was tested in the US Air Force fuel savings advisory system which employed elements of the existing, commercial performance management system in a C-141 flight-trials installation. Lockheed claims that the trials proved the earlier expectation with impressive fuel savings in a variety of operational missions.

The fuel savings advisory system features a remote display unit which shows the appropriate data from the computer over an ARINC 575 data-bus. These displayed parameters are altitude, target and actual engine pressure ratio, target and actual speed (ias or Mach number, depending on aircraft height) and the current flight mode. Presentation of the data is in a slightly different form from that of the commercial system, reflecting military demands for a smaller and lighter unit. The high-intensity displays are alpha-numeric light-emitting diode dot-matrix modules using optical filtering for contrast and readability in high ambient light conditions. The control and display unit has built-in test equipment and a front panel defect monitor light.

Dimensions: (Computer) ½ ATR Long
(Rdu) 3⅛ × 3⅛ × 9 inches
Weight: (Computer) 26·2 lb (Cdu) 2·8 lb

Power: (Computer) 89·3 W
(Rdu) 13·6 W

STATUS: computer is currently in production for the performance management system. More than 300 systems are now on order for the US Air fleet of Boeing KC-135 tankers, and there are options for more than 700 such sets.

Fuel savings advisory and cockpit avionics systems

Delco has now built on its military demonstrator experience to develop a proposal, which has been submitted to the US Air Force, for a comprehensive fuel-savings advisory and cockpit avionics system (fsa cas) intended for the upgrading of the current Boeing C-135 fleet of 600 or so transport and flight-refuelling tanker aircraft.

In this system an additional control and display unit is provided in place of one of the standard ins units to simplify crew management tasks; in particular, all basic data can be entered or interpreted at this point. In addition to the usual engine and navigational infor-

mation a new fuel-management panel and centre of gravity control system has been installed for inputs of fuel state, allocation and usage. The fuel panel is a micro-processor-based unit that transmits the information via the MIL-STD-1553B data-bus.

An additional mode is provided in the fsa to give the crew an indication of critical take-off parameters and to monitor the aircraft performance during take-off roll (the KC-135s are equivalent to very early 707 commercial transports and by today's standards are considerably underpowered). This would give an immediate indication of any incipient problem such as a power loss or failure to accelerate at an acceptable rate.

If the proposal is accepted Delco will act as the prime contractor with Gull Airborne Instruments, Electrospace Systems and Collins Government Avionics Division acting as sub-contractors.

STATUS: awaiting outcome of a proposal submitted to the US Air Force in 1981.

Garrett AiResearch

AiResearch Manufacturing Company, 2525 West 190th Street, Torrance, California 90509

GEMS energy-management system

The Garrett GEMS energy-management system is claimed to be the first airline-standard equipment of its type available for business aircraft. Initially intended for the Gates Learjet 30 Series twin-jets, the device is being offered either as a factory option or as a retrofit on aircraft currently in service.

GEMS is intended to simplify flight planning and ensure optimum conditions for fuel economy, time or cost with the additional advantage of providing engine power-setting trim through the fuel control computers that govern the Garrett TFE731 powerplants. This device

allows accurate and automatic trimming of thrust levels once the power lever angle is manually established within the GEMS trim authority range.

The system comprises an energy-management computer and control and display unit, the latter employing self-cueing for ease of operation. The computer carries the store of information including engine data, derived aircraft data and the cruise control and flight manual data. Five policies or modes are available for flight profile planning and reflect the system's flexibility. They are: minimum time with range assurance, maximum range, minimum overall cost based on fuel and time costs, maximum endurance, and manual speed selection. Once the policy is selected GEMS determines the best altitude, speed and power

setting for climb, cruise and descent modes and displays this information on the control and display unit. Rather typically now for executive aircraft, the company offers an option of using the aircraft weather radar cathode ray tube to display appropriate performance data.

An advantage of GEMS is the provision of colour coding for the control and display unit for ease of operation and unambiguous interpretation.

Dimensions: (Computer) 7½ × 9 × 14½ inches
(Cdu) 4½ × 5¾ × 8 inches
Weight: (Computer) 22·1 lb (Cdu) 4·8 lb

STATUS: in production. Flight trials were conducted in a Learjet Model 36 and were successfully completed at the end of 1981. First deliveries were made in mid-1982.

Lear Siegler

Lear Siegler Inc, Instrument Division, 4141 Eastern Ave SE, Grand Rapids, Michigan 49508

Performance data computer system

Lear Siegler's performance data computer system has been developed jointly by Boeing and the Instrument Division of Lear Siegler for B-727 and B-737 medium- and short-haul jet transports. This collaborative programme involved energy-management studies, ground-based flight simulation and crew inputs from a variety of airlines.

The crew operates the system through a control and display unit, command 'bugs' on the combined asi/Machmeter, engine pressure ratio gauges and the mode-annunciator panel. Modes are provided for take-off, climb, cruise, descent, holding, go-around, loss of an engine, and even the indication of fuel reserves at both the original destination and the alternate airport. Each mode has a variety of options which are available as data 'pages' in the control and display unit. Typically the cruise mode, for example, has four pages detailing conditions applicable to economy, long range, manual (ie crew-selected speed) and thrust limit.

Lear Siegler claims an exclusive function in the 'look-ahead' capability which is the ability of the system to provide a display of information relevant to other flight phases for reference purposes without 'erasing' or disrupting the current commanded 'bug' settings on the asi or engine pressure ratio gauges.

The performance data computer system receives its data from the navigation and central air data systems, as well as from the fuel totaliser unit, autopilot, autothrottle, anti-ice valve control, total air temperature probes, and the cabin air-conditioning and pressurisation mode control. All these inputs are processed in the performance computer and provided in the appropriate form as flight information in the control and display unit.

The system contains its own built-in test circuits which maintain a constant monitor on system operation and fault detection.

STATUS: in production. Trials were initially conducted by Continental Airlines, Lufthansa and VASP and the performance data computer system is currently fitted to the B-737 short haul, twin-jet transports of British Airways and Lufthansa.

Lear Siegler performance data computer system

Performance navigation computer system

Lear Siegler is currently developing a performance data computer system with a navigation function, the new system being known as the performance navigation computer system. It includes a comprehensive en-route and terminal navigation capability in addition to both lateral and vertical guidance.

The system is based on modular design techniques to allow 'building block' additions for aircraft already fitted with a performance data computer system, although it can be equally easily retrofitted to aircraft without any other form of performance or management device.

Crew control and interface is accomplished by means of a 24-character, 10-line cathode ray tube which displays information, data and menu selection and has the added facility of a scratch-patch for crew use, this being the part of the screen into which data may be entered via the alpha-numeric keyboard.

The performance navigation computer system has two processors: the performance computer provides all of the in-flight capabilities available with the original performance data-computer system while the navigation computer has an operational memory capacity of 72 K and a data-base memory expandable to 192 K. The navigation data-base contains information relative to airports, navaids, waypoints, standard instrument departures, en-route airways, standard terminal arrival routes and the various approach and holding patterns. All navigation aids used for position-fixing are automatically tuned by the performance navigation computer system during flight thereby reducing crew workload.

Both lateral and vertical guidance channels are fully coupled into the flight-control system through the autothrottle and the autopilot's pitch and roll channels. With fully automatic mode-switching for altitude capture, speed variations and waypoint sequencing, the performance navigation computer system offers an optimised and largely 'hands-off' flight profile. By coupling with the autothrottle at brake release and the autopilot at 1000 feet (300 metres) a completely automatic 'hands-off' flight may be accomplished until instrument landing system capture at the destination.

STATUS: in development with intended application to the Boeing 737-200 and B-737-300 twin-jets. First production units available in 1983 and certification expected by the end of that year. Lufthansa has already declared its intention to retrofit the performance navigation computer system in its B-737s.

Lear Siegler performance navigation computer system

Safe Flight

Safe Flight Instrument Corporation, White Plains, New York 10602

Speed command of attitude and thrust (scat) system

Safe Flight has been a pioneer company in the development of wing-lift instrumentation and technology and has now produced the speed command of attitude and thrust (scat) system that provides pitch and thrust guidance during the take-off, landing approach and go-around phases of flight. It may be fitted with an autothrottle and under the trade-name Autopower can offer the standard of speed stability appropriate to Category III approach and landing requirements.

Scat drives the pitch-command bar of the flight director during these low-speed phases of flight. If an engine fails just after take-off, for example, the crew is immediately presented with the pitch attitude giving the best rate of climb. It does this by matching the aircraft's polar diagram (its lift/drag characteristic) against the thrust level available.

Emphasis is placed on the system's ability to combine target angle-of-attack and acceleration along the flight-path to optimise the crew's performance during these critical periods, which demand precise attitude and speed control if the situation is to be stabilised. It is particularly useful in restrictive runway operations, noise-abatement procedures, and at times of forecast wind-shear.

STATUS: in production.

Fuel performance computer

Safe Flight produces a fuel performance computer (fpc) which provides cruise control guidance for crews of executive jet aircraft. It computes and displays basic information such as optimum speed for maximum range or long range cruise, along with the optimum en-route altitude.

STATUS: in production.

Safe Flight scat computer

Simmonds

Simmonds Precision, 150 White Plains Road, Tarrytown, New York 10591

Performance Advisory System (pas)

Pas is a performance advisory system that uses a micro-processor to monitor, store and process real-time aircraft performance data. It consists of three items: a display unit, using light-emitting diode elements, a control unit, and a computer.

As part of the pre-flight checks the crew enters all necessary flight-plan data via the control unit keyboard. As this flight plan is followed in the air a crew-member simply turns the control unit selector to each of the seven flight modes to read off the recommended engine pressure rates, pitch attitude and indicated airspeed setting.

Pas receives signals such as pressure, altitude, Mach number, true airspeed, true air temperature, fuel quantity, engine anti-ice and bleed valve position, flap angle and landing gear position. This information is used to update continuously the displayed information.

The display unit has a two-position toggle switch. In the case of the Boeing 727 trijet when the switch is in one position the display window reads centre and pod engine pressure ratios, commanded pitch attitude and commanded indicated air speed. With the switch in the alternative lower position the display reads pod and centre engine pressure ratio limits, optimum and maximum altitude. Simple software changes suffice to adapt the system for any type of aircraft.

The parameters displayed on the upper scale are commands that result in the optimum operation for the selected mode. The lower scale parameters provide information for planning and comparison of various alternatives such as change of flight plan, diversion or extended holding.

The display of optimum altitude and maxi-

Performance advisory system by Simmonds Precision

mum altitude are of particular benefit, allowing the crew to accommodate quickly air-traffic control requests for flight-level changes. Optimum altitude is based on considerations of cost while maximum altitude considers the operational limits of thrust, buffet boundaries, and speed. The control and display unit receives directions from the mode selector on the control panel. There are eight positions: preflight, take-off, climb, cruise, turbulence, descent, holding and approach.

The performance advisory system computer is self-teaching and self-testing. It rejects entry data that does not fall within established parameters and it shuts down automatically on detection of failure.

A dme lock-on facility is offered as a customer option.

A second-generation version of performance advisory system introduces changes in the man/machine interface and the toggle switch is replaced by a push button.

The company stresses that this 'electronic cruise control manual' can be easily modified or updated in the light of changing air-traffic control methods by rapid and inexpensive software changes.

STATUS: operational on Boeing 727 trijet transports.

Sperry

Sperry Flight Systems, Commercial Division, PO Box 21111, Arizona 85036

Flight management system

The advent of the Boeing 757 and 767 short and medium range new-technology airliners boosted the development of all types of avionics, among them flight management systems. Sperry has produced a digital flight management system suitable for these aircraft, and also for Europe's equivalent, Airbus Industrie A310 transport.

The Sperry flight management system features vertical navigation which, in conjunction with inputs from the central air data computer and pre-programmed information, provides climb, cruise and descent guidance options. Lateral navigation produces steering data in conjunction with ins, radio navigation aids and, again, pre-programmed information. It may be coupled to the auto-pilot for ease of operation, though an interlock with the radio altimeter prevents coupling below 1500 feet (460 metres) on some aircraft. Autothrottle may also be coupled. The crew can communicate with the system via a control and display unit and keyboard, and there is a scratch pad capability. The flight management system will then determine the best method of achieving a selected profile and will carry it out via the autopilot and autothrottle. In cruise it controls airspeed and altitude and can present a continuous indication of optimum altitude. When interrogated, it provides an indication of the savings or penalty associated with any proposed altitude change.

A magnetic disc memory is used to store all performance data, navigation aids, route profiles and airport information. Consequently the initial pre-flight checks require only that the crew verifies correct date of data and engine identification and selects the intended route by inserting an identifier number, from on-board documentation, which 'loads' the sector. If the chosen route is not a regular one, or if it is not available in the data bank, then the waypoint co-ordinates may be loaded manually via the keyboard.

The ins inputs may be updated from radio navigation aids, and with the flight-director in operation, command information may be monitored on the attitude deviation indicator. The flight management system also features an automatic tuning facility for the radio navigation equipment.

When this flight management system is used to retrofit older analogue-based aircraft, a data adapter interface is fitted.

Certification programmes have been conducted in the USA with United Airlines DC-10 and 747 wide-body airliners. The system has been chosen by British Airways for retrofit to its Boeing 747s and by early 1982 one of these aircraft with a Sperry flight management system had concluded a number of transatlantic operations with complete success. Full certification of the dual flight management system in the entire fleet was expected by the end of 1982.

STATUS: in production.

Sperry flight management system control and display unit for A310

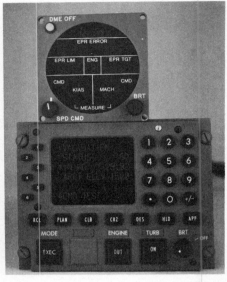

Sperry flight management computer system (two channels) and electronic flight instrument display in company's software validation facility

Sperry performance management system

Performance management system

Sperry has developed a performance management system which is intended for the Boeing 727, 707, 737, and Douglas DC-8 and DC-9 transports. It is said to require few changes in retrofitting and does not need any modification to existing sensors and aircraft instrumentation.

The performance management system is a three-box system consisting of computer, control and display unit and indicator. The computer integrates all performance, autothrottle, thrust rating, air-data and ins functions. The control and display unit provides crew interface and displays performance data on a cathode ray tube with sub-mode selection via a conventional keyboard. The indicator provides the crew with a convenient display of the active thrust and speed commands and may be mounted adjacent to the engine pressure ratio gauges for ease of monitoring by all crew members.

The system provides a means of optimising the aircraft's performance throughout the vertical flight envelope. Four function keys are fitted at the lower edge of the control and display unit and seven adjacent mode-select keys are used to select climb, cruise, descent, hold and approach. These modes provide full flight control including sub-modes for take-off and go-around.

The performance management system features an 'engine out' mode which provides logic to the computer that will sequence on the indicator engine data.

The indicator is based on a Sperry 3-inch cathode-ray tube. It is divided into five data fields with up to 16 characters per field, each field being separated by lines and legends to denote command information. These areas indicate mode, engine pressure ratios, limits and speeds. A manual speed-control knob allows the pilot to select an airspeed to comply with air-traffic control instructions or any other required deviation from the programmed schedule.

The indicator also carries a dme off lamp which warns the crew when the dme-dependent ground-speed calculations are for any reason invalid.

Dimensions: (Computer) ¾ ATR Long (Cdu) 5¾ × 4½ × 10⅞ inches (Indicator) 3 × 7 inches
Weight: (Computer) 28 lb (Cdu) 4 lb (Indicator) 3·3 lb
Power: (Computer) 58–79 W (Cdu) 16 W (Indicator) 12·5 W

STATUS: flight trials began in 1981 in a Western Airlines Boeing 727.

Flight Data Recording Systems

FRANCE

Electronique Serge Dassault

Electronique Serge Dassault, 55 Quai Carnot, 92214 Saint-Cloud

Airframe fatigue recorder

This device measures the steady-state load factors imposed on an aircraft during normal manoeuvres, and the transient loading due to gusts imposed upon its structure. Making use of transistor/transistor logic and large-scale integration technology, the system comprises an accelerometer and a separate counter and storage unit. The accelerometer is a solid-state device, mounted as near to the aircraft centre of gravity as possible so that it responds only to translational movement and not to components of rotation about the centre of gravity.

The accelerometer generates voltages proportional to the imposed loads, and these are compared in the counter and storage unit with reference voltages corresponding to pre-selected g-thresholds. A series of veeder-counters on the face of the storage unit displays the number of counts at each g level.

Accelerometer
Effective measuring range: –2 to +8 g
Linear frequency range: 0 – 10 Hz
Sensitivity to cross-acceleration: 0·02 g/g
Dimensions: 60 × 60 × 33 mm
Weight: 0·22 kg

Counter and storage unit
Number of thresholds: 8
Threshold values: 3,4,5,6 and 8 g; also 0·5, 0·25 and 1·5 g
Accuracy: 2 %
Power: 10 W at 115 V 400 Hz
Dimensions: 180 × 85 × 73 mm
Weight: 1·4 kg

STATUS: in production.

Installation of ESD airframe fatigue recorder behind quick-access panel

Enertec

Enertec, 1 rue Nieuport (BP 54), 78141 Vélizy-Villacoublay Cedex

Enertec is a member of the US-based Schlumberger group of companies, and has specialised in the production of crash and maintenance flight recorders for civil and military aircraft.

Crash recorders impress digital, analogue or voice information continuously on magnetic tape in accordance with FAA Technical Service Order C51a. This standard protects the tape by specifying that the housing and transport mechanism be capable of withstanding the following conditions:
Shock: an acceleration of 10 000 metres a second for 5 milliseconds
Crushing: 22 700 N on 3 axes for 300 seconds
Dynamic penetration: 227 kg dropped from 3·05 metres on an impact area of 0·32 square centimetre
Fire: 1100°C over at least 50 per cent of outer surface for 30 minutes
Immersion of complete enclosure for 36 hours in a mixture of kerosene, Skydrol (hydraulic fluid) and fire-extinguisher fluid
Immersion in sea-water for 30 days
These crash recorders operate continuously, direction of the tape drive being reversed and track-changing being accomplished in 100 milliseconds.

PE 6573 digital flight data accident recorder

The PE 6573 was designed to meet ARINC 573 defining the flight-data and recording system characteristics for modern jet and turboprop transports.

Dimensions: 1/2 ATR Long
Weight: (recorder) 12·6 kg
(anti-vibration mounting) 2·1 kg
Recording capacity and data rate: 25 h, 768 bits/s
Replay data rate: 4608 bits/s
Power supply: 115 V 400 Hz
Accident survival: to FAA TSO C51a

STATUS: in production.

PE 6010 and PE 6011 digital flight data accident recorders

The PE 6010 is a lightweight digital flight data accident recorder in which weight-saving is accomplished by a relaxation of the dynamic penetration requirement. In terms of accident protection however it meets all other requirements of FAA TSO C51a.

The PE 6011 is intended as a replacement for the PE 6010. It is more compact than its predecessor, meets TSO C51a completely with

Enertec PE 6011 crash recorder

Enertec crash recorder: after an accident the tape deck remains intact within its mechanical and thermal protection

Enertec PC 6033 performance/maintenance recorder

only a small weight increase, and has 50 per cent more capacity. The new design of tape deck is simpler, with attendant gains in reliability and maintainability.

Dimensions: (PE 6010) 1/2 ATR Short (length 319 mm)
(PE 6011) 1/2 ATR (length 296 mm)
Weight: (PE 6010) 7·7 kg
(PE 6011) 10·2 kg

Recording capacity and data rate:
(PE 6010) 16 h, 768 bits/s
(PE 6011) 8 h 2308 bits/s
Replay data rate: (PE 6010) 4608 bits/s
(PE 6011) 18 464 bits/s
Power supply: (PE 6010) 18-30·5 V dc
(PE 6011) 12-32 V dc
Accident survival: (PE 6010) TSO C51a
(except penetration is static)
(PE 6011) TSO C51a

STATUS: PE 6010 is in production and service with Dassault Mirage F1 fighters of the French Air Force and the same aircraft for the air forces of Greece, Egypt, Iraq, Kuwait, Morocco, and Spain. It has also been supplied for the HAL Ajeet trainer of India and Super Frelon helicopters of the French Navy. The PE 6011 is in development for the new Dassault Mirage 2000 fighter and 4000 strike aircraft of the French Air Force.

PE 6013 and PE 6015 digital flight data and voice accident recorders

These two recorders are developments of the PE 6010 and PE 6011 and employ their respective technologies.

Dimensions: (PE 6013) ½ ATR Short (PE 6015) ½ ATR (length 296 mm)
Weight: (PE 6013) 11·2 kg (PE 6015) 11·2 kg

Recording capacity and data rate:
(PE 6013) 30 min (voice) 5 h, 2308 bits/s (digital)
(PE 6015) 30 min (voice) 1 h, 2816 bits/s (digital)
Replay rate: (PE 6013) 2308, 11 540 bits/s (PE 6015) 2816 bits/s
Power supply: (PE 6013) 12-32 V dc (PE 6015) 20-30·5 V dc
Accident survival: (PE 6013) TSO C51A (PE 6015) TSO C51a

STATUS: the PE 6013 has application to the Dassault Mirage 2000 fighter and the PE 6015 has been fitted to prototypes of the multinational Tornado bomber.

PC 6033 general-purpose digital cassette recorder

This device records serial data on a continuous basis over a long period of time. Typical applications are performance and maintenance recording, engine health monitoring, aircraft testing, and reconnaissance; the system complies with ARINC 591. Cassettes can be replayed at up to 80 times the recording speed.

Recording capacity and data rate: 50 h, 138 Mbits
50 h at 768 bits/s
25 h at 1536 bits/s
12·5 h at 3072 bits/s
Number of tracks: 12
Recording code: biphase L or M
Power supply: 19-32 V dc or 115 V 400 Hz
Error rate: <1 bit in 10⁵

STATUS: in production for and service with a number of European airlines including Lufthansa (A300, DC-10, B-747), Alitalia (A300, B-747, B-727) KLM (A310), Swissair (A310) and Austrian Airlines (A310) as part of their aids (aircraft integrated data system) equipment.

UNITED KINGDOM

NGL

NGL Electronics Division, Clarence Street, Yeovil, Somerset BA20 2YD

This company produces a range of voice, crash, and maintenance recorders for airborne use.

1203V maintenance data recorder

Claimed by NGL to be the world's smallest airborne maintenance data recorder, the 1203V was chosen by Bendix Flight Systems of Teterboro, New Jersey to become part of that company's data-recording system for the US Navy/McDonnell Douglas F/A-18 Hornet carrier-based air-superiority fighter.

The recorder is a sealed unit to ensure maximum reliability and a data-error integrity of 1 in 10⁷, yet still allowing quick access for replay and data retrieval. Tape tension and increment are independently controlled via precision servo electronics, eliminating the need for pinch wheels and co-belts. Recording-head wear is kept to a minimum by the use of a glass-bonded ferrite/ceramic construction. Control electronics are mounted on 10-layer pcbs, and cmos logic is used for minimum power dissipation. A high standard of built-in test capability is incorporated to isolate faults to sub-assembly level, and by the use of automatic test equipment faults can be diagnosed down to individual components.

STATUS: in production and service.

1403V large capacity data recorder

This is a 42-channel, two-head, write-only machine for the collection of information under severe environmental conditions. A significant feature of the device is that it generates a data-demand clock signal that may be used to select the correct speed automatically to match the data rate to the associated buffer. This results in constant tape density information, so reducing wastage. The recorder has a servo-controlled capstan and spools to ensure constant tape tension during power-off, and prevents spool rotation.

Recorder electronics housed in a separate unit are mounted on pcbs for maintenance and reliability, and are based on cmos technology for reduced power dissipation. The system may be adapted to record fm analogue or digital data, and its capacity of 25 Gbits (25 × 10⁹ bits) doubled by a simple tape change.

STATUS: entering production.

NGL Type 1403V large-capacity recorder

1151V cockpit voice recorder

This unit forms part of a cockpit voice recorder system to meet the requirements of various commercial aviation certification authorities. The system is based on an endless-loop tape of 30 minutes duration on four separate channels. Designed in accordance with FAA TSO 84, ARINC 557, and CAA Specification 11, the unit is housed in a crash-protected ½ ATR Short case with electronics segregated from the mechanical elements at the rear. Self-test facilities and a headphone jack-socket are mounted on the face of the unit. Audio frequency response is flat over a 6 dB range from 350 to 3 kHz.

STATUS: in production and service with commercial airlines.

NGL Type 1151V cockpit voice recorder

1416 maintenance recorder

The 1416 recorder is a development of the Type 1203 unit currently in production for the Bendix flight data recording system chosen by McDonnell Douglas for the US Navy's F/A-18 Hornet fighter. The capacity has been increased from 12 Mbits to 1 Gbits and there are 14 read/write channels with track-switching circuits for series or parallel recording. The system can record digital or analogue information with a bit error rate of better than 1 in 10⁶ under the severe environmental conditions specified by MIL-SPEC-8105 and MIL-STD-5400.

The electronic modules are packaged on multi-layer pcbs with cmos logic for low power dissipation. A single board controls data format and track selection.

STATUS: in development.

NGL Type 1203V miniature recorder

NGL Type 1416 maintenance recorder

1123V digital flight data recorder

NGL's digital flight data recorder Type 1123V is designed to CAA Specification 10 and ARINC 573 for a crash-survivable information system, and can accept 25 hours of continuous serial digital information. Self-test circuits confirm correct operation of the unit, and automatic track-switching is a standard feature.

STATUS: entering production.

Type	Application	Recording mode	Capacity	Data rate	Tracks	Size unit	Power	Environment	Accuracy accuracy
1203	Aircraft recording system	Digital	12 M Bits	30K bits/s	4-Track switching	123 mm × 89 mm × 53 mm	28V dc 13W	MIL-STD-5400 MIL-SPEC-810 B	Better than 1 in 10^7
1403	Aircraft marine data recording system	Fm or digital	25 G Bits or 50 G Bits	630 bit/mm	42	100 mm × 200 mm × 490 mm	14V dc 30W	−20°C to +70°C MIL-SPEC-810 B	Better than 1 in 10^5
1416V	Aircraft recording system	Fm or digital	1 G Bits	30K bit/s	14	203 mm × 178 mm × 178 mm	28V dc 24W	MIL-STD-5400 MIL-SPEC-810 B	Better than 1 in 10^6
1123V	Flight data recording system	Continuous digital	70 M bits	768 bit/s	16	146 mm × 57 mm × 97 mm	10VA at 115V 400Hz	TSO C 51a ARINC 573 BS 2G 100	Better than 1 in 10^5
1151V	Cockpit voice recording system	Continuous direct	30 min 4-channel	Frequency response 350 – 3 KHz	4	146 mm × 57 mm × 97 mm	10VA at 115V 400Hz	CAA Spec. 11 TSO C84 ARINC 557	—

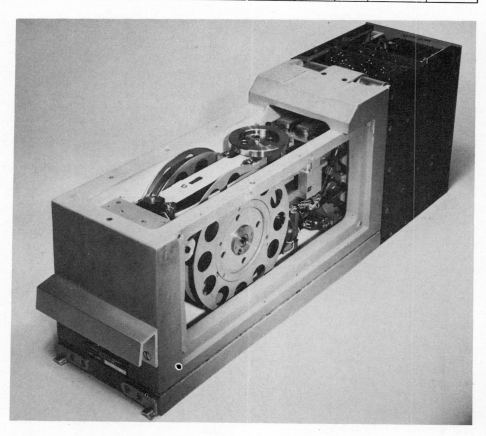

NGL Type 1123V digital flight data recorder

Smiths Industries

Smiths Industries plc, Aerospace & Defence Systems Company, 317 Edgware Road, Cricklewood, London NW2 6JN

Low cycle fatigue counter

In recent years damage known as low cycle fatigue has been recognised as a major limiting factor in the lives of turbine engines. These powerplants benefit from being run at as near constant speed as possible, so that the mechanical stresses set up under changing conditions are as low as possible. With aircraft engines, and particularly those for military types, benign environments and operating usage common to turbines for electrical generation or natural gas pumping is never possible; a typical flight may involve a number of power excursions, or low frequency cycles,

between flight idle and maximum power, each such excursion from the cruise setting representing a cycle. Each time the throttle or power lever is moved, stress variations are set up in the rotating components that produce minute stress cracks, and these propagate in accordance with usage. Cumulative damage so caused is termed low cycle fatigue, and is monitored by devices that record the number of times that an engine has been cycled through specified rpm limits between overhauls. The results are then compared with the damage produced by a 'standard cycle', and factored to allow for the scatter between individual engines of the same type.

The Smiths Industries low cycle fatigue counter, weighing 3·7 lb (1·7 kg) is housed in a ¼ ATR dwarf case, and accepts inputs from tachogenerators or pulse probes in single or two-shaft engines. The functions can be ex-

tended to include temperature inputs so that resulting creep and fatigue damage can be taken into account. Additional inputs can be included to accommodate three-shaft engines.

The counter has four readouts to show selected component lives, for example low-pressure and high-pressure compressors and their corresponding turbines. The unit computes the reduction of low cycle fatigue life using the same basic formula as that used by a stress analyst, but the data-processing uses a micro-processor in conjunction with a memory to compute low cycle fatigue from its stored programs. The damage cycles representing loss of engine life as a result of low cycle fatigue are displayed on electro-mechanical registers to give a running total. The device can handle up to four channels of information and display it on the front face of the unit. Built-in test indicators show input or equipment failures.

STATUS: in service to monitor the Rolls-Royce/Turbomeca Adour engines of the Hawk trainers of the RAF's Red Arrows aerobatic team. Formation aerobatics impose the most severe usage on engines of any form of flying, and as a result of the unavoidable constant throttle movement, low cycle fatigue is the dominant failure mechanism, and so determines overhaul life.

Engine life counter

An extension of the low cycle fatigue counter described above, the micro-processor-based engine life counter developed in conjunction with the UK Ministry of Defence is a considerably more versatile unit, monitoring other parameters as well as low cycle fatigue that influence engine life.

The system can be tailored to the special requirements of particular types of engines, with inputs comprising rotational speeds, operating pressures, temperatures, or vibration. After processing and factoring for the characteristics of different types of engine, the information displayed on the recorder can be related to thermal fatigue, creep, low cycle fatigue, or other limitations appropriate to the engine. The unit therefore provides an accurate prediction of the time to next overhaul or for component or module replacement, so that maintenance costs are substantially reduced.

The processing circuits for all applications are standard, the factoring of information necessary for individual engines or engine types being done on a single replaceable software board.

STATUS: operational trials have been conducted in Lockheed L-1011 TriStar (5000 hours), BAe One-Eleven (3000 hours), and Hawk Trainer.

Smiths Industries low cycle fatigue counter

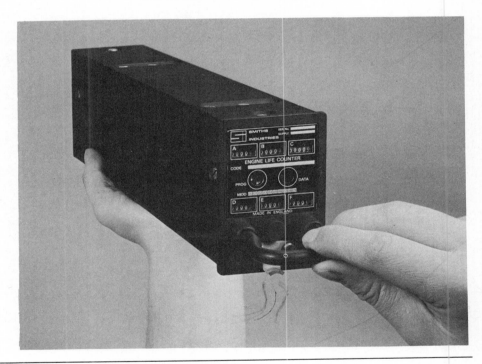

Smiths Industries engine life counter

UNITED STATES OF AMERICA

Edo Western

Edo Western Corporation, 2645 South West Street, Salt Lake City, Utah 84115

Model 655 head-up display television camera recorder

This device is designed for advanced military aircraft to record head-up display symbology superimposed on the outside world, and can also be used as flight-test instrumentation during the development of equipment. It is a substitute for the earlier-generation film-camera recorders, having the advantages of magnetic recording compatibility with many

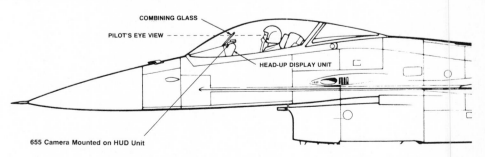

Installation of Model 655 recorder in General Dynamics F-16 fighter

types of reproducer, and instant replay; no time is lost in film processing. For training and reconnaissance, a 'front-seat' presentation can be provided at the rear-seat position.

The system has a number of applications. As a video tape recorder, it can record equipment anomalies and failures, provide a basis for the assessment of weapon delivery and target damage, permit an evaluation of pilot proficiency, act as an intelligence gatherer, and help pilot and crew training. As an airborne video display generator the system provides real-time information to other processing and display equipment – the ability to provide a 'front-seat' picture to the weapons system operator or instructor in the rear seat, can substantially augment the value of a two-seat aircraft for front-line service or training. Again, the system can act as a video communications link, providing in conjunction with an S or L band transmitter a data link to an airborne command post, sensor reporting post, flight-test centre, or ground-based intelligence organisation The Model 655 vidicon unit is detachable from the body of the camera so that it can be used to record cathode ray tube data in other avionic systems such as a radar display or threat-warning panel.

Dimensions (overall): 7½ × 3¾ × 6⁴⁄₇ inches 190 × 95 × 167 mm)
Weight: 3 lb (1·3 kg)
Lens: f/1.8, 18 mm, iris switchable to f/1.8 or f/8
Field of view: 22·5° × 17° at 4:3 image format 20° × 20° at 1:1 image format
Interface: compatible with Marconi Avionics head-up displays, and has a multiplexer that permits input from external video source to be recorded simultaneously with the hud video.
Sensor: ⁵⁄₈ inch FPS (focus, projection, scan) vidicon sulphide target or (optionally) silicon diode target
Bandwidth: Flat to 10 MHz ± 1 dB, 12 MHz –3 dB
Resolution: 700 tv lines at 4:3 image format 800 lines at 1:1 format
Synchronisation generator: EIA RS-170 or CCIR European
Horizontal line rate: 525 US or 625 European
Frame rate: 60 Hz US or 50 Hz European
Video output: to RS-170 or CCIR

STATUS: in production.

Model 679 gunsight television camera

This gunsight television camera is designed for installation in the cockpits of combat aircraft where size and mounting constraints may limit the use of other head-up display camera systems. There are always difficulties with the cockpit installation of 'outside world' cameras because they need to command the same field of view as the pilot but without obstructing his view. To reduce the size of the viewing system, therefore, the Model 679 is configured as two units: the television camera itself, and a camera control unit. The camera can be mounted just in front of the head-up display or above it, while the control unit can be sited in any convenient panel space.

Applications of the Model 679 are identical to those of the Model 655 described above, and the optical and video specifications are the same.

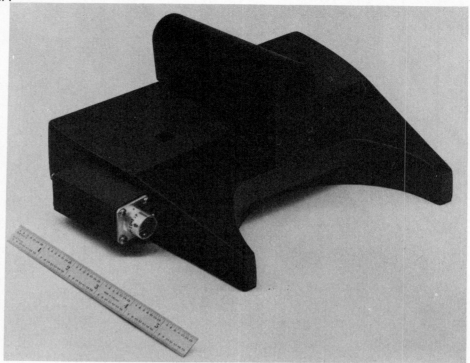

Edo Western Model 655 hud television camera recorder

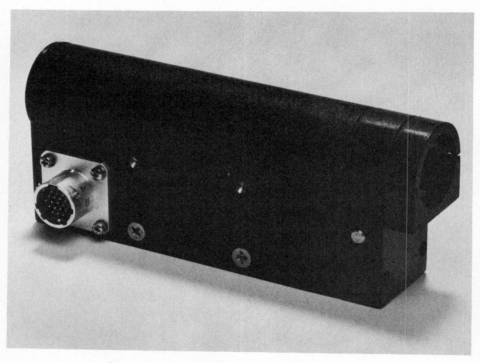

Camera unit for Edo Western Model 679 gunsight recorder

Dimensions: (camera) 5·02 × 2⅙ × 1⅝ inches (127 × 55 × 41 mm) (control unit) 5¾ × 2½ × 6½ inches (146 × 63 × 165 mm)

Weight: (camera) 0·7 lb (0·3 kg) (control unit) 2·7 lb (1·23) **Power supply:** 16 W at 28 V dc

STATUS: in production.

Fairchild Weston

Fairchild Weston Systems Inc, 300 Robbins Lane, Syosset, New York 11791

AN/AXQ-15 Hitmore helicopter installed television monitor recorder

The Hitmore system, based on Fairchild's MV-201 solid-state charge-coupled device camera, increases the proficiency of helicopter TOW missile operators by recording on video his handling of the weapon for immediate monitoring or post-flight playback.

The benefits, according to Fairchild, are that no 'target conditioning' is needed, real-time impact-points can be observed and commented upon; the record can be played back directly on landing; 40 simulated firings can be accommodated on a single cassette; the reusable tape cuts costs; a short-term record of student's performance can be used to demonstrate improvements; and the system is capable of modular growth. The final objective is to improve the gunner's first-round kill probability.

Camera
Frame rate: 30 frames/s
Format: 244 lines, 190 picture elements/line
Camera size: 2½ × 2 × 3¾ inches (63 × 51 × 95 mm)

Video monitor
Display area: 4¼ × 3¼ (108 × 83 mm)
Visual: 525-line scan, 8 shades of grey
Frame frequency: 30 Hz

Fairchild Weston AN/AXQ-15 Hitmore system

Dimensions: 6 × 6 × 14 inches (152 × 152 × 357 mm)
Weight: 10 lb (4·55 kg)

Video recorder
Recording system: rotary two-head helical scan

Power: 30 W at 28 V dc
Dimensions: 9³/₅ × 6 × 13 inches (243 × 151 × 330 mm)
Weight: 23 lb (10·3 kg)

STATUS: in production and service.

Hamilton Standard

Hamilton Standard Division of United Technologies, Windsor Locks, Connecticut 06096

Flight data systems

Hamilton Standard produces examples of both types of information recording system currently operational in military and commercial aircraft. The first is the basic flight data acquisition system (bfdas) to monitor mandatory crash parameters, designed in accordance with ARINC 573 and 717 for mounting and electrical interfaces, and the second is a development with expanded capability to measure other parameters, and referred to as an aircraft integrated data system (aids). The two types of system are built up from a family of building blocks, or modules. These are:
Flight data entry panel (fdep)
Data management unit (dmu)
Flight data acquisition unit (fdau)
Auxiliary data acquisition unit (adau)
Management control unit (mcu)
Digital flight data acquisition unit (dfdau)
Recorders or printers to handle outputs

FDEP 100, 120, 121, 122, 123 flight data entry panels

The fdep is a device that permits flight-crews to feed commands or information such as flight number, gross weight, and aircraft identification into the aids or bfdas recording system. The information may also control various analysis routines in the aids processor. The unit has three configurations: thumbwheel-controlled unit for a bfdas, push-button fdep, and a standard aids unit with display and recall facilities. These units are part of Hamilton Standard's Mk II, III and IV bfdas/aids systems installed on Boeing 737 and 747, Douglas DC-9-80 and DC-10, and Airbus Industrie A300 and A310 transports. Specific applications are FDEP 100: 737, 747 and DC-10; FDEP 120:, 747, DC-9-80, A300; FDEP 123: A310; FDEP 121: 767, 757.

Hamilton Standard basic flight data acquisition system

FDEP 100 for Mk II aids
Dimensions: 5⁴/₅ × 4½ × 9 inches (147 × 114 × 228 mm)
Weight: 7 lb (3·18 kg)
Power: 35 W

FDEP 120 for Mk IV aids
Dimensions: 5³/₄ × 2¼ × 6 inches (146 × 57 × 152 mm)
Weight: 4 lb (1·82 kg)
Power: 4 W

DMU 100, 101 data management unit

This unit is the 'brain' for the aids system, analysing the real-time information from the fdau, the adau, or the dgdau. It controls the digital aids recorder, having program logic that determines what information to record and when. It also provides data to displays including the fdep and airborne printer, and contains extensive built-in test equipment. This dmu is part of the company's Mk II and expanded, Mk III aids system in the following aircraft: DMU 100: 747, DC-10, A300; DMU 101: A310

	DMU 100	**DMU 101**
Dimensions:	1 ATR Long	6 MCU
Weight:	36 lb (16·36 kg)	12·7 lb (5·77 kg)
Power:	180 W	55 W

Hamilton Standard digital flight data acquisition unit

FDAU 100 flight data acquisition unit

Essentially a data gatherer, this unit contains the signal conditioning needed to rationalise the many types of signal from engine, airframe and systems sensors. Signals are multiplexed and digitised so that they can be recorded on the dfdr for accident investigation purposes. The unit is fully compliant with ARINC 573 and meets regulatory agency flight data acquisition requirements. The unit is used on 727, 737, 747, DC-10 and A300

Dimensions: ½ ATR Long
Weight: 18·5 lb (8·41 kg)
Power: 70 W

ADAU 100 auxiliary data acquisition unit

This micro-processor-controlled unit provides the multiplexing, signal conditioning, and digitising of strain-gauge and thermocouple signals for use in engine condition monitoring equipment. The serial output data is sent to a dmu as part of an expanded aids system, and is controlled by software stored in a read-only memory. The unit also provides calibration and cold-junction compensation for the input signals. The adau is part of the company's Mk II aids system for the Boeing 747 and McDonnell Douglas DC-10 transports.

Dimensions: ⅜ ATR Short
Weight: 9 lb (4·1 kg)
Power: 45 W

MCU 110, 111 management control unit

This is the control analysis unit for the Mk IV aids system, providing a reduced or more specialised data-gathering capability. Using state-of-the-art micro-processors, this aids provides some of the benefits of larger systems although packaged into a smaller volume. The computer's job is to read the fdau input data and decide when it is to be passed to the quick-access recorder or optional printer. The mcu writes over data in the fdau serial data stream in unused word locations sequence numbers, flight modes, and other specialised information to aid transcription and analysis on the ground. The unit is part of Hamilton Standard's Mk IV aids system on the 747, DC-10, DC-9-80 (MCU 110) and A300 (MCU 111) transports.

Dimensions: ⅜ ATR Short
Weight: 8 lb (3·6 kg)
Power: 25 W

DFDAU 120 digital flight data acquisition unit

Complying with ARINC 717 and performing the same functions as the fdau for the mandatory flight recording systems on the new-generation transports built to ARINC 700, the unit contains a micro-processor permitting it to record some aids information in addition to its crash recorder functions. This unit is standard equipment on the new Boeing 767 and 757 transports, and is a basic option on the Airbus Industrie A310 derivative of the top-selling European widebody airliner. It is also used as a building block for an expanded aids on the A310.

Dimensions: 6 MCU
Weight: 15 lb (6·8 kg)
Power: 40 W

AIR 100 accident information retrieval system

Hamilton Standard has developed under contract to the US Army a solid-state airborne flight data recorder, a ground-readout unit, and a batch-process computer for data analysis. The recorder receives, conditions and digitises information from aircraft sensors and under micro-processor control stores it in a crash-proof, non-volatile solid-state memory device. The system can record some 40 essential parameters, such as acceleration, air-speed, altitude, attitude, heading, engine power, and fault-warnings to the crew.

Heart of the system is a crash-survivable module housing the solid-state memory device and meeting the requirements of FAA TSO C51a. Depending on requirements, the device can store between 15 minutes and 4 hours of prior flight history. In the event of an accident the unit can be read out directly, or the memory chip can be removed for the subsequent extraction of flight information.

Dimensions: 6½ × 6⁴/₅ × 6½ inches (165 × 172 × 165 mm)

STATUS: in development for small aircraft types including helicopters, business jets, and fighters.

Lockheed

Lockheed Aircraft Service Company, Division of Lockheed Corporation, PO Box 33, Ontario, California 91761

Model 209 digital flight and maintenance data recorder

Since the introduction by Lockheed in 1958 of the first FAA-approved flight recorder on the first US commercial jet transport aircraft, the California company's recorders have become standard equipment on the world's airlines and more than 4000 have been built. The Model 209 digital flight and maintenance recorder, designed to accommodate the increased complexity of the widebody jetliners, was introduced in 1970 with the Boeing 747, and is now in service with most of the world's airlines. The manufacturer says that it has the highest demonstrated reliability of any flight data recorder in current service. Apart from the widebodies, the Model 209 has been chosen by a number of airlines for their new-generation Boeing 767 and 757 transports.

The Model 209 digital flight data recorder was developed expressly for use with the FAA-mandated expandable flight-data acquisition and recording system (efdars). Apart from its application as a mandatory crash recorder, a unique electronic motor speed control and high-speed playback facility permit the device to be used as a data recorder for maintenance purposes. Twenty-five hours of digital information can be copied in a 20-minute turnround, using the companion LAS Model 235 copy recorder. As a crash recorder the device is contained within a titanium housing, and contains mounting facilities for an optional, certificated underwater locator beacon.

Lockheed Model 209 digital flight and maintenance data recorder

Dimensions: ½ ATR Long
4⁹/₁₀ × 7³/₅ × 19½ inches (124 × 193 × 493 mm)
Weight: 22·5 lb (10·23 kg)
Operating temperature: –30 to +50° C
Motor: ac induction type operating at 1/60 rated speed during recording to conserve life
Drive system: V-belt and pulley
Speed control: electronic system using optically coupled tachometer disc for lowest wow and flutter
Input signal: Harvard biphase, 768 bits/s
Tape speed: 0·37 inch/s record and playback
Bit packing density: 2076 bits/inch at 64 words (12 bits each)/s
Recording duration: 25 h, continuously updated

High-speed playback time: < 20 minutes for total record
Power: 12 W at 115 V 400 Hz record, 30 W playback
Qualification: survivability to FAA TSO C51a
Reliability: 6000 h mtbf 15 000 h or 5 years mtbo. Bite and status outputs according to ARINC 573

STATUS: in production and service with more than 700 built. In October 1981 LAS was awarded initial contract for 1000 sets of Model 109 equipment, including cockpit voice recorders, for the US Air Force fleet of Lockheed C-130 Hercules and C-141 StarLifter transports.

Model 280 quick-access recorder

The Model 280 quick-access recorder was designed primarily for the recording and storage of aircraft maintenance data, and has a capacity of more than 51 hours. The system is built to ARINC 591 specifications, and uses a commercially proven cassette modified to accommodate reel brakes when it is removed from the recorder in order to remove any possibility of tape spillage. The system can be easily adapted for use with one, two, three, or four flight-data acquisition units, and can handle two such units without the use of a data-management unit.

Dimensions: ½ ATR Short or ½ ATR Long case
Weight: 17·5 lb (7·95 kg)
Input signal: Harvard biphase, 768, 1536, 2304 or 3072 bits/s
Playback time: < 32 minutes for total record
Bit-packing density: 2400 bits/inch
Power: 18 W

STATUS: in production.

Model 319 flight data recorder system

This crash-recorder system is specially designed for business and commuter aircraft where space and weight considerations are even more important than in commercial transport aircraft. Space is conserved by combining the functions of a data acquisition unit and recorder, normally accommodated in two ½ ATR Long boxes, into a single ½ ATR case with the same capability. The system can playback the entire 25 hours of data in less than 20 minutes at high speed without removing it from the aircraft.

Apart from the saving in volume, Lockheed says that the Model 319 is 30 per cent less heavy and requires 80 per cent less power than equivalent air-transport equipment. The system records the parameters mandated by the FAA's FAR Part 121 airworthiness schedules for this class of aircraft in standard ARINC 573 format.

Dimensions: ½ ATR Long 4⁹/₁₀ × 7³/₅ × 19½ inches (124 × 193 × 493 mm)
Weight: 25 lb (11·36 kg)
Operating temperature: –30 to +50° C
Input signals: 80 analogue, 30 discrete
Sample rate: programmable from 1 sample in 4s to 8 samples/s
Tape speed: 0·37 inch/s for record or low-speed playback, 11·84 inches/s high-speed playback
Bit packing density: 2076 bits/inch at 64 words (12 bits each)/s
Recording duration: 25 h, continuously updated

Lockheed Model 280 quick-access recorder and cassette

Lockheed Model 319 flight data recorder

Power: 25 W at 115 V 400 Hz
Qualification: survivability to FAA TSO C51a
Reliability: 6000 h mtbf, 15 000 h or 5 years mtbo

STATUS: in production.

Sperry

Sperry Flight Systems, Defense and Space Systems Division, PO Box 29222, Phoenix, Arizona 85038

Video recorder for B-52 bomber Offensive Avionics System update

Facing the possibility that its strategic bomber fleet of some 600 Boeing B-52s might have to soldier on to the turn of the century, the US Air Force and industry in the late 1970s instituted a programme to update these early post-war technology aeroplanes. Among the extensive modifications planned was an upgrading of the electronics controlling the weapons and other stores and equipment known as the Offensive Avionics System. In place of many individual instruments and indicators, multi-function cathode ray tube displays are being fitted to improve reliability and reduce congestion on the flight-deck.

Sperry video recorder for the B-52

To maintain the information for post-flight analysis, the cathode ray tube data is put on to a Sperry video recorder, along with indications of the chosen brightness and contrast levels. The system can hold 100 feet of 35 mm film, and recording takes place on command at a rate of up to four frames a second.

Dimensions: 14 × 8¹/₁₀ × 16¹/₂ inches (356 × 206 × 419 mm)
Weight: 39·6 lb (18 kg)
Input format: EIA RS-170 standard
Dynamic range: 10 shades of grey on developed film
Scan density: 875 lines

Exposure: 1/30s with electronic shutter, synchronised with vertical sweep
Power: 70 W at 115 V 400 Hz, 3 W at 28 V dc
Reliability: mtbf 3744 h

STATUS: in production.

Sundstrand

Sundstrand Data Control Inc, Avionic Systems Division, Overlake Industrial Park, Redmond, Washington 98052

Universal flight data recorder

This 25-hour capacity, crash-protected digital flight data recorder is offered in a variety of configurations. Together with optional accessory equipment, the system can be made completely interchangeable with existing flight data recorders to ARINC 542, with digital recorders to ARINC 573/717, or expanded-parameter systems intended for CAA-regulated applications.

The recorder uses Kapton tape for durability, a single motor, and co-planar reel-to-reel tape transport. It can be hard-mounted in the aircraft, and considerable attention has been given to reducing the number of moving (and therefore wearable) parts. High-speed data retrieval permits the entire tape content to be transferred to a copy recorder in less than 30 minutes, so that the operation may be conducted during normal turn-round.

Dimensions: ¹/₂ ATR Long, per ARINC 404
Weight: 30 lb (13·6 kg)
Tape/recording capacity: 450 ft for 25 h at 768 bits/s, 8-track bi-directional sequential
Power: 115 V 400 Hz
Crash survivability: FAA TSO C51a

STATUS: in production and service.

Sundstrand universal flight data recorder

Addenda

Flight Simulation Systems

UNITED STATES OF AMERICA

General Electric

General Electric Company, Electronic Systems Division, Simulation and Control Systems Department, PO Box 2500, Daytona Beach, Florida 32015

Advanced computer-generated visual systems

The Simulation and Control Systems Division is one of the four elements making up GE's Electronic Systems Division, the others being the Ordnance Systems Department, Pittsfield, and the two plants of the Military Electronic Systems Operations at Syracuse.

To maintain and enhance its standing in the simulation field, the company announced three changes in its organisation in late 1979: a name change from Ground Systems Department (a legacy from the Apollo space programme activities) to Simulation and Control Systems Department; plans for an expansion of the testing facilities; and transfer of the department from the Space Division at Valley Forge to the Electronics Systems Division, emphasising the relationship with the electronic, software-intensive business at Syracuse.

The division's work embraces three lines: advanced computer-generated visual systems for flight simulators, digital radar landmass simulation, and mission simulators, especially for the Northrop F-5 light fighter.

Compu-Scene visual simulation systems

Although the international simulation business has been more than 50 per cent military, with an appreciable percentage taken by the offshore operators, there has been a swing towards civil airlines. The spur for this development has come from rapidly increasing fuel costs, which have led regulatory bodies (predominantly the FAA) to sanction the use of flight simulators for greater proportions of regular proficiency training. The goal is 100 per cent, and to assist this progression, considerable refinements in visual simulation are being demanded.

Many airlines have current-generation simulators on six-degree-of-freedom motion bases that cost around $6 million. These systems are typically amortised over 15 years, and such investments cannot be discarded prematurely. As a result, GE policy has been to anticipate sales of advanced visual systems to bring simulators up to the standards required by the new training schemes.

GE's visual system is called Compu-Scene and is a development of the class of equipment largely pioneered by the company itself. GE built the world's first computer-image generation system in 1958, and supplied it to the US Army and Navy for a joint programme called Janip (joint Army Navy instrumentation programme). In 1962 it delivered to NASA the first computer-image generation system to train astronauts in space rendezvous and docking manoeuvres, an essential prerequisite to the subsequent moon landings. This was followed up in 1972 and 1974 when the first such systems were delivered respectively to the US Navy and Air Force. In 1975 GE provided the first full day/night computer-image generation system for commercial flying training.

Compu-Scene is an all-electronic, full-colour, day/dusk/night, variable-visibility 'outside world' display with a broad choice of image display configurations. The system is completely modular, and based on two types of equipment: an image generator and an image display. Integration with existing simulators is accomplished in a straightforward manner, the image displays being typically 'wrap-around' boxed units mounted in front of the simulator enclosure. Among many major capabilities are texture generation, circular feature generation, and curved-surface shading. The first produces texture patterns of selectable colour to provide enhanced motion and distance cues on bounded terrain areas and model face surfaces. Texture algorithms are modified to produce modulated patterns to represent sea, cloud, crops, or various types of terrain. Circular-feature generation software creates circles, ellipses, spheres and ellipsoids of selectable colours to represent clusters of circular objects such as trees, clouds, storage tanks, silos and similar cultural objects. Such features can be generated with curves instead of straight lines, resulting in great economy in computer storage space; a single curve can be generated that might require 50 or so edges. The last-named capability permits the use of continuously variable shade or tone of colour across the face of an object to make it appear curved, even though the object is modelled with flat faces.

High resolution is achieved by suitable choice of raster density between 625, 763, 875 and 1023 lines, and by the capacity of the image-storage system, which can provide up to 8000 edges and the same number of light points.

STATUS: in production and service. One such system is used by Boeing in conjunction with a Rediffusion simulator to train 747 crews.

Mission simulator system for Northrop F-5 fighter

GE's mission simulator for the top-selling F-5 light fighter (the F-5E version was the subject of the 1970 IFA (International Fighter Aircraft) competition sponsored by the US Government to supply a combat aircraft to third world countries) closely reproduces the capability and environment of the tactical fighter to provide training in visual and imc conditions. The system is based on real-time digital simulation and includes a full-colour GE computer-image generation visual system.

The mission simulator system accurately simulates the F-5 instruments, navigation, communications, and fire-control systems, and provides aural cues, realistic stick-force feel, and outside-world scenes. The system can be supplied in either single or dual cockpit configurations; the basic single-cockpit system comprises representative pilot enclosure, instructor station, and three-window continuous-scene visual display, all controlled by a general-purpose digital computer. The advantage of the dual-cockpit configuration is that it permits simultaneous use of a single display by two pilots, in order to practise formation flying, one-versus-one combat, or even two-versus-one if a computer-controlled model is programmed into the scene.

The instructor is located directly behind the student, permitting full communication and also allowing the former to use the same display as the latter. The visual system depicts ground scenes and sky equivalent to an area of 340 × 340 nautical miles and extending up to 60 000 feet. The data-base is three-dimensional and can include hills as well as fields, man-made objects, and bodies of water. The density and distribution of data-base detail is great enough to support speed, altitude, and distance estimation.

Operating costs of the mission simulator system, according to GE, are conservatively estimated at between 10 and 15 per cent of the cost involved in operating an F-5.

STATUS: in production.

GE F-5 mission simulator in formation flight mode, following another, computer-generated F-5

Digital radar landmass simulation

The training of radar operators in conjunction with pilots is now recognised to be a critical part of ensuring the combat-readiness of aircrews. Advanced flight or mission simulators with a digital radar landmass simulation (drlms) system can simulate every aspect of a flight, and particularly the navigational procedures and workload in the approach to the target, weapon-delivery phase, and subsequent escape.

GE's drlms programme began in 1971 with an experimental radar prediction device, and the company now produces several families of radar-image simulation systems for all three US armed services and overseas customers.

One of the most important applications is to be found in a weapon-system trainer for the US Air Force/Boeing B-52 bomber. The drlms provides the required proficiency training without the need to fly these large bombers and the resulting penalties in fuel and maintenance costs. The B-52 simulator programme requires a correlated radar and electro-optical viewing system and outside-world display, and the GE system is the first of a new generation of airborne-radar landmass simulators, incorporating advances in processing speed, accuracy, and the full range of special radar effects.

STATUS: in production. One member of the family has been ordered for the West German version of the multi-national Tornado bomber.

Mission simulator visual system for C-130 transport

As the Viet-Nam war demonstrated, assault transport aircraft need the same skills in low-flying and caution in the battle area as interdiction and strike aircraft. Formation flying in large aircraft, often necessary for accurate para-drops, also calls for precision handling. For these reasons the US Air Force has established at its Little Rock, Arkansas Military Airlift Command training base a C-130 mission simulator equipped with a GE visual system that can provide crew introduction and proficiency training 20 hours a day, seven days a week.

The visual system can generate up to 8000 edges and 4000 point light sources simultaneously, with eight levels of detail, and provides up to seven simultaneous models, including formation aircraft and surface-to-air missiles. To accomplish this the image generator uses one of the most powerful special-purpose computers ever built. An innovative technique known as scene texturing provides the essential visual cues for manoeuvres such as low-altitude parachute extraction, assault landings and contour flying.

The displays are formed from high-resolution, full-colour cathode ray tube images and are presented in six units in front of the aircraft windows on the six-degree-of-freedom moving platform. Each scene is an independent 'snapshot' determined by instantaneous computations of aircraft position and attitude. A unique feature is that two of the side-mounted displays (one each side) can be lowered to give the navigator the view he needs to identify drop references, or raised to show airfield references while banking during a turn. Instructors can call up weather effects such as fog, storm and cloud and can trigger hostile action such as anti-aircraft tracer and shell bursts and the flight of missiles. The instructor can also operate the system in a crash override mode, in which a training mission that would otherwise have been abandoned due to a fatal error on the student's part can be continued. A freeze mode can also be selected so that a particular situation can be discussed, and the previous 7½ minutes of the mission can be replayed for the same purpose.

STATUS: in service.

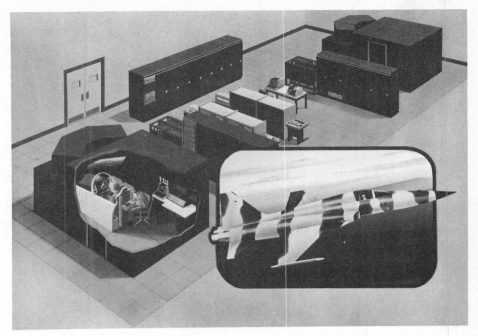

GE mission simulation system for the F-5. Inset, F-5E fighter

Daylight touchdown scene in GE C-130 mission simulator

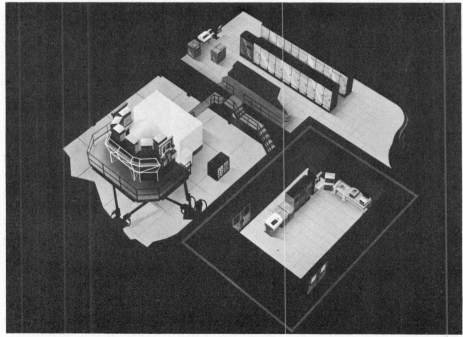

C-130 mission simulator with GE visual system display, image generator and data-base facility

Index